Introductory Probability and Statistics

Applications for Forestry and Natural Sciences

Dedicated to the Hungarian freedom fighters of 1956, without whom this book would never have materialized.

£ 35.00

Introductory Probability and Statistics

Applications for Forestry and Natural Sciences

Antal Kozak
Professor Emeritus, Faculty of Forestry
The University of British Columbia, Vancouver, BC, Canada

Robert A. Kozak
Associate Professor, Faculty of Forestry
The University of British Columbia, Vancouver, BC, Canada

Christina L. Staudhammer
Assistant Professor, School of Forest Resources and Conservation
University of Florida, Gainesville, Florida, USA

Susan B. Watts
Lecturer, Faculty of Forestry
The University of British Columbia, Vancouver, BC, Canada

CABI is a trading name of CAB International

CABI Head Office
Nosworthy Way
Wallingford
Oxfordshire OX10 8DE
UK

Tel: +44 (0)1491 832111
Fax: +44 (0)1491 833508
Email: cabi@cabi.org
Web site: www.cabi.org

CABI North American Office
875 Massachusetts Avenue
7th Floor
Cambridge, MA 02139
USA

Tel: +1 617 395 4056
Fax: +1 617 354 6875
Email: cabi-nao@cabi.org

A catalogue record for this title is available from the British Library, London, UK.

Library of Congress Cataloging-in-Publication Data (Hardcover edition)
Introductory probability and statistics : applications for forestry and natural sciences / Antal Kozak … [et al.].
p. cm. – (Modular texts)
Includes bibliographical references and index.
ISBN 978-1-84593-275-6 (alk. paper)
1. Forests and forestry—Statistical methods. 2. Probabilities. I. Kozak, Antal. II. Title: Applications for forestry and natural sciences. III. Series.

SD387.S73I585 2007
634.901'5192—dc22

2007005704

ISBN-13: 978-1-84593-275-6 (HB)
ISBN-13: 978-1-78064-051-8 (PB)

First published (HB) 2008
First paperback edition 2012

Typeset by Columns Design, Reading, UK.
Printed and bound in the UK by CPI Group (UK) Ltd, Croydon, CR0 4YY

Contents

List of Figures

List of Tables

Preface

What is different about this probability and statistics book from the seemingly endless supply of other introductory texts available? Simply put, our main objective was to develop an elementary statistics text for and about the growing number of people who have (quite wisely) chosen to make careers in the fields of *forestry*, *forest products*, *conservation* and *other natural sciences*. Our approach is unique in that we have used methods, examples and exercises that are particularly relevant to these increasingly important fields.

Notably, there have been several previous attempts at books aimed at these disciplines: *Experimental Design and Analysis in Forest Research* (1959) by J. Jeffers; *Forest Biometrics* (1961) by M. Prodan; and F. Freese's series, *Elementary Statistical Methods for Foresters* (Handbook, 1974), *Elementary Forest Sampling* (Handbook, 1962) and *Linear Regression Methods for Forest Research* (Handbook, 1964). Each of these texts has made an important contribution to the field, but they are outdated, specialized and limited from a pedagogical point of view. To the best of our knowledge, our book is unique in that it is designed for and well suited to a one-semester introductory probability and statistics course in forestry and natural sciences. This book also serves as a useful reference and it is hoped that students, practitioners and researchers will use it to understand and appreciate the important role of statistics in their respective disciplines.

The book starts with the usual topics found in any introductory statistics text – the use of descriptive statistics and a basic introduction to probability, random variables and probability distributions. We use this as the foundation for discussing some of the more commonly used inferential tools in statistics – estimation, hypothesis testing, analysis of variance, goodness-of-fit, test of independence, regression and correlation analyses. Unlike many other introductory texts, we have also provided discussions of statistical topics that are commonplace in a forestry context – design of experiments, sampling methods, non-parametric tests and statistical quality control. It is our opinion that a good background in high-school algebra is sufficient for reading and understanding the material presented here. We have made every attempt to use examples and exercises from the disciplines of forestry, wood science and conservation. However, we have also incorporated some everyday examples (like coin flipping and dice tossing) to make it more readable and understandable, especially where probability theories are concerned.

This book did not occur in isolation and we owe a debt of gratitude to a number of people, most notably our families, friends, peers and co-workers. We would also like to acknowledge everyone who has assisted us in the development of this book. We are particularly grateful to Patsy Quay, Jamie Myers and Denise Allen for their help in the preparation of the several versions of the manuscript. We also appreciate the insightful comments provided by Dr Lisa Zabek and the dozens of Forestry 231 (*Introduction to Biometrics and Business Statistics*) students at the University of British Columbia on one of the earlier drafts. Last, but certainly not least, the many concepts, ideas, thoughts and examples in this book were inspired by and originated from several other texts. These are listed in the Bibliography and the contributions made by these authors within the domain of probability and statistics are very much appreciated.

1 Statistics and Data

What do Numbers have to do with Trees?

In this chapter we define the term **statistics**. We also discuss **data**, the building blocks of statistics, and introduce data collection procedures and measurement scales.

1.1 What is Statistics?

When most people think of *statistics*, they imagine percentages, averages and rates of change, which are displayed in tables, graphs and charts. Statistics give us information about debits and credits, incomes and taxes, births and deaths, home prices, daily temperatures, smokers' mortality rates and so on. In sports, such as hockey, the term statistics refers to the records kept on each player's performance, such as the number of goals, number of shots on goal and number of penalty minutes.

Similarly, in forestry, statistics are collected to summarize tree heights, diameters, volumes, seedling survival rates, bark beetle infestation costs and much more. In sawmills, tables and charts are compiled to indicate the quality of the products produced, such as the distribution of lumber by grade, the strength of the lumber and other important quality characteristics.

We offer two definitions for statistics in this book.

- *Statistics* is the science of collecting, organizing, analysing and interpreting information (in this case, *statistics* is singular).
- *Statistics* are numbers calculated from information (in this case, *statistics* can be singular or plural).

The study of statistics is generally subdivided into two distinct fields: **descriptive statistics** and **statistical inference.**

Consider a large body of information, such as 5000 measurements of tree height collected from a forest management unit. Ordinarily, it is almost impossible to look at a large listing of numbers like this and draw any meaningful conclusions. Using descriptive statistics, we can describe this information with tables, summary numbers, charts and graphs. In this way, an observer (e.g. the forest manager) can very easily and quickly characterize, summarize and communicate the attributes of the forest management unit being measured (tree heights) since it is generally easier to understand the information when it is presented in the form of tables, charts, graphs or summary numbers, the latter often being referred to as *statistics*. The organization or tabulation of such large bodies of information has become a necessary skill for people employed in the forestry sector, from conservation biologists to foresters to wood products manufacturers.

> **Descriptive statistics** deals with the collection, organization and presentation of *information* and the calculations of some measures (*statistics*) that describe the information.

We may also want to use the information on hand to make predictions about the future or make statements about the larger body of information from which our data were taken. Statistical inference, or inferential statistics, uses information contained in a **sample** to reach conclusions about one or more characteristics of the whole **population.**

> A **population** is the entire collection of items/subjects possessing certain common characteristics about which information is being sought. The characteristics of a population are called **parameters** and are usually denoted with Greek letters (e.g. μ, σ).
>
> A **sample** is a portion or subset of the population. The characteristics of a sample are called **statistics** and are usually denoted with Roman letters (e.g. x, p).

It would be ideal if we could obtain information from every item (subject) in a population. However, populations are quite often very large (e.g. all possible trees in a forest type) and, therefore, it is simply not practical, or even possible, to collect the desired information from each item of the population. In other cases, such as the testing of modulus of rupture (strength) in 2 × 4s, it is not tenable to observe every item in the population because the item being observed is destroyed in the gathering of this information.

In these situations, we must collect the desired information from a sample. This portion, or subset, of the population is used to draw conclusions (inferences) concerning the whole population. This type of generalization, based on an incomplete set of information, involves a certain amount of **risk.** Therefore, in studying and using inferential statistics, a considerable amount of time is spent quantifying the associated risk. Some theories in **probability** will help us to properly quantify these risks.

> **Inferential statistics** or **statistical inference** is concerned with generalizing from the information obtained in a sample to an entire population. This generalization involves **estimation, hypothesis testing, determining relationships** and **prediction.**

1.2 Data

Pieces of information collected on subjects or items from a population form the building blocks of statistics and are called data (data is the plural of datum, a piece of information). Data can be collected, organized, analysed and summarized. Table 1.1 shows an example data set, which contains information collected from 50 trees. The trees here are the *elements* (or items or subjects) on which the data were collected. A *variable* is a characteristic of an element that we want to study. Seven *variables* were recorded in this data set: (i) tree identification number; (ii) date of measurement; (iii) species; (iv) crown class; (v) number of neighbouring trees (growing within a 5 m radius); (vi) diameter at breast height (dbh, which is measured at a height of 1.3 m in Canada, or 4.5 ft in the USA); and (vii) height.

Usually, a variable takes on different values from element to element, hence the name. In general, variables whose values are determined by chance are referred to as *random variables.* A set of measurements (such as the seven variables seen in Table 1.1) collected for one element is called an *observation*, and thus Table 1.1 contains 50 observations.

Table 1.1. A data set for 50 trees

Tree number	Date of measurement[a]	Species[b]	Crown class[c]	Number of neighbouring trees[d]	Diameter at breast height (cm)	Total height (m)
1	12	F	C	4	15.3	14.78
2	12	F	D	3	17.8	17.07
3	9	C	D	5	18.2	18.28
4	9	H	S	4	9.7	8.79
5	7	H	I	6	10.8	10.18
6	10	C	I	3	14.1	14.90
7	10	C	C	2	17.1	15.34
8	12	C	D	2	20.6	17.22
9	16	F	C	4	18.2	15.15
10	14	F	I	5	16.1	14.66
11	8	H	D	3	14.2	17.43
12	5	H	D	6	14.8	17.45
13	12	F	I	2	19.1	14.18
14	5	C	I	2	16.7	13.40
15	12	C	S	4	18.9	10.40
16	20	H	S	3	12.4	11.52
17	15	H	C	0	17.3	14.61
18	20	F	D	1	22.7	21.46
19	15	C	C	4	15.1	17.82
20	14	C	I	3	17.7	11.38
21	14	C	S	5	13.4	8.50
22	13	C	I	4	16.2	12.80
23	14	F	D	1	18.5	18.71
24	20	F	I	4	15.0	14.48
25	21	F	C	2	18.8	14.81
26	5	H	I	4	15.8	12.01
27	2	H	I	3	16.1	11.70
28	22	C	C	3	15.4	16.03
29	22	C	I	0	17.8	14.46
30	18	C	S	1	18.5	8.47
31	16	C	I	3	14.1	11.22
32	16	C	C	5	14.8	12.34
33	17	F	C	4	15.5	16.79
34	17	F	I	6	13.8	16.06
35	18	F	S	4	13.0	13.20
36	20	H	C	2	18.2	14.30
37	22	H	C	0	22.3	16.84
38	20	H	I	3	17.8	13.84
39	17	C	I	4	13.1	11.31
40	17	C	I	6	12.8	13.20
41	16	C	C	3	13.3	13.75
42	23	F	C	3	15.6	14.60
43	23	H	C	4	16.6	12.56
44	22	C	I	5	13.0	10.88
45	24	C	I	4	10.2	13.93
46	23	F	I	3	14.4	12.68
47	24	C	S	6	7.7	10.00
48	25	C	S	5	9.9	8.69
49	25	H	D	1	20.4	16.73
50	24	H	D	3	20.9	16.25

[a] Day of the month (March, 2006). [b] C, western red cedar; F, Douglas-fir; and H, western hemlock. [c] D, dominant; C, codominant; I, intermediate; and S, suppressed. [d] Trees within a 5 m radius of the subject tree.

Variables can be classified as either **qualitative** or **quantitative. Qualitative variables** are also known as **categorical variables** because they can be placed into distinct categories according to some characteristic. Species and crown class (Table 1.1) are qualitative variables. Other examples of categorical variables include gender, forest type, level of insect infestation (low, medium and heavy) and field of study (Forestry, Engineering, Agriculture, Education, Arts, Science).

Quantitative variables are numerical and can indicate 'how many' or 'how much' or 'how big' on a numeric scale. For example, dbh, height and number of neighbouring trees (Table 1.1) are quantitative variables. Quantitative variables can be further subdivided into **discrete** and **continuous variables.** Discrete variables, which take on whole numbers only, usually result from counting something such as the number of neighbouring trees (Table 1.1). Continuous variables are those which can take on 'all possible values' over a specific interval and are generally measured, e.g. height and dbh (Table 1.1). Often, 'all possible values' exist only in theory since measurement processes are limited to the precision of measurement devices. For example, current measurement techniques only allow dbh to be measured to the nearest 0.1 cm and tree height to the nearest 0.01 m. This means that a recorded dbh of 15.2 cm includes all possible values between 15.15 and 15.25 cm (not including trees with 15.25 cm dbh).

1.3 Measurement Scales

In analysing variables, the **scale of measurement** refers to the amount of information contained within the variable and indicates what types of statistical analyses are appropriate. Four common scales are used for measurements: **nominal**, **ordinal**, **interval** and **ratio.**

Nominal scale data can be quantitative or qualitative and are used mainly for identification and classification of items. Examples of quantitative nominal scale data include the tree numbers listed in Table 1.1, numbers on hockey jerseys, zip codes in the USA and telephone numbers (note that the use of numbers here is for identification purposes only). Examples of qualitative nominal scale data are the species identified in Table 1.1, gender, marital status and postal codes in Canada (e.g. V6S 1B9). Even if a variable is quantitative, arithmetic operations (addition, subtraction, multiplication and division) and/or ranking the items by their values are not meaningful for nominal scale data.

The **ordinal scale** is similar to the nominal scale, but in an ordinal scale, the order or rank of the values is valid. For example, crown class (Table 1.1) is in an ordinal scale, as it is known that the dominant trees are taller than the codominant trees within a stand. Again, ordinal scale data can be qualitative or quantitative. Examples of qualitative ordinal scale data are letter grades, levels of insect infestation (light, medium and heavy) and ranking of food quality (excellent, good, medium and poor). Examples of quantitative ordinal scale data are addresses in a block on one side of a street and numeric quality rankings (e.g. 1 for excellent, 2 for good, ..., 5 for poor). While the ranking of items in an ordinal scale is valid and meaningful in interpreting data, arithmetic operations (addition, subtraction, multiplication and division) are not.

The **interval scale** has the same properties as the ordinal scale, but interval scale data are always quantitative and differences between data values are meaningful.

Examples of interval scale data are temperature (in Celsius or Fahrenheit), Scholastic Aptitude Test (SAT) scores and measurement date (Table 1.1). When using an interval scale, zero does not indicate an absence of measurement. For instance, zero degrees is set as the icing point on a Celsius temperature scale; however, zero degrees does not indicate an absence of temperature. Similarly, if the temperature on a given day was 20°C in Vancouver and 10°C in Toronto, the difference of 10°C is meaningful. However, it does not mean that it is twice as warm in Vancouver as in Toronto.

The **ratio scale** is similar to the interval scale, but with two main differences. In the ratio scale, zero means 'none' and, therefore, the ratio of two variables becomes meaningful. Height, dbh and number of neighbouring trees (Table 1.1) are measured in a ratio scale; other examples are weight, distance, height and cost. All arithmetic operations (addition, subtraction, multiplication and division) are valid with ratio scale data. For example, we can say that a 20 m tree is twice as tall as a 10 m tree.

1.4 Data Collection

Data can be collected in many ways. Sometimes, data required for a particular application are available from government or company offices where operational data sets have been historically maintained. Data on forest inventory levels, production quantities and imports and exports are often collected by organizations, such as the United Nations Food and Agriculture Organization (UN/FAO). Employment rates, wage rates and other labour force information can usually be obtained from various government agencies.

If the required data are not available from existing sources, we can turn to some well-known statistical tools for data collection, namely **experimental designs** or **sampling designs** (or a combination of the two). These two techniques are frequently referred to as **experimental** and **observational** studies, respectively, and are discussed in more detail in Chapter 13 of this volume.

In experimental studies, one or more factors affecting the variable(s) of interest are controlled. The objectives of the study are to investigate how these controlled factors affect the variable(s) of interest. For example, to investigate the effect of seeding date on burnt and unburnt seedbeds, the dates and preparation of the seedbeds are controlled and the effects on germination are studied.

In observational studies (sampling), no attempt is made to control the variables of interest; we merely observe a given situation. The main purpose of sampling is to collect data from a subset of the population and to use this data to make predictions or inferences about the entire population. For example, if we would like to estimate the average height of a lodgepole pine plantation, we could randomly select 40 trees from the stand, measure their heights and estimate (with some degree of error) the unknown population average of height. These sorts of sampling designs are often referred to as **sample surveys**.

Exercises

Section 1.1

1.1. Define the word 'statistics'.

1.2. Give three examples of how descriptive statistics can be used in your field of interest.

1.3. Give an example in forestry, conservation or wood science where inferential statistics can be applied.

1.4. Two summer students are sent out to measure the dbh and height of 75 randomly selected trees in an experimental plantation, where each of the 10,753 trees in the plantation has been labelled with a number. Seventy-five random numbers between 1 and 10,753 were generated to indicate the trees to be measured.

a. Describe the population.
b. Describe the sample.
c. Will the students be using descriptive statistics, inferential statistics or both in this study?

1.5. A wood science student is working for a particleboard mill during her co-op term. She is asked to pull a single board every 15 min as it comes off the production line and to measure its thickness. These observations will be used to study the quality of the boards being produced as part of a programme for *statistical quality control.*

a. Describe the population.
b. Describe the sample.
c. Will she be using descriptive statistics or inferential statistics in this study?

Section 1.2

1.6. Classify each of the following variables as qualitative or quantitative:

a. Number of trees per hectare.
b. Colour of Douglas-fir flowers.
c. Number of leaders on a weevil-infested Sitka spruce seedling.
d. Outside bark diameter at breast height of a cork oak tree.
e. Fire hazard classification (low, moderate or severe).
f. Thickness of plywood.
g. Grade of lumber (No 1, No 2 or defective).
h. Length of a piece of dimensional lumber.
i. Age of a ponderosa pine tree determined from the number of annual rings.
j. Species.
k. Daily low and high temperatures measured in degrees Celsius.
l. Annual wage (in dollars) earned by 20 foresters working for a large company.
m. Phone number of each of the above 20 foresters.
n. Date when each of the above 20 foresters began working for the company.

1.7. Classify the quantitative variables in Exercise 1.6 as discrete or continuous.

Section 1.3

1.8. Classify the variables listed in Exercise 1.6 by their scale of measurement (nominal, ordinal, interval or ratio).

Section 1.4

1.9. The effect of chemical treatment on the modulus of rupture of 30 pieces of oriented strandboard was studied. Ten boards were treated with chemical A, 10 boards were treated with chemical B and 10 were left untreated. After the treatment, the 30 pieces were tested and their moduli of rupture were measured. Identify and briefly describe the data collection method used in this study.

1.10. Identify and describe the data collection method used in Exercise 1.4.

1.11. Modify the study described in Exercise 1.9 so that both experimental design and sampling design are used to obtain the required information.

This page intentionally left blank

2 Descriptive Statistics

Making Sense of Data

In order to adequately monitor and manage natural resources, such as forests and rangelands, many very large data sets are compiled. The objective of this chapter is to explore the tools used to make data sets more comprehensible. By organizing variables into tables, charts and graphs, and by calculating numbers that best describe the characteristics of a variable of interest, managers can quickly get information about the natural resources for which they are responsible.

2.1 Tables

Data, such as those presented in Table 1.1 (see Chapter 1), are called *raw data*. Even considering only one variable (e.g. diameter at breast height, or dbh), it is difficult to assess this listing of 50 observations of unprocessed data, let alone a larger data set of 5000 or more data points. One of the simplest ways to organize variables is to rank them in ascending or descending order. Ranking observations, as shown with the 50 dbh observations in Table 2.1, does not reduce the size of the data set and is usually used to describe data sets with smaller numbers of observations only.

When the number of observations is large, a more powerful tool, known as the **frequency distribution**, is used to describe a variable. In frequency distributions, observations are grouped into classes, and the frequency of observations in each class are tallied and presented in tabular form. Depending on the nature of the variable being grouped, we distinguish between three types of frequency distributions: **categorical**, **ungrouped** and **grouped frequency distributions**.

Categorical frequency distributions are used to place qualitative, ordinal or nominal level variables into specific categories. Table 2.2 shows the frequency of the trees from Table 1.1 (see Chapter 1) by crown class. Since crown class is a categorical variable, four discrete classes are used and the number of trees in each class is tallied. The relative frequency of each class can be calculated by dividing its frequency (f_i) by the total frequency (n, the number of observations):

Table 2.1. Ranked dbh measurements (in cm) of 50 trees.

7.7	9.7	9.9	10.2	10.8	12.4	12.8	13.0	13.0	13.1
13.3	13.4	13.8	14.1	14.1	14.2	14.4	14.8	14.8	15.0
15.1	15.3	15.4	15.5	15.6	15.8	16.1	16.1	16.2	16.6
16.7	17.1	17.3	17.7	17.8	17.8	17.8	18.2	18.2	18.2
18.5	18.5	18.8	18.9	19.1	20.4	20.6	20.9	22.3	22.7

Table 2.2. Categorical frequency distribution of 50 trees by crown class.

Crown class	f_j (frequency)	R_j (relative frequency)
Dominant	9	0.18
Codominant	14	0.28
Intermediate	19	0.38
Suppressed	8	0.16
Total	50	1.00

$$R_j = \frac{f_j}{n}$$

where j denotes the class.

Often, relative frequencies are expressed as percentages:

$$R_j = 100\frac{f_j}{n}$$

Ungrouped frequency distributions are used to summarize discrete quantitative variables. Table 2.3 shows the frequencies of the number of neighbouring trees from Table 1.1 (see Chapter 1). Because the variable could only take on integer values of 0, 1, ..., 6, seven discrete classes were used and their frequencies and relative frequencies are displayed in Table 2.3.

To summarize continuous (ratio scale) variables, **grouped frequency distributions** are generally used. For grouped frequency distributions, we divide the total range of the observations into a number of *classes* and tally the number of observations that fall into each class. Table 2.4 is a grouped frequency distribution for the 50 dbh measurements from Table 1.1 (see Chapter 1), recorded to the nearest 0.1 (one-tenth) cm. Seven class **intervals** have been used: 7.6–9.8, 9.9–12.1, ..., 19.1–21.3 and 21.4–23.6. The **class limits** are the smallest and largest possible values that can fall into a given class. For the second interval, 9.9–12.1, the **lower class limit** is 9.9 and the **upper class limit** is 12.1.

As the dbh values were recorded to the nearest tenth of a centimetre, the three trees in the second class interval must all be greater than 9.85 cm and less than

Table 2.3. Ungrouped frequency distribution of 50 trees by number of neighbouring trees.

Number of neighbouring trees	f_j (frequency)	R_j (relative frequency)
0	3	0.06
1	4	0.08
2	6	0.12
3	13	0.26
4	13	0.26
5	6	0.12
6	5	0.10
Total	50	1.00

Table 2.4. Grouped frequency distribution of 50 dbh measurements (in cm).

Class limits	Class midpoint	Frequency	Relative frequency
7.6–9.8	8.7	2	0.04
9.9–12.1	11.0	3	0.06
12.2–14.4	13.3	12	0.24
14.5–16.7	15.6	14	0.28
16.8–19.0	17.9	13	0.26
19.1–21.3	20.2	4	0.08
21.4–23.6	22.5	2	0.04

12.15 cm. These values, halfway between the upper class limit of one interval and the lower class limit of the next interval, are known as the **class boundaries** (Table 2.5). Class limits always have the same precision as the original observations. Conversely, class boundaries are always carried to one more decimal place than the original observations (unless, for example, a measuring device has 0.02 mm precision) and are halfway between the upper class limit of one interval and the lower class limit of the next interval (in most cases ending with the digit 5). In this way, no observation can fall on a class boundary and every observation can be uniquely classified into only one class.

The number of observations falling within a particular class is called the **class frequency** (f_j). The **class width** (w) is defined as the difference between the upper and lower class boundaries of a given class. It is convenient to have equal class widths for all classes, as in Table 2.5, although sometimes variable widths are used. It is also sometimes necessary to have **open classes** for the first or last class to accommodate a very few (one or two) extreme observations in the data set. In our example, the first class could be labelled as '9.8 and below' (≤ 9.8) or the last class as '21.4 and above' (≥ 21.4).

The **class midpoint** or **class mark** is defined as the average of the upper and lower class limits or the midpoint between the upper and lower boundaries of a class. When calculating the class midpoints for open classes, it is assumed that they have the same class width as the other classes in the distribution. If the first class is open, a 'false' lower class boundary is estimated by subtracting the class width from the upper class boundary of the first class, and the midpoint is calculated by averaging the two class

Table 2.5. Expanded grouped frequency distribution of 50 dbh measurements.

Class limits	Class boundaries	Class mark	Frequency	Relative frequency	Cumulative frequency	Relative cumulative frequency	Inverse cumulative frequency	Relative inverse cumulative frequency
7.6–9.8	7.55–9.85	8.7	2	0.04	2	0.04	50	1.00
9.9–12.1	9.85–12.15	11.0	3	0.06	5	0.10	48	0.96
12.2–14.4	12.15–14.45	13.3	12	0.24	17	0.34	45	0.90
14.5–16.7	14.45–16.75	15.6	14	0.28	31	0.62	33	0.66
16.8–19.0	16.75–19.05	17.9	13	0.26	44	0.88	19	0.38
19.1–21.3	19.05–21.35	20.2	4	0.08	48	0.96	6	0.12
21.4–23.6	21.35–23.65	22.5	2	0.04	50	1.00	2	0.04

boundaries. Similarly, if the last class is open, the upper class boundary is estimated in the same manner. The class midpoint plays a very important role in the calculation of various statistics from grouped frequency distributions, as each observation in a frequency class is represented by its class midpoint.

The following is a step-by-step procedure for classifying observations into a grouped frequency distribution. Note that this is not the only way of creating a grouped frequency distribution, but the reader cannot go wrong following these guidelines. Our example uses data from the continuous variable dbh listed in Table 1.1 (see Chapter 1).

1. *Decide on the number of classes (c).* Usually, the number of classes is set between 5 and 20. The choice depends on the number of observations in the data set. Too many or too few classes limit the usefulness of the frequency distribution to adequately describe the shape or pattern of the data sets. Although there is no definite method for selecting the number of classes, a rough estimate can be obtained by applying **Sturges' Rule**:

$$c = 3.3 \log_{10}(n) + 1$$

where n is the number of observations in the data set. In our example:

$$c = 3.3 \log_{10}(50) + 1 \approx 6.6066 \cong 7$$

(we round the calculated number to the nearest whole number.)

2. *Determine the class width.* Dividing the range by the number of classes (rounded) produces an approximate class width:

$$w = Range/c = (22.7 - 7.7)/7 \approx 2.143.$$

The class width should be rounded to the same precision as the measurements and the class limits. Whenever possible, the class width should be rounded up to avoid open frequency classes at the extremes of the data set. Some statisticians round up so that the last digit of the class width is an odd number. This ensures that the class mark will always have the same precision as the observations, the class limits and the class width. In our example, this suggests using 2.3 as the class width.

3. *Determine the lower class limit and boundary of the first class.* Unless an open class is used, the lower class limit of the first class should be less than or equal to the smallest observation. For our example, our lower class boundary should be less than or equal to 7.7 (the smallest value in our data set) and, therefore, we select 7.6. To obtain the lower class boundary of the first class, we subtract half of the precision (0.1/2 = 0.05 in our example) from the lower class limit (7.6 – 0.05 = 7.55).

4. *Calculate remaining class limits and boundaries.* First, we calculate the upper class boundary for the first class by adding the class width to the lower class boundary (7.55 + 2.3 = 9.85). The upper class limit of the first class is obtained by subtracting half of the precision (0.05) from the upper class boundary of the first class (9.85 – 0.05 = 9.8). The second class limits and boundaries are calculated by adding the class width to each of the lower and upper class limits and boundaries of the first class. The remaining class limits and boundaries are calculated from their preceding classes in a similar manner. Note that the upper class boundary of one class should be identical to

the lower class boundary of the subsequent class. This does not present a problem in placing each point in a unique class because the precision of the class boundaries is one decimal more than the raw data.

5. *Calculate the class midpoint for each class.* The class mark (midpoint) is calculated as the average of either the upper and lower class limits or the average of the upper and lower class boundaries. For example, our first class midpoint (m_1) is:

$$m_1 = (7.6 + 9.8)/2 = (7.55 + 9.85)/2 = 8.7.$$

6. *Tally the observations.* One practical way to count the number of observations in each class is to consider each observation in turn and make a 'tally mark' in the class where each observation falls. Table 2.6 shows the tally marks which add up to the frequency in each dbh class of our example.

Table 2.4 is the most common way of presenting grouped frequency distributions. If required, it can be extended to have more descriptive capabilities by including further information such as relative frequency, cumulative frequency, relative cumulative frequency, inverse cumulative frequency and relative inverse cumulative frequency (Table 2.5).

While the frequency gives us the number of observations that fall within a particular class, we may be interested in how many observations fall above or below a particular class. The **cumulative frequency** is the frequency of all observations less than the upper class boundary of a given class. It answers the question, 'How many observations fall within a certain class or lower?', and is often referred to as the 'less than frequency'. Conversely, the **inverse cumulative frequency** represents the frequency of all values greater than the lower class boundary of a given class. It answers the question, 'How many observations fall within a certain class or higher?', and is often referred to as the 'more than frequency'.

Table 2.7 shows how the cumulative and the inverse cumulative frequencies are calculated. For the cumulative frequency, the frequency of each class is added to the number of observations that fall below that class, starting with the first class. The same logic is applied to the inverse cumulative frequency, but starting with the last class. Both cumulative frequencies can be expressed as percentages (or proportions) of the total frequencies and, in these cases, are called **relative cumulative frequencies** and **inverse relative cumulative frequencies**, respectively.

We may also be interested in categorizing data from Table 1.1 (see Chapter 1) in terms of the pattern or shape of two variables. Table 2.8 shows a **bivariate frequency**

Table 2.6. Preparation of grouped frequency distribution of 50 dbh measurements.

Class boundaries	Tally	Frequency
7.55–9.85	//	2
9.85–12.15	///	3
12.15–14.45	~~////~~ ~~////~~ //	12
14.45–16.75	~~////~~ ~~////~~ ////	14
16.75–19.05	~~////~~ ~~////~~ ///	13
19.05–21.35	////	4
21.35–23.65	//	2

Table 2.7. Preparation of cumulative and inverse cumulative frequency distributions of 50 dbh measurements.

Class limits	Class mark	Frequency	Cumulative frequency	Inverse cumulative frequency
7.6–9.8	8.7	2	0 + 2 = 2	48 + 2 = 50
9.9–12.1	11.0	3	2 + 3 = 5	45 + 3 = 48
12.2–14.4	13.3	12	5 + 12 = 17	33 + 12 = 45
14.5–16.7	15.6	14	17 + 14 = 31	19 + 14 = 33
16.8–19.0	17.9	13	31 + 13 = 44	6 + 13 = 19
19.1–21.3	20.2	4	44 + 4 = 48	2 + 4 = 6
21.4–23.6	22.5	2	48 + 2 = 50	0 + 2 = 2

distribution, which gives the distribution of trees by species and number of neighbouring trees. In constructing Table 2.8, we use a categorical frequency distribution in one direction and an ungrouped frequency distribution in the other. Any combination of the three frequency distributions (categorical, ungrouped and grouped) can be combined to form bivariate frequency distributions.

Table 2.8. Bivariate frequency distribution of 50 trees.

Neighbouring trees	Cedar	Douglas-fir	Hemlock	Total
0	1	0	2	3
1	1	2	1	4
2	3	2	1	6
3	5	3	5	13
4	5	5	3	13
5	5	1	0	6
6	2	1	2	5
Total	22	14	14	50

When the data are grouped to form univariate or bivariate frequency distributions, we gain very valuable descriptive information and we can begin to see the pattern or the 'shape' of the data. On the other hand, this information is gained at a price; we lose the 'identity' of the original observations. For example, Table 2.7 tells us that there are three observations in the class identified as 9.9–12.1, but we do not know the original or 'measured' values of these observations. It is always good to keep this in mind when we present data in frequency distributions and use the graphical tools introduced in the next section.

2.2 Graphical Tools

Often, a graphic presentation of data can display the essential features of a frequency distribution more readily and comprehensively than a table. Pictorial representation of the information in graphic form often makes the important characteristics of the data more apparent. In this book, we will present several common graphical tools: bar graphs, histograms, pie charts, frequency polygons and ogives.

Bar graphs are used to present information summarized in categorical frequency distributions or ungrouped frequency distributions created for discrete variables. In bar graphs, the horizontal axis is not a continuous random variable and, consequently, the bars do not touch each other. Figure 2.1 presents a bar graph of the categorical variable crown class (from Table 2.2). Figure 2.2 shows a specific type of bar graph – a stick graph of the discrete variable number of neighbouring trees (from Table 2.3). With a stick graph, the 'bars' associated with a given number (label) have meaning only at that number. For example, the stick at two neighbouring trees refers to six occurrences of exactly two neighbouring trees and has no other meaning between one and three. This is why these graphs, done properly, are made using 'sticks' instead of 'bars'.

Histograms are used to present grouped frequency distributions for continuous variables. Therefore, they should not contain spaces between bars. The middle of each bar must be the class midpoint, and the bars touch at the class boundaries. Class boundaries are used because this allows for the graph to be 'continuous' since the upper class boundary of one class is the same as the lower class boundary of the subsequent class. Figure 2.3 presents the frequency distribution of dbh from Table 2.4.

Each of Figs 2.1, 2.2 or 2.3 can be plotted using relative frequencies (in per cent or as a proportion) on the vertical axis instead of actual frequencies. For example, Fig. 2.4 shows the frequency distribution of dbh measurements using relative frequencies.

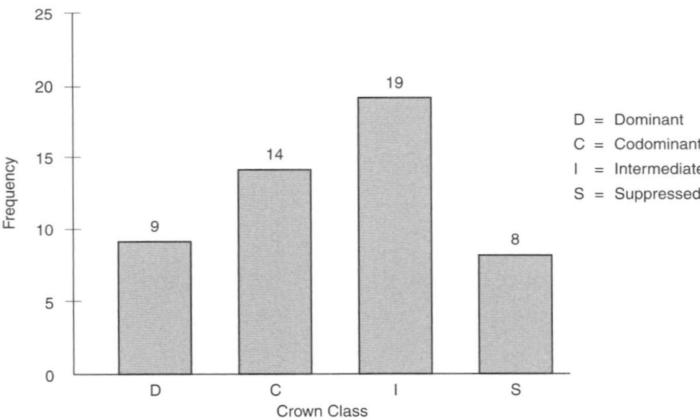

Fig. 2.1. Bar graph for crown class data.

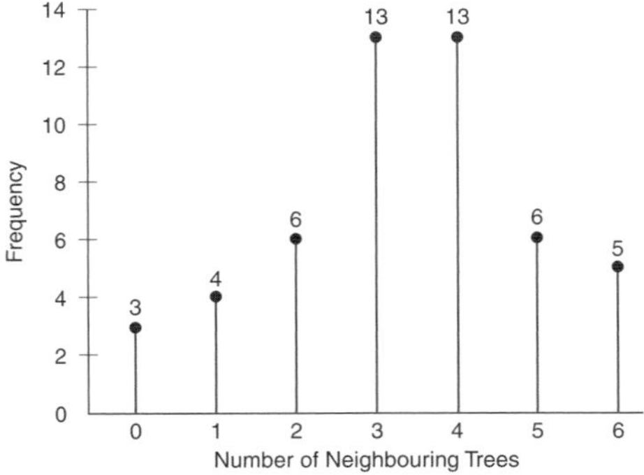

Fig. 2.2. Stick graph for number of neighbouring trees.

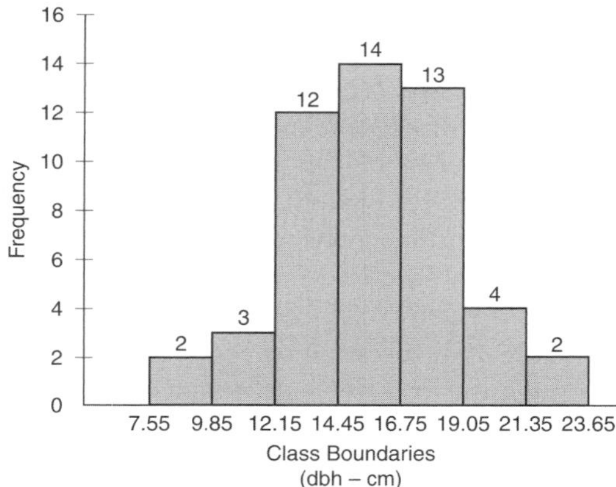

Fig. 2.3. Histogram for dbh data.

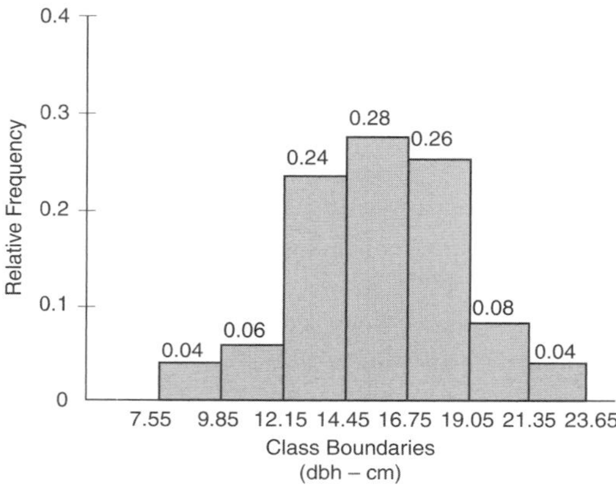

Fig. 2.4. Histogram for dbh data (relative frequencies).

In some cases, the presentation of a variable is best conveyed relative to a totality. Here, **pie charts** are used in place of bar charts. In the pie chart for crown class data (Fig. 2.5), the circle is divided into sections representing each category's frequency proportional in size to the total. In pie charts, frequencies or proportions for each class can be given.

Grouped frequency distributions can also be graphically presented with a **frequency polygon.** Frequency polygons are constructed by plotting class frequencies (or relative frequencies) against class marks and then joining each point by a sequence of line segments. To close the polygon, we add an 'imaginary' class midpoint with zero frequency to both ends of the distribution (e.g. 8.7 – 2.3 and 22.5 + 2.3). Figure 2.6 shows the frequency polygon for the dbh data.

Cumulative or inverse cumulative frequency graphs are called **ogives** and are plotted in a similar manner to polygons. When cumulative frequency graphs are prepared, the cumulative frequencies are plotted against the upper class boundaries and joined by line segments. To close the ogive at the lower end, the graph is extended

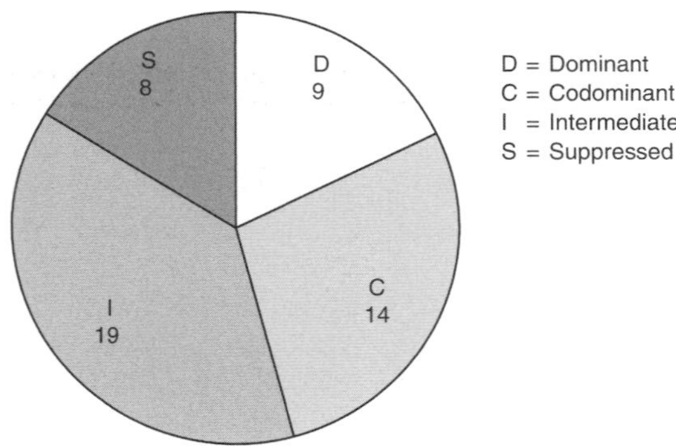

Fig. 2.5. Pie chart for crown class data.

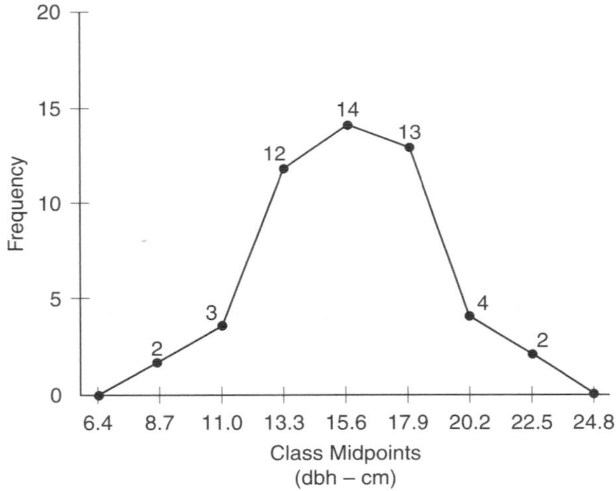

Fig. 2.6. Frequency polygon for dbh data.

to the lower bound of the first class boundary and given a frequency of zero because no points fall below this boundary (Fig. 2.7). This graph indicates the frequency of observations below a given class boundary. For example, in our data set, there are 31 trees below 16.75 cm dbh.

When inverse cumulative graphs are prepared, the inverse cumulative frequencies are plotted against the lower class boundaries and joined by line segments. To close this graph, the upper class boundary of the last class is graphed versus a frequency of zero. The inverse cumulative frequency graph indicates the frequency of observations above a given class boundary. For example, Fig. 2.8 shows that there are 33 trees above 14.45 cm dbh.

Earlier, it was stated that one of the main purposes of frequency distributions and histograms was to show the 'shape' of the distribution of the data. Related to this, we introduce some terms that are frequently used in statistics. In general, distributions can be classified as either **symmetric** or **skewed**. A distribution is said to be symmetric if a vertical axis at the 'centre' of the distribution separates the distribution into two

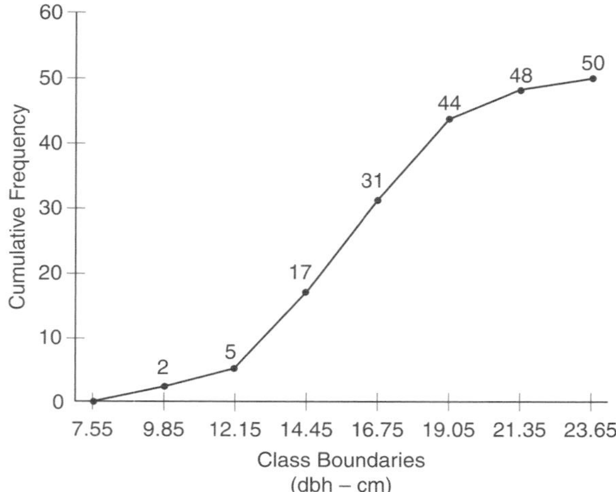

Fig. 2.7. Cumulative frequency graph (ogive) for dbh data.

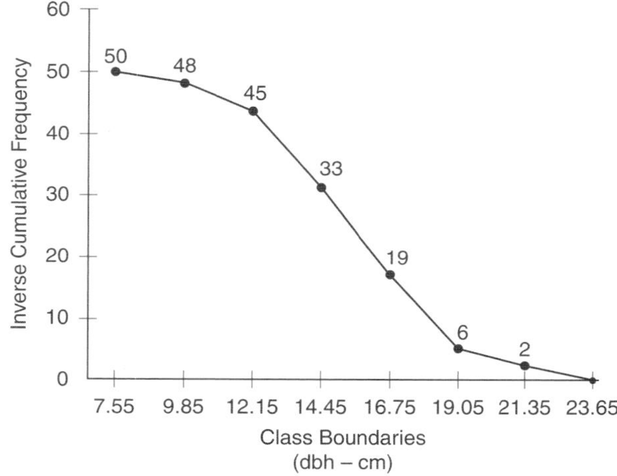

Fig. 2.8. Inverse cumulative frequency graph (ogive) for dbh data.

identical (mirror image) or near-identical parts. A distribution is skewed if it lacks symmetry with respect to this central vertical axis. Figure 2.9 presents two drastically different, but symmetric, distributions.

Figure 2.10 shows a positively skewed and a negatively skewed distribution. A distribution is positively skewed if it has a long right tail (Fig. 2.10a) and negatively skewed if it has a long left tail (Fig. 2.10b). The shapes of distributions will be discussed further in later chapters.

When constructing any of the charts discussed here, keep in mind that the number of classes has a direct effect on the shape of the distribution. If too many frequency classes are used, the histograms or bar charts do not satisfactorily reflect the shape of the distribution, as too few observations will fall into each class. On the other hand,

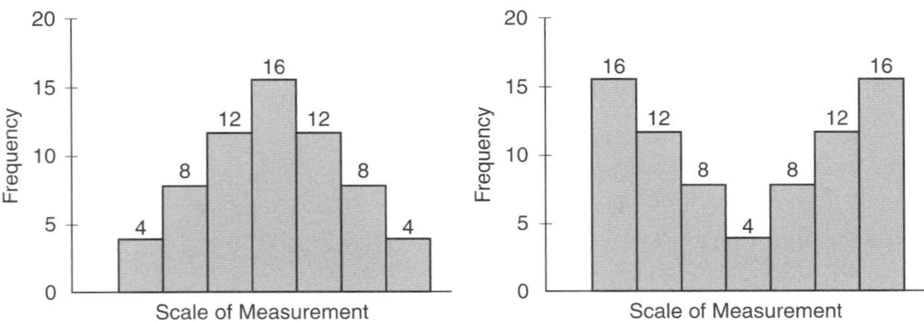

Fig. 2.9. Symmetric distributions.

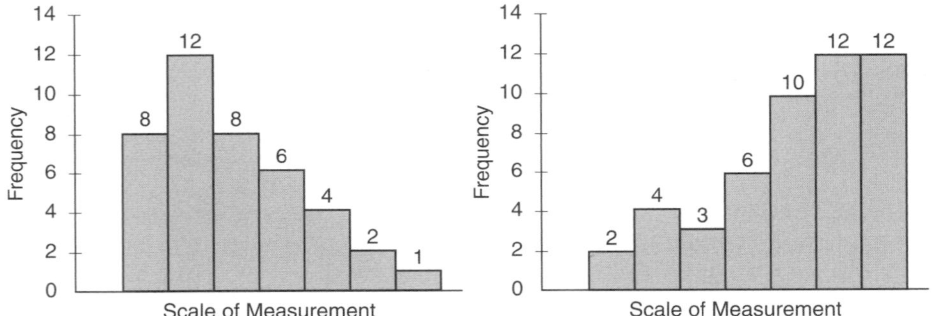

Fig. 2.10. Skewed distributions.

if too few classes are used, each class will have a high number of observations and, again, the pattern of the distribution is lost. The practitioner should be careful to use appropriately constructed classes in making these graphs, bearing in mind that it is easy to interpret graphs in different ways and potentially obscure their true meanings.

2.3 Measures of Central Location

Variables can be described using a range of statistical measures. Perhaps the best place to start is with some measure or measures of central location. Measures of central location are used to define, in some sense, the centre of a set of measurements. The most commonly used measures of central location are the **mean, median, mode** and **midrange.**

The **mean** is a statistic that is often referred to as the *average* or *arithmetic average*. To calculate this, we simply divide the sum of the measurements by the number of measurements. For a brief introduction to symbols and summation notation, see Appendix B. Since the calculation of the mean could be required for several variables, it is customary to use different symbols to represent each variable (such as x, y and z for dbh, height and biomass, respectively). The number of values in a sample is usually denoted by n and the number of values in a population by N. Individual values in a sample or population are symbolized by x_1, x_2, … and x_n or x_N.

The **sample mean** (for a variable, x) is calculated as:

$$\bar{x} = \frac{\sum_{i=1}^{n} x_i}{n} = \frac{x_1 + x_2 + \ldots + x_n}{n} \tag{2.1}$$

with the mean of variables y or z being symbolized by $\bar{y}$ or $\bar{z}$.

The **population mean** is calculated as:

$$\mu = \frac{\sum_{i=1}^{N} x_i}{N} = \frac{x_1 + x_2 + \ldots + x_N}{N} \tag{2.2}$$

Example 2.1. The height measurements of the first 5 Douglas-fir trees in Table 1.1 (see Chapter 1) are 14.78, 17.07, 15.15, 14.66 and 14.18 m. If these observations constitute a sample, then the sample mean height is:

$$\bar{x} = (14.78 + 17.07 + 15.15 + 14.66 + 14.18)/5 = 15.168.$$

In general, 'mean' refers to the *arithmetic mean*. However, in some disciplines, statisticians may be concerned with the *geometric mean* (used for ratio data like population growth, rates of change, economic indicators, etc.) and the *harmonic mean* (used for data where one element remains constant but another changes, like equal monthly contributions to a pension plan that varies in value). Since the use of these means is restricted, by and large, to special situations, they will not be discussed further in this book.

A **weighted mean** is often used when we wish to average a number of values by attaching more importance to some numbers than to others. This is done by assigning different weights (w_1, w_2, …, w_n) to the n observations, where these weights represent measures of their relative contribution to the overall average. The weighted mean ($\bar{x}_w$) is then calculated as:

$$\bar{x}_w = \frac{\sum_{i=1}^{n} w_i x_i}{\sum_{i=1}^{n} w_i} = \frac{w_1 x_1 + w_2 x_2 + \ldots + w_n x_n}{w_1 + w_2 + \ldots + w_n} \tag{2.3}$$

Example 2.2. If the costs of 3 models of chainsaw are US\$487, US\$596 and US\$759, and a company purchased 5, 9 and 11 of these saws, respectively, the average cost of chainsaws using a weighted mean can be calculated as:

$$\bar{x} = [(5)(487) + (9)(596) + (11)(759)]/(5 + 9 + 11) = 645.92.$$

A special application of this weighting procedure is used when finding the overall mean of several data sets when they are combined and the mean of each individual set is known. The equation to calculate the *grand mean* from k individual means is:

$$\bar{\bar{x}} = \frac{\sum_{j=1}^{k} n_j \bar{x}_j}{\sum_{j=1}^{k} n_j} = \frac{n_1\bar{x}_1 + n_2\bar{x}_2 + \ldots + n_k\bar{x}_k}{n_1 + n_2 \ldots n_k} \tag{2.4}$$

where k = number of means to combine.

Example 2.3. Based on sample sizes of 14, 22 and 14 trees, the sample means of tree heights for Douglas-fir, western red cedar and western hemlock are 15.62 m, 12.94 m and 13.87 m, respectively (see Table 1.1, Chapter 1). The mean of the combined data (50 observations) is the weighted mean of the individual means, weighted by their corresponding observations:

$$\bar{\bar{x}} = [(14)(15.62) + (22)(12.94) + (14)(13.87)]/(14 + 22 + 14) \approx 13.95.$$

Another application of the weighting procedure is for the calculation of the mean from a grouped frequency distribution (Table 2.4). In this case, since each individual observation has lost its *identity* (the value of the original observations), each observation is represented by its class midpoint. In the process of calculating the mean, the class midpoints (m_j) are weighted by class frequencies (f_j):

$$\bar{x} = \frac{\sum_{j=1}^{c} f_j m_j}{\sum_{j=1}^{c} f_j} = \frac{f_1 m_1 + f_2 m_2 + \ldots + f_c m_c}{f_1 + f_2 + \ldots + f_c} \tag{2.5}$$

where c = number of frequency classes; m_j = class midpoints; f_j = class frequencies.

$$\sum_{j=1}^{c} f_j = n.$$

Example 2.4. Using the information given in Table 2.4, we can calculate the mean of the dbh measurements as:

$$\bar{x} = [(2)(8.7) + (3)(11.0) + \ldots + (2)(22.5)]/50 = 15.738.$$

Interestingly, the mean in Example 2.4 (15.738 cm) is different from the mean calculated from the raw data in Table 1.1 (see Chapter 1). Using Eqn 2.1 with all 50 dbh observations, we get a mean of 15.794 cm. This difference stems from the fact that, in Eqn 2.5, we have replaced the original measurements with class midpoints (representations or proxies of the original data points). In other words, the true value of the mean is 15.794 cm but, using the grouped frequency distribution, we get 15.784 cm – a close approximation.

Apart from the fact that the mean is easy to calculate and it is a statistic that is familiar to most people, it also has some desirable properties that make it an invaluable tool for the interpretation of data sets (other desirable properties of the mean will be explored in later chapters). For instance, it is a reliable indicator of the centre of the values of a variable and it does not fluctuate much from one sample to another. On the other hand, its value is sensitive to extreme (very small or very large) values. In cases where very small or large values are apparent, the so-called trimmed mean can be used, which is the mean calculated after removing the upper and lower 5% of the ranked data.

The preferred measure of central location in the presence of extreme values is the **median**. The median is the middle value when a set of n measurements is arranged in increasing or decreasing order of magnitude. When n is *odd*, the median is the middle value of the ranked items. When n is *even*, the median is the mean of the two middle values of the ranked items. We usually use $\tilde{x}$ or m as symbols for the median.

Example 2.5. Consider the following two sets of dbh measurements (ranked):

12.4, 13.5, 13.5, [15.8,] 15.9, 18.2, 19.1

and

14.8, 16.3, 17.2, [17.2,] [17.4,] 18.3, 18.3, 19.4

The median of the first set is 15.8, or the fourth point in the ranked set of 7. The median for the second set is the average of the fourth and fifth points in the ranked set of 8, or (17.2 + 17.4)/2 = 17.3.

Although there are arithmetic procedures available to approximate the median from grouped frequency distributions, these procedures are beyond the scope of this book. The main advantage of the median over the mean is that it is not affected by extreme values. For symmetrical distributions, the mean and the median are equal.

Another measure of the central location is the **mode**, which is the most frequently occurring value in a sample or a population. A data set will not necessarily possess a mode, e.g. when all observations occur with the same frequency. Other data sets may have more than one mode, such as when several values occur with the greatest frequency. A population or sample with two modes is referred to as **bimodal**, while one with more than two modes is referred to as **multimodal**. The first set of data in Example 2.5 has one mode (13.5 occurs twice) and the second set has two modes (17.2 and 18.3 each occur twice). The use of the mode is advantageous in that it does not require any calculations and it can be used to study qualitative, as well as quantitative, variables. For example, the mode of the crown class observations in Tables 1.1 (see Chapter 1) and 2.2 is the intermediate trees, which occur with the highest frequency (19).

The **midrange** is another measure of central tendency and is defined as the sum of the minimum and maximum values divided by two. Like the average, its main disadvantage is that it is affected by the occurrence of extreme observations.

Because of its desirable properties, the mean (specifically, the arithmetic average) is the most commonly used measure of central location in statistics. It uses all observations, is easy to calculate and does not change much from one sample to another (taken from the same population). More importantly, when means are calculated from repeated samples from a population, a clear relationship between the sample mean and population mean emerges. Because of this relationship, the sample mean is a good estimator of the population mean. This very important characteristic of the sample mean will be discussed in more detail in later chapters.

2.4 Measures of Variation

An important characteristic of observations is that they are not exactly alike (hence the name **random variable**). The variation or spread of the observations has important properties in statistics. Although the measures of central location discussed in Section 2.3 are important, they do not provide a complete picture of the nature of the data. For example, each of the five Douglas-fir dbh measurements below, taken from natural regeneration and plantation stands, have exactly the same mean: 15.8 cm.

Natural	12.4, 14.5, 15.2, 17.8, 19.1
Plantation	14.8, 15.0, 15.9, 16.5, 16.8

While these two data sets have the same central location, they are very different. The dispersion or spread of the observations from natural regeneration appears to be higher than from the plantation stand. If you were the manager of a sawmill, which stand of trees would you prefer to deal with, all things being equal? For a more complete description of observations, it is often useful to provide some measure of the spread of the data in addition to a measure of the central location. Several such measures, namely the **range, mean deviation, variance, standard deviation** and **coefficient of variation,** will be discussed in this section.

Range, the simplest measure of variation, is the difference between the highest and lowest values in the data sets.

Range = highest value – lowest value, or
$R = \max(x_i) - \min(x_i)$.

Example 2.6. Douglas-fir measurements below, taken from natural regeneration and plantation stands, have exactly the same mean: 15.8 cm.

Natural	12.4, 14.5, 15.2, 17.8, 19.1
Plantation	14.8, 15.0, 15.9, 16.5, 16.8

The range of the natural regeneration dbh data is 19.1 – 12.4 = 6.7 cm, while that of the plantation is 16.8 – 14.8 = 2.0 cm. These numbers confirm that the spread, at least in terms of range, is higher for the natural regeneration stand. In fact, it is more than three times higher.

Although the range is easy to calculate, it usually does not provide a satisfactory measure of spread. Because its calculation involves only two of the observations, it is considered a rough estimate only. Furthermore, the range is affected by the number of observations, as well as by **outliers** (extreme values). As the number of observations increases, the range tends to increase because the chance that the data contains outliers also increases.

Unlike the range, the **mean deviation**, *MD*, utilizes all of the observations. It is calculated as the average of the absolute values of the deviations (Fig. 2.11) of each of the observations from the mean (sample or population).

$$MD = \frac{\sum_{i=1}^{n} \left| x_i - \bar{x} \right|}{n} \tag{2.6}$$

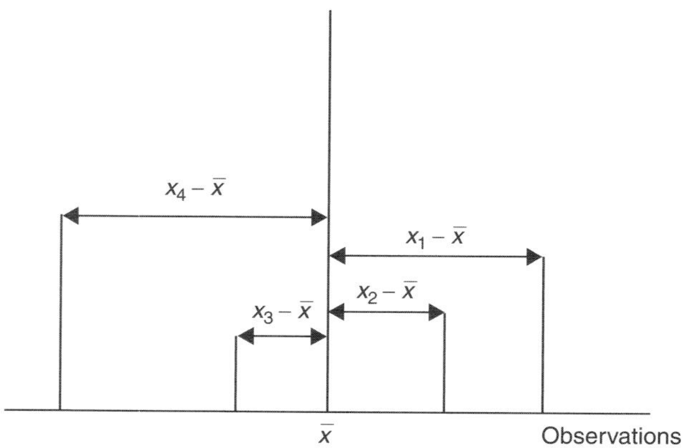

Fig. 2.11. Deviations from sample mean.

Note that this equation uses the absolute value of the deviations because, by definition, the summation of all deviations (positive and negative) would equal zero.

Although the mean deviation is a better descriptor of the variation than the range, its application in practical statistics is limited because theoretical interpretation is difficult. To overcome this disadvantage, another measure of spread was created, the **sum of squares of the deviations from the mean.** These sums of squares can be computed for sample data (SS), or for an entire population (SS_p):

$$SS = \sum_{i=1}^{n} (x_i - \bar{x})^2 = (x_1 - \bar{x})^2 + (x_2 - \bar{x})^2 + \ldots + (x_n - \bar{x})^2 \tag{2.7}$$

$$SS_p = \sum_{i=1}^{N} (x_i - \mu)^2 = (x_1 - \mu)^2 + (x_2 - \mu)^2 + \ldots + (x_N - \mu)^2 \tag{2.8}$$

The sum of squares of the deviations from the mean is also referred to as the **corrected sum of squares**, as each observation is subtracted from, or 'corrected for', the mean before it is squared. Since our natural regeneration and plantation data (above) are samples, we use Eqn 2.7 to compute the sums of squares (note that the units will be in squared terms, i.e. cm^2):

Natural $(12.4 - 15.8)^2 + (14.5 - 15.8)^2 + (15.2 - 15.8)^2 + (17.8 - 15.8)^2 + (19.1 - 15.8)^2 = 28.5$

Plantation $(14.8 - 15.8)^2 + (15.0 - 15.8)^2 + (15.9 - 15.8)^2 + (16.5 - 15.8)^2 + (16.8 - 15.8)^2 = 2.14$

It is clear that the sum of squares reflects the measure of spread, but its size is entirely dependent on the number of observations in the sample. Thus, samples of differing sizes cannot be directly compared. To overcome this, the average sum of squares can be calculated by dividing the sum of squares by the number of observations (N) for populations, or by the number of observations minus one ($n - 1$) for samples. The term

(n – 1) is known as **degrees of freedom** and is defined as the number of unrestricted observations used to calculate a statistic. Since the same observations are used to calculate the sample mean and the sample sum of squares, and since the sample mean is part of the equation to calculate the sample sum of squares, one of the n observations is not independent. In other words, in the equation for the sample sum of squares, *the sample mean defines one of the observations* and this must be accounted for. Many readers may find this concept difficult to grasp and so a simple example is provided for clarification.

Assume there are three observations from which a sample mean is calculated: 12, 14 and 16. The sum of these observations is $\Sigma x_i = 42$ and therefore the mean is 14 (42/3). Let us assume now that we know the mean, the number of observations and the sum of the observations, but we do not know exactly what the observations are. Working backwards to determine the values of the observations, we soon discover that they could be any set of three numbers that add up to 42. When we think about it a little more, we come to the further realization that actually we are free to choose only two out of three observations because the three observations must add up to 42. In other words, if we know that the sum of the observations must be 42 and we choose two of the observations, say 12 and 16, we know that the third observation *must* be 14 in order for the three observations to sum to 42. Thus, in this case, the degrees of freedom (the number of observations that are free to vary) are 3 – 1 = 2.

In most instances involving the sample mean, the degrees of freedom are n – 1. That is the case in calculating the sum of squares from a single variable data set where we can choose only n – 1 observations freely. However, the reader should be cautioned that there are many types of sum of squares used in statistics and each has its own associated degrees of freedom. We will give the appropriate degrees of freedom with any new sum of squares as it arises.

The sum of squares divided by its degrees of freedom is called the **variance** or, less commonly, the **mean square**. Using the sum of squares, the sample and population variances, s^2 and σ^2, are respectively calculated as follows:

$$s^2 = \frac{ss}{n-1} = \frac{\sum_{i=1}^{n}\left(x_i - \bar{x}\right)^2}{n-1} \tag{2.9}$$

$$\sigma^2 = \frac{ss_p}{N} = \frac{\sum_{i=1}^{N}\left(x_i - \mu\right)^2}{N} \tag{2.10}$$

Note that the population variance is divided by N instead of N – 1, because the population mean is a parameter, not an estimated statistic. Using Eqn 2.9, we can now calculate the variances for the two samples (note that the units are in squared terms):

Natural $s^2 = 28.5/(5 - 1) = 7.125\ \text{cm}^2$

Plantation $s^2 = 3.14/(5 - 1) = 0.785\ \text{cm}^2$

Eqns 2.9 and 2.10 are the theoretical equations to calculate variances. If pocket calculators are used to calculate the variance, the following equations – algebraically equivalent to 2.9 and 2.10 – provide the results much more efficiently, but are perhaps more complicated to conceptualize. These are referred to as 'computational' or 'working' or 'machine' equations:

$$s^2 = \frac{\sum_{i=1}^{n} x_i^{\ 2} - \left(\sum_{i=1}^{n} x_i\right)^2 \bigg/ n}{n-1} \tag{2.11}$$

$$\sigma^2 = \frac{\sum_{i=1}^{N} x_i^{\ 2} - \left(\sum_{i=1}^{N} x_i\right)^2 \bigg/ N}{N} \tag{2.12}$$

where

$\sum_{i=1}^{n} x_i^{\ 2} = x_1^2 + x_2^2 + \ldots x_n^2$, the uncorrected sum of squares;

and

$\sum_{i=1}^{n} x_1 = x_i + x_2 + \ldots x_n$, the sum of the observations.

Using Eqn 2.11, we first compute the uncorrected sums of squares:

Natural $12.4^2 + 14.5^2 + 15.2^2 + 17.8^2 + 19.1^2 = 1276.70 \text{ cm}^2$

Plantation $14.8^2 + 15.0^2 + 15.9^2 + 16.5^2 + 16.8^2 = 1251.34 \text{ cm}^2$

Then, the sample variances:

Natural $s^2 = (1276.70 - 79^2/5)/(5 - 1) = 7.125 \text{ cm}^2$

Plantation $s^2 = (1251.34 - 79^2/5)/(5 - 1) = 0.785 \text{ cm}^2$.

As also indicated by the range, the variance of dbh from natural regeneration Douglas-firs is considerably higher than that of the plantation. Thus, the spread of observations is higher for the natural regeneration stand than for the plantation stand.

Although the variance has a number of desirable theoretical characteristics, the **standard deviation**, which is the square root of the variance, has more descriptive power, mainly because the standard deviation is in the same *units* as the original observations (and the mean). From above, the standard deviations are 2.669 cm and 0.886 cm for natural regeneration and plantation stands, respectively.

The standard deviation holds unique properties for describing the spread of a data set. From our two examples of natural and plantation forests, it should be clear that the standard deviation is small if the values cluster closely around their mean, and the standard deviation is large if the values are widely dispersed around their mean. This observation was formalized by the Russian mathematician, P.L. Chebyshev, in **Chebyshev's Theorem.** This theorem, which can be applied to samples or populations of any kind, states that *at least* the fraction $(1 - 1/k^2)$ of the observations must lie within k standard deviations of the mean, *regardless of the shape of the distribution* of the data set (where k is any constant greater than one). For instance, using $k = 2$, we can say that at least 75% of the observations lie within two standard deviations of the mean.

If the data has a symmetrical, bell-shaped distribution (this will be called the *normal* distribution later on in this volume), the variability of the data can better be described by the **Empirical Rule**, which states that approximately 68%, 95% and 99.7% of the observations will lie within one, two or three standard deviations of the mean, respectively (Fig. 2.12).

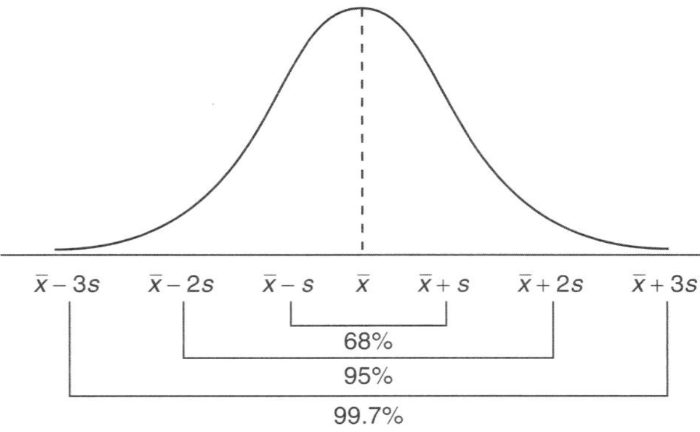

Fig. 2.12. Empirical Rule.

Although the Empirical Rule is more powerful, it is much less robust than Chebyshev's Theorem and should be applied only to distributions that exhibit the normal bell shape. Chebyshev's Theorem will work for any type of distribution (bell-shaped or not), and is thus a much more conservative way to interpret data.

Example 2.7. The mean and the standard deviation of the 50 dbh measurements in Table 1.1 are 15.8 cm and 3.2 cm, respectively. At least what fraction of the trees should be within two and three standard deviations of the mean?

Using Chebyshev's Theorem, at least 75% $[(1-1/2^2)100]$ of the trees should be within 9.4 cm and 22.2 cm, and at least 89% $[(1-1/3^2)100]$ of the trees should be within 6.2 cm and 25.4 cm. However, if we assume that the distribution is symmetrical and bell-shaped, we can say that 95% of the data will lie within 9.4 cm and 22.2 cm, and 99.7% will lie within 6.2 cm and 25.4 cm.

When it is appropriate to calculate a weighted mean as a measure of central tendency, then it is also appropriate to calculate a weighted standard deviation. Equations 2.13 and 2.14 give the theoretical and working equations, respectively, for the weighted sample variance. The standard deviation is computed by taking the square root as above:

$$s^2 = \frac{\sum_{i=1}^{n} w_i (x_i - \bar{x})^2}{\left(\sum_{i=1}^{n} w_i\right) - 1} \tag{2.13}$$

$$s^2 = \frac{\sum_{i=1}^{n} w_i x_i^2 - \left(\sum_{i=1}^{n} w_i x_i\right)^2 \Big/ \left(\sum_{i=1}^{n} w_i\right)}{\left(\sum_{i=1}^{n} w_i\right) - 1} \tag{2.14}$$

where

$$\sum_{i=1}^{n} w_i x_i^2 = w_1 x_1^2 + w_2 x_2^2 + \ldots w_n x_n^2.$$

A special application of the weighting procedure is used to calculate the variance and standard deviation from a grouped frequency distribution (Table 2.4). In the process of calculating the weighted variance, the class midpoints (m_j) are weighted by the class frequencies (f_j). The theoretical and working equations for the variance from a grouped frequency distribution are given in Eqns 2.15 and 2.16, respectively:

$$s^2 = \frac{\sum_{j=1}^{c} f_j\left(m_j - \bar{x}\right)^2}{n-1} \tag{2.15}$$

$$s^2 = \frac{\sum_{j=1}^{c} f_j m_j^2 - \left(\sum_{j=1}^{c} f_j m_i\right)^2 \Big/ n}{n-1} \tag{2.16}$$

where

$$\sum_{j=1}^{c} f_j m_j^2 = f_j m_1^2 + f_2 m_2^2 + \ldots f_c m_c^2.$$

Example 2.8. We can now calculate the variance and standard deviation of the dbh measurements from the grouped frequency distribution given in Table 2.3 as:

$\Sigma f_j m_j^2 = (2)(8.7^2) + (3)(11.0^2) + \ldots + (4)(20.2^2) + (2)(22.5^2) = 12{,}854.09$ cm^2

$\Sigma f_j m_j = 786.9$, as in Example 2.4.

$s^2 = (12{,}854.09 - 786.9^2/50)/(50 - 1) = 9.589$ cm^2

$s = 3.097$ cm.

As in the calculation of the mean, the observations in a grouped frequency distribution have lost their *identity* and therefore each observation is represented by its class midpoint. Thus, this estimate is not as exact as one computed from raw data (Eqns 2.9 and 2.11). Using the raw data and Eqn 2.9 or 2.11, the variance and the standard deviation were 10.414 and 3.227, respectively, compared to 9.589 and 3.097 above. We can say that the former values are correct, while the latter values are rough, but acceptable, approximations.

We now return to the hypothetical example of 5000 dbh measurements introduced in the beginning of the chapter. Let us assume that we know the mean of these measurements is 24.0 cm, the standard deviation is 4.0 cm and that the histogram shows a symmetrical, bell-shaped distribution. We can draw several conclusions very quickly. First of all, the centre of the distribution is 24.0 cm. Second, using the Empirical Rule, about 3400 (68%) of the trees are between 20.0 and 28.0 cm and about 4750 (95%) of the trees are between 16.0 and 32.0 cm.

When two data sets have the same units of measurement, the variances and standard deviations are comparable. For example, the standard deviation of the 50 dbh measurements in Table 1.1 (3.23 cm – see Chapter 1) and our hypothetical example (4.0 cm) are directly comparable, as they are both given in centimetres. Thus, it can be concluded that the 50 dbh measurements are less variable than those of our hypothetical example. On the other hand, if we compare the standard deviations of the height measurements (2.91 m) and dbh measurements (3.23 cm) of Table 1.1 (see Chapter 1), we could come to the wrong conclusion because the units are different. In cases like this, it is appropriate to use the **coefficient of variation** (CV).

The coefficient of variation is the standard deviation expressed as a percentage of the mean:

$$CV = 100\frac{s}{\bar{x}} \tag{2.17}$$

Thus, the *CV* is a standardized measure, meaning that variables measured in different units are directly comparable.

Example 2.9. The mean and standard deviation of the 22 western red cedar heights are 12.92 m and 2.92 m, respectively. The mean and standard deviation of the 22 western red cedar dbh measurements are 14.94 cm and 3.20 cm, respectively. The coefficients of variation are therefore:

Height: $CV = 100(2.92/12.92) = 22.6\%$

Dbh: $CV = 100(3.20/14.94) = 21.4\%$.

This means that the relative variation (relative to the mean) of the dbh measurements is somewhat less than the relative variation of height measurements.

The coefficient of variation can be very informative; however, some caution should be exercised with its use. Like percentages, the coefficient of variation can be misused. Specifically, as the mean of a set of observations approaches zero, the coefficient of variation approaches infinity. Therefore, coefficients of variation calculated for samples having a mean near zero should be avoided.

2.5 Measures of Position

Measures of position, such as **standard scores** and **percentiles**, are used to make statements about the relative position of an observation or observations within a particular set of data.

The relative position of an observation can be expressed in terms of the mean and standard deviation by calculating a standard score. Standard scores can be computed either for samples or for populations using the following:

$$z_i = \frac{x_i - \bar{x}}{s} \tag{2.18}$$

$$z_i = \frac{x_i - \mu}{\sigma} \tag{2.19}$$

A *z*-score indicates how many standard deviations an observation is above or below the mean value. With *z*-scores, items can be compared from two samples, regardless of the units of measurement or the relative variation of the two samples. It should be noted that the *z*-value is unitless and, if all the observations in a sample are transformed into *z*-scores, they will have a mean of 0.0 and a standard deviation (or variance) of 1.0. We will discuss *z*-scores in much more detail in later chapters.

Example 2.10. A student obtained grades of 65% in English and 85% in mathematics. The mean grade of all of the students in the English class was 60% with a standard deviation of 6% and the mean grade of all of the students in the mathematics class was 80% with a standard deviation of 8%. The *z*-scores corresponding to the student's grades are therefore:

English: $z_{65} = (65 - 60)/6 = 0.833$

Mathematics: $z_{85} = (85 - 80)/8 = 0.625$

Because the two z-scores are above zero, we know the student's performance was above the class average in both classes. Comparing these two z-scores enables us to say that, in a relative sense, the student performed better in English (the score is 0.833 standard deviations above the mean) than in mathematics (the score is only 0.625 standard deviations above the mean).

Percentiles indicate the position of an observation within a data set, but they are not the same as percentages. Assume a student scored 76 out of 100 possible points in a test (a score of 76%). This score could be the lowest, or the highest, or somewhere in the middle in the class. However, if the score of 76 corresponds to the 82nd percentile, then he or she performed better than 82% of the students in the class. In general, the pth *percentile* is the value such that p per cent of the items in the data set fall at or below that value:

$$p = 100(\text{number of items in the data set below that value} + 0.5)/n \qquad (2.20)$$

Some commonly used percentile values are:

- **deciles** divide the distribution into ten equal groups and correspond to the 10th, 20th, ..., and 90th percentiles;
- **quartiles** divide the distribution into four equal groups and correspond to the 25th, 50th and 75th percentiles, and
- the **median** divides the distribution into two equal groups and corresponds to the 50th percentile.

2.6 Computers and Statistical Software

Computers can be used for data organization, statistical analyses and arithmetic calculations. In most cases, they provide efficient and numerically accurate results. Recently, the general availability of computers and user-friendly software packages has had a tremendous impact on statistics. Most of the data manipulation and statistical analyses discussed in this book could be carried out with little or no difficulty using some of the more popular programs such as MINITAB, SPSS, BMDP, SAS, R or SYSTAT. Many statistical problems can also be solved in popular spreadsheet programs, like Microsoft Excel. We purposely do not cover computer packages in this book because we strongly believe that solving the exercises presented in this text by pocket calculators will help students to understand and learn the theory and applications of statistical techniques. However, most of the exercises and examples in this book can be duplicated using the above packages. If you are interested in doing so, we have included a few excellent references in the References section at the end of this volume that should help to familiarize you with one or more of these packages. Upon completion of your first statistics course, we would encourage you to explore these software packages – along with knowledge of the theory underlying statistics, they will become powerful tools in your careers.

Exercises

Section 2.1

2.1. The number of accidents per month in a sawmill for the last 20 months are as follows:

0	1	0	2	2	1	4	3	0	1
5	1	2	3	4	0	1	1	3	4

Construct a frequency table and calculate the relative frequencies.

2.2. The tree species in a permanent sample plot on the west coast of British Columbia were recorded as follows (F = Douglas-fir, H = western hemlock, C = western red cedar and A = red alder).

F	H	F	C	F	A	H	F
H	C	A	C	F	H	H	H
F	H	A	C	F	H	H	F

Construct a frequency table and calculate the relative frequencies of each species.

2.3. In a commercial timber cruise, the number of trees per plot were recorded as follows:

5	6	5	5	4	5	4	5	3	6	4	5	6	2	7
2	3	5	5	6	7	8	2	3	4	5	6	4	3	2

Construct a frequency table and calculate the relative frequencies.

2.4. Construct a frequency table using the 50 height measurements given in Table 1.1 (see Chapter 1). Show the class limits, class boundaries, class midpoint, relative frequencies, cumulative frequencies, relative cumulative frequencies, inverse cumulative frequencies and relative inverse cumulative frequencies.

2.5. The following are the amounts, in parts per million (ppm), of a nitrogen compound found in 60 soil samples.

3.6	3.2	3.3	3.6	2.7	3.4	4.5	3.3	2.8	5.4
6.1	3.4	2.9	2.7	4.1	4.7	5.1	4.7	3.2	3.6
5.1	2.6	3.6	3.8	3.8	3.1	3.7	5.5	3.2	3.7
4.2	4.5	4.3	3.7	3.6	3.9	3.5	4.4	2.8	3.3
3.9	4.4	5.1	4.6	3.4	2.6	4.5	3.1	2.5	3.1
3.7	3.4	4.1	2.7	5.7	3.5	4.7	4.4	4.4	5.0

Construct a frequency table and show the class limits, class boundaries, class midpoint, relative frequencies, cumulative frequencies, relative cumulative frequencies, inverse cumulative frequencies and relative inverse cumulative frequencies.

2.6. Consider the following grouped frequency distribution of tree crown lengths (in metres) of trees collected in a young stand.

Class limits	Frequency
< 4.4	2
4.5–6.4	7
6.5–8.4	11
8.5–10.4	13
10.5–12.4	8
12.5–14.4	4

a. What is the precision of the measurements?
b. Calculate the class boundaries.
c. Calculate the class widths.
d. Calculate the class midpoints.
e. Describe the first class.
f. Find the number of trees with less than 12.45 m crown length.
g. What percentage of trees has a crown length of less than 8.45 m?
h. What percentage of trees has a crown length of more than 6.45 m?
i. Find the number of trees with a crown length greater than 8.45 m.
j. Would you have used six frequency classes for these data?

Section 2.2

2.7. Construct a 'stick graph' for Exercise 2.1.

2.8. Construct a bar graph for Exercise 2.2.

2.9. Construct a pie chart for Exercise 2.2.

2.10. What graph or chart would you recommend for Exercise 2.3? Construct your recommended graph.

2.11. Construct a histogram and a frequency polygon for Exercise 2.6.

2.12. Construct a relative histogram and a relative frequency polygon for Exercise 2.5.

2.13. Construct a cumulative frequency graph and an inverse cumulative frequency graph for Exercise 2.4.

2.14. The following frequency distribution was constructed for log lengths (in feet) from trees bucked on a landing:

Class limits	Frequency
12.1–14.0	6
14.1–16.0	23
16.1–18.0	44
18.1–20.0	27
20.1≤	4

Construct a frequency histogram and a frequency polygon.

Section 2.3

2.15. The specific gravity (density) of each of eight coniferous tree species was measured as follows:

0.682 0.357 0.412 0.582 0.556 0.576 0.368 0.381

Find the mean, median and mode of specific gravity.

2.16. The following are the minimum temperatures (in Celsius) of seven cities in Canada recorded on 14 January 2006. Calculate the mean, median and mode of these temperatures.

–12 –5 2 2 0 –3 5

2.17. Find the mean, median and mode for the number of accidents given in Exercise 2.1.

2.18. Find the mean, median and mode for the number of trees per plot given in Exercise 2.3.

2.19. From the data in Table 1.1 (see Chapter 1), calculate the mean of the dbh measurements by species (i.e. separately for Douglas-fir, western hemlock and western red cedar).

a. Find the median and mode of the dbh measurements for western red cedar trees.
b. Calculate the weighted mean of the dbh measurements for all three species.

2.20. Assume that a cutblock consists of three forest types, A, B and C, and their areas are 420, 350 and 210 ha, respectively. If the average volume per ha (in m^3/ha) for each of the three types are 450, 480 and 620, respectively, what is the average volume for the cutblock?

2.21. Find the mean from the 'raw' data and from the grouped frequency distribution you constructed for the 60 soil samples given in Exercise 2.5. Compare the two means.

Section 2.4

2.22. Find the range, mean deviation, variance and standard deviation for the data given in Exercises 2.15 and 2.16.

2.23. Find the variance and standard deviation for the data given in Exercise 2.1.

2.24. Find the range, mean deviation, variance and standard deviation for the data given in Exercise 2.3.

2.25. If it is known that the mean ($\bar{x}$) of 12 observations is 12.5 and the uncorrected sum of squares (Σx_i^2) is 2000, calculate the variance and standard deviation for the observations.

2.26. Calculate the variance and standard deviation from the raw data and from the frequency distribution that you constructed in Exercise 2.5. Compare the results.

a. Apply Chebyshev's Theorem with $k = 2$. Are the results consistent with the theorem? *Hint: at least 75% of the observations should be within two standard deviations of the mean.*
b. Using the Empirical Rule, find the proportion of the observations within one and two standard deviations of the mean. Does this theory apply? Why or why not?

2.27. Calculate the coefficients of variation from the means that you obtained in Exercises 2.15 and 2.16 and the standard deviations that you obtained in Exercise 2.22. Compare the two coefficients of variation and draw some conclusions.

2.28. What does it mean if the standard deviation of a particular data set is zero?

Section 2.5

2.29. Using the data given in Exercise 2.5, find:

a. The standard score for soil samples with 4.5 and 3.7 ppm, respectively: interpret the results.
b. The 75th percentile.
c. The percentile rank for soil samples with 4.5 and 3.7 ppm, respectively.

3 Probability
The Foundation of Statistics

We use statistical information every day to qualify statements and to help us make decisions. For example, we may hear statements like:

- There is an 80% *chance* of rain today.
- The *odds* are one in 13 million that you will win the lottery.

Or we may be confronted with questions like:

- What is the *likelihood* of receiving an A on the first exam in this course?
- What is the *chance* that the Vancouver Canucks will win the next Stanley Cup?

Statistical inference, the generalization from a *sample* to a *population*, involves drawing a conclusion about a population on the basis of available, but incomplete, information. Hence, statistical inference involves a certain amount of uncertainty, and statisticians should not base decisions on statistical inference unless the risk of uncertainty can be reduced to a tolerable minimum.

Problems involving 'uncertainty', 'chance', 'likelihood', 'odds' and other such factors require an understanding and application of the theory of **probability**. Probability is the branch of mathematics that incorporates the most important set of concepts used in the field of statistics. The purpose of this chapter is to introduce the basic theories of probability that are required to appreciate and understand many of the concepts of statistical inference.

3.1 Sample Space and Events

In statistics, we define an *experiment* as a process that produces some data. In Chapter 1, we described an experiment to study the effects of seeding date and seedbed preparation on germination. A wood scientist could be interested in studying the effect of temperature and applied pressure on the strength properties of plywood. Experiments such as tossing a coin, rolling a die, or drawing a card from an ordinary (52 cards) deck of cards will also produce some data. In this chapter, we will deal with some simple experiments in order to make the concept of probability easier to understand.

In most cases, the **outcome** of an experiment (real or simplified) will depend on chance, and the outcome cannot be predicted with certainty. All possible outcomes of an experiment are called the **sample space** and are represented by the symbol S. A single outcome of an experiment is called an **element** or **sample point** of the sample space. When sample spaces are finite, their elements can be listed. The general practice is to list the elements separated by commas and enclosed in brackets. Some examples of sample spaces are:

Tossing a coin	$S = \{H, T\}$	Two possible outcomes where H = heads and T = tails.
Tossing two coins	$S = \{HH, HT, TH, TT\}$	Four possible outcomes.
Rolling a fair die	$S = \{1, 2, 3, 4, 5, 6\}$	Six possible outcomes where the values indicate the number rolled.

In other cases, it is easier to describe sample points rather than list them. When we draw a card from an ordinary deck of cards, the sample space contains all possible 52 cards, 13 of each of 4 suits: hearts (♥), clubs (♣), diamonds (♦) and spades (♠). Each suit contains nine 'numbered' cards (numbered from 2 to 10), three 'face' cards (jack, queen and king) and one ace. This sample space can be described as:

$S = \{\text{1 of 52 possible cards}\}$.

Sample spaces can also be described with coded numbers such as 1 for heads and 0 for tails, or by some other characteristic of the sample points. For example:

Tossing a coin $S = \{1, 0\}$
Rolling a die $S = \{\text{even, odd}\}$.

A sample space can also be described or qualified in general terms. For example, the sample space of all trees of species that are native to British Columbia can be described as:

$S = \{\text{all trees of species } y \mid y \text{ is a species native to British Columbia}\}$,

where the vertical bar '|' is read as 'such that' or 'given'.

Within a sample space, we may be concerned with the occurrence of a particular subset of all possible elements. An **event** is defined as a subset or portion of the elements of a sample space. Events are usually represented by capital letters. *A* might be the event that we have one head in an experiment of tossing two coins; *B* might be that we have the 'same' outcomes for the two coin tosses; and *C* might be that we have at least one tail.

$A = \{HT, TH\}$ $B = \{TT, HH\}$ $C = \{HT, TH, TT\}$

We could describe three events from a deck of cards such as:

A = {the card is black (clubs or spades)}
B = {the card is a 'face' card}
C = {the card is an ace of clubs or an ace of spades}.

We can distinguish four types of events:

- a **sample space** contains all possible outcomes;
- a **simple event** contains only one element;
- a **compound event** contains more than one element; and
- a **null space** contains no elements.

The usual symbol for the null space is $\varnothing$. Readers should note that, in some textbooks, compound events are also referred to as *unions* and *intersections* of events. This definition is ambiguous, as it is possible that unions or intersections of two or more events might not contain more than one element!

In approaching probability problems, it is oftentimes useful and informative to draw a picture or **Venn diagram** of events as they relate to each other within the sample space. This is especially useful where compound (multiple) events are concerned. In a Venn diagram, the sample space is shown as the interior of a rectangle. Events are identified (often as circles) as specified regions inside the rectangle. Figure 3.1 shows the three events *A*, *B* and *C* from the sample space of cards discussed above. It shows that there is some overlap, or intersection, between *A* and *B*, indicating some common elements in the two events. Also, since *C* is completely contained within *A*, we can see that all the elements in *C* belong to *A* as well.

Using events, mathematical operations, known as **unions** and **intersections**, can be carried out and they play an important role in the theory of probability. In this book, we will deal with unions and intersections of two events; however, the procedures can easily be extended to three or more events. Before looking at unions and intersections, the **complement** of an event should be defined. The complement of an event, *B*, is the event containing all the elements of the sample space that are not contained in *B*. The complement of *B* is denoted by B' (the shaded area in Fig. 3.2) and is discussed further in Section 3.3.

The **union** of two events *A* and *B* is the event that contains all the elements in *A* or *B*, including elements common to both. It is denoted by $A \cup B$. The shaded area in the Venn diagram in Fig. 3.3 shows the union of two events.

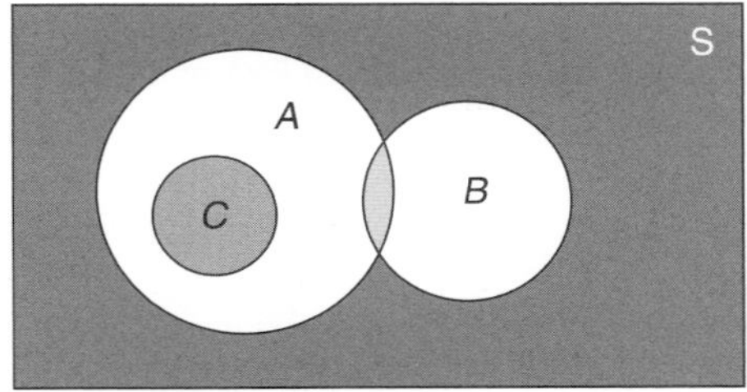

Fig. 3.1. Venn diagram of sample space and events.

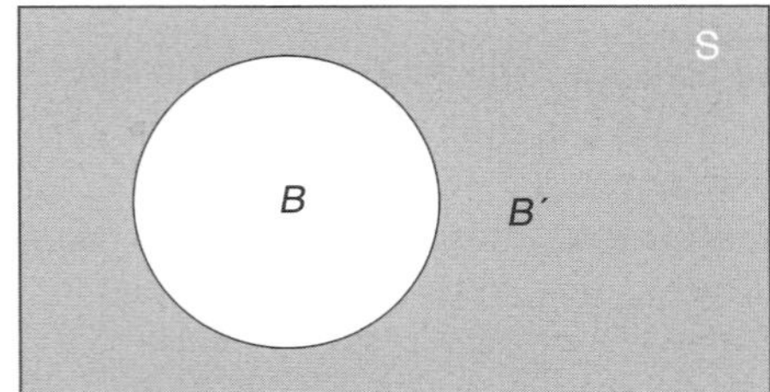

Fig. 3.2. Complement of event *B*.

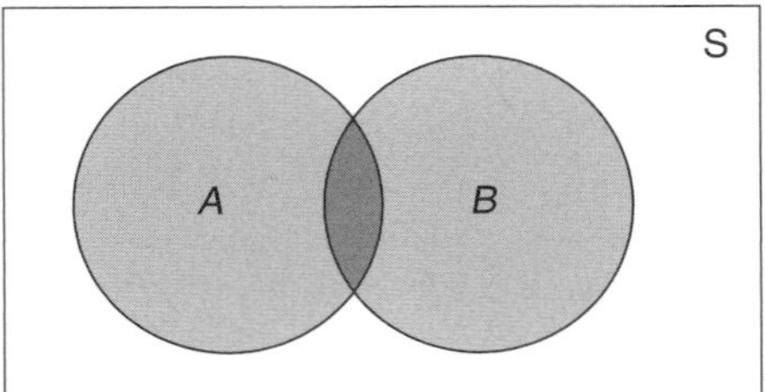

Fig. 3.3. Union of two events, *A* and *B*.

Example 3.1. Two trees are chosen at random from a stand of southern pine, where some trees are infested with southern pine bark beetle. Event *A* is that the first tree chosen is not infested. Event *B* is that the same outcome occurs on two consecutive trees. What is the union of these two events?

A = {NN, NI} B = {NN, II} where N = not infested and I = infested
$A \cup B$ = {NN, NI, II}.

Note that NN, the common element, is not listed twice.

Example 3.2. In an experiment of rolling a die, event A is the even numbers and event B is the numbers less than 3. What is the union of these two events?

$$A = \{2, 4, 6\} \quad B = \{1, 2\} \quad A \cup B = \{1, 2, 4, 6\}.$$

The **intersection** of two events A and B is the event that contains all the elements common to both A and B. It is denoted by $A \cap B$. Figure 3.4 shows the Venn diagram of the intersection of two events.

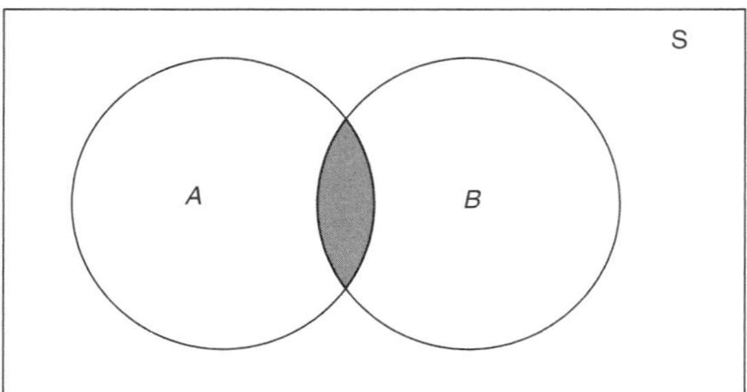

Fig. 3.4. Intersection of two events, A and B.

Example 3.3. The intersection (common element) of the two events from Example 3.1 is:

$$A \cap B = \{\text{NN}\}.$$

Example 3.4. The intersection of the two events from Example 3.2 is:

$$A \cap B = \{2\}.$$

If $A \cap B = \varnothing$, A and B are said to be **mutually exclusive**; that is, they have no common elements. Figure 3.5 shows two mutually exclusive events.

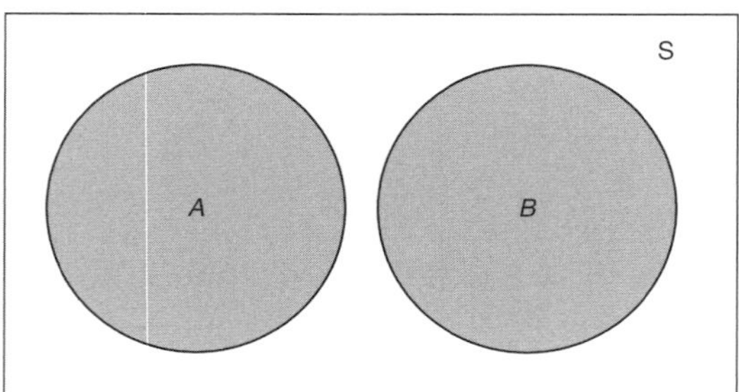

Fig. 3.5. Mutually exclusive events.

Example 3.5. If event C contains all the spades and event D contains all the clubs from a deck of cards, they are mutually exclusive, as they do not have a single common element.

$$C \cap D = \varnothing.$$

Conversely, if all of the elements in event C are also elements of event A (see Fig. 3.1), event C is said to be a **subset** of event A ($C \subset A$).

It is easy to verify the following statements on unions and intersections using Venn diagrams or simple logical arguments:

1. $(A')' = A$
2. $\varnothing' = S$
3. $S' = \varnothing$
4. $A \cup A' = S$
5. $A \cap A' = \varnothing$
6. $A \cup \varnothing = A$
7. $A \cap \varnothing = \varnothing$
8. $(A \cap B)' = A' \cup B'$

3.2 Counting Techniques

The concept of sample and event spaces is essential to understanding classical probability because oftentimes we must list and count the numbers of elements in a sample space and in various events to calculate the chances of those events occurring. Sometimes, determining the number of elements in an event and sample space is simple (e.g. a coin toss), but other times it is more complex (see below). In general, the probability of an event, E, is calculated as:

$$P(E) = \frac{f}{n}$$

where f = number of ways an event can occur; n = total number of outcomes in the sample space; and $P(E)$ = probability of event E occurring.

There are two ways to find the number of outcomes in a sample space and in an event. One is listing and then counting all of the elements in both the sample space and the event. In these cases, a **tree diagram** is a simple tool for listing and counting outcomes. A second method is to use mathematical techniques to calculate the number of ways something can happen. In these generally more complex cases, we can apply the **multiplication rule** and/or use **permutations** and **combinations**. Many times, we must use more than one mathematical technique to determine the number of elements.

A **tree diagram** is a systematic procedure for listing all possible outcomes in a sample space or an event. Example 3.6 illustrates the construction of a tree diagram.

Example 3.6. Assume that we are carrying out a quality control check in a particleboard mill and we have to select 3 sheets from the production line, 1 piece at a time. The mill produces either defective (D) or defect-free (N) boards. Figure 3.6 shows the construction of a tree diagram for sequentially selecting 3 boards, with every choice of board having two paths, either D or N. When all the sheets have been selected (3 in our case), we can list and count all the possible outcomes represented by each path. Generally, tree diagrams move from left to right, following the logical sequence of time.

Our sample space has eight possible outcomes (the number of terminal branches on the tree):

S = {DDD, DDN, DND, DNN, NDD, NDN, NND, NNN}

Once the sample space is identified, listing a given

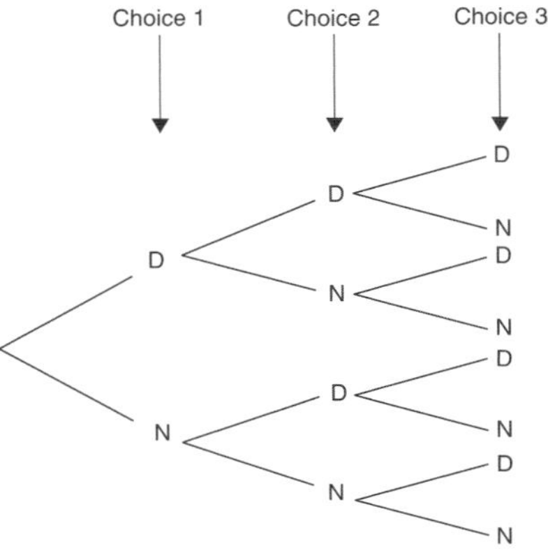

Fig. 3.6. Tree diagram for selecting 3 boards.

event is also simple. For instance, we can now define an event A as selecting only 1 defective board. Event *A* has three outcomes:

A = {DNN, NDN, NND}.

It should be fairly intuitive to the reader that the probability of selecting 1 defective board in a sample of 3 is 3/8. We will discuss this result more formally in the subsequent section.

Tree diagrams can be much more complicated with increasing numbers of outcomes. In some situations, tree diagrams can have a different number of choices (branches) for each successive step.

Example 3.7. We construct an experiment where we first flip a coin. If the outcome is a head (H), we then roll a die. If the outcome is a tail (T), we flip the coin again. This random experiment results in an asymmetric set of branches (Fig. 3.7). The sample space has eight outcomes.

S = {H1, H2, H3, H4, H5, H6, TH, TT}.

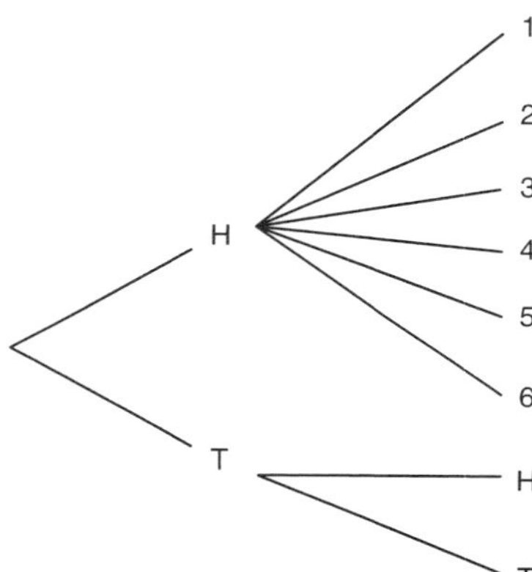

Fig. 3.7. Tree diagram for flipping a coin and rolling a die.

The total number of outcomes for a sample space in a tree diagram can generally also be calculated by what is known as the **multiplication rule**. This mathematical counting rule states that if a random experiment has a sequence of two steps, in which there are n_1 possible outcomes for the first step and n_2 for the second, the total number of outcomes is the product of the two numbers ($n_1 \times n_2$).

Example 3.8. If a restaurant offers 2 soups and 4 main courses on the lunch menu, we can order (2)(4) = 8 unique lunches.

The above definition for the multiplication rule can easily be extended to more than two steps. That is, if a random experiment has a sequence of k steps, in which there are n_1 possible outcomes for the first step, n_2 for the second step, and so on (to n_k possible outcomes on the kth step), then the total number of outcomes is $(n_1)(n_2)\ldots(n_k)$.

Example 3.9. In a home centre, wooden decks can be made using four kinds of wood. Each deck can be stained with 5 types of stain and put together with 3 different types of hardware fasteners. How many different kinds of decks can a customer buy?

(4)(5)(3) = 60.

A further extension of the multiplication rule occurs when each step of the experiment has an equal number of possible outcomes. In these cases, the total number of outcomes is n^k, where n is the number of outcomes in each step and k is the number of steps.

Example 3.10. How many outcomes are possible if we choose 5 trees from a stand infested by southern pine bark beetle, each time noting whether or not the tree is infested?

$2^5 = 32$; or
$(2)(2)(2)(2)(2) = 32$.

Note that for the above two examples, a tree diagram will give exactly the same results.

Another way to calculate the number of outcomes in sample spaces and events is by using a **permutation**, which is the number of arrangements of all or part of a set of n distinct objects. The number of permutations of n objects taken r at a time is:

$$_nP_r = \frac{n!}{(n-r)!} \tag{3.1}$$

Equation 3.1 can be derived from the multiplication rule, as

$$_nP_r = (n)(n-1)(n-2)\ldots\ldots\ldots\ldots(n-(r-1)) \tag{3.2}$$

meaning that the first object can be selected (n) ways and, as one object has been taken, the second object can be selected ($n-1$) ways, and so on until the rth object, which can be selected ($n-(r-1)$) ways (since ($r-1$) objects have been taken already). Equation 3.2 can then be algebraically simplified to Eqn 3.1 using *factorial notation*. By way of example, $n!$ is expressed in factorial notation and indicates the product of consecutive numbers from 1 to n. Most statistical calculators have these functions built in, which allow for the easy computation of factorials like $5! = (1)(2)(3)(4)(5) = 120$. Note that $0!$ is, by definition, equal to 1, although the proof of this is beyond the scope of this text.

Example 3.11. We can plant 4 trees in a row at the front of a house. If we have 6 trees, all different species, how many ways can we plant them in a row?

$$_6P_4 = \frac{6!}{(6-4)!} = 360$$

Example 3.12. If we have 3 unique paintings, how many ways can we hang 2 of them side-by-side in a room?

$$_3P_2 = \frac{3!}{(3-2)!} = 6.$$

If we label the paintings from Example 3.12, *A*, *B* and *C*, these six permutations can easily be listed as *AB*, *AC*, *BA*, *BC*, *CA* and *CB*. This should clearly show that, in a permutation, *the order is important*. Permutations '*AB*' and '*BA*' are considered two distinct arrangements.

When the permutations of *all n* objects are considered (in other words, we are not taking r objects at a time), Eqn 3.1 can be simplified as:

$$_nP_n = n! \quad (3.3)$$

Since $r = n$, we have $_nP_n = n!/(n-n)! = n!/0!$. Because $0! = 1$, $_nP_n = n!$.

Example 3.13. In how many ways can we plant all 6 of the trees from Example 3.11 in a row at the front of a house?

$$_6P_6 = 6! = 720.$$

A special case of a permutation is the **circular permutation**. Consider the problem of seating 5 people at a round table. Their arrangement is not considered different if they each move seats one place to the right or to the left because they are in a connected circle and not a disconnected line. The solution to this problem is to fix the position of one person and find the number of permutations for the remaining 4 persons; that is, $4! = 24$. In general, the number of permutations of n distinct subjects in a circle is denoted P_c and is calculated as:

$$P_c = (n-1)! \quad (3.4)$$

Another special case of a permutation is the **permutation of similar objects.** This occurs when some of the objects, among the n objects, are not distinguishable. For example, if we have 3 Douglas-fir, 2 birch and 4 oak seedlings, we may assume that we cannot distinguish between the trees within each species (i.e. we cannot tell the 2 birches apart). In general, out of n objects, if n_1 objects are of one kind that are indistinguishable, n_2 are of a second kind and so on until n_k, the number of permutations of similar objects, P_s, is:

$$P_s = \frac{n!}{(n_1!)(n_2!)\ldots(n_k!)} \quad (3.5)$$

Example 3.14. The number of ways the above 9 trees can be planted in a row is:

$$P_s = \frac{9!}{(3!)(2!)(4!)} = 1260.$$

Oftentimes, we are not concerned with the order in which r objects are selected from n distinct objects. If, for instance, in Example 3.11 we were merely selecting paintings as opposed to arranging them on a wall, we would not need to distinguish between outcomes 'AB' and 'BA' because they are made up of the same objects. If *order is not important*, we use **combinations** of the objects to describe the number of possible outcomes. The number of combinations of n distinct objects, taken r at a time is:

$$_nC_r = \frac{n!}{r!\,(n-r)!} \quad (3.6)$$

Another common notation for combinations is $\binom{n}{r}$ and is often stated as 'n choose r'.

Notice that Eqn 3.6 can be obtained by dividing the number of permutations (Eqn 3.1) by $r!$, since the *same* objects appear together $r!$ times. For instance, in the painting example (Example 3.12), the two permutations 'AB' and 'BA' are considered to be only one combination. Thus, if order is not important, we can derive the number of combinations by dividing the number of permutations (six, from Example 3.12) by $r!$ (2! in this case) for a total of three combinations.

Example 3.15. Using the 6 species of trees in Example 3.11, how many ways can we randomly select 4 out of the 6 trees (i.e. order is unimportant)?

$$_6C_4 = \frac{6!}{(4!)(6-4)!} = 15.$$

Using the combination equation above, we can see that there are 15 ways to select 4 out of 6 trees without regard to order. However, if we instead planted or arranged the trees, a certain order is implied for each of the 15 combinations and the problem becomes one of permutations. In fact, each of the 15 combinations of 4 trees can be lined up (or arranged) in 24 (4!) distinct ways. In other words, the total number of permutations equals 360 (15 combinations × 24 ways that each combination can be arranged), which agrees with Example 3.11. This example should again illustrate the difference between considering the number of combinations and the number of permutations of objects. However, the reader should also be cautioned that it is sometimes difficult to distinguish between combination and permutation problems. Often, in counting problems, statements about order are not explicitly made, but are implied. We recommend practising these sorts of problems as much as possible.

Example 3.16. How many ways can we select 3 students out of 5 to sit on a university committee? If we have not assigned any 'positions' (chair, secretary, treasurer) to the individuals, the question can be answered by considering the number of combinations.

$$_5C_3 = \frac{5!}{(3!)(5-3)!} = 10.$$

Using the combination equation, we see that there are 10 different committees of 3 that can be set up from 5 students. However, if we look at assigning positions on the committee, the problem becomes one of permutations. Again, the importance of order here is not explicitly stated but implied, and we must recognize that a committee consisting of Black (chair), Jones (secretary) and Smith (treasurer) is different from one consisting of Jones (chair), Black (secretary) and Smith (treasurer). The number of ways that 3 students can be picked from 5 if each student is to have a distinct position then becomes the permutation:

$$_5P_3 = \frac{5!}{(5-3)!} = 60$$

Considering the result of the combination above, this means that every 3-person committee selected can actually form 6 different committees when positions are assigned.

Example 3.17. We have a bag of 9 seedlings, 3 of which are stunted in growth. How many ways can we select 4 seedlings such that exactly 1 out of the 4 selected is stunted? Using the combination equation, consider first how many ways that a normally growing seedling can be selected:

$_6C_3 = (6!)/[(3!)(6–3)!] = 20$ ways.

Now consider the number of ways that a stunted one can be selected:

$_3C_1$ = (3!)/[(1!)(3–1)!] = 3 ways.

Then, using the multiplication rule, multiply the combinations of normal and stunted seedlings together for a total of (20)(3) = 60 ways.

3.3 Probability

Probability is the measure of likelihood of the occurrence or non-occurrence of an event. As defined earlier, an event is usually symbolized by a capital letter, say A, and its probability is symbolized by $P(A)$. Mathematically, a scale ranging from 0 to 1 is used to evaluate the likelihood of occurrence of an event. If an event is very likely to occur, it is assigned a probability close to 1. If an event is very unlikely to occur, it is assigned a probability close to 0. It follows, then, that an event that is 'certain' to occur has a probability of 1, while an event that is 'impossible' has a probability of 0. The probability of the event that the sun will rise tomorrow is 1. The probability of the event that a tossed coin will not land anywhere (stays in the air) is 0. In practical applications, probabilities are often converted to percentages, with the possible values ranging from 0% to 100%, and are frequently referred to as chances. For example, a weather forecaster may say that, 'The chance of showers tomorrow is 80%,' meaning that the probability of rain tomorrow is 0.8.

There are three kinds of probabilities: classical, empirical and subjective.

Classical probability is calculated from the knowledge of the sample space and an event from a random experiment. It is so named because it was the first type of probability studied by mathematicians in the 17th century. As we discussed in Section 3.2, the probability of an event, A, can be calculated from the total number of outcomes in a sample space, n, and the number of ways that event A can occur, f.

$$P(A) = \frac{f}{n}. \tag{3.7}$$

In other words, f is the number of outcomes in event A, whereas n is the number of total outcomes in the entire sample space. Equation 3.7 assumes the total number of outcomes, n, is equally likely; that is, they all have exactly the same probability of occurring.

Example 3.18. Two dice, one red and the other green, are rolled. What is the probability of event A, defined as having the number of dots totalling 7, occurring? All of the 36 outcomes in this sample space are listed, with the event A defined in **boldface**:

1–1	1–2	1–3	1–4	1–5	**1–6**
2–1	2–2	2–3	2–4	**2–5**	2–6
3–1	3–2	3–3	**3–4**	3–5	3–6
4–1	4–2	**4–3**	4–4	4–5	4–6
5–1	**5–2**	5–3	5–4	5–5	5–6
6–1	6–2	6–3	6–4	6–5	6–6

A = {1–6, 2–5, 3–4, 4–3, 5–2, 6–1}

Since $n = 36$ and $f = 6$, $P(A) = \frac{6}{36} = \frac{1}{6} \approx 0.16667$.

Empirical probabilities are based on experiments for which the possible outcomes and the number of outcomes favouring an event are not known exactly, but generally have been observed. If an experiment is repeated n times and f out of the n trials favours event B, the probability can be calculated as:

$$P(B) \approx \frac{f}{n}. \tag{3.8}$$

Here, f is called the *frequency* of event B. The symbol $\approx$ means that the probability is approximately equal to the theoretical value that would be expected (i.e. the classical probability). Empirical probabilities change from one experiment to another for the same event, while classical probabilities remain the same. Take the example of flipping a coin 500 times. Classical probability tells us that the probability of getting a head would be 0.5 and, thus, we would expect 250 of the flips to be heads. In reality, however, this is unlikely to occur. We may get 240 heads in a 500-flip experiment and, thus, the empirical (observed) probability would be 240/500 = 0.48, a value that is close, but not exactly equal to 0.5.

The relative frequencies in frequency distributions are empirical probabilities if the distributions are created from samples. These relative frequencies change if we take a different sample from the same population. Table 2.2 (see Chapter 2) shows the frequency distribution of crown classes for 50 trees taken from a stand. Since 14 codominant trees and 9 dominant trees were observed, the empirical probabilities of trees being codominant and dominant are 0.28 and 0.18, respectively. If we took another sample of 50 trees from the same stand (independently from the first), these two probabilities would very likely change.

Subjective probabilities are based on a person's experiences, or 'educated guesses'. For example, an avian biologist may say that, 'If we log this area, there is a 15% chance that cavity-nesting birds may never return,' or a forester may note that, 'If we plant this logged-over area next spring, 80% of the seedlings will survive.' These statements are not substantiated by exact scientific evidence and are based solely on an individual's experience.

Subjective probabilities are often seen in gambling, sporting events and horse racing, where the term '**odds**' is generally used in lieu of probability. However, before defining what 'odds' means precisely, it is necessary to state some properties of probabilities.

The **properties of probability** can be summarized as follows:

1. For a given event A, $0 \le P(A) \le 1$ (i.e. the probability of an event must be between complete uncertainty and complete certainty).
2. The sum of the probabilities of all possible simple events in a sample space must equal 1.
3. For a given event A,

$$P(A) + P(A') = 1,$$

where A' is called the complement of A and represents an event defined by A *not* occurring. Thus,

$$P(A') = 1 - P(A) \qquad \text{and} \qquad P(A) = 1 - P(A').$$

4. For a given event A, $P(A)$ is the sum of the probabilities of all simple events

corresponding to A. That is, if A consists of several simple events, the sum of the probabilities of all these events will sum to $P(A)$.

Now that we know the mathematical meaning of an event's complement, we can define the term 'odds'. The odds in favour of an event A are:

$$\text{odds}(A) = \frac{P(A)}{P(A')}.$$

Again, odds are often used in gaming events involving subjective probabilities, but this is not always the case. Take, for example, an experiment where two coins are tossed and the sample space is defined as $S = \{\text{HH, HT, TH, TT}\}$. The odds in favour of obtaining at least one head are computed as follows (note that the denominator term is the probability of not obtaining at least one head):

$$\text{odds(at least one head)} = \frac{\frac{3}{4}}{\frac{1}{4}} = 3$$

The way to state these odds is to say, 'there are 3 to 1 odds in favour of obtaining at least one head.' It can also be written as a ratio, 3:1. Alternatively, we may state that 'there are 1 in 3 odds against not obtaining at least one head.'

Odds of $a{:}b$ can be converted back to a probability as follows:

$$P(A) = \frac{a}{a+b}.$$

For example:

$$P(\text{at least one head}) = \frac{3}{3+1} = \frac{3}{4}.$$

3.4 Rules for Probabilities

Two basic and commonly used rules for operations with probabilities are the **addition rule** and the **multiplication rule**.

The **addition rule** is based on the probability of the union of events. For two events A and B, the addition rule is:

$$P(A \cup B) = P(A) + P(B) - P(A \cap B) \tag{3.9}$$

The union of the two events above represents the probability of either event A or event B occurring. It is represented by all of the sample points found in A or B. As shown in Fig. 3.8, as we add $P(A)$ and $P(B)$, we are double-counting the intersection space $P(A \cap B)$ because it is included in both events A and B. Hence, to arrive at $P(A \cup B)$, we must subtract (once) the probability of the intersection from the sum of probabilities of the two events.

Again, the addition rule should be interpreted as the probability of the occurrence of A **or** B. This is why in several texts, the notation of $P(A \text{ or } B)$ is used in place of $P(A \cup B)$. In situations where events A and B are **mutually exclusive** or **disjoint** (that is, they do not have any common intersecting elements), Eqn 3.9 simplifies to:

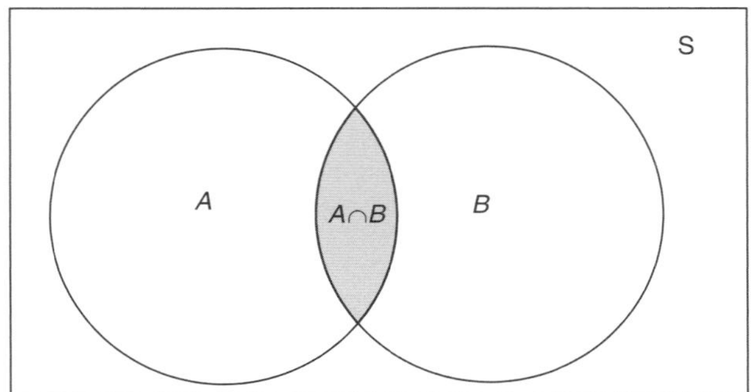

Fig. 3.8. Common elements (intersection) of events *A* and *B*.

$$P(A \cup B) = P(A) + P(B), \tag{3.10}$$

since

$$P(A \cap B) = 0.$$

Equation 3.10 can be extended for *k* mutually exclusive events, such as A_1, A_2, ... and A_k:

$$P(A_1 \cup A_2 \cup \ldots \cup A_k) = P(A_1) + P(A_2) + \ldots + P(A_k). \tag{3.11}$$

Equation 3.9 can also be extended to more than two events. However, the inclusion of multiple intersections creates a rather complicated equation and is beyond the scope of this book.

Example 3.19. What is the probability of getting a total of 3 or 7, when two dice are rolled? Let event *B* be the total of 3 and event *A* the total of 7. From Example 3.18, we already know that

$$P(A) = \frac{6}{36}, \text{ and we can work out that } P(B) = \frac{2}{36}, \text{ since } B = \{1-2,\ 2-1\}.$$

Examining the elements in the two events, we quickly realize that they do not have any common elements. We can therefore say that they are mutually exclusive events and we apply Eqn 3.10:

$$P(A \cup B) = \frac{6}{36} + \frac{2}{36} = \frac{8}{36} = \frac{2}{9} \approx 0.222.$$

It is good practice to work out the meaning of probabilities like 0.222 to get a sense of what they imply for the question in hand. The practical interpretation of 0.222 is that, if we roll the two dice several times, the total of '3 or 7' will occur a little more than one out of five times.

Example 3.20. Assume now that we roll a single die, what is the probability that event *A* or *B* occurs? Where

$A = \{< 3\} = \{1, 2\}$
$B = \{\text{odd}\} = \{1, 3, 5\}.$

From here

$$P(A) = \frac{2}{6} \qquad \text{and} \qquad P(B) = \frac{3}{6}.$$

We can see that A and B are not mutually exclusive events because they have one common element: 1 (see Fig. 3.8). The probability of the intersection of A and B (rolling a '1') is then:

$$P(A \cap B) = \frac{1}{6}.$$

To answer the question about the probability of either event A or B occurring, we must consider the union of two events using Eqn 3.9:

$$P(A \cup B) = \frac{2}{6} + \frac{3}{6} - \frac{1}{6} = \frac{4}{6} = \frac{2}{3} \approx 0.667.$$

If we roll the die several times, a roll of 3 or less or an odd number should occur about twice in every three rolls.

When required, Eqn 3.9 can be rearranged so that the probability of the intersection of two events can be readily calculated:

$$P(A \cap B) = P(A) + P(B) - P(A \cup B).$$

Often, the occurrence of one event will influence the probability of another event. For example, it is well known that drinking alcohol before driving affects the chance of having a car accident, and applying fertilizer affects the height growth of seedlings. In such cases, we are dealing with **conditional probabilities**.

We will introduce the concept of conditional probability by further exploring Example 3.17. In this example, rolling two dice (a red one and a green one) produced 36 outcomes, which are listed below.

1–1	1–2	**1–3**	1–4	**1–5**	1–6
2–1	**2–2**	2–3	**2–4**	2–5	**2–6**
3–1	3–2	**3–3**	3–4	**3–5**	*3–6*
4–1	**4–2**	4–3	**4–4**	*4–5*	***4–6***
5–1	5–2	**5–3**	*5–4*	***5–5***	*5–6*
6–1	**6–2**	*6–3*	***6–4***	*6–5*	***6–6***

Consider event A to be that the total is an even number (in **boldface** in the list above) and event B to be that the total is greater than or equal to 9 (in italics in the list above). Note that some of the sample points meet the criteria of both events (shown in ***boldface-italic***). After counting the outcomes for these two events, we can calculate the probability, of each event as:

$$P(A) = \frac{18}{36} = 0.5$$

and

$$P(B) = \frac{10}{36} \approx 0.278.$$

An example of a conditional probability problem here would be asking the question, 'If two dice are rolled, what is the probability that the outcome is even if we already know that the outcome is greater than or equal to 9?' In this situation, we know that event B has occurred and we are interested in understanding the effect of this information on the probability of event A. We symbolize the conditional probability of event A given that event B has occurred as $P(A|B)$. The conditional probability represents a *redefined sample space*. In our example, out of the 36 outcomes (original sample space) only ten qualify for the new sample space, as ten outcomes have a total greater than or equal to nine (italics), and four out of these ten outcomes are even (boldface-italics). By intuition, the conditional probability should then be:

$$P(A|B) = \frac{4}{10} = 0.4.$$

It is interesting to note that the probability of A changed from 0.5 (unconditional) to 0.4 (conditional) with the additional information. The general equation to calculate conditional probabilities is:

$$P(A|B) = \frac{P(A \cap B)}{P(B)}; \quad P(B) > 0. \tag{3.12}$$

It is also true that

$$P(B|A) = \frac{P(A \cap B)}{P(A)}; \quad P(A) > 0. \tag{3.13}$$

Example 3.21. We draw a card from a deck of 52 cards. Event A is that the card is a spade and event B is that the card is a face card (jack, queen, king). Since 13 cards are spades, 12 cards are face cards and 3 cards are both spade and face cards, the probabilities are:

$$P(A) = \frac{13}{52}, \quad P(B) = \frac{12}{52} \quad \text{and} \quad P(A \cap B) = \frac{3}{52}.$$

If we are told that the outcome is a spade, what is the probability that it is also a face card? This can be calculated using Eqn 3.12:

$$P(B|A) = \frac{\frac{3}{52}}{\frac{13}{52}} = \frac{3}{13} = \frac{12}{52}.$$

On the other hand, if we are told that the outcome is a face card, what is the probability that it is also a spade?

$$P(A|B) = \frac{\frac{3}{52}}{\frac{12}{52}} = \frac{3}{12} = \frac{13}{52}.$$

Interestingly, the conditional probabilities in the previous example for both A and B are the same as the unconditional probabilities. In cases like these, we refer to the events A and B as **independent**. Two events are said to be independent if the probability of one event is not affected by the occurrence or non-occurrence of the other event.

If two events, A and B, are independent, then:

$$P(A) = P(A|B) \qquad \text{and} \qquad P(B) = (B|A).$$

Either of these equalities can be used to prove or disprove the independence of two events.

To extend the idea of conditional probability, we will study a research example with empirical probabilities. A forest scientist has studied the germination of white pine and yellow pine seeds. The results are summarized in the following table.

	Germinated	Not germinated	Total
Yellow pine	60	10	70
White pine	65	15	80
Total	125	25	150

Consider the two events; G, that a seed has germinated and Y, that a seed is yellow pine. The following probabilities can be calculated from the frequencies above:

$$P(G) = \frac{125}{150} \approx 0.833, \qquad P(Y) = \frac{70}{150} \approx 0.467, \qquad P(G \cap Y) = \frac{60}{150} = 0.40.$$

We can also compute the conditional probability that a seed germinates given that it is yellow pine using Eqn 3.12:

$$P(G|Y) = \frac{0.40}{0.467} \approx 0.857.$$

This conditional probability can also be calculated by using intuition. Consider the redefined sample space of 70 yellow pine seeds. In this new sample space, 60 seeds germinated. Therefore, the conditional probability is:

$$P(G|Y) = \frac{60}{70} \approx 0.857.$$

A point that confuses first-time readers is the distinction between a conditional probability and an intersection between two events. A conditional probability refers to the probability of one event occurring, given that another has already occurred, while an intersection refers to the probability of two events occurring simultaneously. This is discussed in further detail below.

Equations 3.12 and 3.13 can be rearranged such that the intersection of two events can be calculated:

$$P(A \cap B) = P(B)\,P(A|B) \tag{3.14}$$

and

$$P(A \cap B) = P(A)\,P(B|A). \tag{3.15}$$

Equations 3.14 and 3.15 are referred to as the **multiplication rule**, which enables us to calculate the probability that two events both occur. Several books use the notation, $P(A$ and $B)$, rather than $P(A \cap B)$, indicating the probability of both A *and* B occurring simultaneously.

The multiplication rule simplifies for independent events, since $P(A|B) = P(A)$ and $P(B|A) = P(B)$:

$$P(A \cap B) = P(B)\,P(A) = P(A)\,P(B). \tag{3.16}$$

Equation 3.16 can be restated in words as: the probability that two *independent* events will both occur (or the probability of the intersection of two independent events) is the product of their corresponding probabilities.

Equation 3.16 can also be extended for k independent events, $A_1, A_2, \ldots, A_k$.

$$P(A_1 \cap A_2 \ldots A_k) = P(A_1)\, P(A_2) \ldots P(A_k). \tag{3.17}$$

Similarly, Eqns 3.14 and 3.15 can be extended to more than two events but, because they take on complicated forms, these equations will not be discussed in this text.

Example 3.22. In a lumber remanufacturing plant, 8 spare parts are kept for a planing machine. It is known that 2 of these spare parts are defective. If the mechanic randomly selects 2 of these 8 spare parts, what is the probability that the 2 spare parts are both functional? Let D represent the event that a defective part is chosen, and N represent the event that a non-defective part is chosen.

Note that this problem introduces a very important statistical concept, **sampling without replacement**. Let N_1 be the event that the first part selected by the mechanic is functional and N_2 be the event that the second part selected is functional. The probability of the first choice being functional is simply the proportion of non-defective parts:

$$P(N_1) = \frac{6}{8}.$$

However, the probability of the second choice is conditional. Since the first choice was a functional part, then only 5 functional parts remain out of a new total of 7 parts:

$$P(N_2|N_1) = \frac{5}{7},$$

But if the first choice was a defective one, 6 functional parts still remain out of the 7 parts:

$$P(N_2|D_1) = \frac{6}{7}.$$

Since the problem relates to 2 functional parts being selected in sequence, we need not worry about the latter probability. The multiplication rule is used to answer this question:

$$P(N_1 \cap N_2) = P(N_1)P(N_2|N_1) = \left(\frac{6}{8}\right)\left(\frac{5}{7}\right) \approx 0.538.$$

Example 3.23. Assume that we are drawing 2 cards from a deck of 52 cards. Cards can be drawn with or without replacement and this affects the probabilities of each subsequent draw. In this case, **sampling with replacement** means that after a card is drawn and observed, it is put back into the deck before the next card is drawn. Sampling without replacement means that after the first card is drawn, it is set aside from the deck when the subsequent card is drawn. We can compute the probability of drawing two spades (our event) for both situations.

When the sampling is done without replacement, the case follows the logic in Example 3.22.

$$P(S_1) = \frac{13}{52} \quad \text{and} \quad P(S_2|S_1) = \frac{12}{51}.$$

From the multiplication rule, the probability of the intersection, that is, that both cards drawn are spades, is:

$$P(S_1 \cap S_2) = \left(\frac{13}{52}\right)\left(\frac{12}{51}\right) \approx 0.0588.$$

However, when sampling is done with replacement, the respective probabilities are:

$$(S_1) = \frac{13}{52} \quad \text{and} \quad P(S_2|S_2) = \frac{13}{52}.$$

When sampling with replacement, the two events S_1 and S_2 are independent because the probability of drawing one spade is the same as the probability of drawing a second spade, given that one has already been drawn. Therefore, the probability of the intersection can be more simply calculated as:

$$P(S_1 \cap S_2) = P(S_1)\,P(S_2) = \left(\frac{13}{52}\right)\left(\frac{13}{52}\right) = 0.0625.$$

This example serves two purposes. It demonstrates the difference between sampling with replacement and sampling without replacement. It also demonstrates the application of the multiplication rule when two events are *dependent* or *independent*.

Example 3.24. One box contains 3 US dimes and 5 Canadian dimes and another box contains 4 US dimes and 2 Canadian dimes. One dime is picked from Box 1 and transferred into Box 2, unseen. Once this is done, what is the probability that a dime that is picked from Box 2 is Canadian?

Tree diagrams (Fig. 3.9) can be very helpful in conceptualizing conditional problems like this. Note how the branches of the tree follow the temporal (time) order of the problem (picking dimes from two boxes, one at a time), which serves to illustrate how each subsequent branch is actually a conditional probability. Based on the information provided in the problem, a tree diagram can be constructed; however, one needs to be mindful of how the conditional probabilities will change along every path. A quick test is to make sure that the probabilities for each set of branches (representing all of the choices at any given step) add up to 1.

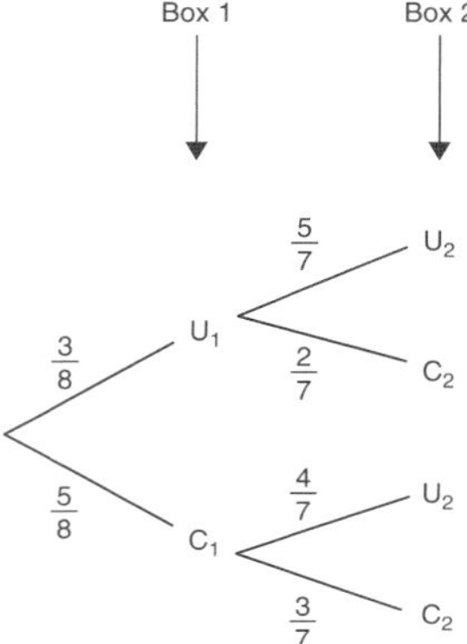

Fig. 3.9. Tree diagram for selecting US and Canadian dimes.

Let us define U as the event of picking a US dime and C as the event of picking a Canadian dime. Subscript 1 indicates that a dime was picked from Box 1, while subscript 2 indicates that it was picked from Box 2.

By following the tree diagram, it can be seen that there are two possible ways of picking a Canadian dime from the second box – one in which the first pick was a US dime and the other in which the first pick was a Canadian dime. Thus, this is actually a union problem and the probability is computed as follows:

$$P[(U_1 \cap C_2) \cup (C_1 \cap C_2)] = \left(\frac{3}{8}\right)\left(\frac{2}{7}\right) + \left(\frac{5}{8}\right)\left(\frac{3}{7}\right) = 0.375,$$

since

$$P(U_1 \cap C_2) = P(U_1)P(C_2|U_1) = \left(\frac{3}{8}\right)\left(\frac{2}{7}\right)$$

and

$$P(C_1 \cap C_2) = P(C_1)P(C_2|C_1) = \left(\frac{5}{8}\right)\left(\frac{3}{7}\right).$$

It is worth noting that this problem is actually a union of two intersecting events. Again, the difference between intersecting probabilities and conditional probabilities should be considered, and tree diagrams are useful in this regard. Note that conditional probabilities occur within the tree at each subsequent branching. In this case, there are four conditional probabilities. There are also four possible intersections, but they are not conveyed directly on the tree diagram. Rather, they can be obtained by multiplying through all of the probabilities for each final outcome (branch).

3.5 Bayes' Theorem

The conditional probabilities that we have seen thus far have followed a logical, temporal order. For example, if we had enough information, it would be possible to work out the conditional probabilities for a student to break a leg given that he or she: (i) went skiing; or (ii) went hiking; or (iii) played soccer; or (iv) stayed at home to study for the statistics midterm. These are referred to as **prior probabilities**, since they are based on previously observed frequencies for the four activities.

Sometimes, though, conditional probabilities can be reversed. For instance, if a student walked into the classroom with a broken leg, we could also work in the reverse order and calculate the probabilities that he or she: (i) went skiing; or (ii) went hiking; or (iii) played soccer; or (iv) stayed at home to study for the statistics midterm. These are referred to as **posterior probabilities**, since they are based on the fact that the student already has a broken leg. **Bayes' Theorem** is used to work out posterior probabilities. This theorem is introduced with a simplified practical example.

Example 3.25. Assume that logs arrive by truck to a sawmill from 3 cutblocks. The probabilities are 0.60, 0.25 and 0.15 that the logs are from cutblocks A_1, A_2 and A_3, respectively. Note that the events for Bayes' Theorem must be mutually exclusive (disjoint) and **collectively exhaustive**. These two conditions are met if the following are true:

1. When one event occurs, the others cannot (the events are mutually exclusive).
2. The sum of the probabilities for all the possible events (three in our example) equals 1 (the events are collectively exhaustive).

Let us further assume that only two species, Douglas-fir and western hemlock, are logged on each of the 3 cutblocks and that the two species are sorted and transported on separate logging trucks. The following table shows the proportions (probabilities) of occurrence for the two species by cutblock.

	Cutblock		
Species	A_1	A_2	A_3
Douglas-fir	0.50	0.40	0.30
Western hemlock	0.50	0.60	0.70

Figure 3.10 shows the sample space with events A_1, A_2, A_3 and B, where A_i refers to the ith cutting block and B is the event that the species is Douglas-fir. If we observe that a truck has arrived with a load of Douglas-fir logs, we can work out the probabilities that it originated from any 1 of the 3 possible cutting blocks by applying Bayes' Theorem:

$$P(A_i|B) = \frac{P(A_i \cap B)}{\sum_{i=1}^{k} P(A_i \cap B)} \tag{3.18}$$

and

$$P(A_i|B) = \frac{P(A_i)\,P(B|A_i)}{\sum_{i=1}^{k} P(A_i)\,P(B|A_i)}. \tag{3.19}$$

$i = 1, 2 \ldots k.$

Equations 3.18 and 3.19 are given for k mutually exclusive and collectively exhaustive events assuming a prior knowledge of event B. This equation is essentially the same formulation as the conditional probability Eqns 3.12 and 3.13, as the denominator function, when expanded, simply reduces to $P(B)$.

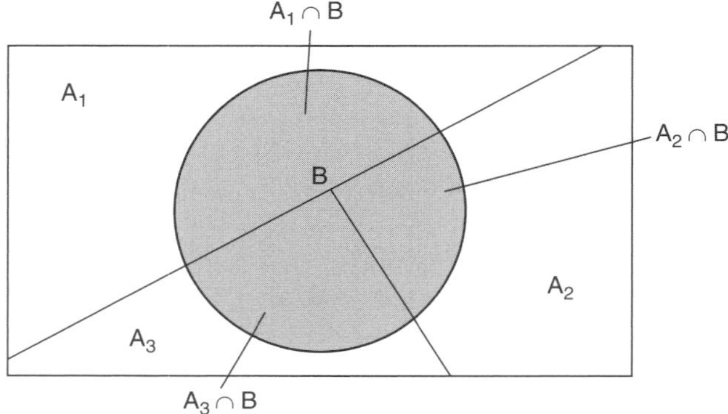

Fig. 3.10. Venn diagram for events A_1, A_2, A_3 and B.

The Venn diagram (Fig. 3.10), a tree diagram (Fig. 3.11) and the following algebraic steps demonstrate the derivation of Eqns 3.18 and 3.19. In order to define event B (Douglas-fir), we can use the Venn diagram (Fig. 3.10) to show that it is actually a union of three mutually exclusive events, each defined by an intersection:

$$B = (A_1 \cap B) \cup (A_2 \cap B) \cup (A_3 \cap B).$$

Since the intersections are disjointed, we can see from Fig. 3.11 that:

$$P(B) = P(A_1 \cap B) + P(A_2 \cap B) + P(A_3 \cap B).$$

Using the multiplication rule (Eqn 3.14) and Fig. 3.10, the probability of B can be calculated as:

$$P(B) = P(A_1)P(B|A_1) + P(A_2)P(B|A_2) + P(A_3)P(B|A_3).$$

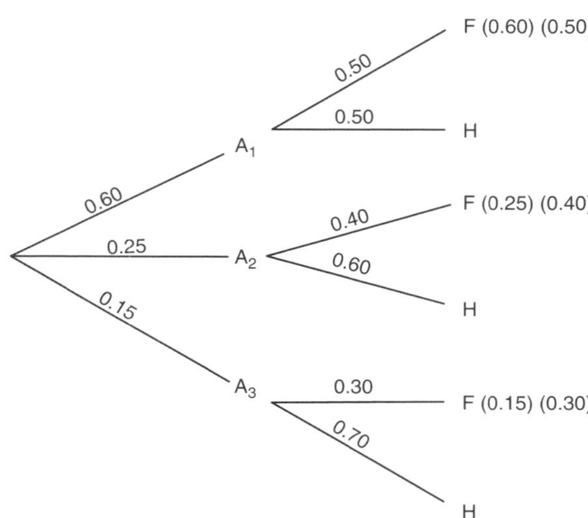

Fig. 3.11. Tree diagram for Bayes' Theorem (F = Douglas-fir and H = western hemlock).

It can be seen from Fig. 3.11 that the ratio of $P(A_1)P(B|A_1)$ divided by $P(B)$ is the probability that a known Douglas-fir load is from cutblock A_1. Numerically,

$$P(A_1|B) = \frac{(0.6)(0.5)}{(0.6)(0.5)+(0.25)(0.40)+(0.15)(0.30)} = \frac{0.30}{0.445} \approx 0.674.$$

Similarly, the probabilities can be calculated for A_2 and A_3 as well.

$$P(A_2|B) = \frac{(0.25)(0.40)}{(0.445)} \approx 0.225$$

$$P(A_3|B) = \frac{(0.15)(0.3)}{0.445} \approx 0.101.$$

These steps can also be used to verify Eqns 3.18 and 3.19.

Note that in using Bayes' Theorem, the numerator term will appear somewhere in the denominator term. This is a useful check, but it also makes the probability easier to conceptualize. In the first case, the conditional probability is defined by one intersection (a load of trees being both Douglas-firs and from cutblock 1) being divided by the sum of all the intersections (which works out to be the total probability of a load being Douglas-firs). In this light, Bayes' Theorem, like all probabilities, is just a proportion. It should also be noted that some statisticians question the validity of Bayes' Theorem because it is disputed whether prior probabilities can reasonably be assigned. However, this debate is beyond the scope of this text.

Exercises

Section 3.1

3.1. Assume a random experiment in which a balanced die is rolled and then a balanced coin is tossed.

a. List the elements in the sample space and count the total number of possible outcomes.
b. Let event A be that the outcome on the die is an even number. List the elements corresponding to event A.
c. Let event B be that the outcome on the coin toss is a head. List the elements corresponding to event B.
d. Draw a Venn diagram of the sample space with events A and B.
e. List the elements in the intersection of events A and B.
f. List the elements in the union of events A and B.
g. List the elements in A'.
h. List the elements in B'.
i. Define a simple event in this random experiment.
j. Define a compound event in this random experiment.

3.2. Suppose that we have 5 cards numbered 0, 1, 2, 3 and 4. Two cards are drawn in such a way that after the first card is drawn, we put the card aside and draw the second card from the remaining 4 (sampling without replacement).

a. List the elements in the sample space.
b. Let event A be that the total on the 2 cards is odd. List the elements corresponding to event A.

c. Let event B be that the total on the 2 cards is even. List the elements corresponding to event B.
d. Draw a Venn diagram of the sample space and events A and B.
e. Let event C be that the total on the 2 cards is less than 3. List the elements corresponding to event C.
f. Draw a Venn diagram of the sample space and events A, B and C.
g. List the elements in $A \cap B$, $A \cap C$, $B \cap C$, $A \cup B$, $A \cup C$ and $B \cup C$.

3.3. Consider the following sample space and events.

S = {Douglas-fir, cedar, hemlock, alder, maple, spruce, birch}
A = {Douglas-fir, hemlock, cedar}
B = {cedar, spruce}
C = {alder, maple, birch}

List the elements corresponding to the following events:

a. $A \cap B$.
b. $A \cup B$.
c. $A' \cap B$.
d. $B \cap C' \cap A$.
e. $(A' \cup B) \cap (A' \cap C')$.

3.4. Three seeds are drawn from a bag of mixed Douglas-fir and ponderosa pine seeds.

a. List the elements in the sample space.
b. Let event A be that all 3 seeds are ponderosa pine. List the elements in event A.
c. What kind of event is A?
d. Let event B be that the first seed drawn is a Douglas-fir. List the elements in event B.
e. What kind of event is B?
f. Describe the relationship between A and B.

Section 3.2

3.5. Find the total number of outcomes in the random experiment described in Exercise 3.2 using:

a. A tree diagram.
b. The multiplication rule.
c. Permutations.

3.6. Find the total number of outcomes in the random experiment described in Exercise 3.4 using:

a. A tree diagram.
b. The multiplication rule.

3.7. Find the total number of outcomes in the random experiment described in Exercise 3.1 using:

a. A tree diagram.
b. The multiplication rule.

3.8. A natural resources conservation student can select his or her second year elective course from a choice of 2 courses, a third year elective from a choice of 4 courses, and a fourth year (final year) elective from a choice of 2 courses. How many ways can the student select electives in the last 3 years of the programme?

3.9. A test consists of 6 multiple-choice questions in which the first 3 questions have 3 possible answers and the last 3 questions have 2 possible answers. If only 1 answer can be selected per question, how many ways can the student choose answers to the questions?

3.10. In how many ways can a true–false test consisting of 12 questions be answered (assuming only 1 answer per question)?

3.11. A telephone company is adding 4 new exchanges to its service area. The 7-digit telephone numbers for these new exchanges must begin with 2, 3, 4 or 5. How many new phone numbers can be created by the company?

3.12. Assuming no ties in a 5 km race with 8 runners, how many ways can medals be awarded for first, second and third place?

3.13. A company that makes tags for trees in research plots produces tags with 3-digit codes. In doing this, they draw from 6 digits, 0 to 5 inclusive, each digit can be used only once and the codes cannot begin with 0.

a. How many 3-digit tags can be formed from these numbers?
b. How many are even numbers?
c. How many numbers are less than 240?

3.14. Three married couples have purchased 6 seats in one row for a play. How many different ways can they be seated:

a. With no restrictions?
b. If each couple sits together?
c. If all the men sit together to the left of all the women?

3.15. In how many ways can a forester select 4 out of 7 available technicians to form a field crew?

3.16. In how many ways can 6 people be seated around a circular table?

3.17. The library's reserve shelf contains 3 copies of a maths textbook, 4 copies of a biology textbook, 2 copies of a chemistry textbook and 3 copies of a physics textbook. How many ways can the librarian arrange these 12 textbooks on a shelf?

3.18. How many ways can the ABC Forestry Company hire 3 co-op students from 7 equally qualified applicants?

3.19. From an ordinary deck of 52 playing cards, how many ways can a poker hand (5 cards) be selected:

a. With no restrictions?
b. With 2 kings, 2 aces and 1 other card (other than a king or an ace)?
c. With a full house (3 cards of one value and 2 cards of another value)?

3.20. A committee of 3 is selected from 4 women and 3 men. How many selections are possible, if the committee consists of:

a. All women?
b. One man and 2 women?
c. Two men and 1 woman?
d. Any combination of men and women?

Section 3.3

3.21. Calculate the probability for events A, B, A' and B' in Exercise 3.1.

3.22. In LOTTO 6/49®, players choose 6 numbers from 1 to 49. On each draw day, 6 randomly generated winning numbers from 1 to 49 are selected.

a. How many different ways can the 6 numbers be generated (the sample space)?
b. What is the probability of winning the jackpot (matching all the 6 winning numbers) if you select one set of 6 numbers?

3.23. The American roulette wheel contains 38 sections: 1 to 36, 0 and 00. Out of these, the 0 and 00 sections are green, while 18 of the 36 remaining sections are black and the other 18 are red. A ball is spun in the direction opposite to the wheel's motion and bets are made on the numbers or colours where the ball stops. Assume that the wheel is balanced (fair).

a. What is the probability that the ball stops in a green section?
b. What is the probability that the ball stops in a red section?
c. What is the probability the ball stops on an even number (ignoring 0 and 00)?
d. What is the probability that the ball stops on a black section?
e. What are the odds against the ball stopping on a black section?

3.24. If the weather forecast states that the odds of rain tomorrow are 6 to 11:

a. What is the probability that it will rain tomorrow?
b. What is the probability that it will not rain tomorrow?

3.25. If the forecast is that the chance for rain tomorrow is 60%:

a. What are the odds in favour of rain?
b. What are the odds against rain?

3.26. A survey indicated that last Sunday 20 out of 75 visitors to a provincial park were German citizens.

a. What is the probability that a park visitor is a German citizen?
b. What kind of probability is this?
c. Will this probability be the same if the survey is repeated on another day?

Section 3.4

3.27. A permanent sample plot contains 15 Douglas-fir, 10 western hemlock and 3 red alder trees. If a tree is selected at random:

a. What is the probability that it is either Douglas-fir or red alder?
b. What is the probability that it is not western hemlock?

3.28. In a forest stand, the probability that a tree is a Douglas-fir is 0.65, the probability that a tree is infested by bark beetles is 0.18 and the probability that a tree is a Douglas-fir and is infested by bark beetles is 0.15. If a tree is selected at random, what is the probability that it is a Douglas-fir or it is infested by bark beetles?

3.29. For a small ski resort to operate, at least 1 of their 2 chair lifts must be functioning. Given the following information, what is the probability of a shutdown?

P(lift 1 works) = 0.90
P(lift 2 works) = 0.85
P(both lifts work) = 0.765.

3.30. If 2 cards are drawn from an ordinary deck of 52 playing cards, what is the probability of getting 2 kings, if:

a. Cards are drawn with replacement?
b. Cards are drawn without replacement?

3.31. In a certain mill, 20% of the lumber cut is grade No 1, 45% is grade No 2 and 35% is grade No 3.

a. What is the probability that a piece selected at random will be grade No 1?
b. What is the probability that a piece selected at random will not be grade No 1?
c. If it is known that a piece selected at random is not grade No 1, what is the probability that it is not grade No 2 either?

3.32. If a student guesses on all 8 questions on a true–false exam, what is the probability that he or she will get them all correct?

3.33. Suppose that a balanced coin is independently tossed twice. List the elements in the following events and their associated probabilities:

A = head appears on the first toss;
B = head appears on the second toss; and
C = both tosses yield the same outcome.

Prove whether:

a. A and B are independent.
b. A and C are independent.
c. All of the three events are independent.

3.34. List the elements in the following events and their associated probabilities if a balanced die is rolled:

A = the outcome is even;
B = the outcome is odd; and
C = the outcome is less than 3.

Prove whether:

a. A and B are independent.
b. A and C are independent.
c. B and C are independent.
d. Make a general statement about $A|B$ and $B|A$ when A and B are mutually exclusive (disjoint) events.

3.35. In a forest stand, the trees were classified by species and insect infestation. The results were:

	Spruce	Pine	Douglas-fir
Infested	18	10	12
Not infested	52	70	58

a. If P is the event that a tree is pine and I is the event that a tree is infested, are P and I independent? Prove why or why not.
b. If F is the event that a tree is Douglas-fir and N is the event that a tree is not infested, are F and N independent? Prove why or why not.

3.36. Two balanced dice are rolled. What is the probability that both show a 6? Show which equation you used to calculate $P(6 \cap 6)$.

3.37. A forest district has 3 helicopters for firefighting. The probability that any one of them is available when needed is 0.95.

a. What is the probability that none of them is available?
b. What is the probability that all of them are available?
c. What is the probability that only one is available?

3.38. Two manufacturers supply a certain filter used in measuring river sedimentation. Manufacturer A supplies 65% and manufacturer B supplies 35%. It is known that 5% of the filters supplied by A are defective and 10% of the filters supplied by B are defective.

a. What is the probability that a filter selected at random is defective?
b. Given that a filter is defective, what is the probability that manufacturer B supplied it?
c. Given that a filter is not defective, what is the probability that it was supplied by manufacturer A?

Section 3.5

3.39. A small sawmill is supplied with 30%, 25% and 45% of its logs by logging companies A, B and C, respectively. It is known that red stain (a pathological defect) is present in 20% of the logs supplied by A, in 5% supplied by B and in 15% supplied by C.

a. Find the probability that a randomly selected log contains red stain.
b. If it is known that a log came from company A, what is the probability of it containing red stain?
c. What is the probability of randomly selecting a log that is both from company A and contains red stain?
d. What is the probability of selecting either a log from company A or a log that contains red stain?
e. If it is known that a log contains red stain, what is the probability that it was supplied by logging company A?
f. If it is known that a log contains red stain, what is the probability that it was supplied by logging company B?
g. If it is known that a log contains red stain, what is the probability that it was supplied by logging company C?

4 Random Variables and Probability Distributions

Outcomes of Random Experiments

Like the theory of probability, random variables and their probability distributions play an important role in *statistical inference*. The main objectives of this chapter are to show how outcomes of random experiments can be described in real (numerical) terms and how probabilities can be assigned to these real numbers. Numerical descriptions of outcomes and their respective probabilities form what are known as probability distributions or probability density functions. We can use these distributions to compute the means and the variances of the random variables that they describe. All of these tools are useful in helping to provide further information for describing populations.

4.1 Random Variables

In Chapter 3 (this volume), we discussed the concepts of random experiments, sample spaces and outcomes. Some random experiments produce outcomes that can be described by letters, symbols or just general descriptions. Other experiments produce outcomes in numerical terms, such as: the number of heads that could occur when a coin is tossed three times; the total number of dots observed when rolling a pair of dice; the number of plants in a 100 m^2 area; or the number of seeds that germinate in a seedbed. A **random variable** is a well-defined *numerical description* of the outcomes in the sample space of a random experiment. We will denote random variables by capital letters, such as X, Y or Z, while small letters, such as x, y or z (usually with subscripts), will denote individual values or outcomes for that random variable.

A sample space associated with a random experiment can be classified as discrete or continuous. A **discrete sample space** is one that contains a finite number of elements, such as the eight possible outcomes from tossing a coin three times. A discrete sample space can also be unending, but countable, such as the sample space associated with tossing a coin until a head appears (the number of tosses necessary to meet this condition is the set of all possible positive whole numbers). Discrete random variables always take the form of data that are counted, such as the number of infested trees or the number of accidents per month in a logging camp.

A **continuous sample space** is one that contains an infinite and uncountable number of outcomes. Any random variable obtained by measurements, like the time to germination, the weight of salmon, the distance between forest dependent communities, or the volume of a tree, can theoretically take on any value in a measurement interval. For instance, for any two given merchantable tree volumes, e.g. 3.1 m^3 and 3.2 m^3, one can always find another value that occurs between them (e.g. 3.17 m^3). Theoretically, this could go on infinitely if measurement instruments were precise enough.

Random variables defined over discrete sample spaces are called **discrete random variables**, while random variables defined over continuous sample spaces are called **continuous random variables.**

4.2 Probability Distributions

A **discrete random variable** can be described by the probabilities that each of its individual values takes on when the random experiment is carried out. The list of all possible numerical outcomes and their associated probabilities is called the **probability distribution** of the random variable. For example, the probability distribution for the number of heads that occur when a coin is tossed three times is as follows (note that the random variable, number of heads, is denoted with an X, while the individual outcomes are denoted with an x):

x	0	1	2	3
$P(X=x)$	$\frac{1}{8}$	$\frac{3}{8}$	$\frac{3}{8}$	$\frac{1}{8}$

Often, it is possible and more convenient to express the outcomes and the probabilities in an equation, called a **probability function.** These equations can often be worked out intuitively using the permutation, combination and multiplication rules from the previous chapter (see Chapter 3); however, their derivation at this time is not as important as the knowledge that probabilities can oftentimes be expressed mathematically. The above probability distribution can be expressed in the following function:

$$P(X=x)=f(x)=\frac{{}_4C_x}{8}=\frac{\frac{4!}{x!(4-x)!}}{8}$$

where $x = 0, 1, 2, 3$.

This function can easily be generalized for any number of coins, n.

$$P(X=x)=f(x)=\frac{{}_nC_x}{2^n}=\frac{\frac{n!}{x!(n-x)!}}{2^n}$$

where $x = 0, 1, 2, \ldots n$.

In these two equations, probability is a function of x, a specific value of the random variable, X. We will denote the probability functions by $f(x)$ or $g(y)$, which can also be expressed in probability terms as $f(x) = P(X = x)$. Numerically, from above, $f(2) = P(X = 2) = 3/8$.

Example 4.1. Consider Example 3.17 (see Chapter 3). We have 9 seedlings, 3 of which are stunted in growth. If we randomly select 4 seedlings without replacement and count the number of normally growing seedlings, we will observe the probability distribution of selecting x normally growing seedlings to be:

x	1	2	3	4
$P(X=x)$	$\frac{6}{126}$	$\frac{45}{126}$	$\frac{60}{126}$	$\frac{15}{126}$

The probability function can be worked out intuitively by using combinations and the multiplication rule (this equation will be generalized in Chapter 5):

$$f(x) = P(X = x) = \frac{{}_6C_x \, {}_3C_{4-x}}{{}_9C_4}$$

where $x = 1, 2, 3, 4$.

Using either the probability function or the probability distribution, one can work out various probabilities for selecting normally growing seedlings:

$$P(odd) = \frac{6}{126} + \frac{60}{126} = \frac{66}{126} \approx 0.524$$

$$P(x \geq 3) = \frac{60}{126} + \frac{15}{126} \approx 0.595$$

$$P(x > 3) = \frac{15}{126} \approx 0.119.$$

It should be noted that this probability function does not have an outcome of $x = 0$, because we have only 3 stunted seedlings – it is impossible not to take a normally growing seedling if 4 seedlings are selected without replacement.

It is often helpful to present probability distributions graphically. In many statistics books, these graphs are incorrectly presented as histograms. For discrete random variables, the probability function and/or histogram should not take on any value between two whole numbers. Stick graphs are more appropriate. Figure 4.1 represents the above probability distribution.

Finally, it should be noted that correctly constructed discrete probability distributions should have a clearly defined domain. In other words, discrete probability distributions must meet two conditions:

1. The probabilities of all possible outcomes must sum to 1.
2. The probabilities of individual outcomes must each be $0 \leq P(X = x) \leq 1$.

With **continuous random variables**, the probability of any exact value, for example $P(x = 2)$, is always zero. Because of this, it is impossible to construct a table similar to the one that we constructed for discrete random variables showing probabilities

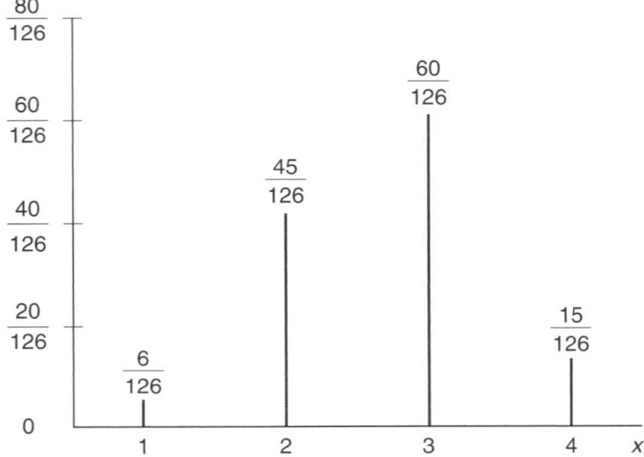

Fig. 4.1. Stick graph of the probability distribution for stunted and normally growing seedlings (from Example 4.1).

associated with individual outcomes. Although an exact value has a zero probability, the probabilities associated with intervals, such as:

$$P(1.9 \leq X \leq 2.1),$$

are possible to calculate. These probabilities must be greater than or equal to 0 and less than or equal to 1. When probabilities are graphed for continuous random variables, they take the form of continuous curves (Fig. 4.2), and are called **probability densities**. From these curves, probabilities for statements like the one above can be obtained by finding the *area under the curve between the two limits*. From the discussion above, we can verify that:

$$P(1.9 \leq X \leq 2.1) = P(1.9 < X < 2.1)$$

since

$$P(X = 1.9) = 0 \text{ and } P(X = 2.1) = 0.$$

For straight-line continuous distributions, the area under the 'curve' can be divided into well-known geometric forms (squares, rectangles, trapezoids and triangles), and the total area can be calculated. For other shapes, probabilities can be determined by integral calculus. These methods are not discussed in this book because easy-to-use tables are available to determine probabilities for the most important continuous functions used in statistics.

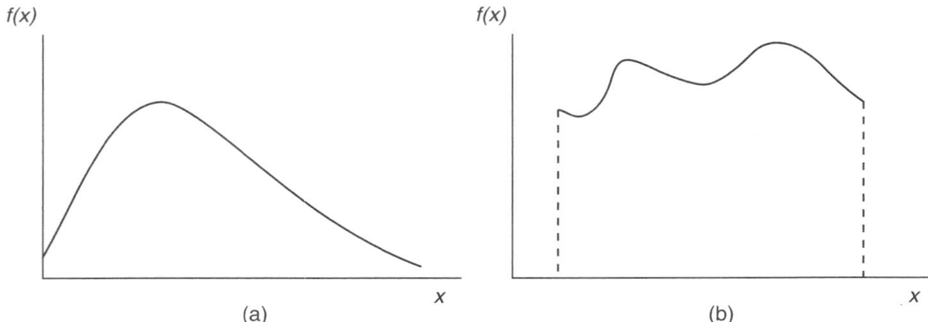

Fig. 4.2. Examples of probability densities.

Example 4.2. Suppose that the probability of the time (in minutes) it takes a tree planter to plant a seedling can be described by a continuous probability function as:

$$f(x) = \frac{x}{6} - \frac{1}{6},$$

where

$3 \leq x \leq 5$ and x = time in minutes.

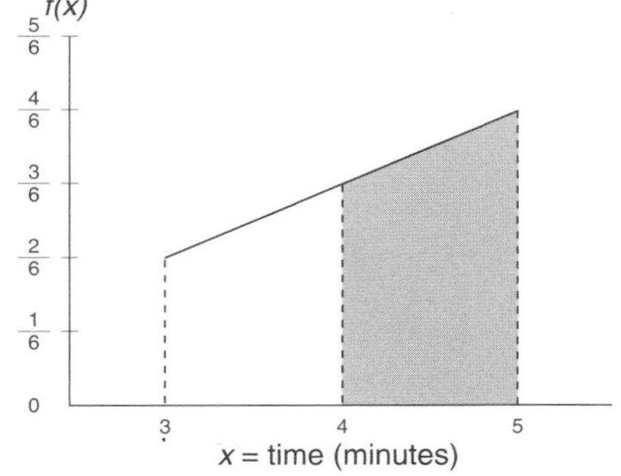

Fig. 4.3. Probability density for the time it takes a tree planter to plant a seedling.

For this function, the plotted probability density curve is a straight line (Fig. 4.3). If we are interested in finding the probability that a tree planter will take between 4 and 5 min to plant a seedling, the area under the curve between $x = 4$ and $x = 5$ can be calculated using the equation of the area of a trapezoid:

$$P(4 < x < 5) = P(4 \le x \le 5) = 1\left(\frac{\frac{3}{6}+\frac{4}{6}}{2}\right) = \frac{7}{12} \approx 0.583.$$

Continuous random variables with well-defined domains have properties similar to those of discrete random variables:

1. The total *area under the curve* between the limits of the domain (the lowest and highest possible outcomes) must add to 1.
2. The probabilities between any two limits, x_1 and x_2, must be $0 \le P(x_1 \le x \le x_2) \le 1$.

Often, it is desirable to study simultaneous outcomes of several random variables. For two discrete random variables, X and Y, a table listing all possible values of x and y with their associated probabilities is called a **joint probability distribution** (also known as a **bivariate distribution**). It is possible to create joint probability distributions for more than two discrete or continuous random variables, but these are complicated and will not be discussed in this book. If joint probabilities are expressed as functions of x and y, they are called **joint probability functions.** Here, $f(x,y)$ represents the probability that a discrete random variable X assumes the value x and, at the same time, another discrete random variable Y assumes the value y. We will demonstrate the construction of a joint probability distribution and joint probability function by modifying Example 4.1.

Example 4.3. Recall that in Example 3.17 (see Chapter 3) we had 9 seedlings, 3 of which were stunted and 6 of which were growing normally. Suppose further that the 6 normally growing seedlings are comprised of 4 spruce and 2 pine seedlings. If we select 3 seedlings at random from the 9 in total, we can construct a joint probability distribution such that X represents the number of normal spruce seedlings and Y represents the number of normal pine seedlings. Table 4.1 shows the tabulated form of the joint probability distribution. Note that since only 3 seedlings are selected, some outcomes, e.g. 2 normal seedlings of each species, are impossible.

Table 4.1. Joint probability distribution.

	Spruce (X)				
Pine (Y)	0	1	2	3	Total $h(y)$
0	$\frac{1}{84}$	$\frac{12}{84}$	$\frac{18}{84}$	$\frac{4}{84}$	$\frac{35}{84}$
1	$\frac{6}{84}$	$\frac{24}{84}$	$\frac{12}{84}$		$\frac{42}{84}$
2	$\frac{3}{84}$	$\frac{4}{84}$			$\frac{7}{84}$
Total $g(x)$	$\frac{10}{84}$	$\frac{40}{84}$	$\frac{30}{84}$	$\frac{4}{84}$	1

The probabilities associated with each combination of (x,y) outcomes can also be found by using the following equation, which is the **joint probability function** for this particular example:

$$P(X = x, Y = y) = f(x, y) = \frac{{}_4C_x \; {}_2C_y \; {}_3C_{3-x-y}}{{}_9C_3}$$

where $x = 0, 1, 2, 3$, $y = 0, 1, 2$ and $x + y \le 3$.

This function can be derived using logic similar to that of Examples 4.1 and 3.17 (see Chapter 3). The denominator is found by calculating the number of ways that 3 seedlings out of 9 can be selected, which is the number of combinations of 9 objects taken 3 at a time. The numerator is found by multiplying together the number of combinations of the 4 spruce seedlings taken x at a time, the number of combinations of the 2 pine seedlings taken y at a time, and the number of combinations of the 3 stunted seedlings taken $(3\text{-}x\text{-}y)$ at a time. The product of these three combinations (multiplication rule) produces the total number of ways x spruces, y pines and $(3\text{-}x\text{-}y)$ stunted seedlings can be taken. The probability is found simply by dividing the number of ways an event of interest can occur (the numerator) by the total number of outcomes in the sample space (the denominator).

Table 4.1 shows the sum of the probabilities for the rows (pines), $h(y)$, and the sum of the probabilities for the columns (spruces), $g(x)$. These are referred to as **marginal probabilities**. If we were to construct discrete probability distributions for X and Y independently, the result would be these one-dimensional marginal probabilities.

From the joint probability distribution, the **conditional distributions** for each of X and Y can also be stated as:

$$P(X = x|Y = y) = f(x|y) = \frac{f(x,y)}{g(x)}$$

$$P(Y = y|X = x) = f(y|x) = \frac{f(x,y)}{h(y)}.$$

These equations can be related to the conditional probability Eqn 3.12 (see Chapter 3), $A|B$, since $f(x,y)$ can be viewed as the intersection of two events, i.e. $P(A \cap B)$. For example, $f(x|y = 1)$ can be calculated as:

x	0	1	2	3
$f(x\|y=1)$	$\frac{6}{42}$	$\frac{24}{42}$	$\frac{12}{42}$	0

;

$$f(2|1) = \frac{\frac{12}{84}}{\frac{42}{84}} = \frac{12}{42} \approx 0.286.$$

Sometimes, it is difficult to interpret the exact meaning of a conditional probability. In the case above, $f(x|y = 1)$ is a function that describes the probabilities of x given the occurrence of y (1 pine seedling being selected). For example, $P(2|1)$ is the probability that 2 spruce seedlings are selected, given (if known) that 1 pine seedling has already been selected.

4.3 Mean of a Random Variable

The **mean** of a random variable can be derived from its probability distribution. It is defined as the *weighted average* of all possible outcomes of the random variable, where the weights are the probabilities of the respective outcomes. For example, from the probability distribution for the number of heads obtained by tossing 3 coins, where the probability distribution is:

x	0	1	2	3
$P(X=x)$	$\frac{1}{8}$	$\frac{3}{8}$	$\frac{3}{8}$	$\frac{1}{8}$

,

the mean (number of heads) is calculated as:

$$\mu = \frac{(0)\left(\frac{1}{8}\right)+(1)\left(\frac{3}{8}\right)+(2)\left(\frac{3}{8}\right)+(3)\left(\frac{1}{8}\right)}{1} = 1.5$$

Since the weights (probabilities) always sum to 1, the denominator in this weighted mean equation can always be omitted.

The above mean is symbolized by the Greek letter μ, or μ_X, and is treated as the **population mean** of the random variable X. It is the theoretical mean of a probability distribution and is often referred to as the **expected value** or the **mathematical expectation** of the random variable X. The expression 'expected value' can be misleading, as oftentimes the value is not expected at all. For example, 1.5 heads in the example above is an impossible outcome. Similarly, the 'expected value' of the random variable describing the number of dots appearing when rolling a die (outcomes are 1, 2, 3, 4, 5 and 6, with a probability of 1/6 each) is 3.5 (see Example 4.4 below). Mathematically, the expected value of a random variable, symbolized as $E(X)$, is the weighted measure of the centre, or weighted mean of all the possible values of a random variable. It is interpreted as the long-term average of the outcomes that are 'expected' if the experiment is conducted repeatedly.

The mean of a discrete random variable is a summation computed as:

$$\mu = E(X) = \sum_{i=1}^{n} x_i f(x_i), \tag{4.1}$$

where $x_1, x_2, \ldots, x_n$ are all possible values of the random variable, X; and $f(x_i)$ are their respective probabilities.

Example 4.4. If we would like to find the mean of the number of outcomes when a die is rolled, we first construct the probability distribution for all possible x values.

x	1	2	3	4	5	6
$P(X=x)$	$\frac{1}{6}$	$\frac{1}{6}$	$\frac{1}{6}$	$\frac{1}{6}$	$\frac{1}{6}$	$\frac{1}{6}$

Then we calculate the mean as:

$$\mu = \sum_{i=1}^{6} x_i f(x_i) = (1)\frac{1}{6} + (2)\frac{1}{6} + (3)\frac{1}{6} + (4)\frac{1}{6} + (5)\frac{1}{6} + (6)\frac{1}{6} = 3.5.$$

Example 4.5. In Example 4.1, we randomly selected 4 seedlings from 6 normally growing seedlings and 3 stunted seedlings. The probability distribution of the number of normally growing seedlings and the mean are as follows:

x	1	2	3	4
$P(X=x)$	$\frac{6}{126}$	$\frac{45}{126}$	$\frac{60}{126}$	$\frac{15}{126}$

$$\mu = (1)\frac{6}{126} + (2)\frac{45}{126} + (3)\frac{60}{126} + (4)\frac{15}{126} \approx 2.67.$$

Again, the expected (or mean) value of 2.67 is an impossible outcome for this random variable (seedlings must occur in whole numbers). To logically interpret this value, we must say that, if we randomly selected 4 seedlings repeatedly and recorded the number of normally growing seedlings for each trial, the average of these recorded numbers would approach 2.67.

Mathematical expectations can be applied to several practical problems, from computing how much an insurance firm should charge for its premiums to how much a casino should pay out in a game of chance. Expectations are particularly useful in the latter case and can be used to determine the degree to which a gambling game is 'fair' (the mathematical expectation will always be in the house's favour). We will demonstrate these applications with a couple of simplified examples.

Example 4.6. You invite a friend to play a game using a deck of 52 cards. Your friend draws a card. You pay US$10 if it is an ace and US$5 if it is a face card, but your friend pays you US$3 if any other card is drawn. What is the expected dollar loss/pay-off for your friend? Before answering this question, ask yourself: Would you play this game? Would you play it once? Would you play it 100 times in a row? Let us define X as the amount of money that your friend earns. Note that the negative number represents a negative earning (loss) for your friend. We can now set up the following probability distribution and compute the expected value.

x	10	5	−3
$P(X = x)$	$\frac{4}{52}$	$\frac{12}{52}$	$\frac{36}{52}$

$$\mu = (10)\frac{4}{52} + (5)\frac{12}{52} + (-3)\frac{36}{52} \approx -0.15$$

The expected value of US$–0.15 indicates that, on average, your friend will lose 15 cents per game. Again, it is impossible to lose 15 cents in a single game as it is played. However, a practical interpretation of this result would be that, if you played this game 1000 times, your friend would lose about (1000)(US$0.15) = US$150. Note that you can estimate the mathematical expectation either for your friend or for yourself. In the latter case, the mathematical expectation is equal to US$+0.15 per game, meaning that the game is in your favour. A fair game refers to one in which the mathematical expectation is zero.

Example 4.7. An insurance company would like to insure a particular type of car, the value of which is US$30,000. The company estimates that a total (100%) loss will occur with a 0.001 probability, a 50% loss with a 0.02 probability and a 25% loss with a 0.1 probability. How much should the company charge (the insurance premium) if they would like to make a 30% profit on each insured vehicle? First, we set up the probability distribution and then we calculate the expected value (mean):

x	–US$30,000	–US$15,000	–US$7500	US$0
$f(x)$	0.001	0.02	0.1	0.879

$$\mu = (-30{,}000)(0.001) + (-15{,}000)(0.02) + (-7500)(0.1) + (0)(0.879) = -1080.$$

Note that, even though a US$0 loss (i.e. no accident) is not explicitly mentioned in the problem, it must be included in the probability distribution since the probabilities of all outcomes must sum to 1. Since the insurance company expects to lose US$1080 per insured vehicle, they would charge US$1404 (US$1080 × 1.30) in order to make a 30% profit. The reader should note that this is an oversimplified example. In real life, loss is a continuous random variable and insurers would cover losses ranging from 0% to 100%. Therefore, a continuous probability density function would be more realistic for this problem.

For continuous random variables, the mathematical expectation or the mean can be obtained by integral calculus only. The general equation is:

$$\mu = E(X) = \int_a^b x f(x)\,dx \tag{4.2}$$

where $a \le x \le b$.

Since problems of this type are beyond the scope this book, no examples are given.

The equation for the expectation can also be applied to many discrete probability problems to find the expectation of some function of the random variable. For example, given a discrete random variable X with a probability function $f(x)$, the expected value of $g(X)$, which is a function of X, is:

$$\mu_{g(X)} = E[g(X)] = \sum_{i=1}^{n} g(x_i) f(x_i) \tag{4.3}$$

Example 4.8. Let the function $g(X) = 4x + 3$, where X is the number of heads (0, 1, 2, 3) when 3 coins are tossed. The expected value of $g(X)$ is:

$$\mu_{g(X)} = (3)\tfrac{1}{8} + (7)\tfrac{3}{8} + (11)\tfrac{3}{8} + (15)\tfrac{1}{8} = 9,$$

since

$g(x)$	3	7	11	15
$f(x)$	$\frac{1}{8}$	$\frac{3}{8}$	$\frac{3}{8}$	$\frac{1}{8}$

The concept of expectation can also be extended to several variables. The mean of a function of two variables uses their joint probability distribution. The mean of the function of $g(X, Y)$ is calculated as:

$$\mu_{g(X,Y)} = E[g(X,Y)] = \sum_{i=1}^{m} \sum_{j=1}^{n} g(x_i, y_j) f(x_i, y_j) \tag{4.4}$$

where $x_i = x_1, x_2 \ldots x_m$ and $y_j = y_1, y_2 \ldots y_n$.

Example 4.9. Find the expected value of $g(X, Y)$ in Example 4.3.

$$\mu_{g(X,Y)} = (0)(0)\tfrac{1}{84} + (0)(1)\tfrac{6}{84} + (0)(2)\tfrac{3}{84} + (1)(0)\tfrac{12}{84} + (1)(1)\tfrac{24}{84}$$
$$+ (1)(2)\tfrac{4}{84} + (2)(0)\tfrac{18}{84} + (2)(1)\tfrac{12}{84} + (3)(0)\tfrac{4}{84} = \tfrac{56}{84} \approx 0.667$$

For joint probability distributions, the mean values of x and y, μ_x and μ_y, can be obtained from the marginal probability distributions as:

$$\mu_x = E(X) = \sum_{i=1}^{m} \sum_{j=1}^{n} x_i f(x_i, y_j) = \sum_{i=1}^{m} x_i g(x_i) \tag{4.5}$$

$$\mu_y = E(Y) = \sum_{i=1}^{m} \sum_{j=1}^{n} y_j f(x_i, y_j) = \sum_{j=1}^{n} y_j h(y_j) \tag{4.6}$$

The means of X and Y from Example 4.9 are then:

$$\mu_x = (0)\frac{10}{84} + (1)\frac{40}{84} + (2)\frac{30}{84} + (3)\frac{4}{84} = \frac{112}{84} \approx 1.333$$

$$\mu_y = (0)\frac{35}{84} + (1)\frac{42}{84} + (2)\frac{7}{84} = \frac{56}{84} \approx 0.667.$$

These two averages indicate that, if the process of randomly selecting 3 seedlings is repeated many times over and the numbers of spruce and pine seedlings are recorded each time, on average we could see 1.333 spruce seedlings and 0.667 pine seedlings. These results look reasonable, since there are twice as many spruce seedlings as pine seedlings to select from.

4.4 Variance of a Random Variable

Like observed data sets, random variables can be described by a centre (the mean or expected value) as well as by a spread or variance. The **variance of a random variable** or the **variance of the distribution of a random variable** is the weighted average of the squares of the differences between the mean and each of the possible outcomes of the random variable, where the weights are the probabilities of the outcomes. For the example of tossing 3 coins, the mean number of heads observed is 1.5. The variance is computed as:

$$\sigma^2 = (0 - 1.5)^2(1/8) + (1 - 1.5)^2(3/8) + (2 - 1.5)^2(3/8) + (3 - 1.5)^2(1/8) = 6/8 = 0.75.$$

The notation σ^2 or σ_x^2 is used to denote the variance of a random variable. The Greek letter σ^2 is used to indicate that it is the population or theoretical variance of the random variable, X.

In general, the variance of a discrete random variable is:

$$\sigma^2 = E\left[\left(x_i - \mu\right)^2\right] = \sum_{i=1}^{n}\left(x_i - \mu\right)^2 f\left(x_i\right), \tag{4.7}$$

where $f(x_i)$ is the probability of x_i.

With some algebraic manipulation, Eqn 4.7 can be modified to:

$$\sigma^2 = \sum_{i=1}^{n} x_i^2 f\left(x_i\right) - \mu^2. \tag{4.8}$$

This equation is called the 'computing', 'working' or 'machine' equation for the variance and is similar to Eqn 2.11 (see Chapter 2). While it is not as easy as Eqn 4.7 to grasp conceptually, it is much easier and quicker to use with a pocket calculator. The **standard deviation** is the square root of the variance and is denoted by σ or σ_x. As with many sample and population standard deviations, the appropriate way to interpret standard deviations for random variables is by using the more conservative Chebyshev's Theorem (see Chapter 2) because, typically, distributions of random variables are not bell-shaped and symmetrical. Using probability theory, Chebyshev's Theorem can be re-stated as:

$$P\left(\mu - k\sigma \le x_i < \mu + k\sigma\right) \ge 1 - \frac{1}{k^2}.$$

This expression means the probability that a random variable takes on a value within k standard deviations of the mean is at least $(1 - 1/k^2)$.

Example 4.10. Using Eqn 4.8, find the variance and standard deviation of the number of dots showing when rolling a die.

$$\sigma^2 = (1^2)\frac{1}{6} + (2^2)\frac{1}{6} + (3^2)\frac{1}{6} + (4^2)\frac{1}{6} + (5^2)\frac{1}{6} + (6^2)\frac{1}{6} - 3.5^2$$
$$\approx 15.167 - 12.25 \approx 2.917,$$
$$\sigma \approx 1.708.$$

Example 4.11. In Example 4.5, we randomly selected 4 seedlings from a group of 9, where 3 were stunted. We then counted the number of normally growing seedlings in our sample of three. Find the variance and the standard deviation of the number of normally growing seedlings.

$$\sigma^2 = (1^2)\frac{6}{126} + (2^2)\frac{45}{126} + (3^2)\frac{60}{126} + (4^2)\frac{15}{126} - 2.67^2$$
$$= \frac{966}{126} - 2.67^2 \approx 0.5378$$
$$\sigma \approx 0.7333$$

Like the mean, the equation to calculate the variance for continuous random variables requires an understanding of integral calculus. The equation is given here for reference; however, an example is not given, as it is beyond the scope of this book.

$$\sigma^2 = E\left[(X-\mu)^2\right] = \int_a^b (x-\mu)^2 f(x)dx = \int_a^b x^2 f(x)dx - \mu^2 \tag{4.9}$$

As in Eqn 4.3 with the mean, we can compute the variance of a function of a random variable. If $g(X)$ is a function of a random variable, X, which has a probability function of $f(x)$, the variance of $g(X)$ is:

$$\sigma^2_{g(X)} = E\left\{\left[g(X) - \mu_{g(X)}\right]^2\right\} = \sum_{i=1}^{n} \left[g(x_i) - \mu_{g(X)}\right]^2 f(x_i) \tag{4.10}$$

and its working formula is:

$$\sigma^2_{g(X)} = \sum_{i=1}^{n} \left[g(x_i)\right]^2 f(x_i) - \mu^2_{g(X)} \tag{4.11}$$

Example 4.12. In Example 4.8, X was a random variable describing the number of heads counted when tossing 3 coins and $g(X)$ was defined as $4x + 3$. Find the variance and the standard deviation of $g(X)$.

$$\sigma^2_{g(X)} = (3^2)\frac{1}{8} + (7^2)\frac{3}{8} + (11^2)\frac{1}{8} + (15^2)\frac{1}{8} - 9^2 = 93 - 81 = 12$$
$$\sigma = \approx 3.464.$$

In the case of joint probability distributions, the measure of joint variation between two random variables is called the **covariance.** If the covariance between two random variables is zero, the two random variables are said to be independent. A positive covariance between two random variables means that as the value of one variable increases, the value of the other variable also increases. If the covariance is negative, as the value of one variable increases, the value of the other variable decreases. Covariances can be calculated from joint probability distributions or joint probability functions as:

$$\begin{aligned}\sigma_{XY} &= \mathrm{COV}(X,Y) = E(X-\mu_X)(Y-\mu_Y)\\ &= \sum_{i=1}^{m}\sum_{j=1}^{n}(x_i-\mu_x)(y_j-\mu_Y)f(x_i,y_j)\\ &= \sum_{i=1}^{m}\sum_{j=1}^{n}x_iy_jf(x_i,y_j)-\mu_X\mu_Y,\end{aligned} \tag{4.12}$$

where $x_i = x_1, x_2 \ldots x_m$, and $y_j = y_1, y_2 \ldots y_n$.

Example 4.13. Using Example 4.3, the covariance between the number of spruce seedlings and the number of pine seedlings is:

$$\begin{aligned}\mathrm{COV}(X,Y) = \sigma_{xy} &= (0)(0)\frac{1}{84} + (0)(1)\frac{6}{84} + (0)(2)\frac{3}{84} + (1)(0)\frac{12}{84}\\ &+ (1)(1)\frac{24}{84} + (1)(2)\frac{4}{84} + (2)(0)\frac{18}{84} + (2)(1)\frac{12}{84}\\ &+ (3)(0)\frac{4}{84} - (1.333)(0.667) \approx 0.667 - (1.333)(0.667) \approx -0.220.\end{aligned}$$

Our negative covariance indicates that, as the number of spruce seedlings increases in the sample of three, the number of pine seedlings decreases.

4.5 Rules of Mathematical Expectations Related to the Mean and Variance

In probability distributions, there are several rules of mathematical expectations that allow for the manipulation of means and variances of random variables. Three important rules for means are:

1. $E(aX \pm b) = a\mu_X \pm b$.
2. $E(X \pm Y) = \mu_X \pm \mu_Y$.
3. $E(XY) = \mu_X\mu_Y$, if X and Y are *independent*.

1. In the derivation of Rule 1, it is assumed that a and b are constants and that X is a random variable with a known probability function, $f(x)$. By applying the equation of the expectation of a function, the proof is as follows:

$$\begin{aligned}\mu_{ax\pm b} &= E(aX \pm b) = \sum_{i=1}^{n}(ax_i \pm b)f(x_i)\\ &= (ax_1 \pm b)f(x_1) + (ax_2 \pm b)f(x_2) + \ldots + (ax_n \pm b)f(x_n)\\ &= a[x_1f(x_1) + x_2f(x_2) + \ldots + x_nf(x_n)] \pm b[f(x_1) + f(x_2) + \ldots + f(x_n)]\\ &= a\sum_{i=1}^{n}x_if(x_i) \pm b\sum_{i=1}^{n}f(x_i).\end{aligned}$$

Since the first summation above is our definition of μ and the second summation equals 1, we get:

$$\mu_{ax+b} = a\mu \pm b.$$

Setting $a = 0$, we see that $\mu_b = b$. Setting $b = 0$, we see that $\mu_{ax} = a\mu$.

2. In the derivation of Rule 2, it is assumed that X and Y are two random variables with a known joint probability function $f(x,y)$:

$$\begin{aligned}\mu_{x\pm y} &= E(X \pm Y)\\ &= \sum_{i=1}^{m}\sum_{j=1}^{n}(x_i \pm y_j)f(x_i, y_j) \qquad \text{(from Eqns 4.5 and 4.6)}\\ &= \sum_{i=1}^{m}\sum_{j=1}^{n}x_i f(x_i, y_j) \pm \sum_{i=1}^{m}\sum_{j=1}^{n}y_i f(x_i, y_j)\\ &= \mu_x \pm \mu_y.\end{aligned}$$

3. In the derivation of Rule 3, it is assumed that X and Y are *independent* random variables with a known joint probability function, $f(x,y)$:

$$\mu_{xy} = E(XY) = \sum_{i=1}^{m}\sum_{j=1}^{n}x_i y_j f(x_i, y_j).$$

Since X and Y are independent, we can state that:

$$f(x,y) = g(x)h(y),$$

where $g(x)$ and $h(y)$ are the marginal probability distributions of X and Y, respectively. Hence:

$$\begin{aligned}\mu_{xy} &= \sum_{i=1}^{m}\sum_{j=1}^{n}x_i y_j g(x_i)h(y_j)\\ &= \left[\sum_{i=1}^{m}x_i g(x_i)\right]\left[\sum_{j=1}^{n}y_j h(y_j)\right] \qquad \text{(from Eqns 4.5 and 4.6)}\\ &= \mu_x\mu_y.\end{aligned}$$

There are also three important rules for variances of random variables:

1. $\sigma^2_{X\pm b} = \sigma^2_X.$
2. $\sigma^2_{aX} = a^2\sigma^2_X.$
3. $\sigma^2_{aX\pm bY} = a^2\sigma^2_X + b^2\sigma^2_Y \pm 2ab\sigma_{XY}$, where $\sigma_{XY} = \text{COV}(X,Y)$.

1. In the derivation of Rule 1, it is assumed that b is a constant and X is a random variable with a known probability function, $f(x)$:

$$\sigma^2_{x\pm b} = E\left\{\left[(X \pm b) - \mu_{x\pm b}\right]^2\right\}.$$

From Rule 1 for means, we get: $\mu_{x\pm b} = \mu \pm b$. Therefore,

$$\sigma^2_{x\pm b} = E\left\{\left[X \pm b - (\mu \pm b)\right]^2\right\}$$
$$= E\left[(X-\mu)^2\right]$$
$$= \sigma^2.$$

2. In the derivation of Rule 2, it is assumed that a is a constant and X is a random variable with a known probability function, $f(x)$:

$$\sigma^2_{aX} = E\left\{\left[(aX - \mu_{aX})\right]^2\right\}.$$

From Rule 1 for means, we get $\mu_{ax} = a\mu$. Therefore,

$$\sigma^2_{aX} = E\left[(aX - a\mu)^2\right]$$
$$= a^2 E\left[(X-\mu)^2\right]$$
$$= a^2\sigma^2.$$

3. In the derivation of Rule 3, it is assumed that a and b are constants and X and Y are random variables with a known joint probability function of $f(x,y)$:

$$\sigma^2_{aX\pm bY} = E\left\{\left[(aX \pm bY) - \mu_{aX\pm bY}\right]^2\right\}$$

From Rules 1 and 2 for means we get $\mu_{aX\pm bY} = a\mu_X \pm b\mu_Y$. Therefore,

$$\sigma^2_{aX\pm bY} = E\left\{\left[(aX \pm bY) - (a\mu_X \pm b\mu_X)\right]^2\right\}$$
$$= E\left\{\left[a(X-\mu_X) \pm b(Y-\mu_Y)\right]^2\right\}$$
$$= a^2 E\left[(X-\mu_X)^2\right] + b^2 E\left[(Y-\mu_Y)^2\right] \pm 2ab\, E\left[(X-\mu_X)(Y-\mu_Y)\right]$$
$$= a^2\sigma^2_X + b^2\sigma^2_Y \pm 2ab\,\sigma_{XY}.$$

Note, for independent X and Y, $\sigma_{XY} = 0$. Therefore,

$$\sigma^2_{aX\pm bY} = a^2\sigma^2_X + b^2\sigma^2_Y.$$

The above rules regarding the means and variances of random variables are frequently used both in statistical analyses and descriptive statistics. For instance, Rule 1 (for the means) and Rule 2 (for the variances) are used when measurement units have been changed. Consider the case of 1500 dbh measurements that were taken in inches in 1957, where the mean, variance and other descriptive measures are currently available. Our task may be to compare these measurements to some current observations in centimetres. According to the rules above, we can very easily convert the mean and variance from inches to centimetres simply by multiplying the mean by 2.54 and multiplying the variance by 2.54^2.

In another example, we may want to measure the weight of some pulped wood, but the only way to do so in a laboratory setting is in a small container. In other words, to obtain the exact weight of the pulp samples, the weight of the container would need to be subtracted from each measurement. Instead, we can calculate the mean and the variance of the combined weights of the pulp and the container, and subtract the weight of the container from the mean. We do not have to change the variance since, from Rule 1, we know that the variance does not change if we add (or subtract) a constant to (from) the observations.

Exercises

Section 4.1

4.1. Give five examples of a random variable.

4.2. Give three examples of a discrete random variable.

4.3. Give three examples of a continuous random variable.

4.4. Classify the following random variables as discrete or continuous:

a. Number of pathological indicators on a standing tree.
b. Volume of a standing tree.
c. Length of a plywood panel.
d. Number of salmon in a stream.
e. Height of a tree.
f. Number of insects found in a pine cone.
g. Number of seeds in a pine cone.
h. Length of a pine cone.

Section 4.2

4.5. Assume that 2 balanced dice (one green and one red) are rolled. Construct the probability distribution and graphs for the random variables described as:

a. x = the sum of the number of dots showing.
b. y = the difference (red minus green) of the number of dots showing.

4.6. A random experiment consists of tossing a coin 4 times. Let the random variable, X, be the number of tails.

a. Construct the probability distribution.
b. Construct a probability graph.
c. Find $P(X < 2)$; $P(X \leq 2)$; $P(2 < X < 4)$.

4.7. Five cards are drawn without replacement from an ordinary deck of 52 playing cards (a poker hand). Construct the probability distribution for the number of aces that can be drawn (X).

a. Find $P(X < 3)$.
b. Find $P(X = 0)$.

4.8. From the following examples, decide if $P(X = x)$ represents a probability distribution. If not, why not?

a.

x	15	16	17	18	19
$P(X = x)$	$\frac{1}{12}$	$\frac{2}{12}$	$\frac{3}{12}$	$\frac{4}{12}$	$\frac{2}{12}$

b.

x	5	7	9
$P(X = x)$	−0.3	0.8	0.5

c.

x	1	2	3	4
$P(X = x)$	1.2	0.3	0.8	0.1

4.9. A coin is tossed until a tail occurs. List the possible outcomes with up to 10 tosses and construct the probability distribution (X = the number of tosses). Find:

a. $P(X = 1)$.
b. $P(1 \leq X \leq 4)$.

4.10. It is known that the probability of a ponderosa pine seed germinating is 0.75. Three such seeds are tested and X = number of seeds that germinated is observed. Construct the probability distribution of the random variable, X, and, if possible, derive the probability function of the random variable. Find:

a. $P(X < 2)$.
b. $P(X = 2)$.
c. $P(1 < X \leq 3)$.

4.11. Out of 5 logs, 2 contain ambrosia beetle damage. Three logs are selected at random to take inside the mill. Construct the probability distribution of the random variable, X, representing the number of infested logs inside the mill. Express your results graphically and in an equation. Find:

a. $P(X = 2)$.
b. $P(X \leq 2)$.

4.12. A logging contractor can make a profit ranging from US\$5000 to US\$7000 on a project, depending on several unexpected problems. Let the random variable, X, represent the contractor's profit in thousands of dollars, with the following probability density function:

$$f(x) = \frac{x}{8} - \frac{2}{8}, \qquad \text{for } 5 \leq x \leq 7.$$

Find the probability that the contractor will make:

a. More than US\$6000 profit.
b. Between US\$5500 and US\$6500 profit.
c. Less than US\$5700 profit.

4.13. A nursery has 3 Douglas-fir, 2 white pine and 3 birch seedlings. Four of these are selected at random without replacement. If X is the number of Douglas-fir and Y is the number of white pine seedlings, find:

a. The joint probability distribution of X and Y.
b. $P(X \leq 2, Y = 1)$.
c. $P(X > 2, Y \leq 1)$.
d. $P(Y > 2)$.
e. The marginal distribution of X.
f. The marginal distribution of Y.

Section 4.3

4.14. Find the mean of each of the random variables described in Exercises 4.5–4.7, 4.10 and 4.11. Explain the meaning of the mean in each case.

4.15. Find the means of the random variables, X and Y, described in Exercise 4.13.

4.16. In a game of chance, 2 dice are rolled and the value of the sum of the dots is paid in dollars when the outcomes are the same (e.g. 1,1 = US\$2; 2,2 = US\$4; 3,3 = US\$6; 4,4 = US\$8; 5,5 = US\$10; or 6,6 = US\$12).

a. How much should a player pay for each roll in order to make the game fair?
b. How much should a player pay for each roll if the 'house' would like to make a 15% profit on average?

4.17. By investing US\$100,000 in a certain mutual fund, a person can make a US\$10,000 profit with probability 0.62 or lose US\$14,000 with probability 0.38. What is the person's expected gain/loss for this investment?

4.18. A customer purchases a US\$400,000 fire insurance policy. According to past information, homes in the area sustain total loss with a probability of 0.0008 and 50% loss with a probability of 0.001. Ignoring all other partial losses, what premium should the insurance company charge for a policy in order to make a 15% profit over and above the break-even point?

4.19. A logging equipment sales person can contact either 1 or 2 customers per day, depending on the distances between customers. The probability of 1 customer contact is 0.3, the probability of 2 customer contacts is 0.7. Each contact will result in either no sale or a US\$60,000 sale, with probabilities of 0.9 and 0.1, respectively. What is the expected value of the sales person's daily sales? *Hint: you may want to start with a tree diagram.*

4.20. Find the expected value of the random variable, $g(X) = (2X + 1)^2$, where X has the probability distribution described in Exercise 4.6.

4.21. Let X be the sum of the dots in Exercise 4.16. If the payment is $g(X) = 2X -$ US\$2 (instead of X dollars):

a. How much should a player pay for each roll to make the game fair?
b. How much should a player pay if the house would like to make a 15% profit on average?

Section 4.4

4.22. Find the variance and the standard deviation of each of the random variables described in Exercises 4.5–4.7, 4.10 and 4.11. Explain the meaning of each standard deviation using Chebyshev's Theorem with $k = 1.5$ and $k = 2$.

4.23. Find the variance and the standard deviation of each of the random variables described in Exercises 4.16, 4.17 and 4.18.

Section 4.5

4.24. Find the covariance of the two random variables X and Y described in Exercise 4.13 and explain its meaning.

4.25. Find the variance and standard deviation for $g(X)$ in Exercises 4.20 and 4.21.

4.26. Calculate the variance for the random variable of daily sales described in Exercise 4.19. Now assume that each visit can result in either no sale or a sale of US$130,000 and find the mean, variance and the standard deviation for this new random variable. *Hint: use the values for means and variances from Exercise 4.19.*

5 Some Discrete Probability Distributions

Describing Data that are Counted

Observations generated by different experiments, games and natural processes often show similar general behaviours and can be described by a particular probability distribution or probability function. In fact, most discrete random variables can be classified into one of about half a dozen of these distributions. This chapter introduces common discrete probability distributions for describing many random events encountered in practice.

Some students may find these types of problems challenging when exposed to them for the first time. The mathematics underlying them are simple; it is a question of being able to recognize what type of discrete distribution we are dealing with and which equation to apply. We strongly recommend practising as many of these types of problems as possible.

5.1 Uniform Distribution

The discrete **uniform distribution** describes a process whereby the probability of every outcome is the same. For a process with k possible outcomes, the probability function is given by:

$$f(x;k) = \frac{1}{k}, \quad \text{for } x = x_1, x_2, \ldots x_k. \tag{5.1}$$

The notation of $f(x;k)$ denotes that the probability function of observed x values is defined by one parameter, k, the number of possible outcomes.

Example 5.1. Suppose that we have 8 pieces of wood in a box, numbered from 1 to 8. We select 1 piece at random. The probability of selecting a particular piece of wood follows a uniform distribution with $k = 8$ and, therefore, the probability function is:

$$f(x;8) = \frac{1}{8}, \quad \text{for } x = 1, 2, 3, \ldots, 8.$$

In other words, the probability of each outcome (x = 1, 2,..., 7, 8) is 1/8. Figure 5.1 shows the graphical presentation of the probability distribution for Example 5.1. All uniform distributions will follow this type of flat shape.

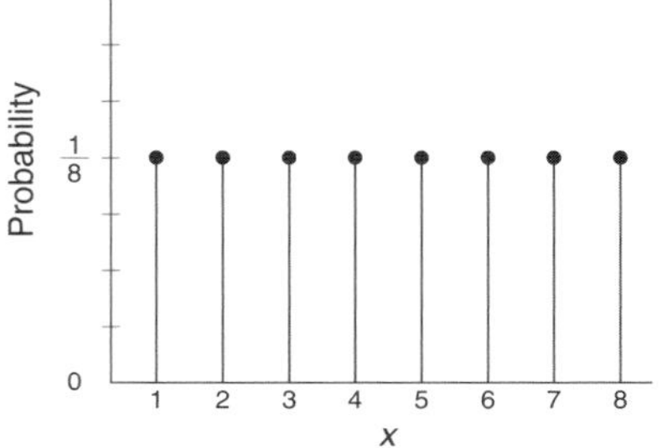

Fig. 5.1. Probability distribution for Example 5.1 (randomly selecting from 8 wood pieces).

Example 5.2. A standard 6-sided die is tossed once. The random variable describing the number of dots showing, x, will follow a uniform distribution with $k = 6$:

$$f(x;6) = \frac{1}{6}, \quad \text{for } x = 1, 2, 3, 4, 5, 6.$$

5.2 Binomial and Multinomial Distributions

Consider an experiment with *two* possible outcomes that is repeated several times. For example, we conduct a germination trial with 3 Douglas-fir seeds. Assume that we know from previous experiments that the probability that a seed will germinate (G) is 0.8 and the probability that it will not germinate (N) is 0.2. If we also assume that the trials are independent (i.e. the same seeds are not used twice) and that the germination of one seed does not affect the germination of another (i.e. it is not a conditional probability), we can expect the eight possible outcomes (permutations) with the following probabilities:

Outcome	x	Probability
NNN	0	$(0.2)(0.2)(0.2) = (0.8)^0(0.2)^3$
GNN	1	$(0.8)(0.2)(0.2) = (0.8)^1(0.2)^2$
NGN	1	$(0.2)(0.8)(0.2) = (0.8)^1(0.2)^2$
NNG	1	$(0.2)(0.2)(0.8) = (0.8)^1(0.2)^2$
GGN	2	$(0.8)(0.8)(0.2) = (0.8)^2(0.2)^1$
GNG	2	$(0.8)(0.2)(0.8) = (0.8)^2(0.2)^1$
NGG	2	$(0.2)(0.8)(0.8) = (0.8)^2(0.2)^1$
GGG	3	$(0.8)(0.8)(0.8) = (0.8)^3(0.2)^0$

If we define the random variable, X, as the number of seeds that will germinate, the probability distribution of X is:

x	$f(x)$
0	$(0.8)^0(0.2)^3$
1	$(3)(0.8)^1(0.2)^2$
2	$(3)(0.8)^2(0.2)^1$
3	$(0.8)^3(0.2)^0$

In the process of creating this probability distribution, we used the multiplication rule, e.g. the probabilities for a given outcome, such as NGG is $P(\text{N}) \times P(\text{G}) \times P(\text{G}) = (0.2)(0.8)(0.8)$. Also implicit in the above probability distribution is that we used the addition rule. For example, to find the probability that one seed germinated, we added probabilities such as GNN + NGN + NNG (we actually multiplied by three, since they are numerically the same). This is possible because these outcomes are mutually exclusive (disjoint).

From the probability distribution above, we can derive a generalized probability function for these experimental trials as:

$$f(x; 3, 0.8) = \binom{3}{x} 0.8^x 0.2^{3-x} = \frac{3!}{x!(3-x)!} 0.8^x 0.2^{3-x}, \quad \text{for } x = 0, 1, 2, 3.$$

In this case, we use two parameters to define the probability function. For this example, the 3 represents the number of trials, and the 0.8 represents the probability that a seed will germinate. Since there are only two possible outcomes, the probability that a seed does not germinate equals $1 - 0.8 = 0.2$ and does not have to be explicitly stated. Again, on the right hand side of the equation, we see evidence of both the multiplication rule and the addition rule. We multiply together the probability of germination, 0.8, x times, with the probability of no germination, 0.2, $3 - x$ times. The expression $\binom{3}{x}$ indicates the number of arrangements (ways) that x seeds can germinate out of the 3 seeds being tested.[1]

This kind of experiment is called a *binomial experiment* because it produces a binomial random variable (i.e. two outcomes). All binomial experiments have the following properties:

1. The experiment consists of a fixed number of trials, n, with each trial resulting in one of two kinds of outcomes: *success* or *failure*.
2. The probability of success, p, is the *same* for each trial.
3. The trials are *independent*.

The above formula can be generalized for n trials, with probability of success, p, and probability of failure, $q = 1 - p$. The term, x, refers to the random variable of interest.

$$f(x;\ n,\ p) = \binom{n}{x} p^x q^{n-x} = \frac{n!}{x!(n-x)!} p^x q^{n-x}, \qquad \text{for } x = 0, 1, 2, \ldots, n. \tag{5.2}$$

One needs to be very careful with the names given to the two kinds of outcomes, as the expressions of 'success' and 'failure' could be misleading. A general rule is that p applies to the probability in question or the probability of interest. If the question is, 'What is the probability that a seed will not germinate?', then the 'success' is actually the outcome that a seed will not germinate, even though this outcome is clearly not a success from a forester's perspective.

Example 5.3. Let us reconsider the above seed germination experiment with 8 seeds. In other words, we have $n = 8$ trials and p is still 0.8.

a. What is the probability that exactly 7 seeds will germinate?
b. What is the probability that more than 6 seeds will germinate?

a. $f(X = 7; 8, 0.8) = \binom{8}{7} 0.8^7 0.2^1 = \frac{8!}{7!(8-7)!} 0.8^7 0.2^1 \approx 0.3355$

b. $P(X > 6) = P(X = 7) + P(X = 8)$. Therefore,

$$P(X > 6; 8, 0.8) = \binom{8}{7} 0.8^7 0.2^1 + \binom{8}{8} 0.8^8 0.2^0 \approx 0.3355 + 0.1678 \approx 0.5033.$$

Example 5.4. Five per cent of the furniture components made by a secondary manufacturer are defective. If we randomly select 10 components from the assembly line:

a. What is the probability that exactly 2 are defective?
b. What is the probability that less than 2 are defective?
c. What is the probability that more than 1 and less than 5 are defective?

[1] Although this equation looks like the number of combinations of three items, it is actually the number of permutations of three items where x are of one kind and (3 – x) are of a second kind (see Eqn 3.5, permutations for similar objects). When there are only *two kinds of similar objects*, the equation for their permutations is mathematically the same as the equation for their combinations.

a. $f(X = 2; 10, 0.05) = \binom{10}{2} 0.05^2 0.95^8 = \frac{10!}{2!\,(10-2)!} 0.05^2\, 0.95^8 \approx 0.0746$

b. $P(X < 2; 10, 0.05) = \binom{10}{1} 0.05^1 0.95^9 + \binom{10}{0} 0.05^0\, 0.95^{10}$

$\approx 0.3151 + 0.5987 \approx 0.9138$

c. $P(1 < X < 5; 10, 0.05) = \binom{10}{2} 0.05^2\, 0.95^8 + \binom{10}{3} 0.05^3\, 0.95^7 + \binom{10}{4} 0.05^4\, 0.95^6$

$\approx 0.0746 + 0.0105 + 0.0001 \approx 0.0852.$

In this example, we chose to define 'success' as selecting a *defective* furniture component, though this is hardly a success from the manufacturer's point of view. Although we could have defined 'success' as selecting a non-defective component, we took our cues from the wording of the questions. The questions ask for the probabilities of *defective* pieces and making these events 'successes' made our calculations easier.

The calculations of binomial probabilities, $f(x; n, p)$, are tabulated for selected p and n values in Table A.1 (see Appendix A). To demonstrate the use of Table A.1, the following probabilities for Example 5.3a and b and Example 5.4a, b and c can be found from the table as functions of p, n and x. All probabilities are from Table A.1.

5.3a. $P(X = 7;\ 8;\ 0.8) = 0.336$

5.3b. $P(X > 6;\ 8;\ 0.8) = P(X = 7) + P(X = 8) = 0.336 + 0.168 = 0.504.$

5.4a $P(X = 2;\ 10;\ 0.05) = 0.075$

5.4b $P(X < 2;\ 10;\ 0.05) = P(X = 1) + P(X = 0) = 0.315 + 0.599 = 0.914$

5.4c $P(1 < X < 5;\ 10;\ 0.05) = P(X = 2) + P(X = 3) + P(X = 4)$

$= 0.075 + 0.010 + 0.001 = 0.086$

The discrepancies between solutions calculated by equations and by table are due to rounding errors in Table A.1 and are usually negligible.

One other short cut that is worth noting requires the knowledge that all of the probabilities of the random variables for a given p and n will sum to 1. This is a helpful time-saving measure in so much as it sometimes allows us to compute probabilities using fewer computations by applying the complement law. Take Example 5.4, where we are randomly drawing 10 furniture components. If we were to pose a question around the probability of obtaining more than 1 defective component, we would need to figure out the individual probabilities of drawing 2, 3, … , 9, 10 components (either using Eqn 5.2 or Table A.1). Rather than going through all of these calculations, we can simply use the complement law to determine the probability of obtaining 0 and 1 defective pieces and subtract these two values from 1.

The term *binomial* originates from the binomial expansion of $(q + p)^n$ for $x = 0, 1, 2, \ldots, n$:

$$(q+p)^n = \binom{n}{0} q^n + \binom{n}{1} p q^{n-1} + \binom{n}{2} p^2 q^{n-2} + \ldots + \binom{n}{n} p^n$$

$$= f(0; n, p) + f(1; n, p) + f(2; n, p) + \cdots + f(n, n, p)$$

and since

$$(q+p) = 1,\ \sum_{x=0}^{n} f(x; n, p) = 1.$$

By applying some algebraic manipulations and some of the rules of expectation (see Section 4.5 in Chapter 4), it can be shown that the mean and variance of a binomial random variable are:

$$\mu = np \qquad \text{and} \qquad \sigma^2 = npq.$$

Example 5.5. Calculate the mean, variance and standard deviation of the binomial random variable defined in Example 5.3 (recall that there were n = 8 trials, with the probability of success, p = 0.8):

$$\mu = (8)(0.8) = 6.4, \qquad \sigma^2 = (8)(0.8)(0.2) = 1.28 \qquad \text{and} \qquad \sigma \approx 1.131.$$

In other words, if we repeated this experiment over and over again, we would expect an average of 6.4. Because this is not a bell-shaped distribution, we need to apply Chebyshev's Theorem to interpret the standard deviation. We will arbitrarily choose k = 2 to find the range where at least 75% of the values of X fall:

$$P(\mu - k\sigma \leq X \leq \mu + k\sigma) = P(4.14 \leq X \leq 8.66) \geq 0.75.$$

To verify, we look at the closest approximation to this range in Table A.1 (see Appendix A), which shows that P(4 < X < 9) = 0.99, a value much greater than 0.75.

Example 5.6. Calculate the mean, variance and standard deviation for the binomial random variable in Example 5.4 (recall n = 10 and p = 0.05):

$$\mu = (10)(0.05) = 0.5, \quad \sigma^2 = (10)(0.05)(0.95) = 0.475 \quad \text{and} \quad \sigma \approx 0.689.$$

When more than two outcomes are possible from each trial, the equation for probabilities associated with binomial experiments can be extended to what is known as a 'multinomial' experiment. Here, a given trial from a random experiment can result in k possible outcomes, $x_1, x_2, \ldots, x_k$, with associated probabilities of $p_1, p_2, \ldots, p_k$, such that:

$$\sum_{i=1}^{k} x_i = n \quad \text{and} \quad \sum_{i=1}^{k} p_i = 1$$

The resultant distribution is referred to as a **multinomial distribution**, for which the probability function is:

$$f(x_1 x_2, \ldots, x_k; p_1 p_2, \ldots, p_k, n) = \binom{n}{x_1, x_2, \ldots, x_k} p_1^{x_1} p_2^{x_2}, \ldots, p_k^{x_k},$$

where (5.3)

$$\binom{n}{x_1, x_2, \ldots, x_k} = \frac{n!}{x_1! x_2!, \ldots, x_k!}.$$

The properties of a multinomial random variable are similar to the ones listed for a binomial random variable, with one modification: *each trial can result in more than two outcomes.*

Example 5.7. On the production line of a particular sawmill, it is known that lumber is classified into grade A, B and cull, in the following proportions: 1/5, 3/4 and 1/20. If we take 6 pieces randomly from the production line, what is the probability that 2 pieces will be grade A, 3 will be grade B and 1 will be a cull?

$$f\left(x_1 = 2, x_2 = 3, x_3 = 1; \frac{1}{5}, \frac{3}{4}, \frac{1}{20}, 6\right) = \frac{6!}{2!\,3!\,1!}\left(\frac{1}{5}\right)^2\left(\frac{3}{4}\right)^3\left(\frac{1}{20}\right)^1 \approx 0.0285.$$

Intuitively, we would expect this probability to be very small. It is indeed a very tall order to obtain exactly two pieces of one grade, three of another, and one of a third. In fact, this is the case for many discrete probability distributions and in answering these sorts of questions, you would be well advised to ask yourself if the result makes sense.

5.3 Hypergeometric and Multivariate Hypergeometric Distributions

The **hypergeometric distribution**, like the binomial distribution, has two kinds of outcomes. However, it is different from the binomial distribution in its underlying assumptions: the probability of success from trial to trial is not constant and the successive trials are not independent. Hypergeometric trials are made *without replacement* from a finite population and thus the binomial distribution does not give correct probabilities. To illustrate the difference between binomial and hypergeometric distributions, we will turn back to Example 3.21 (see Chapter 3). We drew 2 cards from a deck of 52 cards and calculated the probability that they were both spades. As described in Chapter 3, this experiment can be conducted with or without replacement. In an experiment with replacement, a card is drawn, observed and then put back into the deck, which is shuffled before a second card is drawn. This is a binomial experiment, since the probability of drawing a spade in the first draw is the *same* (13/52 = 1/4) as in the second draw, and the outcome in the second draw is *independent* of the first. In an experiment without replacement, one card is drawn, observed, put aside and then the second card is drawn from the remaining 51 cards. This is a hypergeometric experiment, since the probability of the second outcome is affected by the first outcome. The probability that the second draw (from only 51 cards) is a spade depends on whether or not the first card (drawn from a full deck of 52 cards) is a spade. In this example, and in all hypergeometric experiments, the probabilities for the second draw are conditional probabilities. The probabilities associated with drawing 2 cards that are spades (S) or non-spades (N) are illustrated in Fig. 5.2. To calculate the probabilities associated with the hypergeometric experiment, a tree diagram is used; for the binomial experiment, Eqn 5.2 is used. These are tabulated below.

Outcomes	Binomial probabilities (with replacement)	Hypergeometric probabilities calculated from tree diagram below (without replacement)
SS	0.0625	0.0588
SN	0.1875	0.1912
NS	0.1875	0.1912
NN	0.5625	0.5588

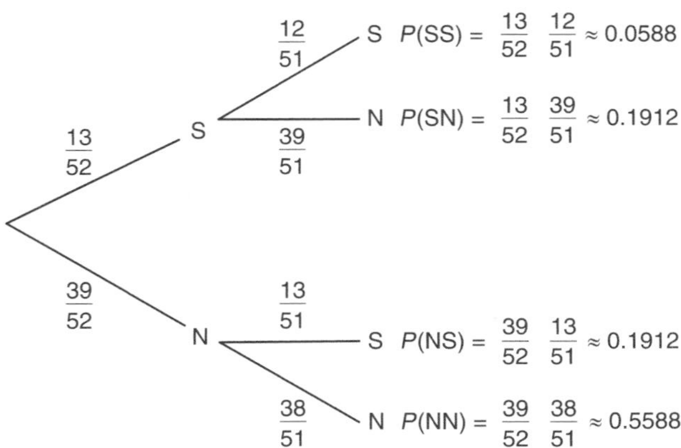

Fig. 5.2. Tree diagram for drawing 2 cards from a deck of 52 cards (without replacement, S = spades and N = other card).

Let N denote the total number of possible outcomes from an experiment, with k of these N possible outcomes labelled a 'success'. If there are two possible outcomes, then $(N - k)$ is the total number of possible outcomes that are 'failures'. The probability function for a random variable, X, representing the number of successes in n trials, follows a hypergeometric distribution and can be written as:

$$f(x;N,n,k) = \frac{\binom{k}{x}\binom{N-k}{n-x}}{\binom{N}{n}} = \frac{\dfrac{k!}{x!(n-k)!}\,\dfrac{(N-k)!}{(n-x)!\left[(N-k)-(n-x)\right]!}}{\dfrac{N!}{n!(N-n)!}}, \qquad (5.4)$$

for $x = 0, 1, 2, \ldots, n$ successes in n samples.

Example 5.8. A technician is sent out to measure 4 Douglas-firs in a sample plot that contains 10 Douglas-firs and 12 western hemlock trees. Unfortunately, the technician has not had a dendrology course and is not very good at distinguishing between species, so he randomly chooses 4 trees to measure (note that while it is not explicitly stated, this is sampling without replacement as he would never choose the same tree twice). What is the probability that exactly 3 of the 4 trees are Douglas-firs?

$$f(X=3;22,4,10) = \frac{\binom{10}{3}\binom{22-10}{4-3}}{\binom{22}{4}} = \frac{(120)(12)}{7315} \approx 0.1969.$$

The mean and the variance of a hypergeometric distribution are given by the following equations, which relate to the mean and variance equations for binomial distributions (proof not given here):

$$\mu = \frac{nk}{N} \quad \text{and} \quad \sigma^2 = \frac{N-n}{N-1}n\frac{k}{N}\left(1-\frac{k}{N}\right).$$

Example 5.9. Calculate the mean, variance and standard deviation for Example 5.8:

$$\mu = \frac{(4)(10)}{22} \approx 1.82$$

$$\sigma^2 = \frac{22-4}{22-1}\, 4\, \frac{10}{22}\left(1-\frac{10}{22}\right) \approx 0.8501$$

$$\sigma \approx 0.9220$$

If the technician in Example 5.9 selects 4 random trees over and over again and records the number of Douglas-firs in each selection, on average there would be 1.82 Douglas-fir trees. Chebyshev's Theorem could then be used to interpret the meaning of the standard deviation (0.92) and draw conclusions about the spread of the outcomes relative to this mean value.

When N is large relative to n, the probabilities in the successive draws do not change much and the hypergeometric distribution approaches the binomial distribution. In this case, the probability of success approaches a constant and, for all practical purposes, can be estimated by $p = k/N$. This approximation is reliable when $n < 0.05N$ and probabilities can reasonably be approximated by Eqn 5.2, or looked up in Table A.1.

Example 5.10. Forty out of 200 logs in a log sort are infested with bark beetles. If 10 logs are selected at random, what is the probability that 2 logs are infested? The exact probability is:

$$f(X = 2; 200, 10, 40) = \frac{\binom{40}{2}\binom{200-40}{10-2}}{\binom{200}{10}} \approx 0.3098.$$

However, since $0.05N > 10$, the binomial approximation should be adequate. Using Table A.1(see Appendix A) with $p = \frac{40}{200} = 0.2$, we obtain $f(x = 2;10,0.2) = 0.302$, which is close to 0.3098 from above.

The hypergeometric distribution can be extended to describe experiments with more than two outcomes. For instance, Example 5.8 could be modified so that the sample plot contains 3 species: 10 Douglas-firs, 12 western hemlocks and 6 western red cedars. If all possible elements, N, in an experiment can be partitioned into k distinct groups with a_1, a_2, …, a_k possible elements in each group, the probability distribution of a sample of size n, taken without replacement, where x_1 are the first kind, x_2 are the second kind, … and x_k are kth kind, is referred to as a **multivariate hypergeometric distribution:**

$$f(x_1, x_2, \ldots, x_k; a_1, a_2, \ldots, a_k, N, n) = \frac{\binom{a_1}{x_1}\binom{a_2}{x_2}\cdots\binom{a_k}{x_k}}{\binom{N}{n}}, \tag{5.5}$$

where

$$\sum_{i=1}^{k} x_i = n \quad \text{and} \quad \sum_{i=1}^{k} a_i = N$$

Example 5.11. An experimental sample plot contains 10 Douglas-firs, 12 western hemlocks and 6 western red cedars, as stated above. What is the probability that, in a sample of 4 randomly selected trees, 1 is a Douglas-fir, 2 are western hemlocks and 1 is a western red cedar?

$$f(x_1 = 1,\ x_2 = 2, x_3 = 1;\ 10,\ 12,\ 6,\ 28,\ 4) = \frac{\binom{10}{1}\binom{12}{2}\binom{6}{1}}{\binom{28}{4}} = \frac{(10)(66)(6)}{20,475} \approx 0.1934.$$

5.4 Geometric and Negative Binomial Distributions

In binomial experiments, the random variable, X, represents the number of successes in a fixed number of trials (n). Geometric experiments possess all the properties of binomial experiments, except that the trials are repeated until the first 'success' occurs. The geometric random variable, X, represents the number of repeated independent trials required to produce the first success, the probability of which is p. As in binomial experiments, the probability of failure is $q = 1 - p$. The probability function of the **geometric distribution** is defined by only one parameter p, as:

$$f(x;p) = pq^{(x-1)} \qquad \text{for } x = 1,2,3,\ldots \tag{5.6}$$

Example 5.12. A bag contains several thousand white pine seeds, 10% of which are empty. What is the probability of finding the first empty seed the fifth time we cut a seed open?

$$f(X = 5; 0.1) = 0.1\ 0.9^{(5-1)} \approx 0.0656$$

An extension of the geometric distribution occurs when trials are repeated until a fixed number of successes, k, occurs. Such a random experiment is called a negative binomial experiment and its probabilities are described by the **negative binomial distribution.** Like a geometric experiment, a negative binomial experiment possesses all the properties of a binomial experiment except that the number of trials is not fixed. The negative binomial random variable, X, represents the number of independent trials required to produce k successes, where p is the probability of success and $q = 1 - p$ is the probability of failure. In developing a general equation, consider that to obtain the kth success on the xth trial, the kth success must be preceded by $k - 1$ successes and $x - k$ failures. The probability of $k - 1$ successes and $x - k$ failures can be arranged in the following number of ways:

$$\binom{x-1}{k-1} = \frac{(x-1)!}{(k-1)!\left[(x-1)-(k-1)\right]!} = \frac{(x-1)!}{(k-1)!\ (x-k)!}$$

The probability of $p^k q^{x-k}$ is multiplied by the number of possible arrangements[2] to obtain the probability of the kth success. The general form of the negative binomial probability distribution function is:

$$f(x;k,p) = \binom{x-1}{k-1} p^k q^{x-k}, \quad \text{for } x = k,\ k+1,\ k+2,\ldots \tag{5.7}$$

The parameters p and k define the negative binomial distribution. Note that, when $k = 1$, the above equation reduces to the geometric distribution because the number of arrangements of one success preceded by $x - 1$ failures is simply 1. In other words, the geometric distribution can be said to be a special case (where $k = 1$) of the negative binomial distribution.

[2] Again, notice that it looks like we are using the number of combinations here. In fact, we are using the number of permutations of *two kinds* of objects.

Example 5.13. Consider the white pine seeds described in Example 5.12. What is the probability that the third empty seed will be found when cutting open the tenth seed?

$$f(X=10;3,0.1)=\binom{10-1}{3-1}0.1^3 0.9^{10-3}=\frac{9!}{2!(9-2)!}0.1^3\,0.9^7\approx 0.0172.$$

Example 5.14. An automated lumber grading system is known to be incorrect 15% of the time. What is the probability that the second incorrect grade will occur when grading the twelfth piece?

$$f(X=12;2,0.15)=\binom{12-1}{2-1}0.15^2 0.85^{10}\approx 0.0487$$

5.5 Poisson Distribution

Experiments where a number of events occur in a fixed time or region (space) with a known average rate are called Poisson experiments. The probability distribution of the number of events in the fixed time (or space), X, is referred to as a **Poisson distribution.** The properties of a **Poisson experiment** are:

1. The probability that an outcome will occur in a small region or during a short time interval is proportional to the size of the region or length of the time.
2. The occurrence of outcomes in a specified space or time interval is independent of the occurrence of outcomes in all other regions or time intervals.
3. The probability that more than one outcome will occur in a small region or in a short time interval is negligible.

For outcomes occurring in a specified space or specified time interval, the probability function is:

$$f(x;\mu)=\frac{e^{-\mu}\mu^x}{x!},\quad \text{for } x=0,1,2,3,\ldots \tag{5.8}$$

where e = 2.71828, base of the natural logarithm, and μ = the average number of outcomes in a specified region or a given time interval.

One parameter, μ, defines the Poisson distribution, and it can be shown that $\sigma^2=\mu$ for Poisson random variables (the proof is not given here). Table A.2 (see Appendix A) can be used to look up Poisson probabilities as a function of μ and x.

Example 5.15. On average, a logging contractor reports 3 accidents per year:

a. What is the probability that exactly 2 accidents will occur in any given year?
b. What is the probability that less than 2 accidents will occur in any given year?
c. Using Chebyshev's Theorem, construct an interval where we expect to find 75% of the occurrences of the number of accidents per year.

a. $f(X=2;3)=\dfrac{2.71828^{-3}\,3^2}{2!}\approx 0.2240$

b. $P(X<2;3)=P(X=0;\ 3)+P(X=1;\ 3)\approx 0.0498+0.1494\approx 0.1992.$

c. Since $\sigma^2=\mu$, $\sigma^2=3$ and $\sigma\approx 1.73$. Applying Chebyshev's Theorem for $k=2$, we get: $P(-0.46\le x_i\le 6.46)\ge 0.75.$

This means that at least 75% of the time, the number of accidents per year will be between 0 and 7. This can be verified with a quick look-up on Table A.2 (see Appendix A):

$P(0 < x < 7) = 0.9881$ (which exceeds 75%).

If n is large and p is small, probabilities associated with binomial distribution can be approximated with the Poisson distribution (Eqn 5.8). The same holds true when n is large and p is close to 1, but the success and failure probabilities must be exchanged. For practical purposes, these approximations are reasonable if

$$n \geq 100 \quad \text{and} \quad np < 10.$$

Example 5.16. From Example 5.4, we know that 5% of a particular secondary manufacturer's furniture components are graded as defective. If 200 furniture components are selected at random, what is the probability that 8 are defective?

If we consider this question in exact terms with a binomial distribution, we use Eqn 5.2 and get:

$$f(X = 8; 200, 0.05) = \binom{200}{8} 0.05^8\, 0.95^{192} \approx 0.1137.$$

To use the Poisson distribution, we assume that the number of defective furniture components occurs over a fixed time at an average rate of 5%. Using the number of trials and the equation for the average number of 'successes' from the binomial distribution, we can calculate the average number of defective components per time period:

$$\mu = np = (200)(0.05) = 10$$

Then, using the equation for a Poisson distribution, the probability of 8 defective components is:

$$f(X = 8;\ 10) = \frac{2.71828^{-10}\, 10^8}{8!} \approx 0.1126$$

This example shows that even when $\mu = 10$, the Poisson distribution gives a good approximation of the binomial distribution. It is advantageous to use this approximation when n is large, as ${}_nC_x$ can be very difficult to calculate.

Exercises

Section 5.1

5.1. A special kind of roulette wheel is divided into 20 equal sections numbered from 1 to 20. Find the equation for the probability distribution of the random variable, X, which describes the number chosen when the wheel is spun. Also find:

a. $P(2 < X < 6)$.
b. $P(X > 15)$.

5.2. A fair die is rolled. Let X be the number of dots showing on the die after it is rolled.

a. Find the equation for the probability distribution of X.
b. Calculate the expected value of the random variable, X.
c. Derive the general equation to calculate the mean of a uniform distribution.
d. Derive the general equation to calculate the variance of a uniform distribution.

Section 5.2

5.3. In a certain sawmill, the probability of a clear grade A board is 0.2. Eight boards are randomly pulled from the green chain. If X = the number of clear grade A boards, use the binomial equation to find the probability that:

a. $P(X < 2)$.
b. $P(1 < X < 3)$.
c. $P(X \leq 4)$.

5.4. Verify the probabilities in Exercise 5.3a, b and c using Table A.1 (see Appendix A).

5.5. Find the mean, variance and standard deviation of the random variable, X, described in Exercise 5.3. Explain the meaning of the standard deviation using Chebyshev's Theorem with $k = 1.5$ and $k = 2$.

5.6. A survey showed that 20% of seedlings died within one year of planting on a given site. Assume that the random variable describing the number of seedlings that survived, X, is a binomial random variable. If 12 seedlings are examined, use the binomial equation to find the probability that:

a. $P(X > 10)$.
b. $P(9 \leq X \leq 12)$.
c. $P(X \leq 10)$.
d. Verify the probabilities in a, b, and c using Table A.1.
e. Find the mean, variance and standard deviation of the random variable, X.
f. What is the meaning of the mean?

5.7. A multiple-choice test is given with 3 choices for each of 10 questions. At least 6 out of 10 correct answers are required to pass the test. If a student guesses on each question, find the probability that the student will pass the test.

5.8. The probability that a logging truck will have no violations, one violation, or two or more violations when it is given a safety inspection is 0.35, 0.42 and 0.23, respectively. If 5 trucks are inspected, what is the probability that 2 will have no violations, 2 will have one violation and 1 will have two or more violations?

5.9. Logs at a sort yard have ambrosia beetle infestation, red stain, or are free of infestation with probabilities of 0.05, 0.07 and 0.88, respectively. If 10 logs are selected at random, what is the probability that 1 will have beetle infestation, 1 will have red stain and 8 will be free of infestation?

Section 5.3

5.10. In a shipment of 25 pieces of furniture, 5 pieces are checked. The shipment is rejected if 1 or more pieces are found to be scratched. Assume that the shipment has 8 scratched pieces.

a. Find the probability that the shipment will be rejected.
b. Find the mean, variance and standard deviation of the random variable for the number of scratched pieces.

5.11. A committee of 3 is selected from 4 foresters and 2 conservation biologists to conduct an environmental impact assessment on logging near a riparian zone. Assume that the members are randomly chosen and the random variable of interest is the number of foresters selected.

a. Find the probability that exactly 2 foresters will be selected.
b. Find the mean and standard deviation of the random variable.
c. Use Chebyshev's Theorem to describe the variability in the number of foresters selected with $k = 1.5$ and $k = 2$.

5.12. In a plantation, there are 400 spruce and 100 pine trees. If a random sample of 8 trees is selected:

a. Find the probability that exactly 3 are pine, using the hypergeometric equation.
b. Find the probability that exactly 3 are pine, using the binomial approximation.
c. Compare the results obtained in a and b.
d. Is the condition required for binomial approximation satisfied?

5.13. A bag contains 10 Douglas-fir, 8 white pine and 6 ponderosa pine seeds. In a random sample of 5 seeds, find the probability that 2 Douglas-fir, 2 white pine and 1 ponderosa pine seeds are selected.

Section 5.4

5.14. The probability that a loaded logging truck will get a flat tyre on its way to the mill is 0.03. Find the probability that a new driver will have his or her first flat tyre during the:

a. First trip.
b. Fifth trip.
c. Twentieth trip.
d. Discuss the results of a, b and c.

5.15. The probability that a log transported to a particular sawmill is infested by ambrosia beetles is 0.1. Find the probability that:

a. The first infested log is the fifth one checked.
b. The second infested log is the fifth one checked.
c. The third infested log is the fifth one checked.

5.16. A fair die is rolled. Find the probability that:

a. The first 6 occurs on the first roll.
b. The first 6 occurs on the sixth roll.
c. The second 6 occurs on the sixth roll.
d. The second 6 occurs on the twelfth roll.

5.17. On any summer day, 12% of the hikers in the Garibaldi Provincial Park in British Columbia are European visitors. If a summer student stationed at the park randomly selects hikers for interviews, find the probability that:

a. The third European visitor is found on the seventh interview.
b. The second European visitor is found on the seventh interview.
c. The first European visitor is found on the seventh interview.

Section 5.5

5.18. In a lodgepole pine stand infested by bark beetles, an average of 2.2 beetle larvae are found under a 10×10 cm^2 sample of bark taken at 1.3 m height from the ground. Find the probability that a randomly selected bark sample has:

a. Exactly 3 larvae.
b. Less than 3 larvae.
c. More than 3 larvae.
d. Find the standard deviation of the random variable for the number of larvae and evaluate Chebyshev's Theorem for $k = 1.5$ and $k = 2$.
e. Use Table A.2 to check the probabilities you calculated in a, b and c.

5.19. A luthier (guitar manufacturer) receives, on average, 3.12 orders per day. Find the probability that on a given day they receive:

a. Exactly 4 orders.
b. Less than 4 orders.
c. More than 4 orders.
d. Find the standard deviation of the variable for the number of orders.

5.20. A young lad in a forestry school is quite popular with his classmates. On average, he receives 2.74 phone calls every night. What is the probability that tomorrow night, the number of calls he receives will:

a. Exceed 4.
b. Be exactly 3.
c. Be less than 3.

6 Continuous Distributions and the Normal Distribution

Describing Data that are Measured

Continuous sample spaces produce continuous random variables. When we deal with quantities measured on continuous scales, such as weights, heights, volumes, time intervals and so on, their probability functions are described by continuous distributions called probability density functions. Like discrete probability distributions, there are several kinds of continuous distributions that describe the numerous sets of data that occur in nature, industry and research. The most important of these continuous distributions is the *normal distribution*, which is a symmetric bell-shaped curve that extends to infinity on both the negative and positive sides. Much of the theory surrounding statistical inference is based on this distribution. The normal distribution will be discussed in this chapter, along with two other continuous distributions: the uniform distribution and the exponential distribution.

6.1 Uniform Distribution

The uniform distribution has both a discrete and a continuous form. For example, the number of dots showing when a single die is tossed (1, 2, ..., 6) is a discrete **uniform random variable** (see Chapter 5). A continuous uniform random variable is similar, but it can take on any real numbered value. It is described by the **uniform probability distribution** and, like its discrete form, the continuous uniform random variable is the simplest description of a continuous random variable. Its probability density function is a horizontal line between two points, a and b:

$$f(x; a, b) = \frac{1}{b-a}, \qquad \text{for } a < x < b. \tag{6.1}$$

Figure 6.1 shows the probability density function for a uniform distribution. If x is not between a and b, $f(x) = 0$, i.e. there is a probability of zero outside the region between a and b. Probabilities between two given values, x_1 and x_2, that are in the region between a and b can be calculated as:

$$P(x_1 < X < x_2;\ a,b) = (x_2 - x_1)\frac{1}{b-a},$$

where $x_2 \ge x_1$, $a \le x_1 \le b$ and $a \le x_2 \le b$.

The mean and variance of the uniform distribution can be calculated as:

$$\mu = E(X) = \frac{a+b}{2}, \text{ and} \tag{6.2}$$

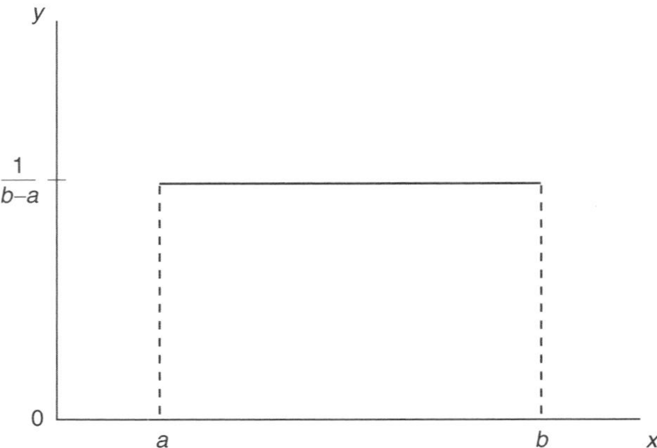

Fig. 6.1. Probability density function for the uniform distribution.

$$\sigma^2 = \frac{(b-a)^2}{12}. \tag{6.3}$$

For any continuous probability distribution, the area under its curve sums to 1, which is to say that the probabilities associated with all possible values for the random variables must sum to 1. In the case of the uniform distribution, the area under the horizontal line segment between a and b is equal to 1. In addition, for any continuous random variable, the probability of an exact x value is 0. For example, $P(x = 1.25) = 0$. Hence,

$$P(x_1 < x < x_2) = P(x_1 \le x \le x_2).$$

Equation 6.1 is more completely stated as:

$$f(x) = \begin{cases} \dfrac{1}{b-a} & a < x < b, \text{ and 0 elsewhere} \\ 0 & \end{cases}$$

Example 6.1. Consider a random variable that represents a logging truck's travelling time from the log sort yard to the highway. The variable can take on any value between 20 and 30 min with equal probabilities and it is therefore represented by a uniform distribution. What is the probability that the truck will get to the highway in more than 22 min and less than 27 min? Calculate the mean, variance and standard deviation of the distribution.

$$P(22 < X < 27) = (27-22)\frac{1}{30-20} = 0.5$$

$$\mu = \frac{20+30}{2} = 25.0$$

$$\sigma^2 = \frac{(30-20)^2}{12} \approx 8.33, \qquad \sigma \approx 2.87$$

It follows that probability associated with a uniformly distributed random variable is proportional to the length of the interval it covers.

Example 6.2. Suppose that the length of a 10 ft piece of dimensional lumber can be trimmed to any length of between 9.9 ft and 10.1 ft with equal probabilities. Therefore, the actual lumber length is described by a uniform distribution. What is the probability that a piece of lumber selected at random is between 9.9 ft and 9.95 ft?

$$P(9.9 < X < 9.95) = (9.95 - 9.9)\frac{1}{10.1 - 9.9} = 0.25$$

6.2 Exponential Distribution

The **exponential distribution** is the continuous counterpart to the Poisson distribution and both depend on the mean. As we learned in Chapter 5, the Poisson distribution is often useful to describe time-related events. While the Poisson distribution describes the probability of events occurring within a time interval, the exponential distribution describes the elapsed times between occurrences of consecutive events. Some examples of exponential distributions are:

- The lifespan of an increment borer.
- Times between arrivals of customers to a supermarket cashier.
- Times between hurricanes in a certain region.
- Times between breakdowns of a woodworking machine centre.
- Time required to load a logging truck.

Suppose that X is a random variable that follows an exponential distribution. Its probability density function is:

$$f(x;\mu) = \frac{1}{\mu}e^{-\frac{x}{\mu}}, \text{where: } x > 0, \text{and } \mu > 0. \tag{6.4}$$

Some typical exponential density functions are shown in Fig. 6.2. As indicated in Eqn 6.4, these functions are defined by μ, the mean of the distribution. The variance of the distribution is a function of the mean, with $\sigma^2 = \mu^2$ (proof not given here).

Since the exponential distribution is continuous, probabilities associated with

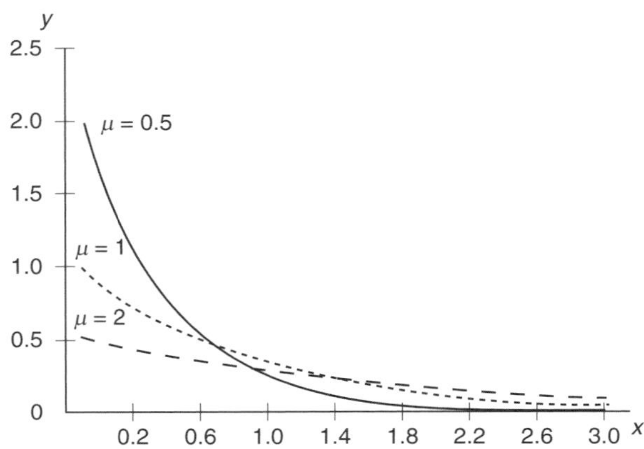

Fig. 6.2. Probability density functions for some exponential distributions.

intervals of the random variable are determined by calculating the area under the density function curve over that interval. The cumulative probability function of the exponential density function can be derived from its probability density function and is given as:

$$F(x;\mu) = 1 - e^{-\frac{x}{\mu}}, \qquad \text{where: } x > 0, \qquad \mu > 0. \tag{6.5}$$

The cumulative density function is used to calculate probabilities between 0 and a given x value.

Example 6.3. Suppose that the time to machine a kitchen cabinet door follows an exponential distribution with a mean of 7 min.

a. What is the probability that a door will be completed in less than 4 min?
b. What is the probability that a door will be completed in more than 4 min and less than 8 min?

a. $P(X > 4) = F(4;7) = 1 - e^{-\frac{4}{7}} \approx 0.435.$
b. $P(4 < X < 8) = P(X > 8) - P(X > 4) \approx 0.681 - 0.435 \approx 0.246.$

6.3 Normal Distribution

Abraham de Moivre (1667–1754), a French mathematician, derived the mathematical equation of the **normal distribution** in 1733. Later, Pierre Laplace (1749–1827) and Karl Gauss (1777–1855) further studied and explored the properties of the normal curve. Because of their independent contributions, the normal curve is often called the *Gaussian distribution*, or in France, the *Laplacian distribution*. Since its discovery more than 360 years ago, it has become the most important distribution, not only in statistics, but in almost every branch of science. This is because the *frequency distributions* of many real, natural events follow a normal distribution. It is also because many of the most important theories in statistical inference are based on the normal distribution. We will begin discussing this in detail in Chapter 7.

A normal distribution is a symmetric, bell-shaped or mound-shaped distribution (Fig. 6.3), and the general equation of its probability density function is given as:

$$f(x;\mu,\sigma) = \frac{1}{\sigma\sqrt{2\pi}} e^{-\frac{1}{2}\left(\frac{x-\mu}{\sigma}\right)^2}, \qquad \text{for } -\infty < x < \infty, \tag{6.6}$$

where $\pi = 3.14159\ldots$ and $e = 2.71818\ldots$

As shown in Eqn 6.6, the probability function is defined by the mean, μ, and the variance, σ^2 (or standard deviation, σ). The two parameters, μ and σ, respectively specify the **position** and the **shape** (or **spread**) of a normal distribution. To illustrate the effect of the position parameter (μ), Fig. 6.4 shows two normal curves with different means (10 and 30), but identical standard deviations (5). To illustrate the effect of the shape parameter (σ), Fig. 6.5 shows two normal curves with the same means (10), but with different standard deviations (3 and 5). Figure 6.6 illustrates the effect of both the shape and position parameters by showing two normal curves with different means (10 and 15) and different standard deviations (5 and 3).

When the means are different and the standard deviations are the same, the curves have the same spread, but are centred at different places (Fig. 6.4). When the means

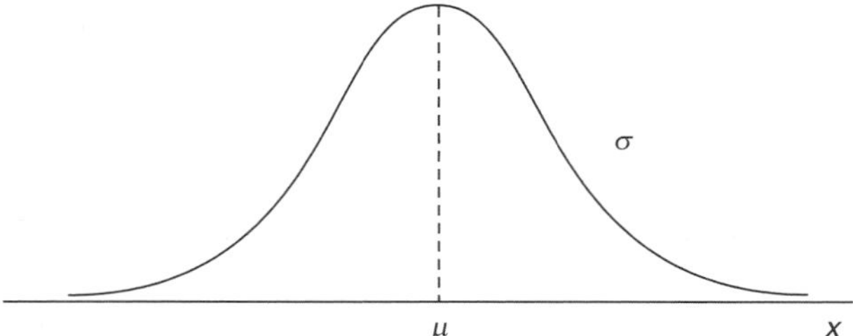

Fig. 6.3. Probability density function for a normal distribution.

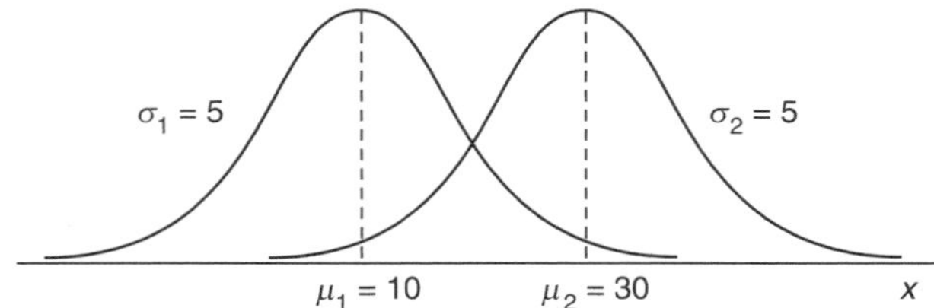

Fig. 6.4. Normal curves for $\mu_1 \neq \mu_2$ and $\sigma_1 = \sigma_2$.

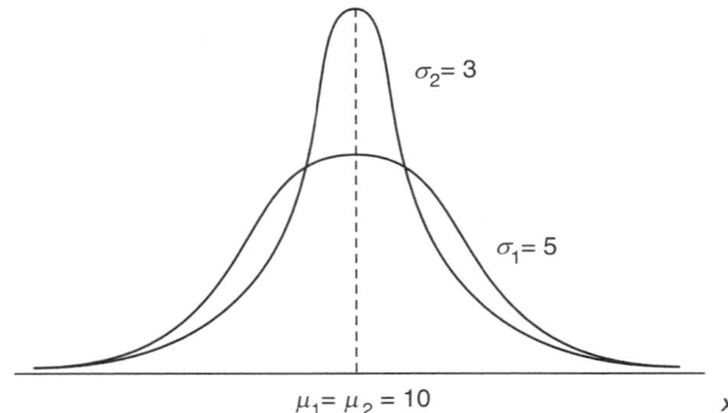

Fig. 6.5. Normal curves for $\mu_1 = \mu_2$ and $\sigma_1 \neq \sigma_2$.

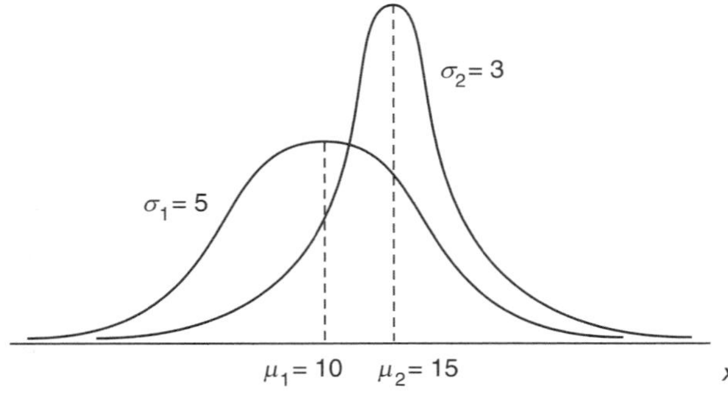

Fig. 6.6. Normal curves for $\mu_1 \neq \mu_2$ and $\sigma_1 \neq \sigma_2$.

are the same and the variances are different, the curves are centred at the same position, but have different spreads (Fig. 6.5). When both the means and standard deviations are different, the two curves are entirely different, centred at different places and having different degrees of spread (Fig. 6.6).

A normal curve has four important properties:

1. The *mode*, *median* and *mean* all coincide at the same point, where the curve is at its maximum.
2. The curve is *symmetric* around the vertical axis drawn through the mean.
3. The curve is *asymptotic* to the x-axis in both the positive and negative directions. In other words, in theory, it never touches the x-axis and extends from $-\infty$ to $+\infty$.
4. The total area under the curve, from $-\infty$ to $+\infty$, is equal to 1.

Other bell-shaped curves may also possess the above properties; however, not all such curves are normal curves. To judge whether or not a frequency distribution is actually normal requires a *statistical test for normality* (introduced in Chapter 10).

As illustrated with the exponential probability distribution, the various probabilities of any continuous density function can be calculated as the area under the curve between two bounds. Figure 6.7 illustrates this concept with bounds x_1 and x_2.

As shown in Figs 6.4 through 6.7, the area under the curve between two fixed bounds depends on both the standard deviation and on the relative positions of the bounds to the mean.

To obtain probabilities between two bounds, the probability density function (Eqn 6.6) must be integrated to obtain its cumulative probability density function. However, because a closed-form expression for the integral of Eqn 6.6 does not exist, other procedures are required to determine the area under the curve. Fortunately, we can rely on tabular values to estimate the value of the cumulative density function. Thousands upon thousands of normal distributions occur in nature, business and science. So that we need use only one table for any normal distribution, we must first transform a normal random variable, X, into another normal random variable, Z, which will always possess two properties: the mean is 0 and the variance is 1. This normal distribution is called the **standardized normal distribution** and is described by:

$$Z = \frac{X - \mu}{\sigma}. \tag{6.7}$$

Note that Eqn 6.7 is similar to the *standard scores* that we introduced briefly in Section 2.5 (see Chapter 2). This equation provides us with z-scores, and their meaning can be interpreted in exactly the same way. A transformed z-score means that the point in question (from the original distribution of the random variable, X) is z standard

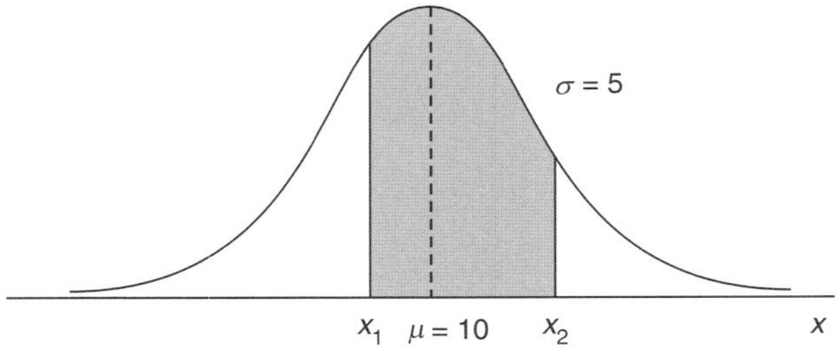

Fig. 6.7. $P(x_1 < X < x_2)$ from a normal curve (shaded area).

deviations away from the mean. The values of z-scores can be either positive or negative, meaning that they are either on the left or the right side of the mean value of 0.

Using the rules of mathematical expectations, it can be shown that the mean of Z is zero:

$$E(Z) = E\left(\frac{X-\mu}{\sigma}\right) = \frac{1}{\sigma}E(X-\mu) = \frac{1}{\sigma}\left[E(x) - E(\mu)\right] = \frac{1}{\sigma}(\mu-\mu) = 0$$

Using the rules of variances, it can be shown that the variance of Z is 1:

$$\sigma_Z^2 = \sigma_{(X-\mu)/\sigma}^2 = \sigma_{X/\sigma-\mu/\sigma}^2 = \sigma_{X/\sigma}^2 = \frac{1}{\sigma^2}\sigma_X^2 = \frac{\sigma^2}{\sigma^2} = 1.0$$

or

$$\begin{aligned}\text{var}(Z) &= \text{var}\big((X-\mu)/\sigma\big) = \text{var}\big(X/\sigma - \mu/\sigma\big) \\ &= \text{var}\big(X/\sigma\big) + \text{var}\big(\mu/\sigma\big) = \frac{1}{\sigma^2}\text{var}(X) + 0 = \frac{\sigma^2}{\sigma^2} = 1.0.\end{aligned}$$

With a normal curve, the general procedure for finding the area (i.e. the probability) of a bounded occurrence, X, between two bounds, x_1 and x_2, is to transform x_1 and x_2 into standard scores or z-values, z_1 and z_2, respectively. These z-values are then used to look up the probabilities in a z-table, which are called p-values and are generated by numerical integration (see Table A.3 in Appendix A). It can be shown that the area under the standardized normal curve is the same as the area under the original (not standardized) curve (Fig. 6.8). That is:

$$P(x_1 < X < x_2) = P(z_1 < Z < z_2).$$

Table A.3 (see Appendix A) gives the area under the standardized normal curve for $P(Z < z)$ for values of z from –3.49 to 3.49. This range of values covers the probabilities required for most practical purposes, and the areas under the curve outside of –3.49 and 3.49 are negligible. It is a cumulative probability distribution, which is similar to a cumulative relative frequency distribution. It provides the probability that the standardized normal variable Z will be less than a given z-value.

To demonstrate the use of Table A.3, let us find the probability that the standard normal variable, Z, is less than 1.58, i.e. $P(Z < 1.58)$. The table is laid out with increments of z in tenths in the leftmost column, and increments in hundredths added to these values across the rows. For z = 1.58, first we find the value of 1.5 in the first column under z, then we move across the row to the column 0.08, where we read the

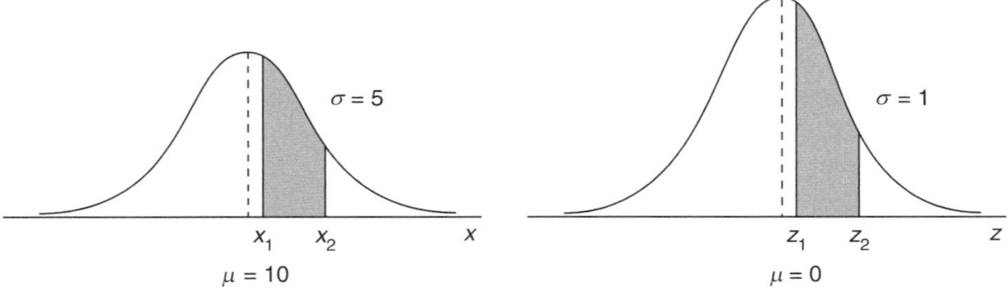

Fig. 6.8. Original and transformed normal distribution.

p-value 0.9429. This means that the probability that a z-value is less than 1.58 is 0.9429. Since the total area under the curve is equal to 1, we could easily find the probability that the standard normal variable Z is greater than 1.58, i.e. $P(Z > 1.58)$. In this case, we look up the value as before (Fig. 6.9) and, using the rule of complements, subtract the p-value from 1:

$$P(Z > 1.58) = 1 - 0.9429 = 0.0571.$$

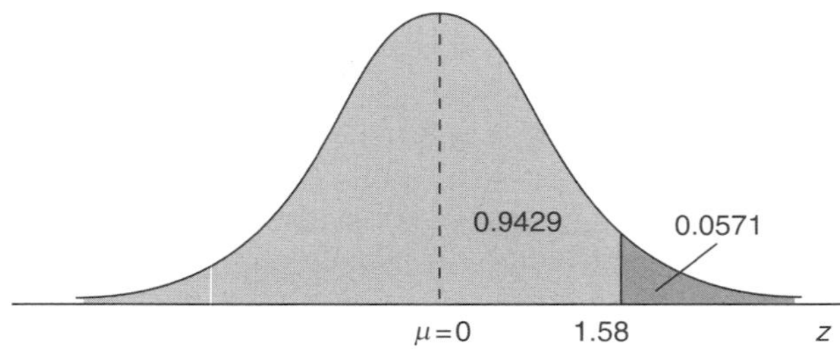

Fig. 6.9. $P(Z < 1.58)$ and $P(Z > 1.58)$.

In Table A.3 (see Appendix A), the values of z can be located for numbers from –3.49 to +3.49 with two decimal places, e.g. 1.58. If a z-value with more than two decimals is required, linear interpolation of the p-values will result in a reasonable approximation (e.g. to find the probability that $Z < 1.585$, we would find the value halfway between $P(Z < 1.58)$ and $P(Z < 1.59)$). The application of the Z transformation changes depending on the context of the probability question being posed. This is demonstrated in the following example.

Example 6.4. Assume that the heights of young Sitka spruce trees in a plantation have a mean of 10 m with a standard deviation of 2.5 m, and that the heights are normally distributed. If you randomly select a single tree, what is the probability of finding a tree that is:

a. Shorter than 6 m?
b. Taller than 12 m?
c. Between 6 m and 12 m in height?
d. Between 12 m and 14 m in height?

Drawing a picture before attempting to solve problems like these is very helpful, and highly recommended (see Figs 6.10 to 6.13). A clear picture of the normal curve will indicate whether the probability may be looked up directly, or needs further mathematical manipulation as a result of the z-table showing only probabilities *less than* a given z-score.

$$P(X < 6) \quad = P\left(Z < \frac{6-10}{2.5}\right) = P(Z < -1.60) \approx 0.0548.$$

$$P(X > 12) \quad = P\left(Z > \frac{12-10}{2.5}\right) = P(Z > 0.80) \approx 1.0 - 0.7881 \approx 0.2119.$$

$$P(6 \le X \le 12) \quad = P(X \le 12) - P(X \le 6) = P\left(Z \le \frac{12-10}{2.5}\right) - P\left(Z \le \frac{6-10}{2.5}\right)$$
$$\approx 0.7881 - 0.0548 \approx 0.7333.$$

$$P(12 < X < 14) \quad = P(X < 14) - P(X < 12) = P\left(Z < \frac{14-10}{2.5}\right) - P\left(Z < \frac{12-10}{2.5}\right)$$
$$= P(Z < 1.60) - P(Z < 0.80) \approx 0.9452 - 0.7881 \approx 0.1571.$$

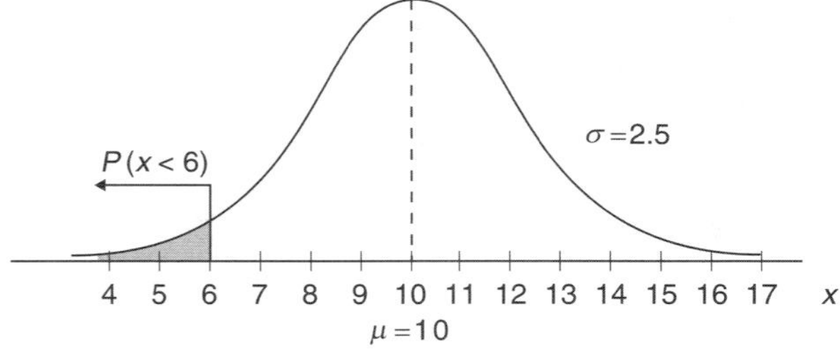

Fig. 6.10. a. $P(X < 6)$.

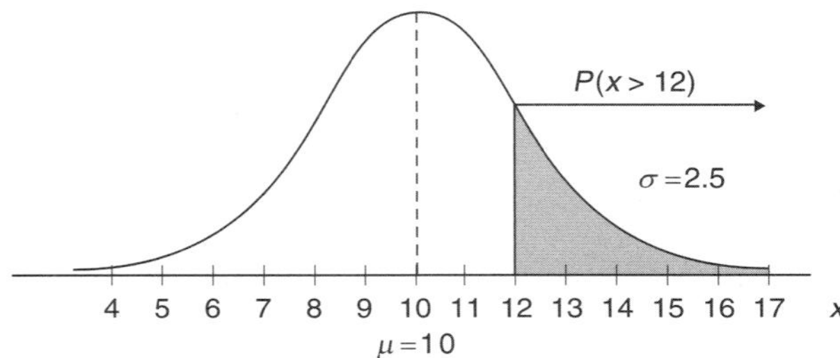

Fig. 6.11. b. $P(X > 12)$.

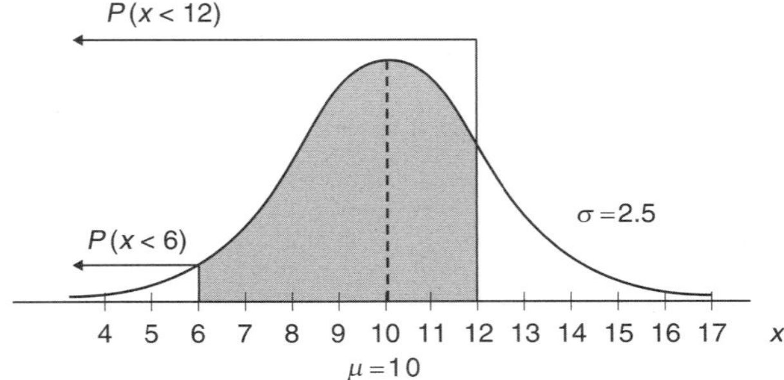

Fig. 6.12. c. $P(6 < X < 12)$.

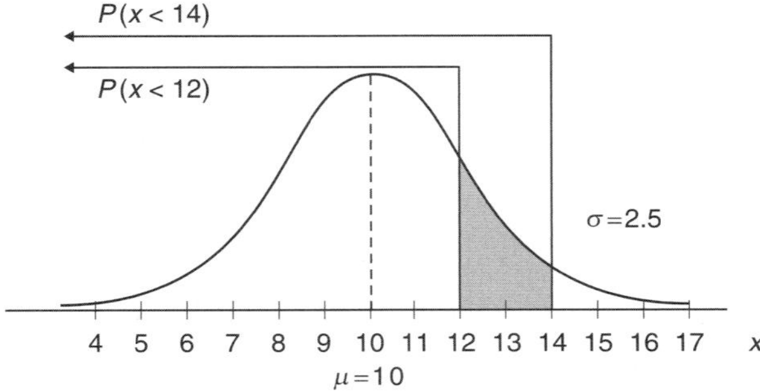

Fig. 6.13. d. $P(12 < X < 14)$.

A word of caution is in order regarding the use of Table A.3 (see Appendix A). This table is not the only kind of standardized normal probability table in common use. Because of the symmetrical nature of the standardized normal curve, some textbooks present only the *p*-values associated with the *positive* half of the *z*-table, expecting the reader to make additional mathematical manipulations when computing probabilities. It is advisable to read the instructions on the use of *z*-tables prior to solving any problems and, again, a diagram is always a useful starting point.

Another use for *z*-tables is for solving probability problems in reverse. For example, we may be interested in knowing where a particular value of a random variable lies in terms of its probability distribution.

Example 6.5. The life of a motor for a chainsaw is a normal random variable with an average of 8 years and a standard deviation of 2.25 years. It is the manufacturer's policy to replace all motors that fail while under guarantee. How long a guarantee should they offer if they are prepared to replace:

a. 2% of all motors that they sell?
b. 4% of all motors that they sell?

Note that this problem is not explicitly about probabilities but, nonetheless, 2% and 4% refer to areas under the curve and follow the exact same logic. In this example, the areas are on the left side of the normal distribution because this side represents lifespans that are below average.

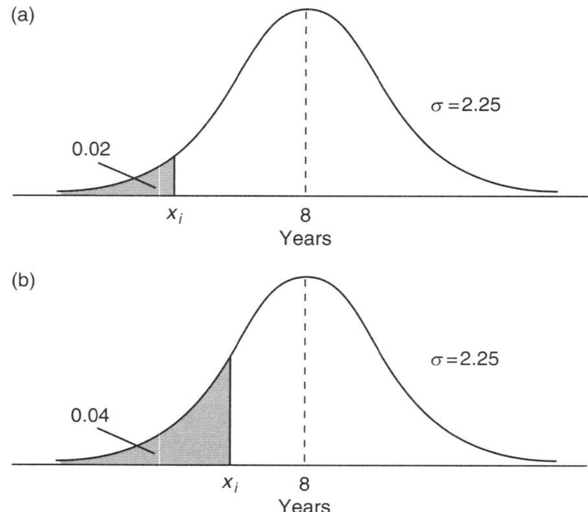

Fig. 6.14. a. Replace motor with guarantee 2% of the time. b. Replace motor with guarantee 4% of the time.

a. Looking at the *p*-values in Table A.3 (see Appendix A), we find 0.0202, which corresponds to a *z*-value of –2.05. We also find 0.0197, which corresponds to a *z*-value of –2.06. We approximate that the *z*-value corresponding to a probability of 0.02 lies halfway between, at –2.055, i.e. $P(Z > -2.055) \approx 0.02$. Now solving for x_i using the Z transformation:

$$-2.055 = \frac{x_i - 8}{2.25}, \text{ and}$$

$$x_i = (-2.055)(2.25) + 8 \approx 3.38 \text{ years.}$$

Since guarantees usually occur in 1-year or 6-month increments, the company could guarantee the motors for 3 or 3.5 years. A guarantee of 3 years would replace less than 2% of the motors, while a guarantee of 3.5 years would replace a little more than 2%.

b. This time, we find the *p*-value 0.0401 at $z = -1.75$, which is very close to 0.04. We then use this *z*-value to solve for x_i:

$$-1.75 = \frac{x_i - 8}{2.25}, \text{ and}$$

$$x_i = (-1.75)(2.25) + 8 \approx 4.06$$

A guarantee of 4 years would replace about 4% of the motors.

Recall that we stated and briefly discussed the **Empirical Rule** in Section 2.4, Fig. 2.12 (see Chapter 2). Using the normal distribution and a *Z* transformation, the Empirical Rule can be verified by working out the probabilities within one, two and three standard deviations of the mean. Using Example 6.5 for one standard deviation, we get:

$$P(8 - 2.25 < X < 8 + 2.25) = P(5.75 < X < 10.25) = P(X < 10.25) - P(X < 5.75)$$
$$= P\left(Z < \frac{10.25 - 8}{2.25}\right) - P\left(Z < \frac{5.7 - 8}{2.25}\right) = P(Z < 1.0)$$
$$-P(Z < -1.0) \approx 0.8413 - 0.1587 \approx 0.6826.$$

The Empirical Rule states that approximately 68% of all observations will be within one standard deviation of the mean, and 0.6826 is very close to 0.68 (68%). Note that any example would give you the same answer, since:

$$Z = \frac{(\mu \pm \sigma) - \mu}{\sigma} = \pm 1.0, \quad \text{and} \quad Z = \frac{(\mu \pm 2\sigma) - \mu}{\sigma} = \pm 2.0.$$

The reader should repeat the above calculations for two and three standard deviations to verify and clearly understand the Empirical Rule.

Sometimes, it is known that a continuous random variable is measured and rounded to a certain precision. For example, the heights of the trees in Example 6.4 were measured to the nearest one-tenth of a metre. In order to improve the probability estimations from the *z*-table for these cases, the random variable is treated as a 'discrete' variable by applying a *continuity correction*, which is equal to one half of the precision of the measurements. Under these conditions, probabilities associated with 'less than' and 'less than or equal to' symbols are treated differently:

$$P(x \le X) = P\left(Z < \frac{(x - h) - \mu}{\sigma}\right)$$

$$P(x < X) = P\left(Z < \frac{(x + h) - \mu}{\sigma}\right)$$

$$P(x \ge X) = P\left(Z > \frac{(x + h) - \mu}{\sigma}\right)$$

$$P(x > X) = P\left(Z > \frac{(x - h) - \mu}{\sigma}\right), \quad \text{where: } h = \text{half of the precision}.$$

This also means that the probability for a single measurement can be estimated. For instance, if heights were measured to the nearest 0.1 m, then h = 0.05 m. Thus, the probability of randomly selecting a tree of height 9.2 m is:

$$P(X = 9.2) = P(9.15 < X < 9.25)$$

$$P\left(\frac{9.15 - 10}{2.5} < Z < \frac{9.25 - 10}{2.5}\right) \approx P(-0.34 < Z < -0.30)$$

$$\approx 0.6331 - 0.6179 \approx 0.0152.$$

Example 6.6. Find the probabilities for Example 6.4a and b, if it is known that measurement precision was to the nearest 0.1 m.

a. $P(X < 6) = P\left(Z < \frac{5.95 - 10}{2.5}\right) \approx P(Z < -1.62) \approx 0.0526.$

In comparison, we found this probability to be 0.0548 in Example 6.4a.

b. $P(X > 12) = P\left(Z > \frac{12.05 - 10}{2.5}\right) \approx P(Z > 0.82) \approx 1.0 - 0.7939 \approx 0.2061.$

In comparison, we found this probability to be 0.2119 in Example 6.4b.

Now, let us reconsider Example 6.4a and b by changing 'shorter than' to 'shorter than or equal to' and 'taller than' to 'taller than or equal to'. These small changes in wording result in small changes in the probabilities:

a. $P(X \le 6) = P\left(Z < \frac{6.05 - 10}{2.5}\right) \approx P(Z < -1.58) \approx 0.0571.$

b. $P(X \ge 12) = P\left(Z > \frac{11.95 - 10}{2.5}\right) \approx P(Z > -0.78) \approx 1.0 - 0.7823 \approx 0.2177.$

All this said, if the precision is reasonably small, these continuity corrections usually do not introduce significant changes to probability values. Thus, they are usually not applied in practical problems. In the above example, the changes are fairly sizeable because 0.1 m rounding is not very precise. The usual practice in forestry is to measure tree heights to the nearest 0.01 m, in which case a continuity correction is generally not necessary.

6.4 Normal Approximation to the Binomial Distribution

As discussed in Chapter 5, binomial probabilities can be difficult and laborious to calculate when n is large. In Chapter 5, we suggested using an approximation with the Poisson distribution. However, under certain circumstances, the normal distribution may yield a better approximation. Consider a binomial distribution with $n = 500$, $p = 0.4$ and the task of finding $P(180 < X \le 200)$.

$$P(180 < X \le 200) = P(X = 181) + P(X = 182) + \cdots + P(X = 200).$$

Since tables are not readily available beyond $n = 20$, the following would have to be solved 20 times, with $x = 181, 182, \ldots, 200$ in order to find $P(180 < X \le 200)$:

$$P(X = 181; 500, 0.4) = \frac{500!}{181!(500 - 181)!} 0.4^{181} 0.6^{319}.$$

Evaluating the above expression with a pocket calculator is not only very time-consuming, but it can lead to large rounding errors. In fact, most calculators cannot compute combinations of this magnitude! When n is large and p is not too far from 0.5, the normal distribution is much easier to use and provides us with a close approximation to the binomial distribution. In other words, binomial probabilities can be estimated from the standard normal distribution in certain instances. Figure 6.15 shows histograms for $p = 0.5$ and $n = 2$, 10 and 20. As n increases, these histograms take on the symmetrical bell-shape characteristic of a normal distribution. For practical purposes, the rule of thumb is that the approximation is said to be acceptable if both np and nq are greater than 5.

In the process of estimating binomial probabilities using the standardized normal distribution, the binomial random variable, X, is transformed using its mean and variance. Recall that the mean and variance of the binomial distribution are np and npq, respectively and thus the transformation is:

$$Z = \frac{X - np}{\sqrt{npq}}. \tag{6.8}$$

Since the binomial random variable is discrete, we should always use a *continuity correction* to find the normal approximation to the binomial probability. Given that the binomial distribution describes counts or whole numbers, a continuity correction of 0.5 is used. For example, the number 7 should be represented by the interval 6.5 to 7.5. This means that the transformations indicated above should be carried out as:

$$P(x_1 \le X \le x_2) \approx P\left(\frac{x_1 - 0.5 - \mu}{\sigma} \le Z \le \frac{x_2 + 0.5 - \mu}{\sigma}\right).$$

Example 6.7. Find the exact probability of getting 8 heads in 20 flips of a balanced coin using the binomial equation. Then, use the standardized normal distribution to approximate this probability.

$$P(X = 8; 20, 0.5) = \frac{20!}{8!(20-8)!} 0.5^8 \, 0.5^{12} \approx 0.1201.$$

In order to use the normal approximation to the binomial distribution, we must first check to see if the conditions for a binomial approximation are met. Indeed, both np (20×0.5) and nq (20×0.5) exceed 5 and thus the normal distribution should provide a good approximation of this binomial probability.

$$\mu = (20)(0.5) = 10 \qquad \sigma = \sqrt{(20)(0.5)(0.5)} \approx 2.24$$

$$P(X = 8) \approx P\left(\frac{7.5-10}{2.24} < Z < \frac{8.5-10}{2.24}\right) \approx P(Z < -0.67) - P(Z < -1.12)$$

$$\approx 0.2514 - 0.1314 \approx 0.1200.$$

It can be seen from this example that the approximation is very good, even for n as low as 20.

Example 6.8. Suppose that a Douglas-fir seedling planted in a given area has a 0.7 probability of surviving its first year after planting. Find the probabilities that out of 200 seedlings checked:

a. More than 150 survived.
b. More than 150 but, at most, 155 survived.

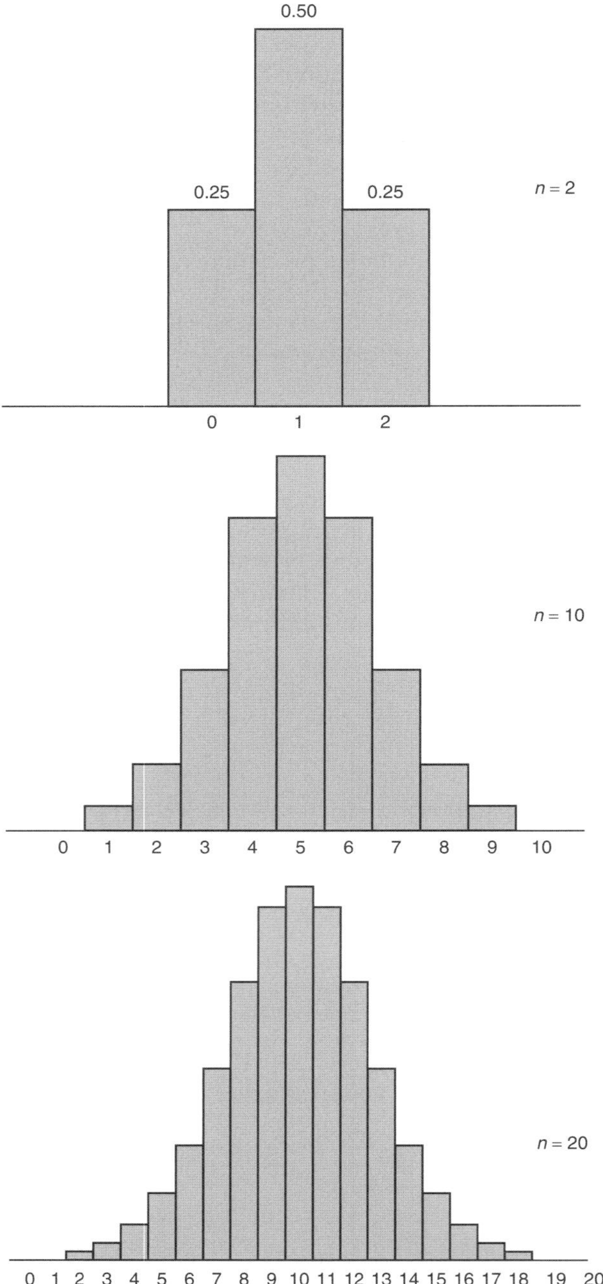

Fig. 6.15. Binomial probabilities for $n = 2$, 10 and 20, $p = 0.5$.

$p = 0.7 \qquad \mu = (200)(0.7) = 140 \qquad \sigma = \sqrt{(200)(0.7)(0.3)} = 6.48$

a. $P(X > 150) \approx P\left(Z > \frac{150.5 - 140}{6.48}\right) \approx P(Z > 1.62) \approx 1 - 0.9474 \approx 0.0526.$

b. $P(150 < X \le 155) \approx P\left(\frac{150.5 - 140}{6.48} < Z < \frac{155.5 - 140}{6.48}\right)$

$\approx P(Z < 2.39) - P(Z < 1.62) = 0.9916 - 0.9474 \approx 0.0442.$

For cases where either the normal or Poisson approximations (see Chapter 5) can be used, the Poisson approximation is recommended when probabilities associated with single values of the binomial random variable are required (e.g. Example 6.7 with $X = 8$), whereas the normal approximation is recommended for intervals of the binomial random variable (e.g. Example 6.8 with $150 < X \le 155$). This recommendation is based largely on the amount of work required by the two different procedures and does not have any theoretical basis.

Exercises

Section 6.1

6.1. Suppose that annual rainfall is uniformly distributed with values ranging from 400 mm to 700 mm. Find the probability that, for a randomly selected year, the annual rainfall will be:

a. Less than 500 mm.
b. Greater than 500 mm.
c. Between 500 mm and 600 mm.

6.2. The time it takes to completely assemble a kitchen cabinet in a wood products factory is uniformly distributed between 15 min and 22 min. Find the probability that assembly for the next cabinet will take:

a. More than 20 min.
b. Between 18 min and 20 min.
c. Less than 16 min.

6.3. Find the mean, variance and the standard deviation for the random variables described in Exercises 6.1 and 6.2.

6.4. Assume a function $f(x) = (5/6)x$ for $2 < X < b$. If f is a probability density function for the random variable, x, find the value of b.

Section 6.2

6.5. Suppose the operating time (in years) of an industrial battery is exponentially distributed with a mean life of 2.5 years.

a. Find the probability that a randomly selected battery of this type will have an operating time of:
   i. Less than 2 years.
   ii. Longer than 3 years.
   iii. Between 2 and 3 years.

b. Find the standard deviation of the random variable for operating time and evaluate Chebyshev's Theorem for $k = 1.5$ and $k = 2$. Why is it preferable to use Chebyshev's Theorem in this case?

6.6. A manufacturer of weather station anemometers wants to determine the warranty period for one type of anemometer they make. Assume that the lifetimes of these anemometers are exponentially distributed with a mean of 4.1 years.

a. Find the warranty period (to the nearest month) if they are prepared to replace:
   i. 5% of the anemometers.
   ii. 2% of the anemometers.
b. Find the standard deviation of the random variable for anemometer lifetimes.

6.7. A system used in a pulpmill for quality control contains 8 integrated circuits. Each circuit has a lifespan that is exponentially distributed with a mean of 7.2 years. Can we assume that at least 2 will still be working after 10 years?

6.8. The time in minutes between successive logging truck arrivals at a government weigh station is exponentially distributed with a mean of 27 min. Find the probability that a truck will arrive within 5 min of another.

Section 6.3

6.9. The number of hours that forestry students study per week is normally distributed with a mean of 20 h and a standard deviation of 6 h.

a. What percentage of the students study:
   i. Less than 10 h?
   ii. More than 25 h?
   iii. At least 25 h?
   iv. Between 10 h and 25 h?
b. How many students in a class of 120 will study more than 25 h?

6.10. The widths (X) of a dimensional lumber product are normally distributed with a mean of 8.9 cm and a standard deviation of 0.2 cm. Find:

a. $P(X < 8.8)$.
b. $P(8.7 < X < 9.1)$.
c. $P(X \geq 9.2)$.

6.11. In a very young plantation, tree heights, X, are normally distributed with a mean of 2.20 m and a standard deviation of 0.42 m.

a. Find the following probabilities:
   i. $P(X \leq 1.8)$.
   ii. $P(X < 3.0)$.
   iii. $P(2.8 \leq X < 3.1)$.
b. Find the probabilities in part a if it is known that the trees are measured to the nearest 0.01 m precision.

6.12. The diameter at breast height (dbh) measurements in a Douglas-fir stand are normally distributed with a mean of 25 cm and a standard deviation of 4 cm.

a. If a forest products company wishes to buy 45% of the biggest trees (as measured by dbh), what is the minimum diameter that the company should specify in its purchase contract?
b. The above forest products company uses its excess chips and sawdust as furnish for the manufacture of particleboard. The particleboard plant has a variable forming machine that discharges furnish in amounts that are normally distributed with a standard deviation of 2.57 kg. The mean, μ kg, is set by the user. The company wishes to use the forming machine to fill a forming belt that holds a maximum of 200 kg of furnish and wants to overfill the forming belt only 1% of the time. What value of μ should the company set for its forming machine?

6.13. The lifespan of a certain type of feller buncher is normally distributed with a mean of 6.2 years and a standard deviation of 1.6 years. Find the time (to the nearest month) before 95% of these feller bunchers fail.

6.14. Verify the Empirical Rule for 1, 2 and 3 standard deviations around the mean.

Section 6.4

6.15. Of the red cedar seedlings planted, 90% survive their first year. Three hundred red cedar seedlings are examined.

a. If x denotes, the number of seedlings that survive, find the probabilities that:
   i. $P(X \leq 280)$.
   ii. $P(X < 280)$.
   iii. $P(265 < X < 275)$.
   iv. $P(265 \leq X \leq 275)$.
b. Find the probability associated with the area within one standard deviation of the mean and evaluate how well the Empirical Rule applies to a binomial random variable.

6.16. It is known that 5% of manufactured particleboard panels are defective and marked as rejects. If 200 of these boards are examined, and X denotes the number of defective boards, what is the probability that:

a. $P(X < 12)$.
b. $P(X < 6)$.
c. $P(6 < X < 12)$.
d. $P(6 \leq X \leq 12)$.
e. $P(X = 6)$, using the binomial equation.
f. $P(X = 6)$, using normal approximation for a binomial variable.

6.17. On any given summer day, 40% of the visitors in a certain provincial park in British Columbia are from outside the province of British Columbia. If 250 visitors are interviewed on a summer day and X denotes the number of non-British Columbian residents, find:

a. $P(X < 90)$.
b. $P(X > 120)$.
c. $P(90 \leq X \leq 120)$.

This page intentionally left blank

7 Sampling Distributions
The Foundation of Inference

As discussed earlier, one of the main purposes of statistics is to obtain information about a population by looking at partial or incomplete evidence that is provided by a subset of the population. It is simply too costly and time-consuming to consider every element of a population. Thus, we usually estimate one or more unknown characteristics (parameters) of the population by observing only a subset of the population (a sample) and by computing the appropriate statistics for characterizing the population from this sample. Since a sample is only a portion of a population, sample values (statistics) will change from sample to sample. In other words, the value of any statistic calculated from a sample is expected to vary. Despite this uncertainty, generalizations from a sample statistic to an unknown population parameter can be made with confidence if the probability distribution of the sample statistic is known. In this chapter, we will study the sampling distributions of means, proportions, differences of two means, differences of two proportions, variances and ratios of variances. These distributions will provide the basic tools to understand subsequent chapters of this book in which we will be dealing with the two most important practical applications of statistics: estimation and hypothesis testing.

7.1 Sampling and Sampling Distributions

Most statisticians agree that statistics calculated from **simple random sampling** usually provide sound, logical and reliable generalizations about population parameters. We will provide a formal definition of simple random sampling shortly, but for all intents and purposes, it can be thought of as randomly drawing sample elements from a hat. For reasons of simplicity, most procedures discussed in this book will be based on simple random sampling. However, it should be noted that in many practical situations, simple random sampling is not the preferred procedure; in fact, it is oftentimes undesirable. A brief discussion of common sampling techniques, such as **systematic sampling, stratified random sampling** and **two-stage sampling**, is provided in Chapter 13.

Prior to introducing the concept of **sampling distributions**, it is important to understand that two types of error can occur when sample values (statistics) are used to estimate population characteristics (parameters): sampling error and non-sampling error. **Sampling error** is a natural consequence of taking samples. As different elements of a population appear in different samples, a sample statistic calculated from one sample will not be the same as those calculated from other samples. This tendency of values to deviate from one sample to another is called sampling error. Sampling error is closely associated with sampling distributions and will be discussed in more detail later.

Faulty equipment, poor measurement techniques and incorrect methods of sampling are all causes of what is referred to as **non-sampling error**. Such errors could consistently under or overestimate real parameter values and, in these instances, estimations are said to be **biased**. Biased estimation occurs, for example, when equipment used for recording measurements is not calibrated properly. Non-sampling error could also occur if diameter at breast height (dbh) is consistently measured at less than 1.3 m height from the ground, or if sample plots are selected so that most are close to roads or are at the base of a slope. Non-sampling errors can usually be minimized and/or completely avoided with proper care of equipment and a random selection of sample elements.

In sampling, we must also distinguish between **finite** and **infinite populations**. Finite populations consist of a fixed number of elements which can be counted and, if necessary, listed. Examples of finite populations include: the number of trees in a permanent sample plot, the number of students registered at a university for the 2007/08 academic year, the number of 2 × 4s coming off the production line in a specified shift, the number of salmon spawning in a stream on a given day, or the number of logging trucks travelling on a section of a logging road during a specified time period. A population is called infinite if, in theory, there is no limit to the number of possible observations (or measurements). In sampling, the word 'infinite' is used rather loosely and, if a population has a large number of possible measurements, it is usually referred to as an infinite population. In light of this definition, some examples of infinite populations include: the number of lodgepole pine trees in British Columbia, the number of trout in a large lake, or the number of Douglas-fir seeds produced in a large forested area.

As we have seen in previous chapters, samples from both finite and infinite populations can be selected either **with replacement** or **without replacement**. If a sample is selected with replacement, each element of the population can appear in the sample as often as it is selected. If a sample is selected without replacement, each element can be selected only once. A **simple random sample** of size n from a *finite* population of size N is a sample chosen randomly, such that each possible sample of size n has the same probability of being chosen (we cannot say the same thing of infinite populations, simply because N cannot be defined). If a sample is selected *without replacement* (a more commonly used method), there are ${}_NC_n$ possible samples. On the other hand, if a sample is selected *with replacement* (something rarely used in practice with finite populations), there are N^n possible samples.

Example 7.1. If we randomly select 4 trees (a sample) out of 10 possible trees (the population) without replacement, there are ${}_{10}C_4 = 210$ possible samples. If the selection is made with replacement, there are $10^4 = 10{,}000$ possible samples. In the latter case (sampling with replacement), the multiplication rule is used to calculate the number of all possible samples: for each of the 4 trees selected, there are 10 trees available to choose from. One correct approach for randomly selecting one sample from the 210 or 10,000 possible samples would be to list each one of the 210 or 10,000 samples on identical pieces of paper, mix the papers up and then randomly draw one. As N and/or n increases, this process becomes more cumbersome and practically impossible.

The above process can be simplified for most practical situations by incorporating the use of **random numbers**. At any stage of sampling, it is vital that the selection of one sampling unit is independent of the selection of other units. This can be accomplished by assigning a number to each sampling unit in the population and then drawing n random numbers from a table of random numbers, such as Table A.4 (see Appendix A). Random numbers can also be generated by using a random number generator, which is available in many computer software packages. A segment of Table A.4 is reproduced in Table 7.1. Ten possible digits consisting of 0, 1, 2, ..., 8 and 9 are roughly equally (randomly) distributed in this table. It is simple to generate a table like this; one need only write the ten digits on ten slips of paper, mix the pieces thoroughly and then randomly select slips with replacement. After recording the first digit, the slip of paper is replaced before the next digit is selected (sampling with replacement). This process is repeated until the desired number of digits is obtained. For convenience, these digits are grouped in columns of five digits, but other tables may be set up differently. The use of the table of random numbers is demonstrated in Example 7.2.

Table 7.1. Partial table of random numbers.

10480	15011	01536	02011	81647	91646	69179	14194	62590	36207	20969	99570	91291	90700
22368	46573	25595	85393	30995	89198	27982	53402	93965	34095	52666	19174	39615	99505
24130	48360	22527	97265	76393	64809	15179	24830	49340	32081	30680	19655	63348	58629
42167	93093	06243	61680	07856	16376	39440	53537	71341	57004	00849	74917	97758	16379
37570	39975	81837	16656	06121	91782	60468	81305	49684	60672	14110	06927	01263	54613
77921	06907	11008	42751	27756	53498	18602	70659	90655	15053	21916	81825	44394	42880
99562	72905	56420	69994	98872	31016	71194	18738	44013	48840	63213	21069	10634	12952
96301	91977	05463	07972	18876	20922	94595	56869	69014	60045	18425	84903	42508	32307
89579	14342	63661	10281	17453	18103	57740	84378	25331	12566	58678	44947	05584	56941
85475	36857	43342	53988	53060	59533	38867	62300	08158	17983	16439	11458	18593	64952
28918	69578	88231	33276	70997	79936	56865	05859	90106	31595	01547	85590	91610	78188
63553	40961	48235	03427	49626	69445	18663	72695	52180	20847	12234	90511	33703	90322

Example 7.2. A permanent sample plot contains 82 trees, numbered 1 through 82. Our task is to randomly select 6 trees without replacement. To do so, we will need to select six 2-digit numbers from Table 7.1 (or Table A.4 – see Appendix A) because 82 is a 2-digit number. First, we must decide on a starting place within the table, perhaps by pointing to a number on the table without looking. We then take a series of 2-digit numbers starting from that point, row-wise or column-wise (both rows and columns are randomly generated numbers). In our example, the starting point 5 is boxed in Table 7.1 in row 10 and column 4. By following a row-wise selection of 2-digit numbers, we would generate the following sequence:

53 98 85 30 60 59 53 33 88 67

We have chosen more than six numbers because we can use only 53, 30, 60, 59, 33 and 67 (numbers greater than 82 or less than 01 are ignored). Additionally, because sampling is done without replacement, 53 cannot be used twice, even though it occurred twice in our generated sequence. In general, it can be stated that n (the required sample size) *distinct* numbers must be selected within the range of numbers used to identify the items in the population. If our task had been to select the trees with replacement, the following numbers would have been selected: 53, 30, 60, 59, 53 and 33 (thereby using the observation for tree No. 53 twice).

For infinite populations, where the items in a population cannot be counted, listed or numbered, the definition of a *simple random sample* is modified such that each item selected is independent. Establishing reliable random samples from infinite populations can be a difficult task. For example, allocating n plots randomly in a large forested area is accomplished by drawing a pair of random numbers, which serve as a random point in a two-dimensional coordinate system (i.e. a grid system made up of column and row numbers superimposed on a map of the forest). If we want to take random samples over a period of time (e.g. taking a sample of minutes within 8-h shifts over a period of 30 working days), we could again do so by generating random pairs of numbers. The first random number would define the day and the second one would define the minute within the 8-h shift.

There are many random samples available from any given population and since the elements included in each of these samples are different, every statistic (mean, median, variance, proportion, etc.) will vary from sample to sample, even though they are all estimates of a single population parameter. This means that a statistic is a **random variable**, because its value depends on the elements included in the sample. Consequently, on the rare occasion when we talk about a statistic in general, we denote it with a capital letter. More commonly, we refer to a given value from a number of possible statistics and we denote this by a lower case letter. For instance, we use $\bar{X}$ to talk about the mean in general terms, whereas we use $\bar{x}$ when we refer to a single occurrence of the mean from a given sample.

Table 7.2 presents the most important statistics and the parameters that they estimate. These will be discussed in detail in the following sections. However, one important point is worth noting. Since these statistics are random variables, there will be a probability distribution associated with each of them. The probability distribution of a statistic is called the **sampling distribution of the statistic**. For example, the probability distribution describing means is called the **sampling distribution of the mean**. These sampling distributions vary in different ways and the spread of the random variable can be described by a standard deviation. However, as a convention in statistics, the standard deviation of a sampling distribution is called the **standard error of the statistic**. It follows, then, that the standard deviation of the sampling distribution of the mean is called the **standard error of the mean**. In practice, it is a common mistake to use the term 'standard error' by itself when referring to the 'standard error of the mean'. The term 'standard error' should be qualified by indicating the standard error 'of what'.

Table 7.2. Parameters and statistics.

	Population parameter	Random variable	Sample statistic
Mean	μ	$\bar{X}$	$\bar{x}$
Difference of means	$\mu_1 - \mu_2$	$\bar{X}_1 - \bar{X}_2$	$\bar{x}_1 - \bar{x}_2$
Proportion	P	$\hat{P}$	$\hat{p}$
Difference of proportions	$P_1 - P_2$	$\hat{P}_1 - \hat{P}_2$	$\hat{p}_1 - \hat{p}_2$
Mean of differences of paired observations	μ_d	$\bar{D}$	$\bar{d}$
Variance	σ^2	S^2	s^2
Ratio of variances	σ_1^2/σ_2^2	S_1^2/S_2^2	s_1^2/s_2^2

7.2 Sampling Distribution of the Mean

As discussed in Chapter 2, the mean has favourable characteristics that make it the most common measure of central tendency in samples and populations. In order to understand the relationship between the sample mean and the population mean, let us assume that we take a sample from a small population of 0, 2, 4 and 6. Figure 7.1 graphically shows the probability distribution for this uniform population. The mean, variance and the probability distribution of this small population are:

$$\mu = \frac{0 + 2 + 4 + 6}{4} = 3.0$$

$$\sigma^2 = \frac{(0-3)^2 + (2-3)^2 + (4-3)^2 + (6-3)^2}{4} = 5.0$$

X	0	2	4	6
$f(x)$	$\frac{1}{4}$	$\frac{1}{4}$	$\frac{1}{4}$	$\frac{1}{4}$

Now let us assume that we are taking samples $n = 2$ from this population. Table 7.3 lists all possible samples of size 2 taken with replacement ($4 \times 4 = 16$), while Table 7.4 lists all possible samples of size 2 taken without replacement (${}_4C_2 = 6$). In either case, the respective means of each sample are also given.

Beginning with the case of sampling with replacement, the mean and the variance of all possible sample means from Table 7.3 are:

$$\mu_{\overline{X}} = \frac{\sum_{i=1}^{16} \overline{x}_i}{16} = 3.0$$

$$\sigma_{\overline{X}}^2 = \frac{\sum_{i=1}^{16} (\overline{x}_i - \mu_{\overline{x}})^2}{16} = \frac{5}{2} = 2.5.$$

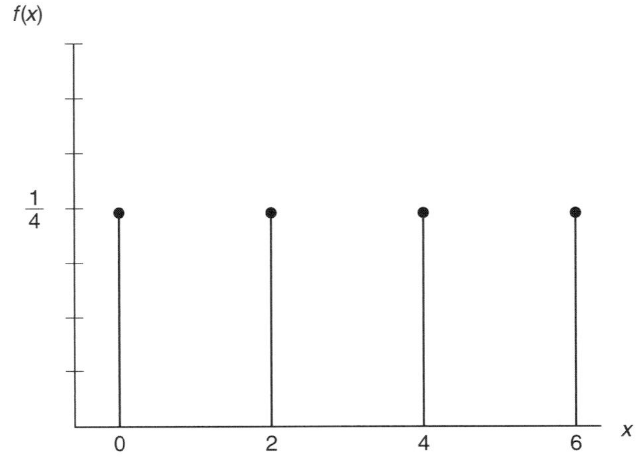

Fig. 7.1. Probability distribution of the population 0, 2, 4 and 6.

Table 7.3. All possible samples of size 2 taken with replacement and their means.

Sample	$\bar{x}$	Sample	$\bar{x}$
0,0	0.0	4,0	2.0
0,2	1.0	4,2	3.0
0,4	2.0	4,4	4.0
0,6	3.0	4,6	5.0
2,0	1.0	6,0	3.0
2,2	2.0	6,2	4.0
2,4	3.0	6,4	5.0
2,6	4.0	6,6	6.0

Table 7.4. All possible samples of size 2 taken without replacement and their means.

Sample	$\bar{x}$	Sample	$\bar{x}$
0,2	1.0	2,4	3.0
0,4	2.0	2,6	4.0
0,6	3.0	4,6	5.0

Figure 7.2 illustrates the frequency distribution for all of these possible means. From the results above, it can be seen that:

1. The mean of the sample means is equal to the population mean.
2. The variance of the sample means is equal to the population variance divided by the size of the sample, σ^2/n (here, 5/2).
3. The probability distribution of all possible means has a symmetrical (approaching a bell) shape.

If we sampled with replacement from the above population at a higher rate of $n = 3$, there would be 64 possible means, with the probability distribution seen in Fig. 7.3

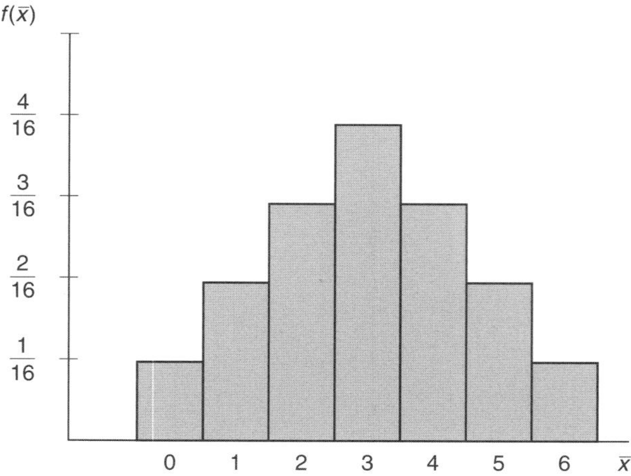

Fig. 7.2. Probability distribution of all possible means from sampling with replacement ($n = 2$).

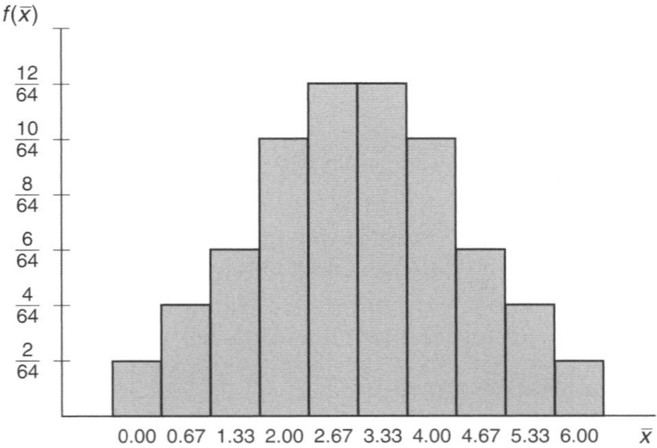

Fig. 7.3. Probability distribution of all possible means from sampling with replacement ($n = 3$).

(samples and individual means are not shown here). Note that this distribution is much closer to being bell-shaped than the distribution for sample size 2.

Moving to the case of sampling without replacement, the mean and the variance of all possible sample means from Table 7.4 are computed as:

$$\mu_{\overline{X}} = \frac{\sum_{i=1}^{6} \overline{x}_i}{6} = 3.0$$

$$\sigma_{\overline{X}}^2 = \frac{\sum_{i=1}^{6} \left(\overline{x}_i - \mu_{\overline{x}}\right)^2}{6} = \frac{10}{6} = \frac{5}{3}$$

Figure 7.4 shows the probability distribution of the means from Table 7.4. From the results above we can see that, like the case of sampling with replacement, the mean of

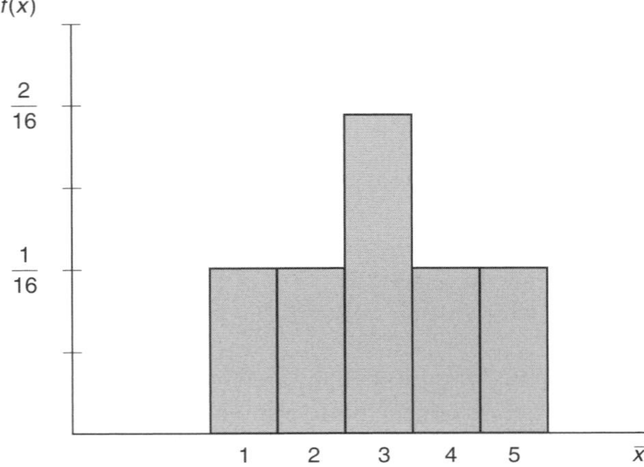

Fig. 7.4. Probability distribution of all possible means from sampling without replacement ($n = 2$).

all possible sample means is equal to the population mean. However, the variance of all possible sample means in this case is computed from the population variance somewhat differently:

$$\sigma_{\overline{X}}^2 = \frac{\sigma^2}{n}\frac{N-n}{N-1}$$

Also, the probability distribution shown in Fig. 7.4 is neither normal nor bell-shaped. However, if the population size was larger, and larger samples were taken, the probability distribution of the means taken from sampling without replacement would be very similar to the one seen in Fig. 7.3 (sampling with replacement).
To summarize sampling with replacement and without replacement:

The mean of the sample means is: $\mu_{\overline{X}} = \mu$. (7.1)

The standard error of the sample means is:

$$\sigma_{\overline{X}} = \frac{\sigma}{\sqrt{n}}, \text{ when sampling with replacement} \qquad (7.2)$$

or

$$\sigma_{\overline{X}} = \frac{\sigma}{\sqrt{n}}\sqrt{\frac{N-n}{N-1}}, \text{ when sampling without replacement} \qquad (7.3)$$

where

$$\sqrt{\frac{N-n}{N-1}} = \text{finite population correction factor.}$$

As defined earlier, the **standard error of the mean** is the standard deviation describing the sampling distribution of the means for a given sample size. In practice, the **finite population correction factor** (fpc) above is omitted when $n \le 0.05N$ (since $(N - n)/(N - 1) \to 1$ as N gets larger); that is, if the sample size is less than 5% of the population, we can omit the fpc. Therefore, if $n \le 0.05N$, Eqn 7.3 reverts to Eqn 7.2. For infinite populations, Eqn 7.2 always applies, regardless of the sampling procedure.

At this stage, we need to introduce the **Central Limit Theorem** to formalize the relationship between a specific parameter of a population and its estimate (statistic). This is, by far, the most important theorem in statistics. It will be introduced in relation to the sample mean, but it will be reiterated for several other statistics. While definitions of the Central Limit Theorem vary, for the purposes of this book it will be defined by the following four points:

1. The sample mean ($\overline{X}$) is a random variable since its value changes from sample to sample.
2. The mean of all possible sample means is equal to the population mean, which can be expressed as an expectation: $E(\overline{X}) = \mu$.
3. The spread of all possible sample means, $\sigma_{\overline{X}} = \frac{\sigma}{\sqrt{n}}$ or $\sigma_{\overline{X}} = \frac{\sigma}{\sqrt{n}}\sqrt{\frac{N-n}{N-1}}$, is the standard error of the mean. Note that there are two ways to compute the standard error, depending on whether or not the finite population correction factor is included.
4. The probability distribution of all possible sample means is normal when the parent population is normally distributed, or is approximately normal when the sample size (n) is sufficiently large ($n \ge 30$). Regardless of the distribution of the parent

population, as n approach infinity, the probability distribution of all possible means becomes normal.

So, why is the Central Limit Theorem so important? Simply stated, this theorem makes it possible to make inferences about a population mean, μ. This is true even when the shape of the population distribution is unknown, as probabilities associated with the statistic, $\overline{X}$, can be easily obtained by standardizing particular values of the sample mean ($\bar{x}$) using the equation:

$$z = \frac{\bar{x} - \mu}{\sigma_{\overline{X}}} \tag{7.4}$$

Example 7.3. In a large plantation of trees, dbh measurements are normally distributed with a mean of 12 cm and a standard deviation of 1.6. If n = 16 trees are selected without replacement, what is the probability that their sample mean is greater than 12.7 cm?

This example is very similar to examples we discussed in Chapter 6. Since the parent population is normally distributed (Fig. 7.5), we assume that the means follow a normal distribution, regardless of the size of the sample, and the probability is simply computed using a Z transformation (standardized normal distribution). The only difference here is that the random variable is the sample mean instead of an individual observation. Also, rather than using the standard deviation, we use the standard error of the mean – which is, of course, a standard deviation. In this case, it is the *standard deviation of all possible sample means.*

$$\sigma_{\overline{X}} = \frac{1.6}{\sqrt{16}} = 0.4$$

$$P(\overline{X} > 12.7) = P\left(Z > \frac{12.7 - 12.0}{0.4}\right) \approx P(Z > 1.75) \approx 1.0 - 0.9599 \approx 0.0401.$$

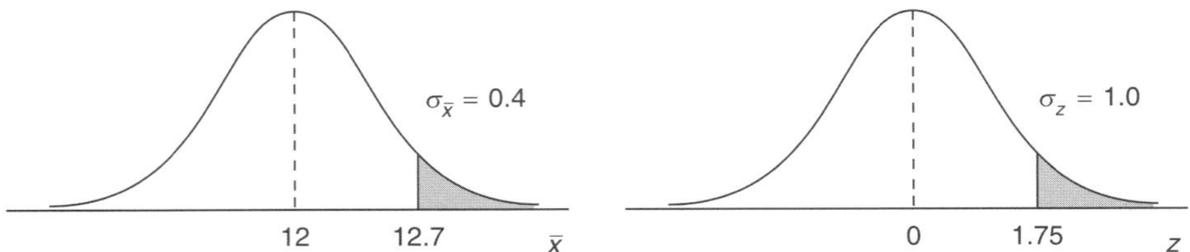

Fig. 7.5. Areas for Example 7.3.

Using the z-table, the resulting probability of 0.04 indicates that there is a small chance (4%) that a sample of 16 trees will have a sample mean of greater than 12.7 cm. In the solution above, we have omitted the finite population correction factor because it was assumed that the number of trees (N) in the stand was very large, so that $n < 0.05N$. Some textbooks suggest using the *continuity correction* (see Chapter 6), depending on the number of decimals used to calculate the sample means. We have not used this correction in our example, as it is unnecessary in most practical applications (the gain in precision will be negligible).

It should be noted that the standard deviation of a population is an inherent characteristic of the spread of the observations within that population. It cannot be manipulated or changed. Although the standard error of the mean is related to the population standard deviation, and behaves in exactly the same way, the size of the standard error of the mean can be manipulated by changing the sample size. The standard error of the mean, which is usually referred to as the sampling error, can be reduced by increasing the sample size, n. When we have 30 or more samples, we can apply the **Empirical Rule** to the standard error of the mean. As such, we know that 68% of all possible means from a given sample size will lie within one standard error of the mean, and 95% of all possible means will lie within two standard errors of the mean.

Example 7.4. Suppose that we have only 50 trees in the population ($N = 50$) from Example 7.3. As before, samples of size = 16 are taken without replacement. Since the sample size of 16 is much higher than 5% of the population size, we must use the finite population correction factor.

$$\sigma_{\overline{X}} = \frac{1.6}{\sqrt{16}}\sqrt{\frac{50-16}{50-1}} \approx 0.333$$

$$P(\overline{X} > 12.7) = P\left(Z > \frac{12.7-12.0}{0.333}\right) \approx P(Z > 2.10) \approx 1.0 - 0.9821 \approx 0.0179.$$

The finite population correction factor essentially serves to 'correct' the standard error downwards as a result of having more information about the population. In other words, when a sample makes up a sizeable portion of a population, the distribution of sample means is less dispersed. Notice that the spread of the probability distribution of the means is much smaller here than in Example 7.3, because more information is available from the population (32% of the population is included in the sample). Consequently, the probability of obtaining a very high or very low sample mean is much lower.

Example 7.5. The specific gravity for green Douglas-fir wood pieces is claimed to be 0.45, with a standard deviation of 0.082. What is the probability that the mean specific gravity of 35 randomly selected Douglas-fir wood specimens is less than 0.435, if the samples are selected without replacement?

$$\sigma_{\overline{X}} = \frac{0.082}{\sqrt{35}} \approx 0.0139$$

$$P(\overline{X} < 0.435) = P\left(Z < \frac{0.435-0.45}{0.0139}\right) \approx P(Z < -1.08) = 0.1401.$$

The distribution of data and corresponding Z transformation are shown in Fig. 7.6. A probability of 0.14 means that about one in every seven samples of size 35 will have an average specific gravity that is less than 0.435. In this example, there was no statement given about the distribution of the parent population; however, since $n \geq 30$, no such statement was necessary.

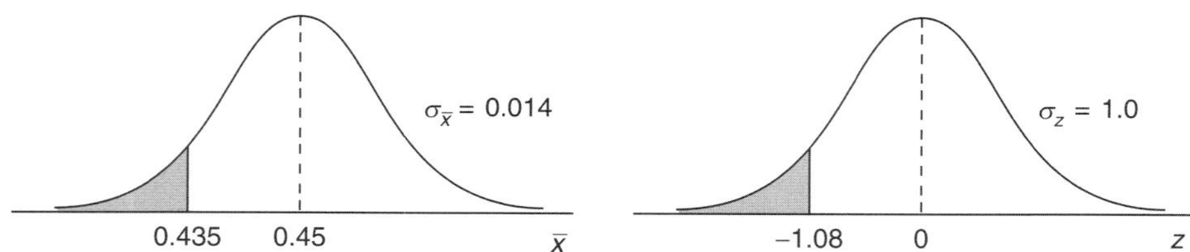

Fig. 7.6. Areas for Example 7.5.

In most real life situations, the population variance (σ^2) is not known (nor is any population parameter for that matter). However, when n is larger than or at least equal to 30, σ^2 can be replaced by its sample value, s^2, in Eqn 7.4. In such cases, s^2 is a good estimate of σ^2, as it does not fluctuate much from one sample to another and the Central Limit Theorem applies.

On the other hand, if the sample size is small ($n < 30$), s^2 varies considerably from sample to sample and Eqn 7.4 does not provide a reasonable standard normal distribution. Instead, Eqn 7.5 is used to transform (standardize) the sample means into another value, the so-called 't' statistic, computed as follows:

$$t_{(v)} = \frac{\overline{x} - \mu}{s/\sqrt{n}} \tag{7.5}$$

where $s_{\overline{X}} = \frac{s}{\sqrt{n}}$ is the standard error of the mean (based on the sample variance), and v = degrees of freedom.

The probability distribution of the t statistic is often referred to as **Student's *t* distribution**, a somewhat cryptic name given that it was derived in 1908 by British chemist, W.S. Gosset. Gosset used the pseudonym 'Student' because he was forbidden to publish by his employer, a popular Irish brewery.

Like the standard normal curve, the t distribution is a symmetrical (about zero), bell-shaped curve. Its standard deviation depends on the sample size and will always be somewhat higher than 1:

$$\sigma_t = \sqrt{\frac{n-1}{n-3}} \quad \text{for } n > 3.$$

From the equation above, it can be seen that as n gets larger, σ_t approaches 1, and thus the distribution of t approaches the standard normal distribution (Z). As shown in Fig. 7.7, the spread of the t distribution is a function of the **degrees of freedom**, which in turn is a function of sample size ($v = n - 1$; for a formal definition, see Chapter 2).

Like the standard normal curve, the probability that a random variable results in a t-value falling between any two specified values is equal to the area under the curve between those two values. A table similar to Table A.3 (see Appendix A) could be produced for every degree of freedom. However, even if we considered only 1 to 30 degrees of freedom, this would result in 30 pages of tables. To save space and simplify

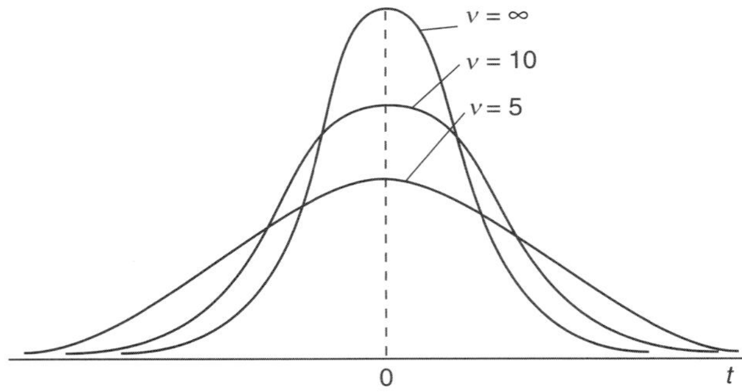

Fig. 7.7. Distribution of t for 5, 10 and ∞ degrees of freedom.

matters, only the most frequently used t-values and their probabilities are generally listed (see Table A.5 in Appendix A).

Table A.5 (Appendix A) looks somewhat different from the normal distribution table (see Table A.3, Appendix A). The first column gives the degrees of freedom, while the first row specifies cumulative probabilities below a specified t-value. Only positive t-values are listed inside the table. In most t-tables, unless otherwise specified, t_α represents the t-value below which (from $-\infty$ to t_α) the area under the curve is equal to α. This makes problems like $P(t_9 < 1.83)$ easy to solve with a simple look-up on the top row of Table A.5 (0.95). But what about the case of $P(t_9 > 1.83)$? Our knowledge that the total area under a probability distribution curve is equal to 1 means that we can solve this problem by taking the complement of the probability: $1 - 0.95 = 0.05$. Additionally, because the curve is symmetrical around zero, negative t-values which occur below the mean need not be listed. Their cumulative probabilities can again be found based on our knowledge that the distribution is symmetrical. For example, $P(t_9 < -1.83)$ is equal to $(t_9 > +1.83)$, both of which equal 0.05.

Because the values of t do not change much between 30 and ∞ degrees of freedom, many tables found in other textbooks do not provide values for more than 30 degrees of freedom. To further complicate matters, many t-tables are organized differently; the reader should be cognizant of this when using other textbooks. The following example demonstrates the use of the t-table presented in this textbook.

Example 7.6. Suppose that the standard deviation of the population in Example 7.3 is not known, but is estimated from the 16 observations to be 2.09 cm.

a. Find the probability that a mean based on 16 observations will be greater than 12.7 cm.
b. Find the probability that a mean based on 16 observations will be less than 11.55 cm.

The distribution of data and corresponding t transformation are shown in Fig. 7.8.

a. $s_{\overline{X}} = \dfrac{2.09}{\sqrt{16}} \approx 0.5225$

$$P(\overline{X} > 12.7) = P\left(T_{(15)} > \frac{12.7 - 12.0}{0.5225}\right) \approx P(T_{(15)} > 1.34) \approx 1.0 - 0.90 \approx 0.10.$$

b. $$P(\overline{X} < 11.55) = P\left(T_{(15)} < \frac{11.55 - 12.0}{0.5225}\right) \approx P(T_{(15)} < -0.861) \approx 1.0 - 0.80 \approx 0.20.$$

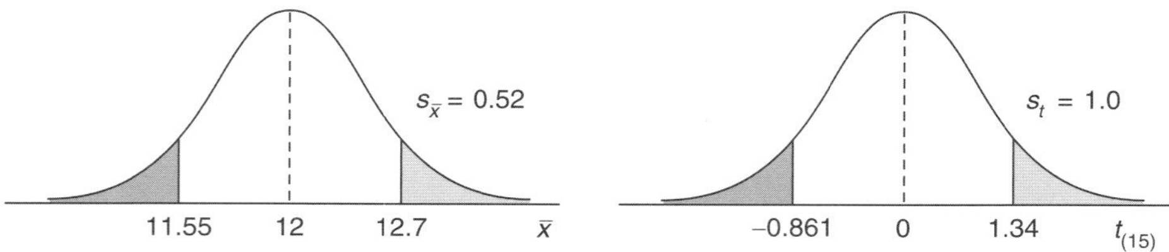

Fig. 7.8. Areas for Example 7.6.

Because of limited entries in the t-table, linear interpolation is often required to obtain proper results. Failing that, estimates based on the proximity to the t-values should suffice. Note that in Example 7.6b, we had to find the probability below a negative t-value. To do this, we used our knowledge of the symmetry of the t

distribution to find the equivalent probability above its positive equivalent. In other words, $P(T_{(15)} < -0.861) = P(T_{(15)} > 0.861) = 1 - P(T_{(15)} < 0.861)$.

7.3 Sampling Distribution of the Sample Proportion

Statistical investigators are often interested in being able to draw conclusions about the *proportion* of objects that possess some specified characteristics in a population, e.g. the proportion of seedlings surviving 2 years after planting, the proportion of defective components in a production line, the proportion of a certain species in a stand, or the proportion of logs infested by ambrosia beetles in a log storage area. Usually, the object that possesses the specified characteristic or property is referred to as a 'success', and the object that does not possess this property is labelled a 'failure'. In other words, we are describing a binomial random variable, and the population follows a binomial distribution (see Chapter 5). In order to make inferences about the usually unknown population proportion, p, the relationship between the proportion based on samples, $\hat{p}$, and the population proportion must be understood.

If n observations are selected from a binomial population and a value of 1 is assigned to each observation (x_i) called a success, while a value 0 is assigned to each observation called a failure, the mean of these observations will be the sample proportion:

$$\bar{x} = \hat{p} = \frac{\sum_{i=1}^{n} x_i}{n} = \frac{\text{number of successes}}{n}.$$

The variance, σ^2, of a population consisting of zeros and ones can be found, using mathematical expectation, as:

$$\sigma^2 = p(1 - p) = pq.$$

It turns out that a distribution of sample proportions behaves in exactly the same way as a distribution of sample means. The general equations for the sample proportion and the standard error of the proportion are:

$$\hat{p} = \frac{X}{n} = \frac{\text{number of successes}}{n}, \tag{7.6}$$

$$\sigma_{\hat{p}} = \frac{\sqrt{pq}}{\sqrt{n}} = \sqrt{\frac{pq}{n}}, \text{ since } \sigma_{\hat{p}}^2 = \frac{pq}{n}, \tag{7.7}$$

or

$$\sigma_{\hat{p}} = \sqrt{\frac{pq}{n}}\sqrt{\frac{N-n}{N-1}}. \tag{7.8}$$

Again, if N is large ($n < 0.05N$), or if samples are taken with replacement, the finite population correction factor is omitted and Eqn 7.8 reduces to Eqn 7.7.

The rules for the sampling distribution of all possible proportions, based on n observations, are as follows:

1. Since the sample proportion, $\hat{P}$, changes from sample to sample it is a random variable.

2. The mean of all possible sample proportions is equal to the true population proportion, p, which can be expressed as an expectation: $E(\hat{P}) = p$.
3. The spread of all possible proportions, or standard error of the proportion, can be calculated from the parameters of the binomial distribution and the sample size:

$$\sigma_{\hat{P}} = \sqrt{\frac{pq}{n}} \quad \text{or} \quad \sigma_{\hat{P}} = \sqrt{\frac{pq}{n}}\sqrt{\frac{N-n}{n}}$$

4. As the sample size (n) increases, the sampling distribution of all possible proportions approaches the normal distribution and, if $n \geq 30$, the sampling distribution is approximated by the normal curve.

Probabilities associated with the random variable, $\hat{P}$, can be obtained by standardizing specific values of $\hat{p}$ using a Z transformation such that:

$$z = \frac{\hat{p} - p}{\sqrt{\frac{pq}{n}}}. \tag{7.9}$$

If the parameter, p, is unknown, as is most often the case, the population variance will also be unknown and the standard error term (as stated in Eqn 7.9) in the denominator cannot be computed. It would make sense to use the t distribution in cases like this, but this is not recommended because the parent population is generally not normal. In other words, the Z transformation is used almost exclusively when $n > 30$ for computing probabilities associated with sample proportions (see Examples 7.7 and 7.8). When p is unknown, therefore, the standard error of the proportion is calculated using the following approximation:

$$s_{\hat{P}} = \sqrt{\frac{\hat{p}\hat{q}}{n}}. \tag{7.10}$$

Example 7.7. The proportion, p, of defective components for a given product is 0.04. What is the probability that, in a sample of 45 components taken without replacement, the proportion of defectives will be greater than 0.07?

The distribution of data and corresponding Z transformation are shown in Fig. 7.9.

$$\sigma_{\hat{P}} = \sqrt{\frac{(0.04)(0.96)}{45}} \approx 0.02921.$$

$$P(\hat{P} > 0.07) = P\left(Z > \frac{0.07 - 0.04}{0.02921}\right) \approx P(Z > 1.03) \approx 1.0 - 0.8485 \approx 0.1515.$$

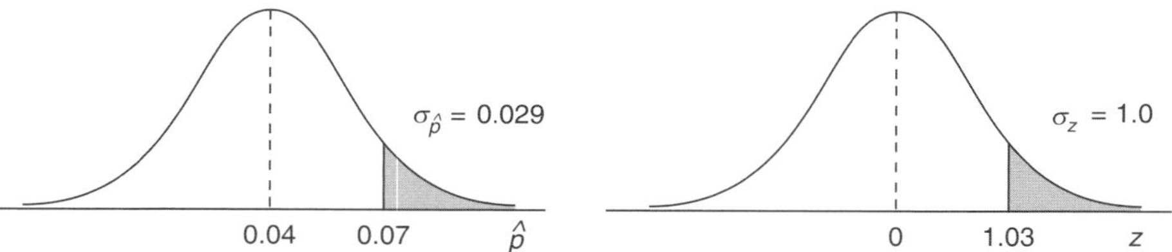

Fig. 7.9. Areas for Example 7.7.

Example 7.8. A forester tells you that the rate of seedling survival in a particular plantation is 0.9. If a sample of 150 seedlings is taken from this population, how likely would it be to obtain a survival rate of less than 0.86?

The distribution of data and corresponding Z transformation are shown in Fig. 7.10.

$$\sigma_{\hat{p}} = \sqrt{\frac{(0.90)(0.10)}{150}} \approx 0.02449$$

$$P(\hat{P} < 0.86) = P\left(Z < \frac{0.86 - 0.90}{0.02449}\right) \approx P(Z < -1.63) \approx 0.0516$$

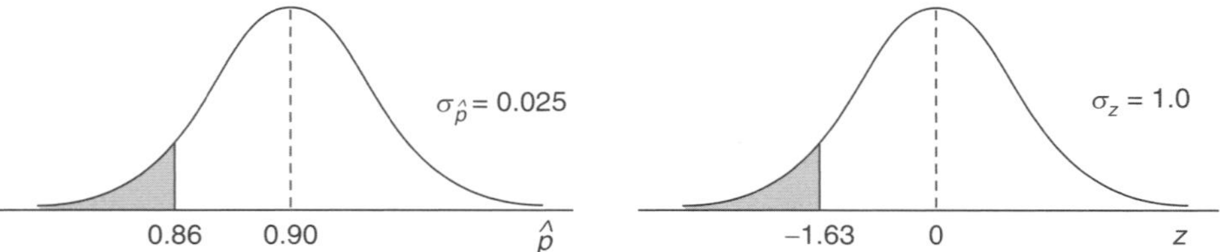

Fig. 7.10. Areas for Example 7.8.

Since a sample proportion can easily be converted into a binomial random variable, $X = n\hat{p}$, Examples 7.7 and 7.8 could be solved using the procedures described in Section 6.4: Normal Approximation of the Binomial Distribution (see Chapter 6). For Example 7.8, we need to find the probability that the number of defects, x, is less than $n\hat{p} = (150)\ (0.86) = 129$:

$$\sigma_X = \sqrt{npq} \approx 3.67 \qquad \mu = np = 135$$

$$P\left(\frac{X}{150} < 0.86\right) = P(X < 129) \approx P\left(Z \le \frac{128.5 - 135}{3.67}\right) \approx P(Z < -1.77) \approx 0.0384$$

Since the product of $n\hat{p}$ is a whole number, the two solutions are not exactly the same. Although small discrepancies exist in some cases, we usually do not use the *continuity correction* for proportions, as this degree of accuracy is not warranted in most real life situations. Note that this procedure is valid only when both np and nq are greater than 5.

7.4 Sampling Distribution of the Differences between Two Means

Oftentimes, we are concerned with comparing means from two different populations in order to answer questions like, 'Is the strength of one species greater than another?' or, 'Do the readings from one measuring instrument vary significantly from another?' In order to answer such questions, we need to discuss two cases of sampling distributions of the differences between two means: (i) from **independent populations** and (ii) from **dependent populations.**

Independent populations

In this section, we will examine samples from two **independent** populations, with known population means, μ_1 and μ_2, and variances, σ^2_1 and σ^2_2, respectively.

Suppose that two sets of all possible sample means are calculated based on samples of size n_1 drawn from the first population and size n_2 from the second population. The sample means from the two populations are random variables, $\overline{X}_1$ and $\overline{X}_2$, and their two sampling distributions behave in exactly the same way as we discussed in Section 7.2. The theory discussed here applies for both finite and infinite populations and for both sampling with and without replacement.

It can be shown that, by calculating *all possible differences* of $\overline{x}_1 - \overline{x}_2$ between the two sets of independent sample means, a new random variable, $\overline{X}_1 - \overline{X}_2$, is created. This new random variable, describing differences between sample means, has a distribution called the **sampling distribution of the differences between two means.**

In order to calculate all possible differences, each sample mean from population 2 is subtracted from each sample mean from population 1 (we can also work the other way if we wish). Suppose that all possible samples of size 2 are taken from a finite population of size 3 (a total of 3^2 samples) and all possible samples of size 2 are taken from another population of size 4 (a total of 4^2 samples). While it would be cumbersome, all possible differences can be calculated for a total of $(3^2)(4^2) = 144$ differences. A careful examination of the resulting sampling distribution of the differences between two means would lead us to the following conclusions:

1. Since the difference of two sample means, $\overline{X}_1 - \overline{X}_2$, changes from sample to sample, it is a random variable.
2. The mean of all possible differences is equal to the difference of the two population means, which can be expressed as an expectation: $E(\overline{X}_1 - \overline{X}_2) = \mu_1 - \mu_2$.
3. The spread of all possible differences, or the standard error of the differences of means, is:

$$\sigma_{\overline{X}_1-\overline{X}_2} = \sqrt{\frac{\sigma_1^2}{n_1}+\frac{\sigma_2^2}{n_2}}. \tag{7.11}$$

4. As the sample sizes, n_1 and n_2, increase, the sampling distribution of all possible differences approaches the normal distribution. If both n_1 and n_2 are greater than 30, the sampling distribution is well approximated by the normal curve.

Figures 7.11a and b show a few points of the sampling distributions of the means for two distinct populations (1 and 2), while Fig. 7.11c shows the distribution of the differences between two means for four arbitrarily selected points from Fig. 7.11a and b.

As before, the random variable, $\overline{X}_1 - \overline{X}_2$, can be transformed into a standard normal variable, Z, allowing probability statements to be made about the differences of two sample means.

$$z = \frac{\left(\overline{x}_1 - \overline{x}_2\right)-\left(\mu_1 - \mu_2\right)}{\sigma_{\overline{X}_1-\overline{X}_2}}. \tag{7.12}$$

The mean and variance of the differences between two means can be derived by using some of the rules covered in Section 4.5 (see Chapter 4) regarding means and variances. Using Rule 2 for means, the mean of the differences between independent sample means is:

$$\mu_{\overline{X}_1-\overline{X}_2} = \mu_{\overline{X}_1} - \mu_{\overline{X}_2} = \mu_1 - \mu_2.$$

Using Rule 3 for variances, the variance of the differences of independent sample means is:

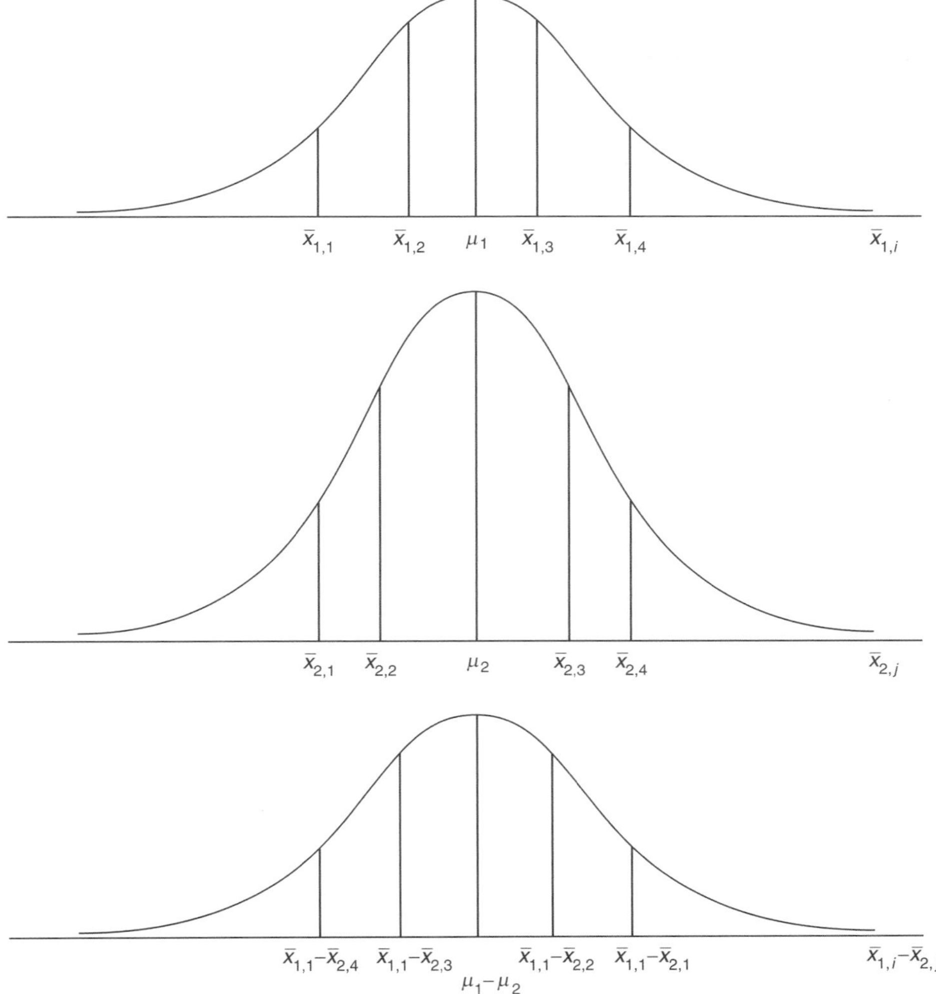

Fig. 7.11. Sampling distribution of the differences between two means.

$$\sigma^2_{\overline{X}_1-\overline{X}_2} = \sigma^2_{\overline{X}_1} + \sigma^2_{\overline{X}_2} = \frac{\sigma_1^2}{n_1} + \frac{\sigma_2^2}{n_2}.$$

Note:

$$\sigma_{\overline{X}_1,\overline{X}_2} = \mathrm{COV}\left(\overline{X}_1, \overline{X}_2\right) = 0, \text{ since the two samples are independent.}$$

If finite population correction factors are required, the standard error of the difference between two means changes to:

$$\sigma_{\overline{X}_1-\overline{X}_2} = \sqrt{\frac{\sigma_1^2}{n_1}\frac{N_1-n_1}{N_1-1} + \frac{\sigma_2^2}{n_2}\frac{N_2-n_2}{N_2-1}}. \quad (7.13)$$

If necessary, the correction factor can be used for only one of the two populations.

Example 7.9. The population mean dbh in one forested area is 16.5 cm with a standard deviation of 0.9 cm, while the population mean dbh in another forested area is 16.0 cm with a standard deviation of 1.2 cm. If samples of size 32 and 36 trees are taken randomly from the two populations, what is the probability that the sample mean from the first stand will be less than the sample mean from the second stand? The information can be summarized as follows:

Population 1	Population 2
$\mu_1 = 16.5$	$\mu_2 = 16.0$
$\sigma_1 = 0.9$	$\sigma_2 = 1.2$
$n_1 = 32$	$n_2 = 36$

Note that we can assume infinite populations in this case. Thus, a finite population correction factor need not be applied in computing the standard error. As shown in Fig. 7.12, the difference in sample means is set to 0 because we are merely asking whether the mean dbh of one species is less than another (i.e. we are not specifying a difference between the two species).

$$\sigma_{\bar{X}_1-\bar{X}_2} = \sqrt{\frac{0.9^2}{32} + \frac{1.2^2}{36}} \approx 0.2556.$$

$$P\left(\left(\bar{X}_1 - \bar{X}_2\right) < 0\right) = P\left(Z < \frac{0-(16.5-16.0)}{0.2556}\right) \approx P(Z < -1.96) \approx 0.025.$$

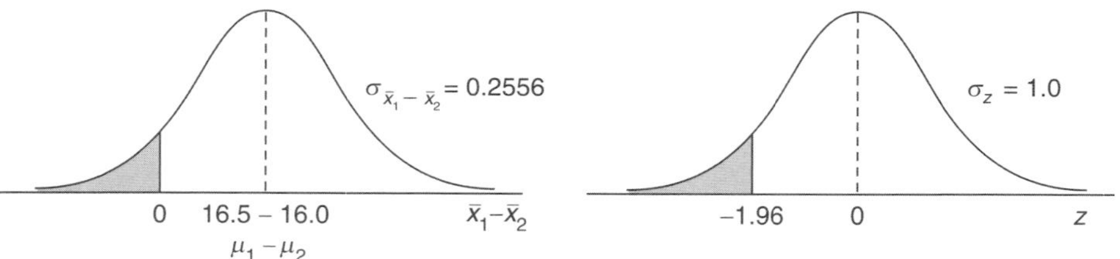

Fig. 7.12. Areas for Example 7.9.

Like before, if the population variances, σ_1^2 and σ_2^2, are unknown, and if samples are taken from two approximately normal populations, Eqn 7.12 is replaced by a t distribution and Eqn 7.14:

$$t_{(v)} = \frac{\left(\bar{x}_1 - \bar{x}_2\right) - \left(\mu_1 - \mu_2\right)}{s_{\bar{X}_1 - \bar{X}_2}} \tag{7.14}$$

where $s_{\bar{X}_1-\bar{X}_2}$ = standard error of the difference between two means, calculated from sample variances based on n_1 and n_2 observations and v = degrees of freedom.

For cases where the population variances are unknown, things become a little more complicated in that there are three distinct cases for which the standard error of the difference between two means and the degrees of freedom are computed differently.

Case 1. Both n_1 and $n_2 \geq 30$ (large samples).

$$s_{\bar{X}_1-\bar{X}_2} = \sqrt{\frac{s_1^2}{n_1} + \frac{s_2^2}{n_2}} \tag{7.15}$$

and $v = \infty$.

Note that with large sample sizes, sample variances are good estimates of the population variances. In such cases, Eqn 7.14 is equivalent to Eqn 7.12, as the degrees of freedom approach infinity and the probability distribution of t approaches the probability distribution of Z. In other words, a Z distribution can be used in place of a t distribution.

Case 2. n_1 or n_2 or both < 30 (small samples) and it can be assumed that $\sigma_1^2 = \sigma_2^2$.

$$s_{\overline{X}_1-\overline{X}_2} = s_p\sqrt{\frac{1}{n_1}+\frac{1}{n_2}} \tag{7.16}$$

where

$$s_p = \sqrt{\frac{s_1^2(n_1-1)+s_2^2(n_2-1)}{n_1+n_2-2}} \tag{7.17}$$

and $\nu = n_1 + n_2 - 2$.

Later on, we will cover a 'statistical test' that tells us whether or not we can assume $\sigma_1^2 = \sigma_2^2$. In the above equation, s_p^2 is the *pooled variance* and can be thought of as a 'weighted' variance term for both populations. The pooled variance has $n_1 + n_2 - 2$ degrees of freedom, two less than the total number of independent observations. Because two statistics, $\overline{x}_1$ and $\overline{x}_2$, are used in the calculation, we lose two degrees of freedom (see Section 2.4 in Chapter 2 for a review of degrees of freedom).

Case 3. n_1 or n_2 or both < 30 (small samples) and we assume that $\sigma_1^2 \neq \sigma_2^2$.

$$s_{\overline{X}_1-\overline{X}_2} = \sqrt{\frac{s_1^2}{n_1}+\frac{s_2^2}{n_2}} \tag{7.18}$$

and

$$\nu = \frac{\left[\frac{s_1^2}{n_1}+\frac{s_2^2}{n_2}\right]^2}{\frac{(s_1^2/n_1)^2}{n_1-1}+\frac{(s_2^2/n_2)^2}{n_2-1}} \tag{7.19a}$$

or

$$\nu = \text{the smaller of } n_1 - 1 \text{ and } n_2 - 1. \tag{7.19b}$$

Note that Eqns 7.15 and 7.18 are the same. However, for Case 1, probabilities can be looked up using the z-table (or in the last line of the t-table) but, for Case 3, only the t-table can be used with the specified degrees of freedom. Note also that there are two ways of obtaining degrees of freedom for Case 3. While both Eqns 7.19a and b result in appropriate degrees of freedom, Eqn 7.19a (a more complicated method) will give a more reliable result. When using Eqn 7.19a, the nearest whole number should be used for degrees of freedom.

Example 7.10. Suppose that the population standard deviations in Example 7.9 are not known, but have been estimated to be 1.1 cm and 0.9 cm from two respective samples of size 32 and 36. Because the sample sizes are both greater than 30, Eqn 7.15 and infinite degrees of freedom (Case 1) can be used to determine if the sample mean from the first stand will be less than the sample mean from the second stand.

$$s_{\bar{X}_1-\bar{X}_2} = \sqrt{\frac{1.1^2}{32} + \frac{0.9^2}{36}} \approx 0.2456.$$

$$P\left(\left(\bar{X}_1 - \bar{X}_2\right) < 0\right) = P\left(Z = T_{(\infty)} < \frac{0-(16.5-160)}{0.2456}\right) \approx P(Z < -2.04) \approx 0.0207.$$

The distribution of data and corresponding Z transformation are shown in Fig. 7.13. As expected, the results from Examples 7.9 and 7.10 are similar.

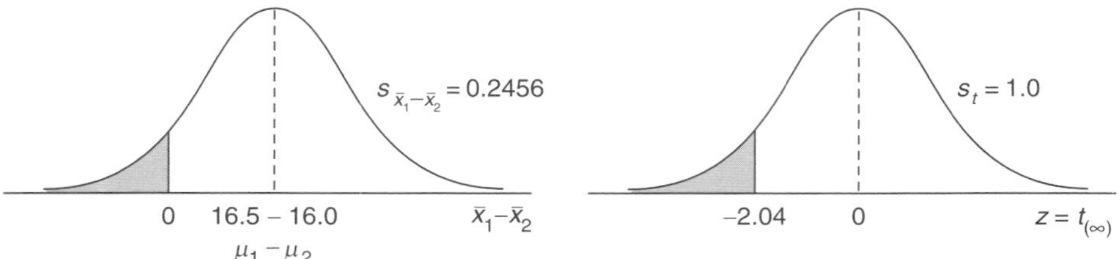

Fig. 7.13. Areas for Example 7.10.

Example 7.11. As stated in Example 7.5, the average specific gravity of green Douglas-fir grown in British Columbia is 0.45. Suppose that it is also known that the average specific gravity for green western red cedar grown in British Columbia is 0.36. Assume that the variances of the two populations are unknown but equal. A sample of 8 Douglas-fir specimens is taken and a sample of 10 western red cedar specimens is taken, and their specific gravities are measured. The sample standard deviations are estimated to be 0.076 for Douglas-fir and 0.068 for western red cedar. What is the probability that the sample mean of Douglas-fir will be greater than the sample mean of western red cedar by at least 0.12? Because our samples sizes are each less than 30 and we have assumed that $\sigma_1^2 = \sigma_2^2$, Case 2 is used. Our information can be summarized as follows:

Douglas-fir	Western red cedar
$\mu_1 = 0.45$	$\mu_2 = 0.36$
$S_1 = 0.076$	$S_2 = 0.068$
$n_1 = 8$	$n_2 = 10$

One should be mindful of the fact that, in this question, we are not asking whether one species will have a higher or lower specific gravity than the other (in other words, the probability of the differences in sample means being less than or greater than 0). Here, we are asking whether the mean specific gravity of one species exceeds the other by at least 0.12 and it is this value that should be included in the t transformation.

The distribution of data and corresponding t transformation are shown in Fig. 7.14.

$$s_p = \sqrt{\frac{0.076^2(8-1)+0.068^2(10-1)}{8+10-2}} \approx 0.0716.$$

$$s_{\bar{X}_1-\bar{X}_2} \approx 0.0716\sqrt{\frac{1}{8}+\frac{1}{10}} \approx 0.0339.$$

$$P\left(\left(\bar{X}_1 - \bar{X}_2\right) > 0.12\right) = P\left(T_{(16)} > \frac{0.12-(0.45-0.36)}{0.0339}\right) \approx P\left(T_{(16)} > 0.8849\right) \approx 0.20.$$

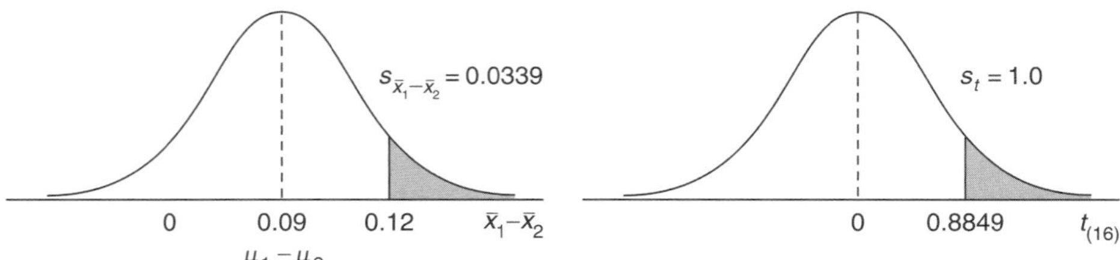

Fig. 7.14. Areas for Example 7.11.

Example 7.12. The mean annual biomass production of spruce seedlings under normal lighting is 4.52 g, while it is 2.61 g under artificial lighting. Suppose that the population variances are unknown. Twelve randomly selected seedlings were exposed to normal lighting and ten were exposed to artificial lighting, resulting in sample standard deviations of 1.2 g and 0.5 g, respectively. What is the probability that the mean biomass of the seedlings grown under normal light exceeds the mean biomass of those grown under artificial light by at least 1.40 g if we assume that the population variances are not equal? Summarizing the information, we get:

Normal	Artificial
$\mu_1 = 4.52$	$\mu_2 = 2.61$
$S_1 = 1.20$	$S_2 = 0.50$
$n_1 = 12$	$n_2 = 10$

Because we assume $\sigma_1^2 \neq \sigma_2^2$ we use Case 3. The distribution of data and corresponding t transformation are shown in Fig. 7.15.

$$s_{\bar{X}_1-\bar{X}_2} = \sqrt{\frac{1.20^2}{12} + \frac{0.50^2}{10}} \approx 0.381.$$

$$v = \frac{\left[\frac{1.2^2}{12} + \frac{0.5^2}{10}\right]^2}{\frac{\left(1.2^2/12\right)^2}{12-1} + \frac{\left(0.5^2/10\right)^2}{10-1}} \approx 15.25 \approx 15.$$

$$P\left(\left(\bar{X}_1 - \bar{X}_2\right) > 1.4\right) = P\left(T_{(15)} > \frac{1.40-1.91}{0.381}\right) \approx P\left(T_{(15)} > -1.34\right) \approx 0.90.$$

If we used the simpler procedure for estimating degrees of freedom (the minimum of $n_1 - 1$ or $n_2 - 1$), the result would be a more 'conservative' estimate of the probability as

$$P\left(T_{(9)} > -1.34\right) \approx 0.89$$

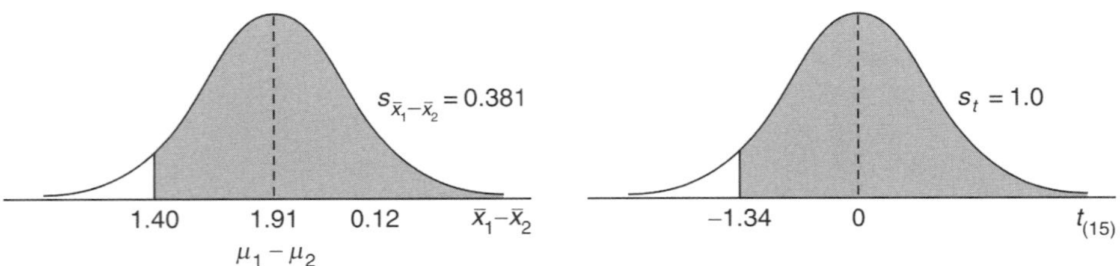

Fig. 7.15. Areas for Example 7.12.

Dependent populations

The sampling distribution for the differences between two means varies when samples come from two different populations that are not independent. Samples are considered **dependent** when the observations occur in pairs, or when the objects are matched by some special characteristics. Take the example of measuring the heights of 15 trees in a particular stand. Pairing or matching occurs if their heights are measured on two separate occasions (say 1 year apart) or by two different people, or if two different instruments are used. Such observations are obviously related (i.e. dependent) as each tree is measured twice.

From the examples above, consider the case when the same trees are measured on two separate occasions. The differences, d_i, between the two corresponding measurements for each tree become the individual values of a new random variable, D. In other words, we create a *new population* comprised of the elements, d_i, which are calculated as the differences of the two sets of corresponding elements of two random variables, X_1 and X_2, from two populations. This new single population has a mean of μ_d (it can be shown that $\mu_d = \mu_1 - \mu_2$) and a variance of σ_d^2, and the means of all possible samples of size n from this population produce the statistic, $\bar{D}$, which behaves in exactly the same way as the sample mean, $\bar{X}$, discussed in Section 7.2. Specifically, four observations are worth noting:

1. Since its value changes from sample to sample, the mean of paired observations, $\bar{D}$, is a random variable.
2. The mean of all possible means of sample size n is equal to the population mean, which can be expressed as an expectation: $E(\bar{D}) = \mu_d = \mu_1 - \mu_2$.
3. The spread of all possible means, or the standard error of differences, is $\sigma_{\bar{d}} = \dfrac{\sigma_d}{\sqrt{n}}$.
4. The probability distribution of all possible means is normal when the parent population is normally distributed or is approximately normal when the sample size (n) is sufficiently large ($n \geq 30$), regardless of the distribution of the parent population. As n approaches infinity, the probability distribution of all possible means becomes normal.

Like the previous cases of sampling distributions of means, the probability distribution of $\bar{D}$ can be standardized as:

$$z = \frac{\bar{d} - \mu_d}{\sigma_{\bar{d}}} \tag{7.20}$$

where

$$\sigma_{\bar{d}} = \frac{\sigma_d}{\sqrt{n}} \quad \text{or} \quad \sigma_{\bar{d}} = \frac{\sigma_d}{\sqrt{n}}\sqrt{\frac{N-n}{N-1}}, \tag{7.21}$$

$$\bar{d} = \frac{\sum_{i=1}^{n} d_i}{n}. \tag{7.22}$$

Example 7.13. Suppose that we ask two students to measure the heights of 5 young trees in a plantation. Even if the population means are unknown, under normal circumstances, $\mu_d = 0.0$ m. Assuming that $\sigma_d = 0.15$ m, what is the probability that the mean of measurements taken by student A will be at least 0.1 m higher than the mean of measurements taken by student B?

The distribution of data and corresponding Z transformation are shown in Fig. 7.16.

$$\sigma_{\bar{d}} = \frac{0.15}{\sqrt{5}} \approx 0.067.$$

$$P(\bar{D} > 0.1) \approx P\left(Z > \frac{0.1 - 0.0}{0.067}\right) \approx P(Z > 1.49) \approx 1.0 - 0.9306 \approx 0.0694.$$

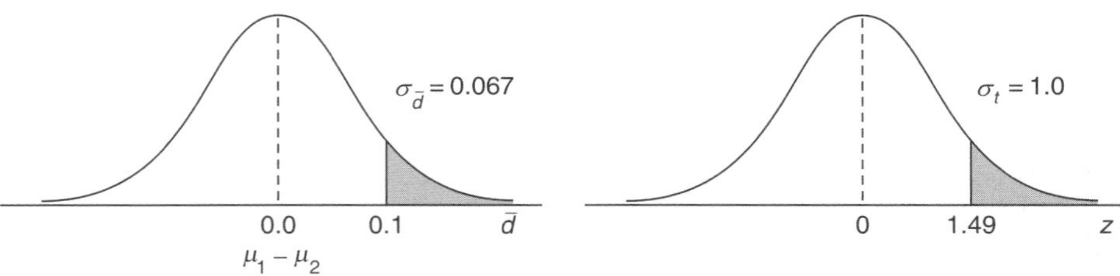

Fig. 7.16. Areas for Example 7.13.

When the population variance for paired observations is unknown, the t distribution can be used to calculate probabilities of $\bar{d}$:

$$t_{(n-1)} = \frac{\bar{d} - \mu_d}{s_{\bar{d}}} \tag{7.23}$$

where

$$s_{\bar{d}} = \frac{s_d}{\sqrt{n}}, \tag{7.24}$$

$$s_d^2 = \frac{\sum_{i=1}^{n} (d_i - \bar{d})^2}{n-1} = \frac{\sum_{t=1}^{n} d_i^2 - \left(\sum_{i=1}^{n} d_i\right)^2 \Big/ n}{n-1}. \tag{7.25}$$

Example 7.14. If the variance (standard deviation) is unknown in Example 7.13, we can estimate it from the raw data:

Student A	Student B	d_i
12.5	12.3	0.2
18.4	18.5	−0.1
11.1	10.8	0.3
12.6	12.6	0.0
15.8	15.7	0.1

What is the probability that the mean of measurements by student A will be at least 0.15 m higher than the mean of measurements by student B?

The distribution of data and corresponding t transformation are shown in Fig. 7.17.

Using the data above and applying Eqns 7.24 and 7.25:

$$s_d \approx 0.158 \quad \text{and} \quad s_{\bar{d}} = \frac{0.158}{\sqrt{5}} \approx 0.071$$

$$P(\bar{D} > 0.15) \approx P\left(T_{(4)} > \frac{0.15 - 0.0}{0.071}\right) \approx P(T_{(4)} > 2.11) \approx 0.05$$

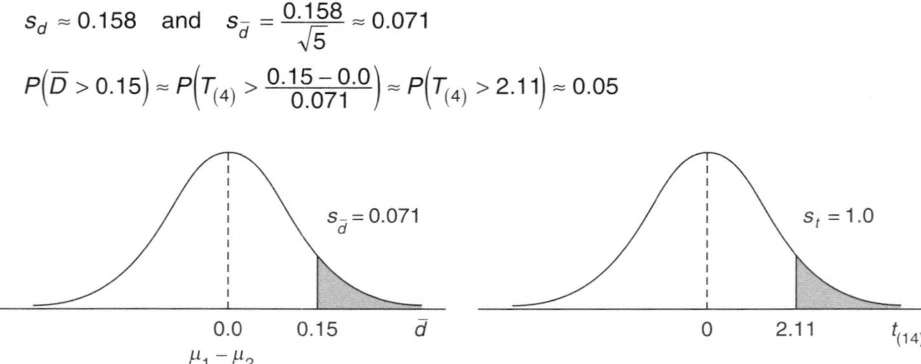

Fig. 7.17. Areas for Example 7.14.

It can be shown that the variance of $\bar{D}$ for paired observations is less than the variance of $\bar{x}_1 - \bar{x}_2$ for independent samples from two populations. This can be an effective way of obtaining more precise results in experiments where pairing is applied.

7.5 Sampling Distribution of the Differences between Two Proportions

It is shown in Section 7.3 that the sampling distribution of proportions behaves in a similar manner to that of the mean. This is because when zeros and ones are assigned to the two kinds of outcomes of the binomial random variable, the proportion of the 'successes' can be calculated in the same way as a sample mean. It follows, then, if independent samples of size n_1 and n_2 are drawn from two binomial populations, the differences of the two sample proportions, $\hat{P}_1 - \hat{P}_2$, will behave very much like differences between two sample means, $\bar{x}_1 - \bar{x}_2$. As such, the sampling distribution of the differences of proportions can be summarized as follows:

1. Since the difference of two sample proportions, $\hat{P}_1 - \hat{P}_2$, changes from sample to sample, it is a random variable.
2. The mean of all possible differences of two proportions is equal to the difference of the two population proportions, which can be expressed as an expectation: $E(\hat{P}_1 - \hat{P}_2) = p_1 - p_2$.
3. The spread of all possible differences, or the standard error of the differences of two proportions, is $\sigma_{\hat{p}_1 - \hat{p}_2} = \sqrt{\frac{p_1 q_1}{n_1} + \frac{p_2 q_2}{n_2}}$.

4. As the sample sizes of n_1 and n_2 increase, the sampling distribution of all possible differences approaches the normal distribution and, if both n_1 and n_2 are greater than 30, the sampling distribution is well approximated by the normal curve.

The random variable, $\hat{P}_1 - \hat{P}_2$, can be transformed into the standard normal variable, Z, for making probability statements about the differences of two sample proportions. Note that the two populations from which sample proportions are obtained are assumed to be independent.

$$z = \frac{(\hat{p}_1 - \hat{p}_2) - (p_1 - p_2)}{\sigma_{\hat{p}_1 - \hat{p}_2}}. \quad (7.26)$$

The mean and the variance of the differences between two proportions can be derived by using the rules for means and variances covered in Section 4.5 (see Chapter 4). Using Rule 2 for the means, the mean of the differences of independent sample proportions is:

$$\mu_{\hat{p}_1 - \hat{p}_2} = \mu_{\hat{p}_1} - \mu_{\hat{p}_2} = p_1 - p_2.$$

Using Rule 3 for the variances, the variance of the differences of independent sample proportions is:

$$\sigma^2_{\hat{p}_1 - \hat{p}_2} = \sigma^2_{\hat{p}_1} + \sigma^2_{\hat{p}_2} = \frac{p_1 q_1}{n_1} + \frac{p_2 q_2}{n_2}.$$

The covariance between $\hat{p}_1$ and $\hat{p}_2$ is zero, since the two samples are independent. If finite population correction factors are required, they are applied in the same way as shown in Eqn 7.13 for the differences between two means.

Example 7.15. The proportions of reject pieces of dimensional lumber in two sawmills (A and B) are 0.05 and 0.07, respectively.

a. If 50 pieces are inspected in mill A and 70 in mill B, what is the probability that the proportion of rejects in mill A will exceed the proportion in mill B?
b. If 150 pieces are inspected at each mill, what is the probability that the proportion in mill A will exceed the proportion in mill B?

The distribution of data and corresponding *Z* transformation are shown in Fig. 7.18.

a. $\sigma_{\hat{p}_1 - \hat{p}_2} = \sqrt{\frac{(0.05)(0.95)}{50} + \frac{(0.07)(0.93)}{70}} \approx 0.0434.$

$$P\left(\left(\hat{P}_1 - \hat{P}_2\right) > 0\right) = P\left(Z > \frac{0.0 - (-0.02)}{0.0434}\right) \approx P(Z > 0.46) \approx 1.0 - 0.6772 \approx 0.3228.$$

b. $\sigma_{\hat{p}_1 - \hat{p}_2} = \sqrt{\frac{(0.05)(0.95)}{150} + \frac{(0.07)(0.93)}{150}} \approx 0.0275.$

$$P\left(\left(\hat{P}_1 - \hat{P}_2\right) > 0\right) = P\left(Z > \frac{0.0 - (-0.02)}{0.0275}\right) \approx P(Z > 0.73) \approx 1.0 - 0.7673 \approx 0.2327.$$

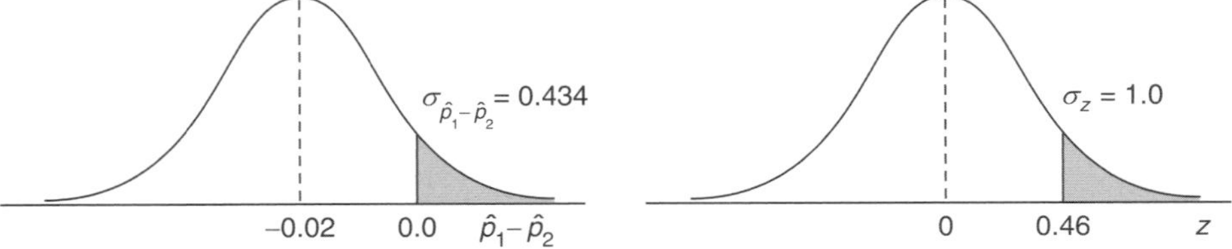

Fig. 7.18. Areas for Example 7.15a.

Note how the probability is considerably smaller as the sample sizes increases from Example 7.15a to 7.15b. This is because higher sample sizes result in smaller standard errors, which in turn decrease the probability in the tail(s) of the sampling distribution.

If the population variance is not known for the sampling distribution of the differences between two proportions, it can be estimated from the sample proportions.

$$s_{\hat{p}_1-\hat{p}_2} = \sqrt{\frac{\hat{p}_1\hat{q}_1}{n_1} + \frac{\hat{p}_2\hat{q}_2}{n_2}}, \tag{7.27a}$$

or

$$s_{\hat{p}_1-\hat{p}_2} = \sqrt{\hat{p}_c\hat{q}_c\left(\frac{1}{n_1} + \frac{1}{n_2}\right)} \tag{7.27b}$$

where $\hat{p}_c$ = pooled or combined proportion = $\dfrac{n_1\hat{p}_1 + n_2\hat{p}_2}{n_1 + n_2}$ and $\hat{q}_c = 1 - \hat{p}_c$.

Note that the pooled or combined proportion in the calculation is essentially the proportion of successes taking both samples into account.

Although the standard error of the difference between two proportions can be estimated using Eqns 7.27a or b, replacing Eqn 7.26 by the t distribution is not recommended unless the parent populations can be considered normal (see Section 6.4 in Chapter 6).

7.6 Sampling Distribution of the Variance

The sample variance, s^2, is used to draw inferences about the unknown population variance, σ^2. Like the sample mean, the sample variance is calculated from elements included in a subset of the population and therefore changes from sample to sample. In this section, we will study the relationship between sample variances and the population variance.

Suppose sample variances, s^2, are calculated from random samples of size n selected repeatedly from a population with a variance, σ^2. Using the different sample variances, the probability distribution of the random variable, S^2, can be constructed (Fig. 7.19a). Unlike a normally shaped sampling distribution of means, though, the sampling distribution of S^2 is actually positively skewed, stretching from zero to infinity with a mean of σ^2.

The sampling distribution of the random variable, S^2, has the following characteristics:

1. Since the sample variance, S^2, changes from sample to sample, it is a random variable.
2. The mean of all possible sample variances of size n is equal to the population variance, which can be expressed as an expectation: $E(S^2) = \sigma^2$.
3. If s^2 is the variance from a random sample of size n taken from a normally distributed population with variance σ^2, it will follow what is known as an χ^2 distribution (chi-squared) with $\nu = n - 1$ degrees of freedom:

$$\chi^2_{(n-1)} = \frac{(n-1)s^2}{\sigma^2}. \tag{7.28}$$

The χ^2 curve is positively skewed (but approaching a normal distribution at larger sample sizes) and describes the sampling distribution of the variances (Fig. 7.19a), as

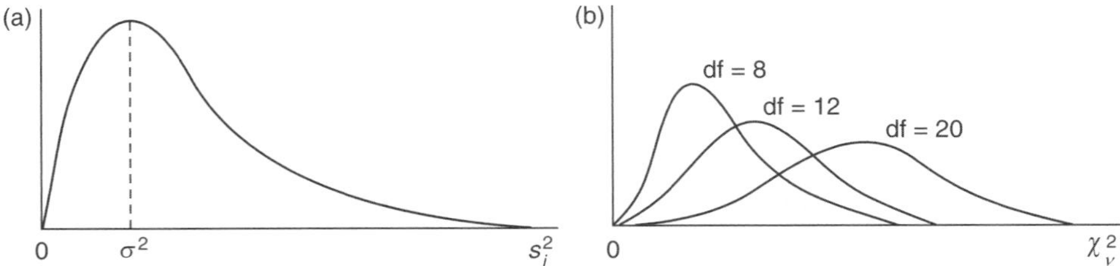

Fig. 7.19. Probability distribution of (a) sample variances and (b) some χ^2 curves.

indicated in Fig. 7.19b. The χ^2 curve starts at zero and extends (in theory) to infinity, with a mean of $n - 1$. Its mathematical formula is rather complicated and beyond the scope of this book. However, it is known that the total area under the curve is one, and areas between any two χ^2-values, which provide probabilities between these two values, can be obtained by integration. For our convenience, however, cumulative probabilities of χ^2-values are listed in Table A.6 (see Appendix A). Like the *t*-table, the first column of Table A.6 contains the degrees of freedom (ν); the first row shows the cumulative probabilities up to a specified χ^2-value, and the table entries are the χ^2-values. For example, for nine degrees of freedom, the probability (area) from 0 to the χ^2-value of 16.9 is 0.95. It follows then that the probability between 16.9 and ∞ is $1.0 - 0.95 = 0.05$. Like the *t* distribution, not every probability can be shown in the χ^2-table. Unlike the *t* distribution, however, the χ^2 distribution is not symmetric. Therefore, the χ^2_ν-value corresponding to a probability of 0.05 cannot be directly found using a probability of 0.95; it must be looked up in the table. A handy feature of the χ^2 distribution is that, for each probability listed, its corresponding complement is also listed: a piece of information to keep in mind for future chapters. One of the main purposes of the χ^2 transformation is to make probability statements about sample variances obtained with a random sampling procedure. However, as we shall see in future chapters, it is also a commonly occurring distribution for several important statistical tests.

Example 7.16. The breaking loads of a certain type of wood beam are normally distributed with a standard deviation of 60 PSI. If a random sample of 12 beams is tested, what is the probability that the standard deviation will be greater than 75 PSI? What is the probability that the standard deviation will be less than 32 PSI?

Since we do not have a sampling distribution for standard deviations, the standard deviations must be changed to variances and then transformed into χ^2 units in order to solve these problems (Fig. 7.20).

$$P\left(S^2 > 75^2\right) = P\left(S^2 > 5625\right) = P\left(\chi^2_{(11)} > \frac{(12-1)5625}{3600}\right)$$

$$\approx P\left(\chi^2_{(11)} > 17.2\right) \approx 0.1.$$

$$P\left(S^2 < 32^2\right) = P\left(S^2 < 1024\right) = P\left(\chi^2_{(11)} < \frac{(12-1)1024}{3600}\right)$$

$$\approx P\left(\chi^2_{(11)} < 3.13\right) \approx 0.01.$$

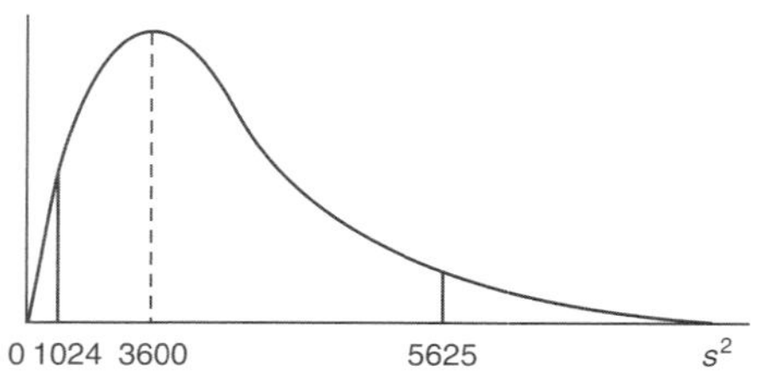

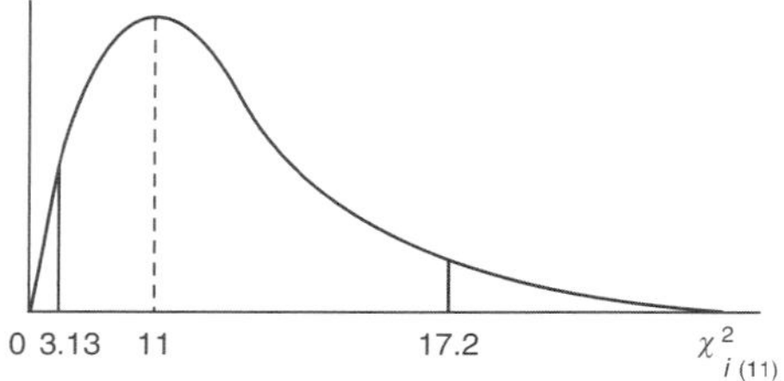

Fig. 7.20. Areas for Example 7.16.

7.7 Sampling Distribution of the Ratios of Two Variances

If repeated independent samples of size n_1 and n_2 are taken from normal populations with variances, σ_1^2 and σ_2^2, and the sample variances of s_1^2 and s_2^2 are computed, then all possible ratios of these sample variances form a random variable in the form of a ratio, s_1^2/s_2^2. The probability distribution of s_1^2/s_2^2 is known as **the sampling distribution of the ratios of two variances** (Fig. 7.21a).

The most important characteristics of the sampling distribution of the ratios of variances can be summarized as:

1. Since the ratio of two variances, s_1^2/s_2^2, changes from sample to sample, it is a random variable.
2. If s_1^2 and s_2^2 are sample variances from independent samples of size n_1 and n_2 from two normal populations with variances of σ_1^2 and σ_2^2, the ratio of the two sample variances can be transformed into an ***F* distribution** (see Fig. 7.21b) using Eqn 7.29:

$$f_{(v_1,v_2)} = \frac{s_1^2/\sigma_1^2}{s_2^2/\sigma_2^2} = \frac{s_1^2\sigma_2^2}{s_2^2\sigma_1^2} \qquad (7.29)$$

where $v_1 = n_1 - 1$ and $v_2 = n_2 - 1$.

In its transformed or standardized form (using Eqn 7.29), the F distribution (named after British statistician, Ronald Fisher) is one of the most important distributions in applied statistics and is used in many of the statistical tests that we will see in later chapters. Theoretically, the F distribution can be defined as the ratio of two independent χ^2-values, where each is divided by its degrees of freedom:

$$F_{(v_1,v_2)} = \frac{\chi^2_{(v_1)}/v_1}{\chi^2_{(v_2)}/v_2} = \frac{\dfrac{(n_1-1)s_1^2}{(n_1-1)\sigma_1^2}}{\dfrac{(n_2-1)s_2^2}{(n_2-1)\sigma_2^2}} = \frac{s_1^2\sigma_2^2}{s_2^2\sigma_1^2}.$$

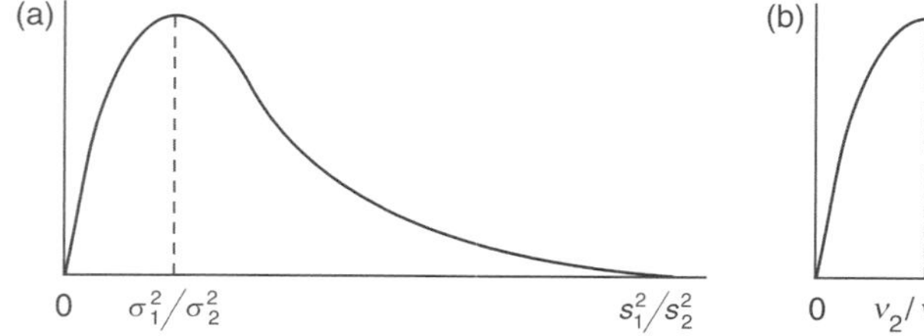

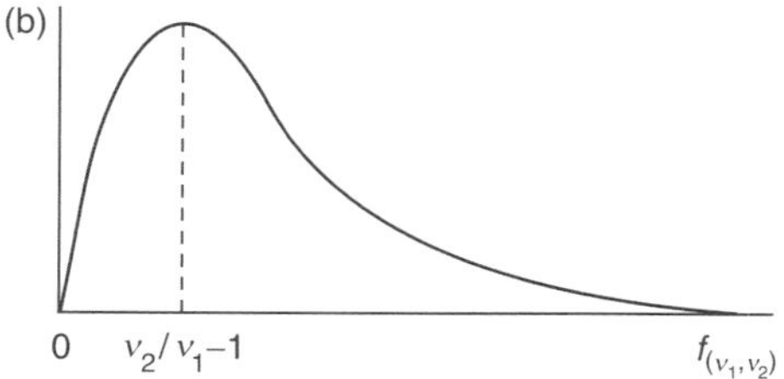

Fig. 7.21. Sampling distribution of (a) the ratio of variances and (b) the *F* distribution.

This is a complicated distribution to visualize for many reasons, not the least of which is that we are dealing with two sets of degrees of freedom! For each set of v_1 and v_2 degrees of freedom, there is a different curve (Fig. 7.22), which is positively skewed and extends from zero to infinity. The total area under each of these curves is equal to one. Their mathematical form, the integration of which would provide probabilities between two specified *F*-values, is beyond the scope of this book. Table A.7 (see Appendix A), however, lists *F*-values as a function of degrees of freedom, v_1 and v_2, and inverse cumulative probabilities. The degrees of freedom associated with the sample variance in the numerator for Eqn 7.29 are always stated first, followed by the degrees of freedom associated with the sample variance in the denominator. In Table A.7 (see Appendix A), the first row contains the degrees of freedom for the numerator, while the first column contains the degrees of freedom for the denominator. The second column shows the probabilities, and the *F*-values are in the body of the table. Note that it can be shown that $E(F) = v_2/(v_2 - 2)$, which approaches 1 as v_2 approaches ∞ (see Fig. 7.21b).

In contrast to Table A.6 (see Appendix A), Table A.7 contains inverse cumulative probabilities; that is, it shows only those probabilities on the right tail of the distribution, where $f > 1.0$. For values of $f < 1.0$, on the left tail of the distribution, probabilities can be calculated by applying the following equation:

$$f_{(v_1,v_2)_{1-\alpha}} = \frac{1}{f_{(v_2,v_1)_\alpha}}. \tag{7.30}$$

The above equation can be verified by taking the reciprocal of Eqn 7.29, which, by convention, requires an exchange of position of the degrees of freedom. If $f > 1$ in Eqn 7.29, then $f < 1.0$ from Eqn 7.30 (or the other way round) and the probability will change from α to $1 - \alpha$ (Fig. 7.23). It may be worth noting at this point that it is

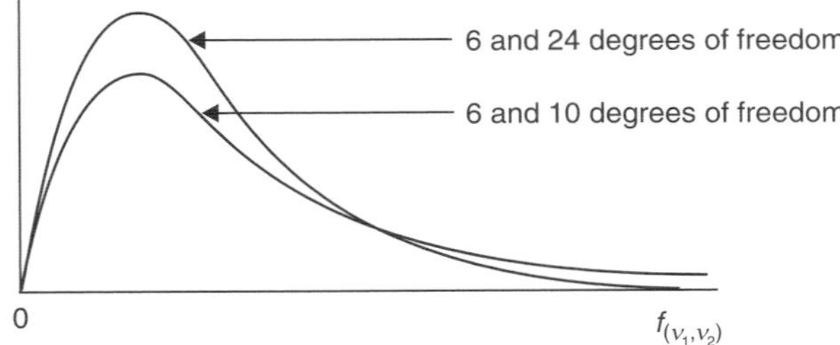

Fig. 7.22. Typical *F* curves.

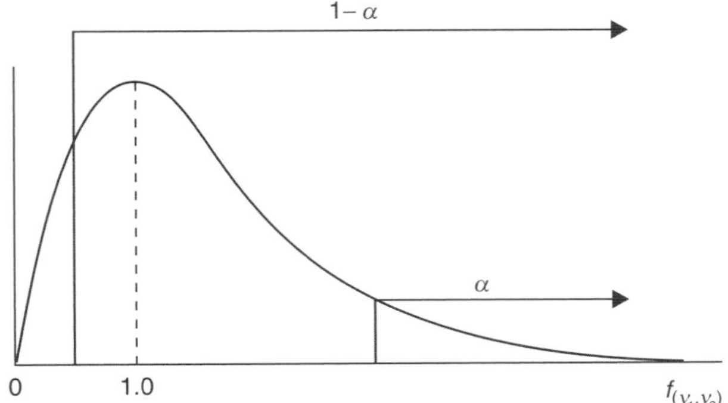

Fig. 7.23. The two tails of the *F* distribution.

common practice in dealing with these sorts of problems to put the higher variance in the numerator. As we shall see in future chapters, this makes the need for this reciprocal calculation unnecessary and simplifies matters greatly.

Example 7.17. In a study of tree diameter measurements, sample sizes of $n_1 = 10$ and $n_2 = 8$ were taken from forest types I and II, respectively. Assume that we know the population variance of the two forest types: $\sigma_1^2 = 36$ cm^2 and $\sigma_2^2 = 25$ cm^2.

a. Find the probability that the ratio of sample variances (S_1^2/S_2^2) will exceed 5.3.
b. Find the probability that the ratio of sample variances (S_1^2/S_2^2) will be less than 0.26.

In order to solve this probability problem, we must first transform the ratio of variances into an F distribution (Fig. 7.24).

a. $P\left(S_1^2/S_2^2 > 5.3\right) = P\left(F_{(9,7)} > 5.3\frac{25}{36}\right) \approx P\left(F_{(9,7)} > 3.68\right) \approx 0.05.$

b. $P\left(S_1^2/S_2^2 < 0.26\right) = P\left(F_{(9,7)} < 0.26\frac{25}{36}\right) \approx P\left(F_{(9,7)} < 0.181\right)$

$\approx P\left(F_{(7,9)} > \frac{1}{0.181}\right) = P\left(F_{(7,9)} > 5.52\right) \approx 0.01.$

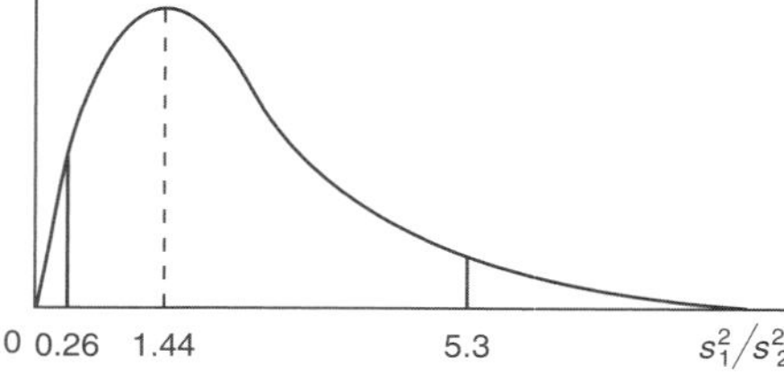

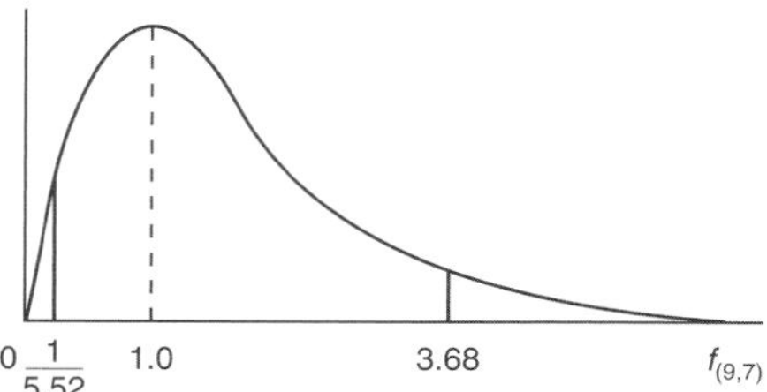

Fig. 7.24. Areas for Example 7.17.

Since table look-ups for F-values can be rather complicated, it is highly recommended that readers practice looking up various values from the F-table. In comparison to the other (Z, t and χ^2) tables, the entries in the F-table are even more limited. However, entries provided in Table A.7 (see Appendix A) are sufficient for most practical applications. A general rule of thumb for using the F-table: if the degrees of freedom for either the numerator or the denominator are not listed in the table, take the next lowest degree of freedom (in either case), as this will give a slightly more conservative result.

7.8 Some Concluding Remarks about Sampling Distributions

One of the main purposes of this chapter is to introduce the student to different sampling distributions and the methods used to solve probability problems associated with these sampling distributions. Four tables for solving these probability problems are introduced. In some ways, the methods for solving these probability problems may seem 'backward' compared to how we will be applying them in later chapters, where we will typically be using the probabilities from the tables in various formulations, as opposed to solving them. That said, it is important to understand the elements that comprise these tables, and this is another main purpose of this chapter.

Exercises

Section 7.1

7.1. A random sample of 15 trees is selected from a permanent sample plot of 148 trees, numbered from 1 to 148. Using Table A.4 (see Appendix A), list the tree numbers you would select for:

a. Sampling with replacement.
b. Sampling without replacement.

7.2. A random sample of 4 soil types is to be selected from a population of 12 soil types. Indicate the number of all possible samples, if the sampling is to be carried out:

a. With replacement.
b. Without replacement.

7.3. A random sample of 25 pieces of lumber is to be selected without replacement from the green chain in a sawmill. Boards come off the green chain every 0.5 min and the samples will be selected from production during the next 10 working days between 8:00 am and 4:00 pm. Use Table A.4 (see Appendix A) to indicate when (both day and time within day) the 25 samples are to be taken.

7.4. Describe how you would randomly allocate 32 fixed radius (r = 5 m) plots in a 20 ha natural forest. Assume that a map is available for the area and that the sampling is done without replacement.

Section 7.2

7.5. List all possible samples of size 2 taken with replacement from the population of 1, 3, 5, 7 and 9. Calculate the population mean and population standard deviation and compare these to the mean of all possible sample means and the standard deviation of all possible sample means. Construct the probability distribution for the population and for the sample means and draw some conclusions. What would happen to the sample mean and the standard deviation of the sample means if the sample size increased from 2 to 3?

7.6. Repeat Exercise 7.5 for the case of sampling without replacement.

7.7. Determine whether the application of the finite population correction factor is required for the following cases:

a. $n = 20, N = 420$.
b. $n = 32, N = 150$.
c. $n = 15, N = 2768$.
d. $n = 5, N = 420$.
e. $n = 40, N = 12{,}000$.

7.8. The annual precipitation in a certain part of British Columbia is normally distributed with a mean of 850 mm and a standard deviation of 95 mm. Assume that the population is infinitely large and that the next n years can be considered a random sample.

a. What is the probability that the average annual rainfall over the next 12 years will be between 700 mm and 825 mm?

b. What is the probability that the average annual rainfall over the next 20 years will be between 700 mm and 825 mm?
c. What is the probability that the average annual rainfall over the next 12 years will be greater than 650 mm?
d. Assume that the population standard deviation is unknown but has been estimated as 91 mm based on the past 10 years' observations. What is the probability that the average annual rainfall over the past 10 years exceeded 830 mm?
e. Calculate the standard error of the mean based on 7.8d and explain its meaning.

7.9. What sample size should be taken from a normal population with a mean of 15 and a variance of 48 to ensure that the probability of the sample mean exceeding 20 will be less than 0.05 (5%)?

7.10. The specified weight limit of an elevator in an office building is 1800 kg. If the mean weight of all people ($N = 200$) ever to use this elevator is 68 kg with a standard deviation of 12 kg, what is the probability that 25 people selected at random will exceed the weight limit? Can this sampling be done with replacement? Calculate the standard error of the mean and explain its meaning.

7.11. Solve Exercise 7.10 for the case when the population standard deviation is unknown but estimated to be 10.3 kg from the 25 randomly selected people.

Section 7.3

7.12. Thirty per cent of all the trees in a certain lodgepole pine stand (N is very large) are attacked by mountain pine beetles.

a. If 50 trees are randomly selected from this stand, what is the probability that more than 41% of the trees are infested? Calculate the standard error of the proportion and discuss its meaning.
b. If 50 trees are randomly selected from this stand, what is the probability that more than 20 trees are infested?
c. How many trees should we sample so that the probability of more than 41% of the trees being infested will be 0.05 or less? Discuss your results.
d. Compare a and b.

7.13. The proportion of grade A pieces in a particleboard plant is 0.15. Based on a random sample of 40 boards, find the probability that:

a. The proportion is less than 0.1.
b. The proportion is greater than 0.25.
c. The proportion is between 0.1 and 0.25.

Calculate the standard error of the proportion and discuss its meaning.

7.14. In a kitchen cabinet plant, 4% of the doors cut to a certain size are defective. Find the probability that a random sample of 55 doors will have:

a. More than 1.8% defectives.
b. Less than 3.6% defectives.

Calculate the standard error of the proportion and explain its meaning.

Section 7.4

7.15. The service lives of two types of logging truck tyres are normally distributed. Brand A tyres have a mean service life of 85,000 km with a standard deviation of 8000 km. Brand B tyres have a mean service life of 90,000 km with a standard deviation of 10,000 km. If samples of 12 and 16 tyres are respectively selected from A and B at random and tested for service life, what is the probability that the difference of two means ($\bar{x}_A - \bar{x}_B$) is:

a. Greater than 2000 km?
b. Greater than 0 km?
c. Less than –2000 km?

Find the standard error of the differences of means, and explain its meaning.

7.16. In Exercise 7.15, assume that the population variances are unknown but equal. The variances were estimated to be 10,200 km^2 and 9800 km^2 by testing 12 and 16 tyres for brands A and B, respectively. What is the probability that:

a. The mean of brand A will exceed the mean of brand B?
b. The mean of brand B will exceed the mean of brand A by at least 2850 km?

Find the standard error of the difference between two means and explain its meaning.

7.17. To test the effect of controlled grazing versus continuous grazing on weight gain, 13 and 12 steers were treated for a period of time. The scientists assumed that the mean difference between the two grazing treatments was zero ($\mu_1 - \mu_2 = 0$). Since the population variances were unknown, they estimated the standard deviations of weight gain and obtained 25 lbs for controlled grazing and 17 lbs for continuous grazing. Based on these estimates, the scientists assumed that the population variances were different. What is the probability that:

a. The mean weight of the cows subjected to controlled grazing will be 15 lbs higher than the mean weight of the cows subjected to continuous grazing?
b. The mean weight of the cows subjected to controlled grazing will be 18 lbs less than the mean weight of the cows subjected to continuous grazing?

Calculate the standard error of the difference between two means and explain its meaning.

7.18. The population means of the tensile strength of dimensional lumber is 3200 lbs per square inch (psi) for Douglas-fir and 3000 psi for western hemlock. The population standard deviations are not available for these species but were estimated to be 400 psi and 370 psi from 35 and 40 specimens of Douglas-fir and western hemlock, respectively. If these specimens are tested, what is the probability that the sample mean of western hemlock will exceed the sample mean of Douglas-fir? Calculate the standard error of the difference between two means and explain its meaning.

7.19. The volume (m^3/ha) in a mature Sitka spruce stand was observed independently by two crews on each of 10 plots. Their observations were:

Plot	1	2	3	4	5	6	7	8	9	10
Crew 1	875	959	475	589	925	1200	971	421	892	728
Crew 2	910	878	480	495	1021	980	1002	410	850	620

What is the probability that the average difference between the two sample means is:

a. Greater than 40 m^3/ha?
b. Less than –66 m^3/ha?

Calculate the standard error of the difference between two means and explain its meaning.

Section 7.5

7.20. Two microprocessor manufacturers, A and B, claim that the proportion of defectives they produce are 0.07 and 0.05, respectively. If a sample of 100 microprocessors are tested from each manufacturer, what is the probability that:

a. The sample proportion of defectives from manufacturer B will exceed the sample proportion of defectives from manufacturer A?
b. The sample proportion of defectives from manufacturer A will exceed the sample proportion of defectives from manufacturer B?
c. The sample proportion of defectives from manufacturer A will exceed the sample proportion of defectives from manufacturer B by at least 0.1?

Calculate the standard error of the difference between two proportions and explain its meaning.

7.21. In two forest types, I and II, 25% and 30% of the trees are infested by bark beetles, respectively. If we examine a sample of 40 trees from forest type I and 50 trees from forest type II, what is the probability that:

a. The proportion of infested trees from forest type II exceeds the proportion of infested trees from forest type I?
b. The proportion of infested trees from forest type II exceeds the proportion of infested trees from forest type I by at least 0.15?

Calculate the standard error of the difference between two proportions and explain its meaning.

Section 7.6

7.22. In Exercise 7.8 we observed that the annual precipitation in a certain part of British Columbia is normally distributed with a mean of 850 mm and a standard deviation of 95 mm. Based on a sample size of 10, what is the probability that:

a. The standard deviation will be greater than 120 mm?
b. The standard deviation will be less than 58 mm?
c. The standard deviation will be between 58 mm and 120 mm?

7.23. The variance of length of dimensional lumber cut to 10-ft lengths is 0.25 ft^2. If 16 randomly selected 10-ft pieces are measured, what is the probability that the variance will be:

a. Less than 0.15 ft^2?
b. Greater than 0.38 ft^2?
c. Between 0.15 and 0.42 ft^2?

Section 7.7

7.24. For an *F* distribution find:

a. $f_{0.05}$ with $v_1 = 8$ and $v_2 = 10$.
b. $f_{0.025}$ with $v_1 = 7$ and $v_2 = 11$.
c. $f_{0.1}$ with $v_1 = 11$ and $v_2 = 10$.
d. $f_{0.005}$ with $v_1 = 15$ and $v_2 = 28$.

7.25. Let s_1^2 and s_2^2 represent the variances of two independent random samples of size $n_1 = 10$ and $n_2 = 16$, taken from two normal populations with variances $s_1^2 = 20$ and $s_2^2 = 15$.

a. What is the probability that $s_1^2 > 32$?
b. What is the probability that $s_2^2 < 6$?
c. Find the ratios of s_1^2/s_2^2 outside of which on each side you will find:
   i. 2.5% of all ratios of the two variances.
   ii. 1.0% of all ratios of the two variances.
d. What is the probability that:
   i. $s_1^2/s_2^2 > 3.458$.
   ii. $s_1^2/s_2^2 < 0.249$.

7.26. Referring back to Exercise 7.15, if two independent random samples of size 12 and 16 are taken for brands A and B, respectively, and the sample variances, s_A^2 and s_B^2, are calculated, what is the probability that:

a. $s_A^2/s_B^2 < 0.235$.
b. $s_A^2/s_B^2 > 1.32$.

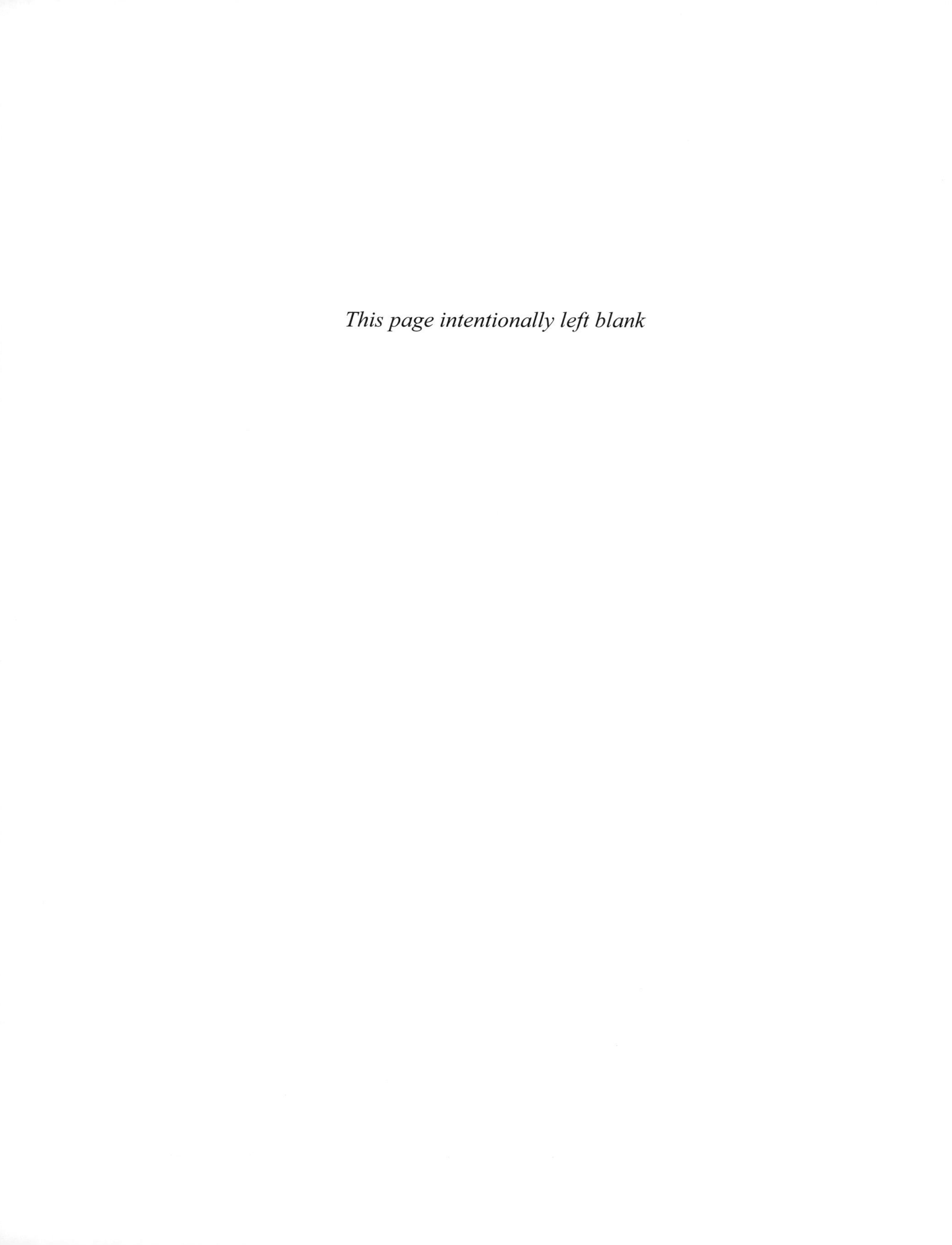
This page intentionally left blank

8 Estimation

Determining the Value of Population Parameters

Statistical inference is the process of drawing a conclusion about a population parameter from information obtained in a sample. We can make **decisions** concerning the value of a parameter or we can **estimate** the value of a parameter. Decision making or **tests of hypothesis** will be introduced in Chapter 9. **Statistical estimation**, which can be classified as either **point estimation** or **interval estimation**, is discussed in this chapter. A point estimate is a single numeric value calculated from the information in a sample. An interval estimate yields two numeric values, between which we can reliably expect to find the target parameter.

8.1 Point Estimation

A **point estimate** of a given population parameter is numeric: our 'best guess' of the true value. For instance, the point estimate of the population mean, μ, is $\bar{x}$ (the sample mean), which is the sample value of the statistic $\bar{X}$ computed from a sample of size n from the population. The statistic $\bar{X}$ is called an **estimator** and the single value, $\bar{x}$, is called an **estimate**. For example, if the mean height of a sample of 10 western hemlock trees is 15.6 m, 15.6 is the point estimate of the unknown population mean. Similarly, the sample proportion, $\hat{P}$, is an estimator of the population proportion, p, while $\hat{p}$ (the sample proportion) is an estimate. The sample variance, S^2, is the estimator of the population variance, σ^2, while s^2 is the estimate.

For any population parameter θ, the quality of a point estimator ($\hat{\theta}$) is judged according to the following characteristics:

1. The estimator should be **unbiased**. In other words, the mean of the sampling distribution is equal to the population parameter. Thus, when we sample, on average we expect to estimate the true population parameter:

$$E(\hat{\theta}) = \theta.$$

Based on the Central Limit Theorem (discussed in Chapter 7), we can conclude that the sample mean, sample proportion and sample variance are all unbiased estimators. Although the sample standard deviation, s, is not an unbiased estimator, if the sample size, n, is greater than 30, the bias is negligible (the mathematical proof for this is beyond the scope of this text).

2. The estimator should be **efficient**. When there are several unbiased estimators of a given parameter, θ, the one with the smallest variance is called the most efficient estimator. The relative efficiency of one estimator versus another is calculated as:

$$\text{relative efficiency} = \frac{\text{variance of } \hat{\theta}_1}{\text{variance of } \hat{\theta}_2} \tag{8.1}$$

where $\hat{\theta}_1$ and $\hat{\theta}_2$ are two different unbiased estimators.

For example, it is known that the variance of the median is about 1.57 times larger than the variance of the sample mean. From this, we can calculate the relative efficiency of the sample median compared to the sample mean:

$$\text{relative efficiency} = \frac{\sigma^2/n}{1.57\,\sigma^2/n} \approx 0.6369.$$

This means that the sample median is only 64% as efficient as the sample mean in estimating the population mean based on the same number of observations.

3. The estimator should be **consistent**. This means that as the sample size, n, approaches infinity, the value of the estimator approaches the value of the population parameter. Therefore, an unbiased estimator is consistent if:

$$\text{as } n \to \infty, \ \hat{\theta} \to \theta.$$

In other words, as n increases, the variance of $\hat{\theta} \to 0$.

Even if an estimator is unbiased, efficient and consistent, it will not estimate the population parameter without error unless n approaches infinity. As discussed in Chapter 7, based on a reasonable sample size, it is unlikely that the numeric value of an estimate will exactly match the population parameter. For this reason, we should keep in mind that point estimates have their limitations. They do not reveal anything about the potential size of their errors, and they do not tell us how much information they are based on. Due to these limitations, it is generally more desirable and even more effective to determine not just a point estimate, but an interval within which we would expect to find the unknown population parameter. This process is referred to as **interval estimation**, and the interval is called the **interval estimate**.

8.2 Interval Estimation

If $\hat{\theta}$ is a point estimate of θ, then $|\hat{\theta} - \theta|$ is the *sampling error* (see Chapter 7), or *error of estimate*. Since we usually do not know the true value of a population parameter, the actual size of the sampling error is unknown. However, we can make probability statements, expressed as 'confidence', concerning the error of estimate. For instance, we can construct an interval, from some lower bound to an upper bound, where the probability of finding the true parameter is set at some value between 0 and 1:

$$P(LCL < \theta < UCL) = 1 - \alpha. \tag{8.2}$$

We call these lower and upper bounds **confidence limits**. The LCL refers to the lower confidence limit and the UCL refers to the upper confidence limit. The value $1 - \alpha$ is the probability (between 0 and 1) that we will find the true population parameter lying somewhere between LCL and UCL. Thus, Eqn 8.2 is read and interpreted as *the probability of finding the unknown population parameter between LCL and UCL is $1 - \alpha$*. If the sampling distribution of $\hat{\theta}$ is symmetric, this implies that the maximum error of estimate, E, is:

$E = (\text{UCL} - \text{LCL})/2.$

E is sometimes called the maximum allowable margin of error. The interval between LCL and UCL is called the $(1 - \alpha)100\%$ **confidence interval**. The quantity $(1 - \alpha)100\%$ is called the **confidence level** or **degree of confidence**. By convention, the most frequently used values of α are 0.10, 0.05 and 0.01, resulting in 90%, 95% and 99% confidence intervals, respectively. The correct interpretation of confidence intervals is sometimes tricky and is discussed in detail in the next section.

8.3 Estimating the Mean

The most frequently used point estimator of the population mean, μ, is the sample mean, $\overline{X}$, and the point estimate is the numeric value of the sample mean, $\bar{x}$, which is unbiased, efficient and consistent.

We can derive an equation for the interval estimate of the unknown population mean, μ, based on the Central Limit Theorem, if the sample is selected from a normal population or if the sample size n is greater than 30. Having more than 30 samples allows us to assume that $\overline{X}$ is approximately normally distributed with a mean and standard deviation of:

$$\mu_{\overline{X}} = \mu \text{ and } \sigma_{\overline{X}} = \frac{\sigma}{\sqrt{n}} \text{ or } \sigma_{\overline{X}} = \frac{\sigma}{\sqrt{n}}\sqrt{\frac{N-n}{N-1}}.$$

Let us assume that we are working with a population with a known variance, σ^2. Using the notation in Fig. 8.1, we describe an interval between $-z_{\alpha/2}$ and $z_{\alpha/2}$, where the probability of finding Z is $1 - \alpha$:

$$P\left(-z_{\alpha/2} < Z < z_{\alpha/2}\right) = 1-\alpha.$$

$\overline{X}$ is normally distributed with mean, μ, and standard deviation, $\sigma/\sqrt{n}$. Therefore, we can substitute the transformation of $\overline{X}$ into a standard normal distribution, Z:

$$Z = \frac{\overline{X}-\mu}{\sigma/\sqrt{n}}$$

hence

$$P\left(-z_{\alpha/2} < \frac{\overline{X}-\mu}{\sigma/\sqrt{n}} < z_{\alpha/2}\right) = 1-\alpha.$$

Rearranging the inequality inside the parentheses, we have:

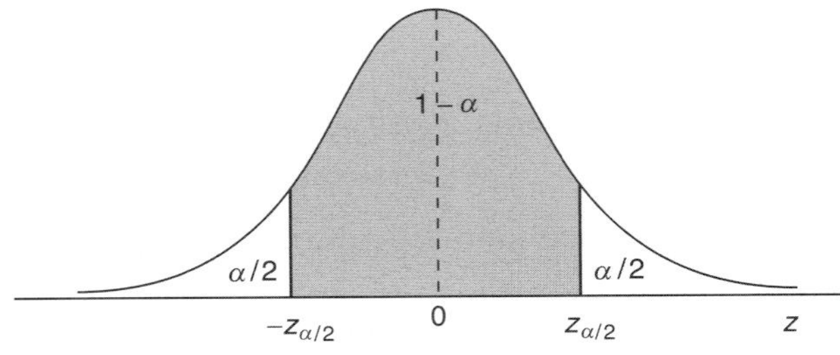

Fig. 8.1. The *Z* distribution.

$$P\left(\overline{X} - z_{\alpha/2}\frac{\sigma}{\sqrt{n}} < \mu < \overline{X} + z_{\alpha/2}\frac{\sigma}{\sqrt{n}}\right) = 1 - \alpha.$$

Replacing the estimator, $\overline{X}$, with the estimate, $\overline{x}$, the final equation is:

$$P\left(\overline{x} - z_{\alpha/2}\frac{\sigma}{\sqrt{n}} < \mu < \overline{x} + z_{\alpha/2}\frac{\sigma}{\sqrt{n}}\right) = 1 - \alpha. \qquad (8.3)$$

Equation 8.3 gives the $(1 - \alpha)100\%$ confidence interval for the unknown population mean, μ, with known variance, σ^2. The LCL and UCL are therefore:

$$\text{LCL} = \overline{x} - z_{\alpha/2}\frac{\sigma}{\sqrt{n}}$$

$$\text{UCL} = \overline{x} + z_{\alpha/2}\frac{\sigma}{\sqrt{n}}$$

and the margin of error is then $E = z_{\alpha/2}\frac{\sigma}{\sqrt{n}}$.

Note that, if samples are taken without replacement, the standard error of the mean can be multiplied by the finite population correction factor (see Chapter 7).

For given values of $\alpha/2$ and n and a known σ, the margin of error is constant. However, since $\overline{x}$ is a random variable, LCL and UCL are both random variables whose values are a function of $\overline{x}$. In other words, if we repeatedly draw n observations taken from a population with a mean, μ, and standard deviation, σ, the values for the LCL and UCL will vary randomly. In Fig. 8.2 we made 20 such draws, each time calculating the 95% confidence interval for μ. Most (19 out of 20) of the intervals contain or bound the population mean, μ. In general, $(1 - \alpha)100\%$ of the means will be within $\overline{x} \pm z_{\alpha/2}\sigma/\sqrt{n}$ and, therefore, $(1 - \alpha)100\%$ of the intervals will contain μ. Restated, the meaning of a 95% confidence interval is that, if 100 samples of size n are taken and intervals are constructed around each of the 100 sample means, then 95% of the intervals will contain the true population parameter, μ (or 19 out of 20 samples). Likewise, in a 99% confidence interval, only 1 out of 100 intervals constructed would not contain μ (implying that the 99% confidence intervals must be wider than the 95% confidence intervals).

Example 8.1. In a large plantation, diameter at breast height (dbh) measurements are normally distributed with σ = 1.6 cm. A random sample of 16 trees was selected without replacement and measured. The sample mean of these measurements was 12.6 cm. Find the 95% and 99% confidence intervals for the unknown population mean dbh and their associated margins of error.

$$\sigma_{\overline{X}} = \frac{1.6}{\sqrt{16}} = 0.4 \qquad z_{0.025} = 1.96 \qquad z_{0.005} = 2.58$$

$$P\left(12.6 - (1.96)(0.4) < \mu < 12.6 + (1.96)(0.4)\right) = 0.95$$

$$P(11.82 < \mu < 13.38) = 0.95 \qquad \text{E} = (13.38 - 11.82)/2 = 0.78.$$

There is a 0.95 probability that the unknown population mean of all possible dbhs will be found between 11.82 cm and 13.38 cm.

$$P\left(12.6 - (2.58)(0.4) < \mu < 12.6 + (2.58)(0.4)\right) = 0.99$$

$$P(11.57 < \mu < 13.63) = 0.99 \qquad E = (13.63 - 11.57)/2 = 1.03.$$

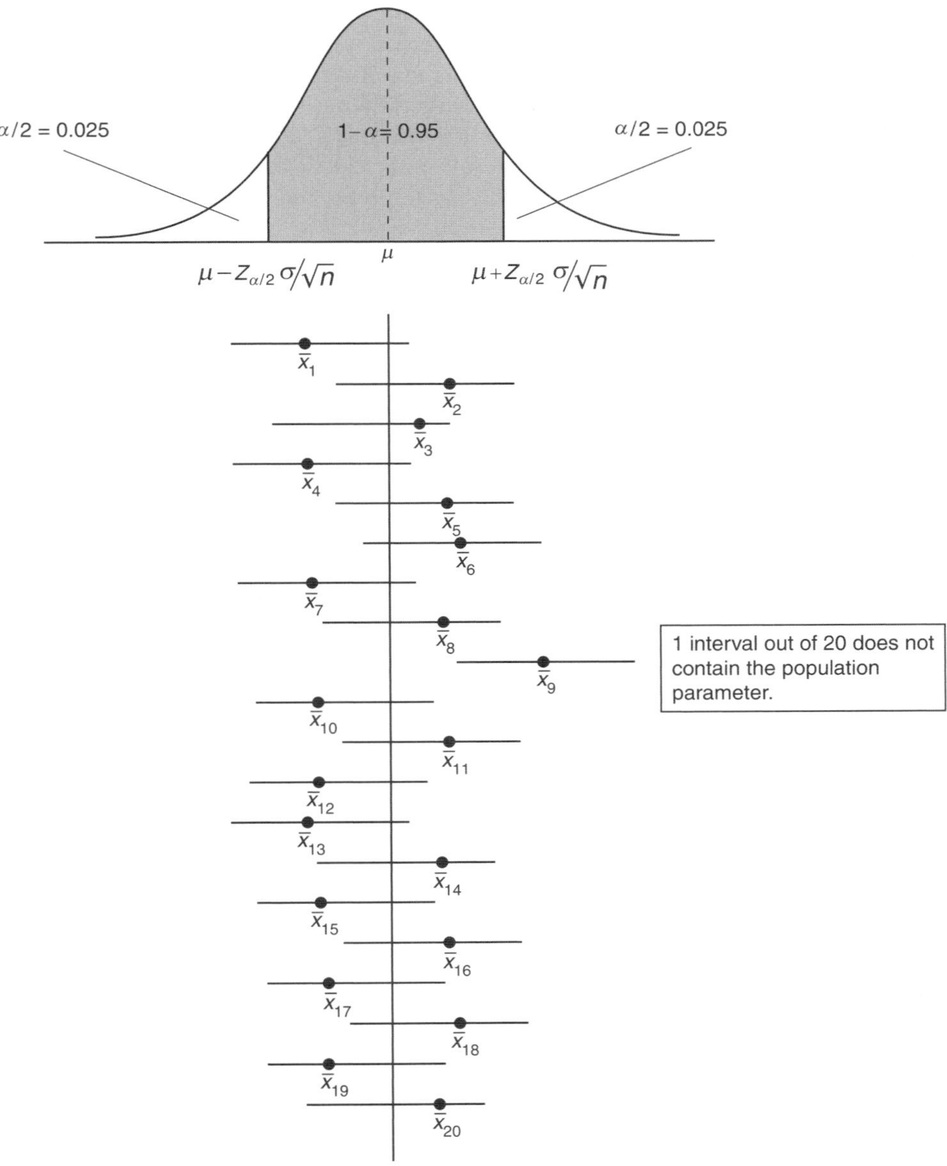

Fig. 8.2. 95% confidence intervals calculated from 20 sample means.

There is a 0.99 probability that the unknown population mean of all possible dbhs will be found between 11.57 cm and 13.63 cm.

Note from the solutions that a wider interval is required for a higher degree of confidence. For a given sample, the width is the function of α. As the value of α decreases, the value of $z_{\alpha/2}$ increases and so does the level of confidence. In other words, there is a trade-off between the degree of confidence and the margin of error.

In order to have a higher confidence level for a population estimate, one must sacrifice precision in the form of a wider confidence interval.

In the previous discussion and example, the population variance, σ, was known. However, for most practical problems, σ is unknown and, therefore, Eqn 8.3 is not applicable. When the original population values, x_i (not $\bar{x}_i$), are approximately normally distributed, the confidence interval can be calculated from the t distribution. However, this is a somewhat restrictive assumption and it can be shown that t distribution works well for many non-normal populations.

Using the notation in Fig. 8.3, our confidence limits are now constructed using values from the t distribution:

$$P\left(-t_{\alpha/2(n-1)} < T < t_{\alpha/2(n-1)}\right) = 1 - \alpha$$

$\bar{X}$ has a t distribution with mean μ and sample standard deviation $S/\sqrt{n}$. Therefore, we can substitute the transformation of $\bar{X}$ into a standard Student's T:

$$T = \frac{\bar{X} - \mu}{S/\sqrt{n}}.$$

Hence

$$P\left(-t_{\alpha/2(n-1)} < \frac{\bar{X} - \mu}{S/\sqrt{n}} < t_{\alpha/2(n-1)}\right) = 1 - \alpha.$$

Rearranging the inequality inside the parentheses, we have:

$$P\left(\bar{X} - t_{\alpha/2(n-1)} \frac{S}{\sqrt{n}} < \mu < \bar{X} + t_{\alpha/2(n-1)} \frac{S}{\sqrt{n}}\right) = 1 - \alpha.$$

Replacing the estimator, $\bar{X}$, with the estimate, $\bar{x}$, the final equation is then:

$$P\left(\bar{x} - t_{\alpha/2(n-1)} \frac{s}{\sqrt{n}} < \mu < \bar{x} + t_{\alpha/2(n-1)} \frac{s}{\sqrt{n}}\right) = 1 - \alpha. \quad (8.4)$$

Equation 8.4 is structurally very similar to Eqn 8.3. There are two differences: the population standard deviation is replaced by the sample standard deviation and the z-values are replaced by t-values with $n - 1$ degrees of freedom (see Chapter 2 for further explanation). The meaning and interpretation of the confidence intervals calculated from Eqn 8.4 are identical to those discussed in reference to Eqn 8.3.

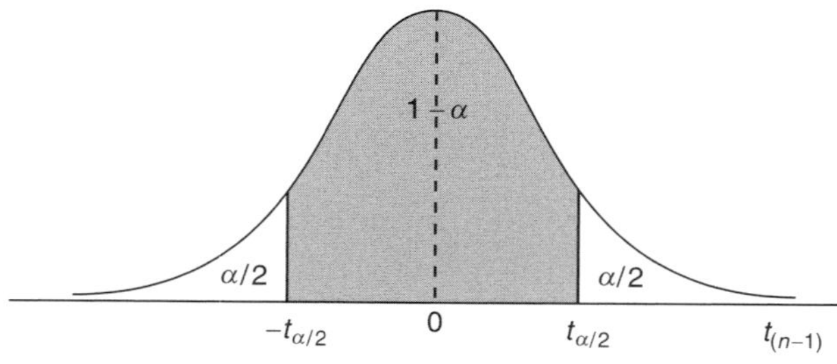

Fig. 8.3. The *t* distribution.

Example 8.2. Suppose that in Example 8.1, the population standard deviation for dbh is unknown, but is estimated to be 2.09 cm using the 16 observations taken. Find the 95% and 99% confidence intervals for the unknown population mean dbh and the associated margins of error.

$$s_{\overline{X}} = \frac{2.09}{\sqrt{16}} \approx 0.5225 \qquad t_{0.025(15)} = 2.13 \qquad t_{0.005(15)} = 2.95$$

$$P\left(12.6 - (2.13)(0.5225) < \mu < 12.6 + (2.13)(0.5225)\right) = 0.95$$

$$P(11.49 < \mu < 13.71) = 0.95 \qquad E = 1.11.$$

Based on a sample of 16 trees, there is a 0.95 probability that the unknown population mean of all possible dbhs will be found between 11.49 cm and 13.71 cm.

$$P\left(12.6 - (2.95)(0.5225) < \mu < 12.6 + (2.95)(0.5225)\right) = 0.99$$

$$P(11.06 < \mu < 14.14) = 0.99 \qquad \text{E} = 1.54.$$

There is a 0.99 probability that the unknown population mean of all possible dbhs will be found between 11.06 cm and 14.14 cm.

Again, notice that the confidence interval is wider for the 99% level of confidence. As the size of α decreases, we create wider confidence limits and therefore have more confidence in our interval estimates. The 95% confidence interval from a t distribution is also wider than that derived from a Z distribution (see Example 8.1), simply because, in the latter case, we have more information at our disposal (in this case, the population variance).

In general, the size of the margin of error for the estimation of the population mean depends on three factors:

- the population or sample standard deviation;
- the level of confidence $(1 - \alpha)$, which determines the size of the z and t; and
- the sample size, n, which affects the size of the standard error of the mean and the size of t.

Of these, we can control only the level of confidence and the sample size, since the standard deviation (population parameter or sample statistic) is a natural characteristic of the population. It can be seen from the equation that the margin of error decreases if we increase the sample size. Hence, a required sample size can be calculated if the desired margin of error (E), the level of confidence and the population/sample standard deviation are known.

If we know the population standard deviation, we start with the equation for margin of error from Eqn 8.3:

$$\text{E} = z_{\alpha/2} \frac{\sigma}{\sqrt{n}}$$

Algebraically rearranging terms to isolate n, we have an expression for the sample size necessary to achieve a desired margin of error when σ is known:

$$n = \frac{z_{\alpha/2}^2 \sigma^2}{E^2} = \left(\frac{z_{\alpha/2}\sigma}{E}\right)^2 \qquad (8.5)$$

Example 8.3. We want to be 95% confident that the margin of error used to estimate the confidence interval around the population mean, μ, in Example 8.1 is 0.6 cm (in other words, μ is bounded by an interval that is ±0.6). How large a sample is required from this stand?

$$E = 0.6; \qquad \sigma = 1.6; \qquad z_{0.025} = 1.96$$

$$n = \frac{(1.96^2)(1.6^2)}{0.6^2} \approx 27.3$$

We should take 27 or 28 samples (rounding up will give a more conservative, but costly, result).

Since we rarely know the population standard deviation for most practical problems, we cannot use Eqn 8.5 in these situations. If we do not know the population standard deviation, we start with Eqn 8.4:

$$E = t_{\alpha/2(n-1)} \frac{s}{\sqrt{n}}$$

Rearranging terms to isolate n, we have an expression for the sample size necessary to achieve a desired margin of error when σ is unknown:

$$n = \frac{t^2_{\alpha/2(n-1)}\, s^2}{E^2} = \left(\frac{t_{\alpha/2(n-1)}\, s}{E}\right)^2 \tag{8.6}$$

Use of Eqn 8.6 is more complicated because there are two unknowns in the equation: n and $t_{\alpha/2(n-1)}$. Furthermore, t is a function of n, which creates a circular equation. To overcome this, iteration is used to find the required sample size, beginning with the first iteration: n_1 = infinity. This is best illustrated by an example.

Example 8.4. We would like to be 95% confident that the margin of error used to estimate the confidence interval around the population mean, μ, in Example 8.2 is 0.8 cm (in other words, μ is bounded by an interval that is ±0.8). How large a sample is required from this stand? Repeat the calculation for $E = 0.6$.

Recall $s = 2.09$. We start with $n_1 = \infty$ degrees of freedom and therefore $t_{0.025(\infty)} = z_{0.025} = 1.96$ (using either the z-table or the t-table).

Using Eqn 8.6, we have:

$$n_1 = \frac{(1.96^2)(2.09^2)}{0.8^2} \approx 26.22 \approx 26$$

We then recalculate Eqn 8.6 using $n_1 = 26$. This means that we use 26 – 1 = 25 degrees of freedom and $t_{0.025(25)} = 2.06$ (we must use the t-table because we do not have infinite degrees of freedom):

$$n_2 = \frac{(2.06^2)(2.09^2)}{0.8^2} \approx 28.96 \approx 29$$

Now, we recalculate Eqn 8.6 using $n_2 = 29$. This means that we use 29 – 1 = 28 degrees of freedom and $t_{0.025(28)} = 2.05$:

$$n_3 = \frac{(2.05^2)(2.09^2)}{0.8^2} \approx 28.68 \approx 29$$

We can stop at this iteration because convergence has been reached in two consecutive solutions: the n-value input into this equation matches the n-value output by it. For a confidence interval that has a margin of error of 0.8 cm on either side of the population mean, a sample size of 29 is required.

For $E = 0.6$, we again start with $n_1 = \infty$ and therefore $t_{0.025(\infty)} = z_{0.025} = 1.96$:

$$n_1 = \frac{(1.96^2)(2.09^2)}{0.6^2} \approx 46.61 \approx 47$$

Because the t-values do not change considerably between 30 and ∞ degrees of freedom (see Table A.5 in Appendix A), we do not need to continue iterating and conclude that the required sample size is 47. In general, when the input to Eqn 8.6 is ∞ and the output is > 30, we stop the iteration process. Of course, if the calculated sample size is not economically feasible, either the desired margin of error or the level of confidence – or both – must be manipulated.

Although more complicated, Eqns 8.5 and 8.6 can also be derived for cases of sampling without replacement from a finite population.

$$n = \frac{1}{\frac{E^2}{z^2_{\alpha/2}\,\sigma^2} + \frac{1}{N}} \tag{8.5a}$$

$$n = \frac{1}{\frac{E^2}{t^2_{\alpha/2(n-1)}\,s^2} + \frac{1}{N}} \tag{8.6a}$$

where N = size of the finite population.

As a final note on sample size calculation, sometimes it is the case that we do not know the population standard deviation. Moreover, we do not have an estimate of the sample standard deviation. In these cases, the sample standard deviation can be crudely approximated as (range)/4, provided that the distribution of the population is not too skewed. One can also look to previous research on the variable in question to come up with a reasonable estimate of the standard deviation.

8.4 Estimating Proportions

The point estimator of the unknown population proportion, p, is the statistic, $\hat{P}$, and the point estimate is the numeric value of a sample proportion, $\hat{p} = x/n$, where x is the number of successes in n trials in a binomial experiment. An interval estimator of the unknown population proportion, p, can be derived from the sampling distribution of the proportions, provided that n is greater than 30 and that p is not very close to 0 or 1. In Chapter 7, we gave the equations for the mean and variance of proportions from a binomial experiment:

$$E(\hat{p}) = \mu_{\hat{p}} = p \qquad \text{and} \qquad \sigma_{\hat{p}} = \sqrt{\frac{pq}{n}} \text{ or } \sigma_{\hat{p}} = \sqrt{\frac{pq}{n}}\sqrt{\frac{N-n}{N-1}}.$$

Using the notation in Fig. 8.1, the confidence interval for a standard Z variable is:

$$P\left(-z_{\alpha/2} < Z < z_{\alpha/2}\right) = 1 - \alpha$$

From Chapter 7, we know that when n is greater than 30, the binomial distribution approaches the normal distribution. The normal transformation, therefore, is:

$$Z = \frac{\hat{P} - p}{\sqrt{\frac{pq}{n}}}.$$

Hence

$$P\left(-z_{\alpha/2} < \frac{\hat{P} - p}{\sqrt{\frac{pq}{n}}} < z_{\alpha/2} \right) = 1 - \alpha.$$

Rearranging the inequality inside the parentheses and replacing the estimator, $\hat{P}$, with the sample proportion, $\hat{p}$, we have:

$$P\left[\hat{p} - z_{\alpha/2}\sqrt{\frac{pq}{n}} < p < \hat{p} + z_{\alpha/2}\sqrt{\frac{pq}{n}} \right] = 1 - \alpha$$

When p and q are unknown (which is usually the case), we use their estimates, $\hat{p}$ and $\hat{q}$. This gives a good approximation and the final equation is:

$$P\left[\hat{p} - z_{\alpha/2}\sqrt{\frac{\hat{p}\hat{q}}{n}} < p < \hat{p} + z_{\alpha/2}\sqrt{\frac{\hat{p}\hat{q}}{n}} \right] = 1 - \alpha \tag{8.7}$$

From Eqn. 8.7, the margin of error is:

$$E = z_{\alpha/2}\sqrt{\frac{\hat{p}\hat{q}}{n}}.$$

Example 8.5. In a random sample of 60 panels in a plywood mill, 27 are grade B. Find the 90% and 95% confidence intervals for the unknown population proportion of grade B panels in production.

$$\hat{p} = \frac{27}{60} = 0.45 \qquad s_{\hat{p}} = \sqrt{\frac{(0.45)(0.55)}{60}} \approx 0.064$$

$$z_{0.05} = 1.645 \qquad z_{0.025} = 1.96$$

$$P\big(0.45 - (1.645)(0.064) < p < 0.45 + (1.645)(0.064)\big) = 0.90$$

$$P(0.344 < p < 0.556) = 0.90 \qquad E = 0.106.$$

There is a 0.90 probability that the unknown population proportion of all plywood panels graded B will be found between 0.344 and 0.556.

$$P\big(0.45 - (1.96)(0.064) < p < 0.45 + (1.96)(0.064)\big) = 0.95$$

$$P(0.324 < p < 0.576) = 0.95 \qquad E = 0.126.$$

There is a 0.95 probability that the unknown population proportion of all plywood panels graded B will be found between 0.324 and 0.576.

The sample size necessary for a desired margin of error and a specified level of confidence can be calculated in the same manner as the sample mean.

From Eqn 8.7, the margin of error is:

$$E = z_{\alpha/2}\sqrt{\frac{\hat{p}\hat{q}}{n}}$$

Rearranging terms to isolate n, we have an expression for the sample size necessary to achieve a desired margin of error when sampling for proportions:

$$n = \frac{z^2_{\alpha/2}\ \hat{p}\hat{q}}{E^2} \tag{8.8}$$

This equation implicitly assumes that we have a sample, $n \geq 30$. In practice, this is usually the case for proportions.

Example 8.6. We want to be 99% confident that the margin of error used to estimate the confidence interval around the population proportion, p, in Example 8.5 does not exceed 0.1. How large a sample is required? Note that in this problem, we use the phrase, 'does not exceed', which is tantamount to an equality in the context of sample size problems.

For $\hat{p}$ and $\hat{q}$ we use our estimates from Example 8.5 and $z_{0.005} = 2.58$.

$$n = \frac{(2.58^2)(0.45)(0.55)}{0.1^2} \approx 164.75 \approx 165$$

We should take 165 samples to maintain a maximum margin of error of 0.1 with a 0.99 confidence level.

For Eqn 8.8, we must have an estimate of $\hat{p}$, the sample proportion. In some situations, it may be very difficult, or impossible, to obtain an estimate of $\hat{p}$. In these cases, a value of $\hat{p} = 0.5$ can be used to provide a conservative estimate, but may overestimate n, since the product of $\hat{p}\hat{q}$ is maximized when $\hat{p} = \hat{q} = 0.5$.

8.5 Estimating the Difference between Two Means

Independent samples

The point estimator of the difference between two unknown population means, $\mu_1 - \mu_2$, is the statistic, $\bar{X}_1 - \bar{X}_2$. The estimate is the numeric value of the difference of two sample means, $\bar{x}_1 - \bar{x}_2$, calculated from two *independent* samples of sizes n_1 and n_2, respectively, from the two populations.

If the independent samples are selected from normal populations with known population variances (σ_1^2 and σ_2^2), the interval estimator or confidence interval can be derived from the sampling distribution of the differences between two sample means. Again, using the notation in Fig. 8.1, a confidence interval for a standard Z random variable is:

$$P\left(-z_{\alpha/2} < Z < z_{\alpha/2}\right) = 1 - \alpha.$$

$\bar{X}_1 - \bar{X}_2$ is normally distributed with mean, $\mu_1 - \mu_2$, and standard deviation, $\sqrt{\left(\sigma_1^2/n_1\right) + \left(\sigma_2^2/n_2\right)}$. We can, therefore, substitute in the transformation of $\bar{X}_1 - \bar{X}_2$ into a standard normal distribution, Z:

$$Z = \frac{\left(\bar{X}_1 - \bar{X}_2\right) - \left(\mu_1 - \mu_2\right)}{\sqrt{\left(\sigma_1^2/n_1\right) + \left(\sigma_2^2/n_2\right)}}.$$

Substituting for Z:

$$P\left[-z_{\alpha/2} < \frac{(\bar{X}_1 - \bar{X}_2) - (\mu_1 - \mu_2)}{\sqrt{(\sigma_1^2/n_1) + (\sigma_2^2/n_2)}} < z_{\alpha/2}\right] = 1 - \alpha.$$

Rearranging the inequality inside the parentheses and substituting the estimate, $\bar{x}_1 - \bar{x}_2$, for the estimator, $\bar{X}_1 - \bar{X}_2$, we get the final equation:

$$P\left[(\bar{x}_1 - \bar{x}_2) - z_{\alpha/2}\sqrt{\frac{\sigma_1^2}{n_1} + \frac{\sigma_2^2}{n_2}} < \mu_1 - \mu_2 < (\bar{x}_1 - \bar{x}_2) + z_{\alpha/2}\sqrt{\frac{\sigma_1^2}{n_1} + \frac{\sigma_2^2}{n_2}}\right] = 1 - \alpha. \quad (8.9)$$

Since Eqn 8.9 contains the two population variances, its practical application is very limited. Instead, the confidence interval for the difference of two unknown population means, $\mu_1 - \mu_2$, is generally calculated using the sample variances and the t distribution.

Starting with the notation from Fig. 8.3, the confidence interval for a standard T random variable is:

$$P\left(-t_{\alpha/2(v)} < T < t_{\alpha/2(v)}\right) = 1 - \alpha$$

In this case, $\bar{X}_1 - \bar{X}_2$ has a t distribution with mean, $\mu_1 - \mu_2$, and standard deviation, $s_{\bar{X}_1 - \bar{X}_2}$. We can therefore substitute the transformation of $\bar{X}_1 - \bar{X}_2$ into a Student's t distribution, T:

$$T = \frac{(\bar{X}_1 - \bar{X}_2) - (\mu_1 - \mu_2)}{S_{\bar{X}_1 - \bar{X}_2}}.$$

Substituting for T:

$$P\left[-t_{\alpha/2(v)} < \frac{(\bar{X}_1 - \bar{X}_2) - (\mu_1 - \mu_2)}{S_{\bar{X}_1 - \bar{X}_2}} < t_{\alpha/2(v)}\right] = 1 - \alpha.$$

Rearranging the inequality inside the parentheses and substituting the estimate, $\bar{x}_1 - \bar{x}_2$, for the estimator, $\bar{X}_1 - \bar{X}_2$ we get the final confidence interval formula for the difference between two means when the population variances are unknown:

$$P\left[(\bar{x}_1 - \bar{x}_2) - t_{\alpha/2(v)} s_{\bar{X}_1 - \bar{X}_2} < \mu_1 - \mu_2 < (\bar{x}_1 - \bar{x}_2) + t_{\alpha/2(v)} s_{\bar{X}_1 - \bar{X}_2}\right] = 1 - \alpha \quad (8.10)$$

To use Eqn 8.10 above, we must distinguish between three cases, which vary according to sample sizes and our assumptions regarding the equality of the two population variances (see Chapter 7). For each case, the standard error of the difference of two means and the degrees of freedom are calculated differently. As a reminder, the three cases are:

Case 1. Both n_1 and $n_2 \geq 30$ (large sample sizes).

$$s_{\bar{X}_1 - \bar{X}_2} = \sqrt{\frac{s_1^2}{n_1} + \frac{s_2^2}{n_2}}$$

and $v = \infty$.

Case 2. n_1 or n_2 or both < 30 (small sample sizes) and it is assumed that $\sigma_1^2 = \sigma_2^2$.

$$s_{\bar{X}_1-\bar{X}_2} = s_p\sqrt{\frac{1}{n_1}+\frac{1}{n_2}}$$

where $s_p = \sqrt{\dfrac{s_1^2(n_1-1)+s_2^2(n_2-1)}{n_1+n_2-2}}$ and $\nu = n_1 + n_2 - 2.$

Case 3. n_1 or n_2 or both < 30 (small sample sizes) and it is assumed that $\sigma_1^2 \neq \sigma_2^2$.

$$s_{\bar{X}_1-\bar{X}_2} = \sqrt{\frac{s_1^2}{n_1}+\frac{s_2^2}{n_2}}$$

and

a. $\nu = \dfrac{\left[\dfrac{s_1^2}{n_1}+\dfrac{s_2^2}{n_2}\right]^2}{\dfrac{\left(s_1^2/n_1\right)^2}{n_1-1}+\dfrac{\left(s_2^2/n_2\right)^2}{n_2-1}}$, or

b. ν = smaller of $n_1 - 1$ or $n_2 - 1$.

Again, the calculation in a is the more precise alternative; however, b suffices for most practical applications.

Example 8.7. Sawmill managers are interested in knowing the dimensions of logs arriving at their mills. Of particular interest are the diameters of the butt and top ends of the logs. A study of the top diameter on two lots of logs was carried out and the measurements from 42 randomly selected logs in Lot 1 and 36 in Lot 2 are summarized below. Find the 95% confidence interval for the difference of the two unknown population means.

	n	Mean	*s*
Lot 1	42	15.9	2.8
Lot 2	36	18.6	3.4

We use Case 1, since n_1 and $n_2 > 30$.

$$s_{\bar{X}_1-\bar{X}_2} = \sqrt{\frac{2.8^2}{42}+\frac{3.4^2}{36}} \approx 0.7126 \qquad t_{0.05(\infty)} = 1.96$$

$$\bar{x}_1 - \bar{x}_2 = 15.9 - 18.6 = -2.7$$

$$P\left(-2.7-(1.96)(0.7126) < \mu_1 - \mu_2 < -2.7+(1.96)(0.7126)\right) = 0.95$$

$$P\left(-4.10 < \mu_1 - \mu_2 < -1.30\right) = 0.95 \qquad E = 1.40.$$

There is 0.95 probability that the unknown difference of the two population means is between –4.10 and –1.30. The practical interpretation of this result is as follows. Since zero is not included in the interval, we can conclude that $\mu_1 \neq \mu_2$ with 95% confidence. Also, since the intervals are negative, we can conclude that $\mu_1 < \mu_2$ with 95% confidence.

Example 8.8. From the two lots analysed above, log volume was measured on a subset of samples. The mean volume of 12 sample logs from Lot 1 was 1.72 m^3 with a standard deviation of 0.56 m^3, and the mean volume of 8 sample logs from Lot 2 was 1.93 m^3 with a standard deviation of 0.61 m^3. Find the 90% and the 95% confidence interval for the difference between the two unknown population means.

Since n_1 and $n_2 < 30$, we either use Case 2 or Case 3. Let us assume that $\sigma_1^2 = \sigma_2^2$ (this assumption will be confirmed in Example 8.13) and, therefore, we use Case 2.

$$\bar{x}_1 - \bar{x}_2 = 1.72 - 1.93 = -0.21 \qquad v = 12 + 8 \;\; 2 = 18$$

$$s_p = \sqrt{\frac{(12-1)(0.56^2) + (8-1)(0.61^2)}{12+8-2}} \approx 0.58 \qquad t_{0.05(18)} = 1.73$$

$$s_{\bar{x}_1 - \bar{x}_2} = 0.58\sqrt{\frac{1}{12} + \frac{1}{8}} \approx 0.265 \qquad t_{0.025(18)} = 2.10$$

$$P(-0.21 - (1.73)(0.265) < \mu_1 - \mu_2 < -0.21 + (1.73)(0.265)) = 0.90$$

$$P(-0.668 < \mu_1 - \mu_2 < 0.248) = 0.90 \qquad E = 0.458.$$

There is a 0.90 probability that the unknown difference of the two population means is between −0.668 and 0.248.

$$P(-0.21 - (2.10)(0.265) < \mu_1 - \mu_2 < -0.21 + (2.10)(0.265)) = 0.95$$

$$P(-0.767 < \mu_1 - \mu_2 < 0.347) = 0.95 \qquad E = 0.557.$$

There is a 0.95 probability that the unknown difference of the two population means is between −0.767 and 0.347.

Since zero is included in both intervals, we can conclude that $\mu_1 = \mu_2$ with 90% (or 95%) confidence.

Example 8.9. In Example 7.12 (see Chapter 7), we described a study where biomass production was measured on groups of seedlings under normal and artificial lighting. The observations are summarized below. Find the 95% confidence interval for the difference between the two unknown population means.

	Lighting	
	Normal	Artificial
Mean	4.65	2.57
s	1.20	0.50
n	12	10

Since n_1 and $n_2 < 30$, we either use Case 2 or Case 3. Let us assume that $\sigma_1^2 \neq \sigma_2^2$ (this assumption will be confirmed in Example 8.14) and, therefore, we use Case 3.

$$s_{\bar{x}_1 - \bar{x}_2} = \sqrt{\frac{1.2^2}{12} + \frac{0.5^2}{10}} = 0.381$$

Using the simpler procedure for determining degrees of freedom, (b): $v = 9$, the smaller of 12 − 1 and 10 − 1. Then,

$$t_{0.025(9)} = 2.26 \qquad \bar{x}_1 - \bar{x}_2 = 4.65 - 2.57 = 2.08$$

$$P(2.08 - (2.26)(0.381) < \mu_1 - \mu_2 < 2.08 + (2.26)(0.381)) = 0.95$$

$$P(1.257 < \mu_1 - \mu_2 < 2.903) = 0.95 \qquad E = 0.823.$$

Since zero is not included in the interval, we can conclude that $\mu_1 \neq \mu_2$ with 95% confidence. Furthermore, since the difference between μ_1 and μ_2 is positive, we can conclude that $\mu_1 > \mu_2$ with 95% confidence.

The meaning and interpretation of confidence intervals for the difference of two population means are very similar to those for one population mean and for one population proportion. The width of the confidence interval depends on three factors that we can control: the level of confidence and the two sample sizes, n_1 and n_2. Although sample size equations can be derived from the desired margin of error, they are much more complicated than those for single means and proportions and, therefore, are not discussed here.

Dependent samples

In Chapter 7, we discussed a special case of the sampling distribution of the differences of two means, where the samples taken from two populations are not independent. Samples are considered to be **dependent** when observations occur in pairs or are matched by some special characteristic. As a result, the topic is discussed under the heading *paired observations* in some basic statistics books.

In the case of paired or dependent observations, the point estimator of the difference of two unknown population means, $\mu_1 - \mu_2$, is $\bar{D}$, and the point estimate is the numeric value, $\bar{d}$, calculated from the two dependent samples (see Chapter 7).

Since we are dealing with one random variable, $\bar{D}$, the derivation of the confidence interval is identical to that of the population mean from one population. The final equation to calculate a confidence interval for the difference of two means from dependent populations, where $\sigma_{\bar{d}}$ is known or $n \geq 30$, is:

$$P\left(\bar{d} - z_{\alpha/2(n-1)}\sigma_{\bar{d}} < \mu_D < \bar{d} + z_{\alpha/2(n-1)}\sigma_{\bar{d}}\right) = 1 - \alpha \tag{8.11}$$

If $n < 30$ or $\sigma_{\bar{d}}$ is not known, the equation is as follows:

$$P\left(\bar{d} - t_{\alpha/2(n-1)}\frac{s_d}{\sqrt{n}} < \mu_D < \bar{d} + t_{\alpha/2(n-1)}\frac{s_d}{\sqrt{n}}\right) = 1 - \alpha, \tag{8.12}$$

where n = number of pairs.

Because the population variance is hardly ever known, Eqn 8.11 has mainly theoretical value. Thus, we present an example for Eqn 8.12 only.

Example 8.10. A study was conducted to find out whether there is a systematic difference in the dry weights (in grams) of seedlings measured with two different scales. Eight seedlings were weighed:

Seedling	Scale 1	Scale 2	d_i
1	12.15	12.17	–0.02
2	9.34	9.35	–0.01
3	14.23	14.26	–0.03
4	10.16	10.21	–0.05
5	16.87	16.92	–0.05
6	11.15	11.13	0.02
7	16.15	16.23	–0.08
8	22.60	22.66	–0.06

Find the 99% confidence interval for the difference of the two unknown population means.

From Eqns 7.24 and 7.25 (see Chapter 7), we have:

$$s_d \approx 0.0316 \qquad s_{\bar{d}} \approx \frac{0.0316}{\sqrt{8}} \approx 0.011$$

$$\bar{x}_1 - \bar{x}_2 = \bar{d} \approx -0.035 \qquad t_{0.005(7)} = 3.50$$

$$P\left(-0.035 - (3.50)(0.011) < \mu_D < -0.035 + (3.50)(0.011)\right) = 0.99$$

$$P\left(-0.074 < \mu_D < 0.004\right) = 0.99$$

Since zero is included in the interval, we can conclude that there is 0.99 probability that the two unknown population means are equal. In other words, the accuracy of the two scales is the same.

8.6 Estimating the Difference of Two Proportions

The point estimator of the differences of two unknown independent population proportions, $p_1 - p_2$, is the statistic, $\hat{P}_1 - \hat{P}_2$. The point estimate is the numeric value of the difference between two independent sample proportions taken from the two binomial populations, $\hat{p}_1 - \hat{p}_2$ with sample sizes of n_1 and n_2, respectively.

Like the difference between two population means, the interval estimator of the difference between two unknown population proportions, $p_1 - p_2$, can be derived from the sampling distribution of the differences between two sample proportions, provided that both n_1 and n_2 are greater than 30 and that both p_1 and p_2 are not very close to 0 or 1.

Using the notation in Fig. 8.1, the confidence interval for a standard Z variable is:

$$P\left(-z_{\alpha/2} < Z < z_{\alpha/2}\right) = 1 - \alpha.$$

From Chapter 7, we know that when both n_1 and n_2 are greater than 30, the difference of two binomial variables is approximately normal. We also know, from Chapter 7, the equations for the mean and variance of the difference of two proportions:

$$\mu_{\hat{p}_1 - \hat{p}_2} = p_1 - p_2 \qquad \text{and} \qquad \sigma^2_{\hat{p}_1 - \hat{p}_2} = \frac{p_1 q_1}{n_1} + \frac{p_2 q_2}{n_2}.$$

The normal transformation is:

$$Z = \frac{\left(\hat{P}_1 - \hat{P}_2\right) - \left(p_1 - p_2\right)}{\sqrt{\left(p_1 q_1 / n_1\right) + \left(p_2 q_2 / n_2\right)}}.$$

Substituting for Z, we get:

$$P\left[-z_{\alpha/2} < \frac{\left(\hat{P}_1 - \hat{P}_2\right) - \left(p_1 - p_2\right)}{\sqrt{\left(p_1 q_1 / n_1\right) + \left(p_2 q_2 / n_2\right)}} < z_{\alpha/2}\right] = 1 - \alpha.$$

Rearranging the inequality inside the parentheses and replacing the estimator, $\hat{P}_1 - \hat{P}_2$, with the estimate, $\hat{p}_1 - \hat{p}_2$ the final equation is:

$$P\left[(\hat{p}_1 - \hat{p}_2) - z_{\alpha/2} s_{\hat{P}_1-\hat{P}_2} < p_1 - p_2 < (\hat{p}_1 - \hat{p}_2) + z_{\alpha/2} s_{\hat{P}_1-\hat{P}_2}\right] = 1 - \alpha, \tag{8.13}$$

where

$$s_{\hat{P}_1-\hat{P}_2} = \sqrt{\hat{p}_c \hat{q}_c \left(\frac{1}{n_1} + \frac{1}{n_2}\right)}, \quad \hat{p}_c = \frac{n_1 \hat{p}_1 + n_2 \hat{p}_2}{n_1 + n_2} \quad \text{and} \quad \hat{q}_c = 1.0 - \hat{p}_c.$$

Example 8.11. We would like to compare two stands of mixed forest, in terms of the proportion of western hemlock trees present. In stand 1, 15 out of 60 trees are western hemlock, while in stand 2, 10 out of 50 trees examined are western hemlock. Find the 99% confidence interval for the difference between the two unknown population proportions of western hemlock trees.

$$n_1 = 60 \qquad \hat{p}_1 = 0.25 \qquad \hat{p}_c = \frac{(60)(0.25) + (50)(0.20)}{60 + 50} \approx 0.227$$

$$n_2 = 50 \qquad \hat{p}_2 = 0.20 \qquad \hat{q}_c = 1.0 - 0.227 \approx 0.773$$

$$s_{\hat{p}_1-\hat{p}_2} = \sqrt{(0.277)(0.773)\left[\frac{1}{60} + \frac{1}{50}\right]} = 0.0802$$

$$z_{0.005} = 2.58 \qquad \hat{p}_1 - \hat{p}_2 = 0.25 - 0.20 = 0.05$$

$$P(0.05 - (2.58)(0.0802) < p_1 - p_2 < 0.05 + (2.58)(0.0802)) = 0.99$$

$$P(-0.157 < p_1 - p_2 < 0.257) = 0.99 \qquad E = 0.207.$$

There is a 0.99 probability that the difference of the two unknown population proportions is between −0.157 and 0.257. Since zero is included in the interval, we can conclude that the difference between the two unknown population proportions is likely equal to zero.

The meaning and interpretation of the confidence interval for the difference of two population proportions is very similar to that of the difference between two population means. Its width depends on three factors that we can control: the level of confidence and the two sample sizes, n_1 and n_2. Although a sample size equation can be derived from the margin of error, it is more complicated than the ones used for single means and proportions. Therefore, it is not discussed in this book.

8.7 Estimating the Variance

The point estimator of the unknown population variance, σ^2, is the statistic, S^2, and the point estimate is the numeric value of the sample variance, s^2.

An interval estimator of the unknown population variance can be derived from the sampling distribution of the sample variances (see Chapter 7) and its standardized form, the χ^2 (chi-squared) distribution (Fig. 8.4).

Using the notation in Fig. 8.4, the confidence interval for a standard χ^2 variable is:

$$P\left(\chi^2_{1-\alpha/2(n-1)} < \chi^2 < \chi^2_{\alpha/2(n-1)}\right) = 1 - \alpha.$$

From Chapter 7 (see Example 8.12), we know that S can be transformed into a standard χ^2 random variable with the following equation:

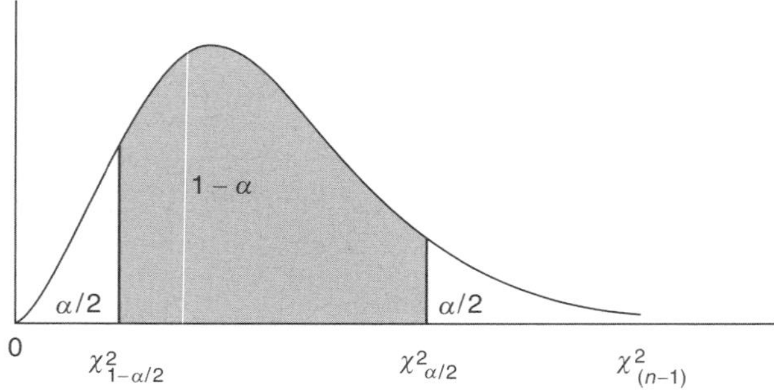

Fig. 8.4. The χ^2 distribution.

$$\chi^2 = \frac{(n-1)S^2}{\sigma^2}.$$

hence, substituting for χ^2, we have:

$$P\left(\chi^2_{1-\alpha/2(n-1)} < \frac{(n-1)S^2}{\sigma^2} < \chi^2_{\alpha/2(n-1)}\right) = 1-\alpha.$$

Rearranging the inequality inside the parentheses, we have:

$$P\left[\frac{(n-1)S^2}{\chi^2_{\alpha/2(n-1)}} < \sigma^2 < \frac{(n-1)S^2}{\chi^2_{1-\alpha/2(n-1)}}\right] = 1-\alpha$$

Replacing the estimator S^2 with the sample variance s^2, the final equation is:

$$P\left[\frac{(n-1)s^2}{\chi^2_{\alpha/2(n-1)}} < \sigma^2 < \frac{(n-1)s^2}{\chi^2_{1-\alpha/2(n-1)}}\right] = 1-\alpha. \tag{8.14}$$

Example 8.12. The sample standard deviation of the 16 dbh measurements described in Example 8.2 is 2.09 ($s^2 \approx 4.37$). Find the 90% and 95% confidence intervals for the unknown population variance.

$s^2 \approx 4.37$ $\quad \chi^2_{0.05(15)} = 25.0 \quad \chi^2_{0.95(15)} = 7.26$

$\chi^2_{0.025(15)} = 27.5 \quad \chi^2_{0.975(15)} = 6.26$

$$P\left[\frac{(16-1)4.37}{25.0} < \sigma^2 < \frac{(16-1)4.37}{7.26}\right] = 0.90$$

$$P(2.62 < \sigma^2 < 9.03) = 0.90$$

There is a 0.90 probability that the unknown population variance is between 2.62 and 9.03.

$$P\left[\frac{(16-1)4.37}{27.5} < \sigma^2 < \frac{(16-1)4.37}{6.26}\right] = 0.95$$

$$P(2.38 < \sigma^2 < 10.47) = 0.95$$

There is a 0.95 probability that the unknown population variance is between 2.38 and 10.47.

The meaning of the confidence interval is that there is a $(1 - \alpha)$ probability that the unknown population variance is contained in the interval between the LCL and UCL. In other words, if the sample variance is calculated from repeated samples of size n from a population with an unknown variance, σ^2, and the $(1 - \alpha)100\%$ confidence interval is calculated using Eqn 8.14, then $(1 - \alpha)100\%$ of the intervals will contain σ^2.

Since the χ^2 distribution is not symmetric like the Z or t distributions, the margin of error and the sample size calculation for estimating the population variance are not readily available and will not be discussed in this book.

8.8 Estimating the Ratio of Two Variances

The point estimator of the ratio of two unknown population variances, σ_1^2/σ_2^2, is the statistic, S_1^2/S_2^2. The point estimate is the numeric value of the ratio of two sample variances, s_1^2/s_2^2, based on n_1 and n_2 samples from the two populations, respectively.

To derive the confidence interval for σ_1^2/σ_2^2, we use the sampling distribution of the ratio of two variances and its standardized form (Fig. 8.5), the F distribution (see Chapter 7).

Using the notation in Fig. 8.5, the confidence interval for a standard F variable is:

$$P\left[F_{1-\alpha/2(v_1,v_2)} < F < F_{\alpha/2(v_1,v_2)}\right] = 1 - \alpha.$$

From Chapter 7, we know that S_1^2/S_2^2 can be transformed into a standard F random variable with the following equation where:

$$F_{(v_1,v_2)} = \frac{S_1^2}{S_2^2}\frac{\sigma_2^2}{\sigma_1^2}.$$

Hence, substituting for F and rearranging the inequality inside the parentheses we have:

$$P\left[\frac{S_1^2}{S_2^2}\frac{1}{F_{\alpha/2(v_1,v_2)}} < \frac{\sigma_1^2}{\sigma_2^2} < \frac{S_1^2}{S_2^2}\frac{1}{F_{1-\alpha/2(v_1,v_2)}}\right] = 1 - \alpha.$$

To simplify the table look-up, we substitute:

$$F_{1-\alpha/2(v_1,v_2)} = \frac{1}{F_{\alpha/2(v_2,v_1)}}$$

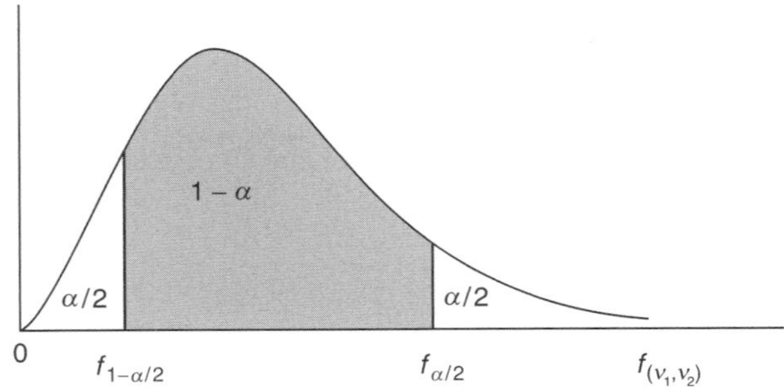

Fig. 8.5. The *F* distribution.

Thus, we get:

$$P\left[\frac{S_1^2}{S_2^2}\frac{1}{F_{\alpha/2(v_1,v_2)}} < \frac{\sigma_1^2}{\sigma_2^2} < \frac{S_1^2}{S_2^2}F_{\alpha/2(v_2,v_1)}\right] = 1-\alpha.$$

Replacing the estimator, S_1^2/S_2^2, with the ratio of two sample variances, s_1^2/s_2^2 the final equation is:

$$P\left[\frac{s_1^2}{s_2^2}\frac{1}{F_{\alpha/2(v_1,v_2)}} < \frac{\sigma_1^2}{\sigma_2^2} < \frac{s_1^2}{s_2^2}F_{\alpha/2(v_2,v_1)}\right] = 1-\alpha \tag{8.15}$$

Example 8.13. In Example 8.8, the variances of log volumes from two lots were $s_1^2 \approx 0.314$ and $s_2^2 \approx 0.372$, based on $n_1 = 12$ and $n_2 = 8$ observations. Find the 95% confidence interval for the ratio of the two unknown population variances, σ_1^2/σ_2^2.

$n_1 = 12$ $s_1^2 = 0.314$ $F_{0.025(7,\ 11)} = 3.76$

$n_2 = 8$ $s_2^2 = 0.372$ $F_{0.025(11,\ 7)} = 4.72$ (by interpolation)

$$P\left[\frac{0.314}{0.372}\frac{1}{4.72} < \frac{\sigma_1^2}{\sigma_2^2} < \frac{0.314}{0.372}3.76\right] = 0.95$$

$$P\left[0.179 < \frac{\sigma_1^2}{\sigma_2^2} < 3.174\right] = 0.95$$

There is a 0.95 probability that the ratio of the two unknown population variances is between 0.179 and 3.174. Since 1.0 is included in the interval, we can conclude that $\sigma_1^2 = \sigma_2^2$ with a probability 0.95 and that the assumption that we made regarding the equality of variances in Example 8.8 is correct. Note that, to make this determination, we relate the interval to 1 instead of 0 (as in the previous cases), because we are now dealing with a ratio and not differences.

Example 8.14. In Example 8.9, the variances of biomass production under two types of lighting were $s_1^2 = 1.44$ and $s_2^2 = 0.25$, based on $n_1 = 12$ and $n_2 = 10$ observations. Find the 90% and 95% confidence intervals for the ratio of the two unknown population variances, σ_1^2/σ_2^2.

$n_1 = 12$ $s_1^2 = 1.44$ $F_{0.025(9,\ 11)} = 3.59$ $F_{0.025(11,\ 9)} = 3.82$ (by interpolation)

$n_2 = 10$ $s_2^2 = 0.25$ $F_{0.05(9,\ 11)} = 2.90$ $F_{0.05(11,\ 9)} = 3.10$ (by interpolation)

$$P\left(\frac{1.44}{0.25}\frac{1}{3.82} < \frac{\sigma_1^2}{\sigma_2^2} < \frac{1.44}{0.25}\ 3.59\right) = 0.95$$

$$P\left(1.508 < \frac{\sigma_1^2}{\sigma_2^2} < 20.678\right) = 0.95$$

There is a 0.95 probability that the ratio of the two unknown population variances is between 1.508 and 20.678.

$$P\left(\frac{1.44}{0.25}\frac{1}{3.10} < \frac{\sigma_1^2}{\sigma_2^2} < \frac{1.44}{0.25}\ 2.90\right) = 0.90$$

$$P\left(1.858 < \frac{\sigma_1^2}{\sigma_2^2} < 16.704\right) = 0.90$$

There is a 0.90 probability that the ratio of the two unknown population variances is between 1.858 and 16.704. Since 1.0 is not included in either the 95% or 90% confidence intervals, we can conclude that $\sigma_1^2 \neq \sigma_2^2$ with a probability of 0.95 (or 0.90) and that the assumption that we made regarding the inequality of variances in Example 8.9 is correct. Note that the 95% interval is wider than the 90% interval, which means we have more confidence, but have traded this off against a larger range.

The meaning of the confidence interval is that there is a $(1 - \alpha)$ probability that the ratio of the two unknown population variances, σ_1^2/σ_2^2, is contained in the interval between the LCL and UCL. In other words, if the sample variances are calculated from repeated samples of sizes n_1 and n_2 from two populations with unknown variances, and the $(1 - \alpha)100\%$ confidence interval is calculated from each ratio, $(1 - \alpha)100\%$ of these intervals will contain the real ratio, σ_1^2/σ_2^2.

Like the χ^2 distribution, the F distribution is not symmetric as in the cases of the Z or t distributions. Therefore, the margin of error and the sample size calculations are not readily available and will not be discussed in this book.

Exercises

Section 8.1

8.1. From the sampling distribution constructed in Exercise 7.5 (see Chapter 7), show that $\bar{X}$ is an unbiased estimator of μ.

8.2. From the information given in Exercise 7.5 (see Chapter 7), show that S^2 is an unbiased estimator of σ^2.

8.3. If s'^2 is calculated as:

$$s'^2 = \frac{\sum_{i=1}^{n} \left(x_i - \bar{x}\right)^2}{n}$$

use the procedure in Exercise 8.2 to show that $E(s'^2) \neq \sigma^2$. What can you conclude from this?

8.4. Repeat Exercise 8.2 for the standard deviation, S. Can you conclude that S is an unbiased estimator of σ?

Section 8.2

8.5. A certain type of light bulb has a normally distributed life length with a known standard deviation of 45 hours. A test of 25 randomly selected bulbs resulted in a mean life length of 975 hours.

a. Calculate the 95% and 99% confidence intervals for the unknown population mean and compare the two intervals.
b. Calculate and compare the margins of error in Exercise 8.5a.
c. How large a sample is required if we want to be 95% confident that the sample mean will be within 15 h of the true mean?
d. Assume that the sample size of light bulbs is increased from 25 to 50. Find the 95% confidence interval and compare it to the 95% confidence interval found in Exercise 8.5a.

8.6. Prices per thousand board feet of spruce–pine–fir lumber are normally distributed. A random sample of 15 prices was taken from the past year:

492	621	521	561	518
571	594	629	603	608
538	562	546	532	576

a. Find the 90% and 95% confidence intervals for the unknown population mean of the price per thousand board feet. Compare the two intervals.
b. Calculate the margins of error in the two intervals calculated in Exercise 8.6a.
c. How large a sample should we take to be 99% confident that the sample mean will not differ from the true mean by more than US$10?

8.7. The population of diameter at breast height (dbh) measurements in a certain forest stand is approximately normal. The following data are dbh measurements (in cm) from randomly selected trees in this stand:

25.6, 28.9, 19.8, 36.9, 40.9, 33.2, 30.1, 40.7, 35.5, 27.3, 23.7, 29.0

a. Find the 95% confidence interval for the mean dbh.
b. How large a sample should be taken to be 95% confident that the sample mean will not differ from the population mean by more than 3.0 cm?
c. Repeat the calculation in Exercise 8.7b for the case when 99% confidence is required and compare your results.

Section 8.3

8.8. A random sample of 80 independent wood specimens was treated with a fire-retardant chemical and the treatment was effective on 68 of them.

a. Find the 95% and 98% confidence intervals for the unknown population proportion of effective treatments of the chemical. Compare the two intervals.
b. How large a sample should you take if you want to be 95% certain that the sample proportion of effective treatments will be within 0.06 of the true proportion?

8.9. In a Douglas-fir plantation, 78 of 105 randomly selected seedlings survived 1 year after planting.

a. Find the 90% and 95% confidence intervals for the unknown population proportion of survival. Compare your results.
b. Find the sample size needed if we want to be 95% confident that the sample proportion will be within 0.04 of the real proportion.

Section 8.4

8.10. A logging company is trying to decide whether to purchase brand A or brand B tyres for their trucks. The population standard deviations supplied by the manufacturer are 8500 km for brand A and 9800 km for brand B. The logging company wishes to compare the two brands and conducts an experiment based on prior records. They look at the service lives (in km) of 12 tyres from brand A and 18 tyres from brand B and find the sample means to be 68,543 km and 61,230 km, respectively. Find the 95% and 99% confidence intervals for the difference of the unknown population means, $(\mu_A - \mu_B)$, and compare the two intervals. Draw some practical conclusions from the results.

8.11. A study was conducted on profits derived from the primary and secondary wood products sectors. The results (in millions of dollars) are summarized below:

	n	$\sum_{i=1}^{n} x_i$	$\sum_{i=1}^{n} x_i^2$
Primary sector	10	31.0	103.3
Secondary sector	9	47.7	262.5

a. Find the 95% and 99% confidence intervals for the difference of the two unknown population means. Assume that $\sigma^2_1 = \sigma^2_2$ and that the populations are approximately normally distributed. Compare the two intervals. *Hint: data from the table above need to be manipulated to solve this problem.*
b. Draw some practical conclusions from the results in Exercise 8.11a.
c. How large a sample must be taken (assume $n_1 = n_2$) from each of the two populations if we want to be 95% confident that the difference of the two sample means will not differ by more than US$1,000,000 from the true difference of the two population means?

8.12. The tensile strengths of two types of commercial fishing lines were compared. The mean tensile strength of brand A was 25.5 kg with a standard deviation of 3.6 kg based on 45 samples, while 50 samples from brand B had a mean tensile strength of 28.9 kg with a standard deviation of 4.7 kg. Find the 90% and 95% confidence intervals for the difference of the two unknown population means. Draw some conclusions.

8.13. To test the effect of controlled grazing versus continuous grazing, 16 and 15 steers were subjected to each of these treatments, respectively. Weight gains (in kg) for each animal are given below. Assume that $\sigma^2_1 \neq \sigma^2_2$ and that the populations are approximately normally distributed. Find the 95% and 99% confidence intervals for the difference of the two unknown population means and draw some conclusions.

Controlled grazing				
130	120	61	111	93
56	85	128	73	56
65	71	109	122	85
131				

Continuous grazing				
44	62	77	58	88
61	42	57	70	38
66	82	81	54	81

8.14. The height increments (in cm) of 8 spruce trees were measured at the end of the growing season in 2002 and in 2003, as shown below. Assume that the two populations are approximately normally distributed. Find the 95% confidence interval for the difference of the two unknown population means and draw some conclusions.

Tree No.	2002	2003
1	31.0	26.3
2	29.5	25.1
3	28.7	24.9
4	30.5	25.9
5	27.0	23.3
6	32.3	26.7
7	27.7	24.2
8	29.9	26.5

Section 8.5

8.15. Two lots of logs in the storage area of a sawmill are infested by bark beetles. In random samples of 45 logs from Lot 1 and 40 logs from Lot 2, 10 and 12 logs were infested, respectively. Find the 95% and 99% confidence intervals for the difference of the two unknown population proportions. Draw some conclusions.

8.16. Regeneration surveys were performed in two areas, with 80 and 110 randomly selected plots, respectively. In the first area, 75% of the plots showed satisfactory regeneration, while in the second, 80% of the plots showed satisfactory regeneration.

a. Construct 90% and 95% confidence intervals for the difference of the two unknown population proportions. Draw some conclusions.
b. How large a sample should we take (assume $n_1 = n_2$) from each of the two binomial populations to be 95% confident that the difference of the two sample percentages does not vary by more than 10% from the true difference?

Section 8.6

8.17. Find the 95% and 99% confidence intervals for the population variance of dbh measurements in Exercise 8.7. Draw some conclusions.

8.18. Construct the 90% and 95% confidence intervals for the population variance of the lumber price measurements in Exercise 8.6. Draw some conclusions.

8.19. A study was carried out to investigate the variation of rainbow trout weights in a certain creek. The weights (in kg) of 10 randomly selected fish are listed below:

0.78, 0.45, 0.35, 0.76, 0.57, 0.42, 0.33, 0.68, 0.66, 0.42

Assume that the population is approximately normally distributed. Find the 95% confidence interval for the unknown population variance. Draw some conclusions.

Section 8.7

8.20. Construct 95% and 99% confidence intervals for the ratio of the two unknown population variances for the primary and secondary wood products sectors, as described in Exercise 8.11. Draw some conclusions. Was the assumption of equal variances justified?

8.21. Find the 90% and 95% confidence intervals for the ratio of the two unknown population variances for the weights of steers given controlled versus continuous grazing treatments, as described in Exercise 8.13. Draw some conclusions. Was the assumption of unequal variances justified?

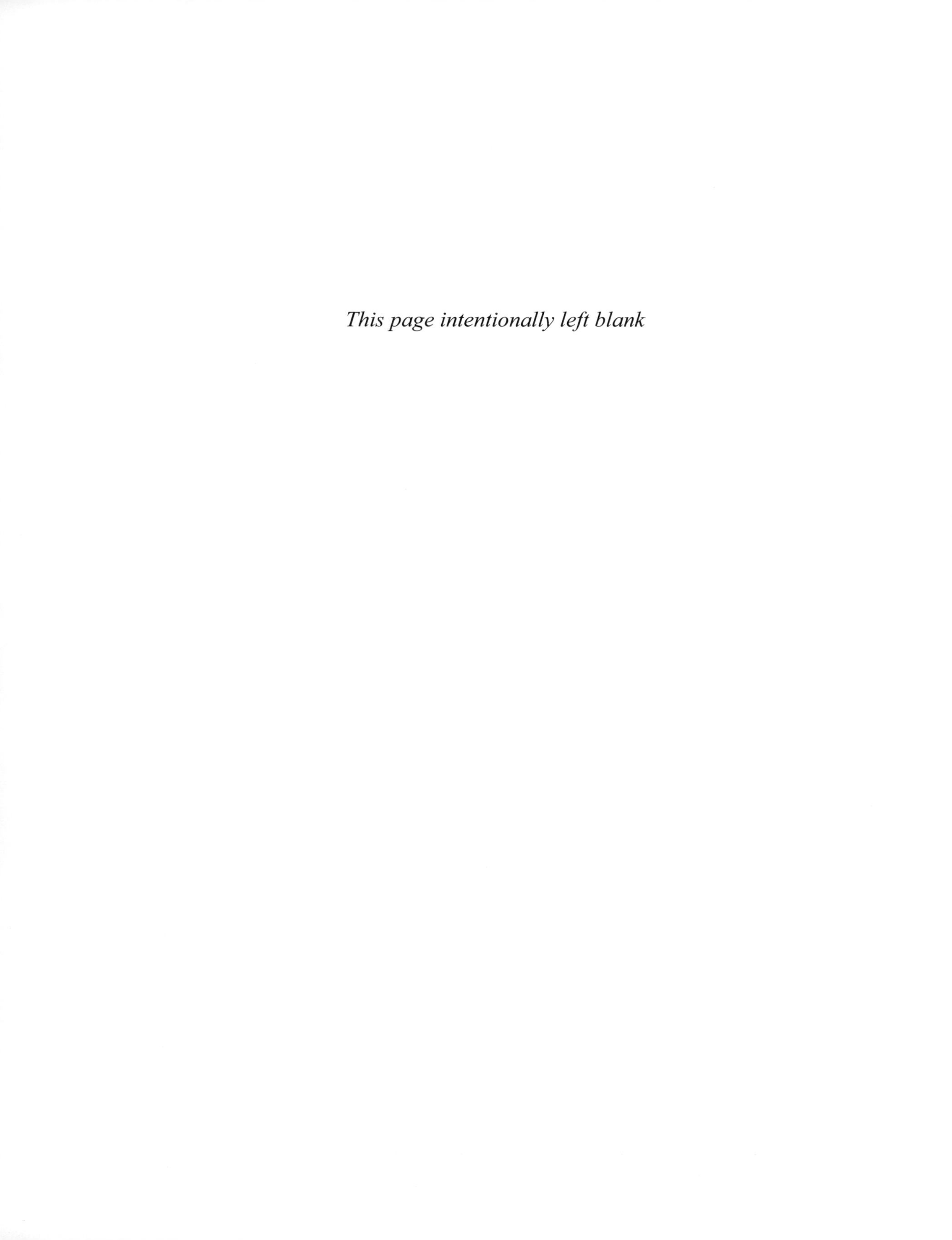

This page intentionally left blank

9 Tests of Hypotheses

Making Claims about Population Parameters

In the previous chapter (Chapter 8), we discussed procedures to estimate unknown population parameters based on a sample from the population. In this chapter, we will learn how to decide whether a statement or claim made about a parameter or a certain characteristic of a population is plausible, based on some sample data from the population. For example, a manufacturer might claim that the mean life expectancy of their chainsaws is at least 5 years. This claim is a **hypothesis** about the mean of the population of the chainsaws they produce. Sample data can be collected from this population to **test** this claim. This is called **hypothesis testing**, and it is one of the most commonly used procedures in applied statistics. We will introduce the general concept of hypothesis testing and apply this theory to various tests concerning means, proportions, differences between two means, differences between two proportions, variances and ratios of two variances.

9.1 Statistical Hypothesis and Test Procedures

A **statistical hypothesis** is a claim or a statement about some characteristic of a population. For example, a silviculturist may claim that the average height of a young Douglas-fir stand is 11.5 m. A park manager may say that 80% of the visitors in a national park in Alberta are Canadian. A wood scientist may claim that the specific gravity measurements of dry western hemlock wood are normally distributed, or that there is a simple linear relationship between the specific gravity and the modulus of rapture in Sitka spruce wood specimens. These are all claims about characteristics of populations. Some are concerned with a parameter – for example, means or proportions – while others are concerned with the distribution of the measurements or the relationship between two (and possibly more) random variables.

Whether a statistical hypothesis is false or true will never be known for certain unless all elements of the population are examined. In many cases, this is impossible or, more often than not, simply impractical. Thus, the decision about whether a statistical hypothesis is true or false must be based on a sample from the population. In decision making or statistical testing, we always have two contradictory hypotheses. One of these is the **null hypothesis**, denoted by H_0, which is a statement about a characteristic of the population assumed to be true. The other one, which is contradictory to the null hypothesis, is called the **alternative hypothesis**, and is denoted by H_1.

In common parlance of science, if H_0 is *rejected*, we *accept* H_1. On the other hand, because the truth of the null hypothesis is not known to us unless we examine the entire population, we can *never really* accept H_0. Nonetheless, as in many other books, the term 'accept H_0' will be used here to connote that the statement being made is, in fact, true. However, it should be noted that acceptance of H_0 implies that we do not have enough evidence to reject it. Thus, in many statistical texts, the phrase, 'we do not reject H_0' is used in place of 'accept H_0'.

The above situation is analogous to a criminal trial. In most judicial systems, a defendant is innocent until proven guilty. In other words, a hypothesis for a criminal trial can be written as:

H_0: the defendant is innocent

H_1: the defendant is guilty.

Sample information in the form of evidence is provided during the trial. If it is consistent with the assumption of innocence, H_0 cannot be rejected. A lack of evidence, however, does not necessarily prove that the defendant is innocent and therefore H_0 can never truly be accepted. On the other hand, if the evidence provided is inconsistent with the assumption of innocence, H_0 is rejected and H_1 is accepted; the defendant is convicted.

We will demonstrate the general process used in testing statistical hypotheses with an example. Suppose that a manufacturer claims that the life expectancy of its brand of 9-volt batteries is 240 days, with a standard deviation of 30 days. These numbers are the manufacturer's specifications (or claims) and are assumed to be the population mean, μ, and population standard deviation, σ. It is also assumed for the purposes of this example that the life expectancy of these batteries is approximately normally distributed. Let us assume that we have enough of these batteries installed in smoke alarms at our plant that we can test the manufacturer's claim regarding the life expectancy. Our H_0 and H_1 will be:

H_0: $\mu = 240$ (the manufacturer's claim)

H_1: $\mu < 240$.

In order to carry out the test properly, H_0 must always be equality; that is, a population parameter must be set equal to a hypothetical or assumed constant. H_1 is always an inequality and can take one of three different forms: $\mu < 240$, or $\mu > 240$, or $\mu \neq 240$. Note that only one of these alternative hypotheses can be used in a single test, but we selected the first one because, as buyers of the batteries, we would prefer to reject H_0 if the chosen sample indicates that the mean life expectancy is considerably less than 240 days. In other words, practically speaking, we would not really mind if the average battery life exceeded 240 days, so we do not test this. On the other hand, if the manufacturer were to carry out this test, the second alternative hypothesis would be selected ($\mu > 240$) because they would prefer to reject H_0 if the sample indicates that the mean is considerably more than 240 days. If the batteries lasted for significantly longer than 240 days, this would indicate that the product is considerably better than the specifications say and that the batteries potentially could sell for a higher price. The first two hypotheses tests are called **one-tailed** tests and the last is called a **two-tailed** test. Each will be discussed in further detail later on.

As stated above, the alternative hypothesis is always an inequality, while the null hypothesis is always an equality. This allows us to construct the sampling distribution with a fixed centre of the statistic being tested (Fig. 9.1).

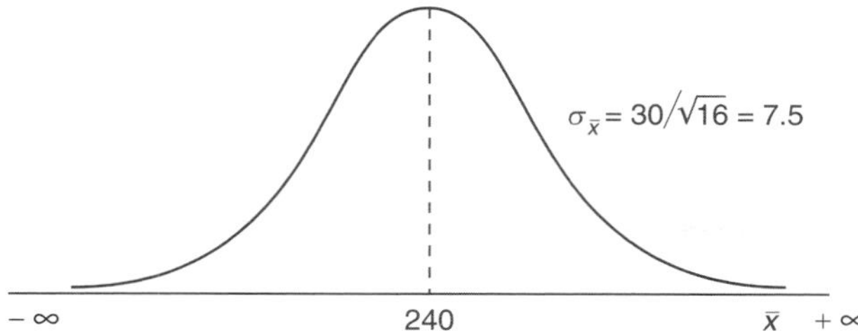

Fig. 9.1. Sampling distribution of the means of battery lives based on 16 observations.

Assume that a sample of 16 batteries was selected and tested to observe their life expectancy. After calculating the sample mean from the 16 observations, the question is: 'How low a value do we need to observe to rightfully reject our H_0?' An examination of the sampling distribution of all possible sample means based on 16 observations from a population with a mean $\mu = 240$ and $\sigma = 30$ (Fig. 9.1) does not answer our question because, in theory, the sample mean (in a sampling distribution) is distributed such that it can take on any value between $-\infty$ and $+\infty$. Yet, we need to make a decision. To do so, an artificial or arbitrary value between $-\infty$ and 240 (the centre of the distribution) is selected, below which the null hypothesis is rejected. This artificial value is called the **critical value.** The region between $-\infty$ and the critical value is called the **critical region**, or **rejection region**, and the region between the critical value and $+\infty$ is called the **acceptance region**, or in other texts, the **non-critical region** (Fig. 9.2). It should be clear from Fig. 9.2 that when H_0 is rejected, H_1 is accepted, and a certain amount of error is committed (since the mean can actually take on any value in this distribution). This error is equal to the area under the curve in the rejection region (Fig. 9.2) and is called **type I error.**

Type I error is the probability of rejecting H_0 when it is true.

The size of type I error is arbitrary in so much as it is selected by the person carrying out the statistical hypothesis test (or decision). It is oftentimes referred to as the **level of significance**, or **significance level**, and statisticians generally agree on using 0.1, 0.05 or 0.01 (more on this decision later). However, if desired, any other level can be used.

While making a decision about the rejection or acceptance of H_0, it is possible to commit another type of error, known as **type II error.**

Type II error is the probability of accepting H_0 when it is false.

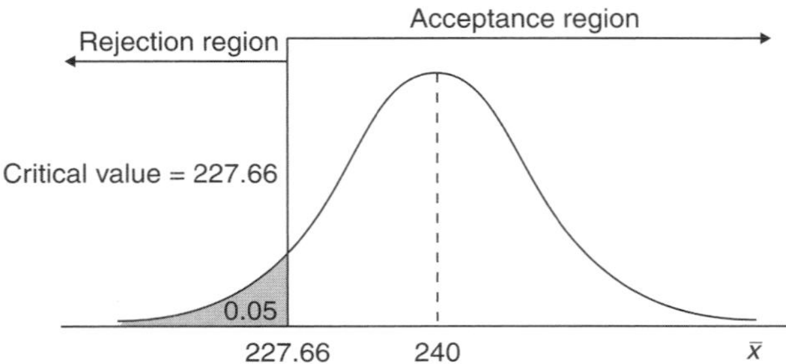

Fig. 9.2. Acceptance and rejection regions for $\alpha = 0.05$, one-tailed test.

To explain type II error, we must make an assumption about knowing the real population mean (in reality, this would never be the case). For instance, assume that the real mean is not 240 (μ_0) days as the battery manufacturer claims, but is actually only 220 (μ_1) days, with the same standard deviation of 30 days (false advertising!). With this information in hand, we can construct another sampling distribution centred at 220. From Fig. 9.3, it can be seen that a part of this new, 'true' sampling distribution overlaps with the acceptance region of H_0 to the right of the critical value in the original sampling distribution, centred at 240. The probability between the critical value and $+\infty$ (using the true sampling distribution) is about 0.1539, which is calculated as follows:

The standard error of the mean of the sampling distribution (Fig. 9.1) is:

$$\sigma_{\bar{x}} = 30/\sqrt{16} = 7.5$$

Using a significance level of $\alpha = 0.05$, the $z = -1.645$ value needs to be converted to the units of the mean, x, where 240 is the claimed population mean and 7.5 is the standard error of the mean:

$$-1.645 = \frac{x - 240}{7.5},$$

$x \approx 227.66$ is the critical value (Fig. 9.2).

Now, we can calculate β, the probability of getting a higher value than x, assuming we know the real value of $\mu = 220$:

$$P\left(\overline{X} > 227.66\right) = p\left(Z > \frac{227.66 - 220}{7.5}\right)$$

$$P\left(Z > 1.02\right) \approx 0.1539,$$

from Table A.3 (see Appendix A).

If 240 days is, in fact, the true mean and the sample of 16 batteries yields a sample mean of 225.2 days, we would reject H_0 and accept H_1 because the value falls in the critical region. Thus, we would be committing type I error. On the other hand, if 220 days is, in fact, the true mean and the sample results in a mean value of 234.9 days, we would 'accept' H_0 (or we would say that we do not have enough evidence to reject it). Thus, we would be committing type II error. In either case, there is a probability that our decision may lead to the wrong conclusion (see Fig. 9.4).

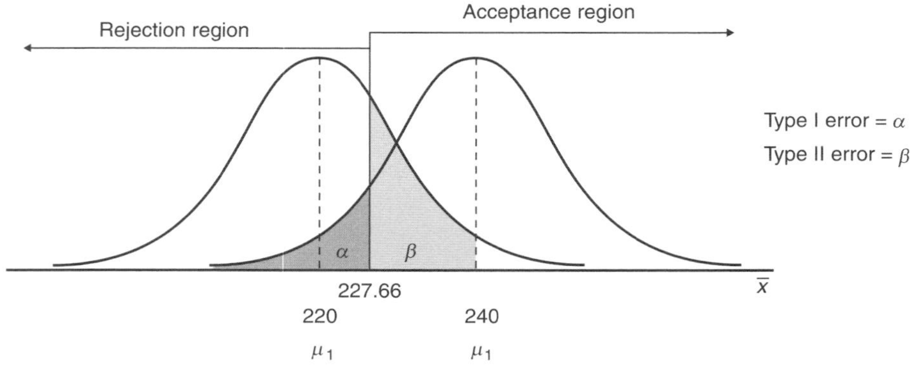

Fig. 9.3. Type I and Type II errors.

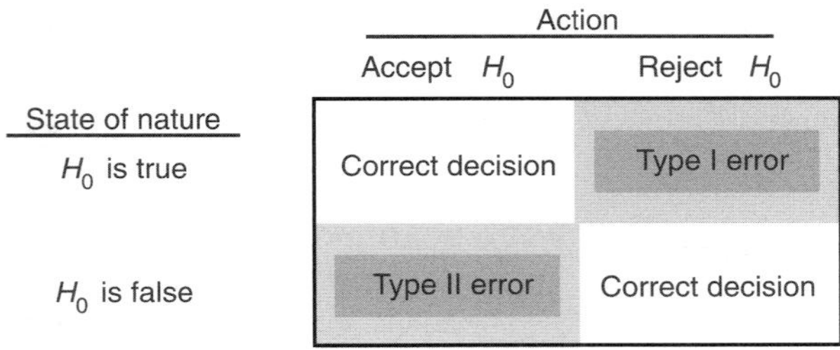

Action
Accept H_0
Reject H_0
State of nature
H_0 is true
Correct decision
Type I error
H_0 is false
Type II error
Correct decision

Fig. 9.4. The nature of Type I and Type II errors.

Type I error (or level of significance) is usually denoted by α, while type II error is denoted by β (see Fig. 9.3), and the value of $1-\beta$ is called the **power of the test.** While the value of α is decided on by a researcher or statistician, the value of β is rarely known to us because its value depends on knowledge that we generally do not possess: the value of an alternative hypothesis (220 days in our example). Even though the value of β is not available in most cases, we should at least be aware of the factors that affect the size of β error:

1. α: if we reduce the size of α, the value of β increases. This statement can be verified with Fig. 9.3 as a shift of the critical value to the left would increase the value of β. This trade-off of errors is the reason why reduction of type I error to a very small level is not advisable.
2. n: assuming a constant distance between μ_1 and μ_0, an increase of the sample size, n, decreases the standard error of the mean ($\sigma_{\bar{x}}$), or the spread of both sampling distributions (the ones around μ_0 and μ_1). This will reduce the overlap between the two sampling distributions. An increase of sample size reduces the area on the right hand side of the critical value of the sampling distribution around μ_1, and therefore reduces β.
3. Distance between μ_1 and μ_0: from Fig. 9.3, we see that, as this distance increases, the value of β decreases.

While the last factor cannot be controlled by the decision maker, the decisions regarding the first two play a very important role in statistical hypothesis testing.

Example 9.1.

a. For the above example, calculate type II error using $\alpha = 0.01$.

If $\alpha = 0.01$, we use $z_{0.01} = -2.33$. Then, the critical value can be calculated as:

$$-2.33 = \frac{x - 240}{7.5}$$

This gives $x = 222.525$. To calculate β, we must find the probability of getting a sample mean $> x$:

$$P(\overline{X} > 222.525) = P\left(Z > \frac{222.525 - 220}{7.5}\right)$$

$$P(Z > 0.34) \approx 0.3669$$

This tells us that the β error is more than double (from 0.1537 to 0.3669) when α is changed from 0.05 to 0.01.

b. Repeat the above calculation of type II error for $\alpha = 0.05$ and $n = 100$. Note that the standard error of the mean will be different.

If $n = 100$ instead of 16 and α remains at 0.05, the critical value is $-1.645 = \frac{x - 240}{30/\sqrt{100}}$.

From here, $x = 235.065$. Note that as we increase the sample size from 16 to 100, the standard error of the mean reduces from 7.5 to 3.0.

Calculation of β:

$$P(\bar{X} > 235.065) = P\left(Z > \frac{235.065 - 220}{3.0}\right)$$

$$P(Z > 5.02) \approx 0.0.$$

Thus, with an increased sample size, the β error of 0.1539 is reduced to approximately zero. In other words, the overlap of the two sampling distributions becomes negligible.

The results in Example 9.1 show that a change of sample size and a change of α affect the size of β. In some statistics books, the values of β are calculated for a range of values of μ_1, n and α, and their plotted form is called the **operating characteristic**, or (**OC**) **curve**, a useful tool for understanding how these factors interact. OC curves are not widely used in natural resources and forestry applications; therefore, we will not discuss their construction in this book.

As stated earlier, the alternative hypothesis can take on one of the three forms:

$$H_1: \mu < 240, \text{ or } \quad \mu > 240, \text{ or } \quad \mu \neq 240.$$

In the first case, the entire critical region (α) lies in the left tail of the sampling distribution, while for the second case, the entire critical region (α) lies in the right tail. For the third case, the critical region is divided into two equal parts ($\alpha/2$) and they lie in each tail of the sampling distribution (Fig. 9.5).

In practice, testing is not carried out in the units of the parameter (see Fig. 9.2), as it is faster and more convenient to convert the estimate (in this case, the sample mean) into z-, t-, χ^2- or F-values and then make the decision regarding H_0 using Z, t, χ^2 or F distributions (Fig. 9.6). In our example, the sample mean of 225.2 would be converted to z-values, such that the equivalent z-value for 225.2 is: $z = \frac{225.2 - 240}{7.5} \approx -1.97$.

This value would then be compared to the standard z-value corresponding to the α level. For $\alpha = 0.05$, this value is $z_{0.05} = -1.645$. Therefore, we would reject H_0 since -1.97 is in the rejection region. This calculated z-value (-1.97) above is called a **test statistic.**

The following is a step-by-step guide for formulating and testing statistical hypotheses. We recommend that students follow these steps, especially as they are learning how to conduct hypothesis tests:

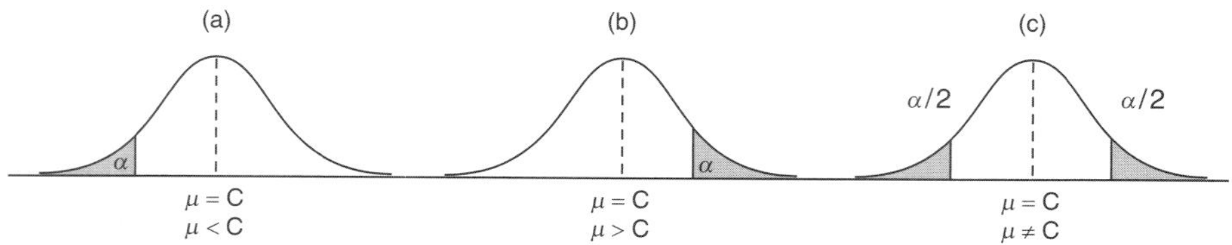

Fig. 9.5. Critical regions for (a) and (b) one-tailed and (c) two-tailed tests.

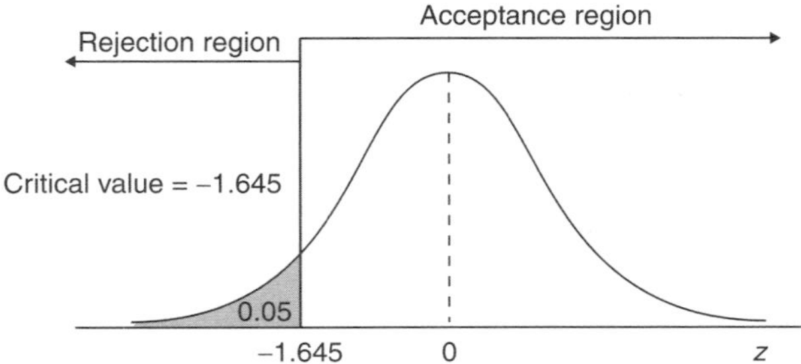

Fig. 9.6. Critical regions in *z* units, one-tailed test.

1. Formulate H_0. This must always be an *equality* statement.
2. Formulate H_1. Use only *one* of the three possible inequalities.
3. Set level of significance. One of 0.1, 0.05 or 0.01 is recommended.
4. Select the equation to be used to calculate the test statistic (e.g. we used Z in our example; others will be introduced in the following sections).
5. Take sample(s) and calculate estimate(s), if not already provided.
6. Find the critical value(s). This will be a z-, t-, χ^2- or F-value(s).
7. Compute the test statistic.
8. Make a decision regarding H_0. Usually, this is more easily understood with a picture.
9. Draw some conclusions.

In the following sections, we will present the most important hypotheses tests used in statistics. We will present these tests by way of some practical examples and follow the step-by-step guide presented above.

Note that several of the equations presented will contain the standard error of the statistic being tested. While these standard errors are presented without their finite population correction factors (to make computation easier), if a sample used in a given test is taken from a finite population without replacement, a correction of the standard error of the statistic is necessary.

9.2 Tests Concerning Means

In this section, we will discuss two tests concerning population means. In both tests, the unknown population mean is assumed to be equal to a constant, c, against an alternative hypothesis that takes on one of the three different forms stated in the previous section.

H_0: $\mu = c$,

H_1: $\mu \neq c$; or $\quad \mu < c$; or $\quad \mu > c$.

If the population variance is known, the test statistic is calculated as:

$$z = \frac{\overline{x} - c}{\sigma/\sqrt{n}} = \frac{\overline{x} - c}{\sigma_{\overline{X}}} \tag{9.1}$$

Since the population variance is known, the critical values can be obtained from the z-table (see Table A.3, Appendix A). Note that the last line of the t-table (corresponding to infinite degrees of freedom in Table A.5, Appendix A) is equivalent to the z-table and is often used to save time when looking up the critical z-values in hypothesis testing.

If the population variance is unknown, it must be estimated from the sample and the test statistic is then calculated as:

$$t_{(n-1)} = \frac{\bar{x} - c}{s/\sqrt{n}} = \frac{\bar{x} - c}{s_{\bar{X}}} \tag{9.2}$$

The critical values are obtained from the t-table (see Table A.5, Appendix A) with $(n - 1)$ degrees of freedom.

The use of Eqn 9.1 assumes that the sample is selected from a normal population or that the sample size is greater than 30, while the use of Eqn 9.2 assumes that the sample is selected from an approximately normal population.

Example 9.2. Suppose that, in a large plantation, the diameter at breast height (dbh) measurements are normally distributed with a standard deviation of 1.6 cm (see Example 8.1, Chapter 8). Can it be assumed that the unknown population mean is equal to 14.0 cm if the sample mean calculated from 16 measurements (taken without replacement) is 12.6 cm? Use $\alpha = 0.05$.

From the description of this problem, we do not know whether the alternative hypothesis would be accepted if the sample mean was considerably less than 14.0 cm or considerably more than 14.0 cm. In cases like these, we use a two-tailed test.

1. H_0: $\mu = 14.0$.
2. H_1: $\mu \neq 14.0$.
3. $\alpha = 0.05$.
4. Use Eqn 9.1, since σ^2 is known.
5. $n = 16$; $\bar{x} = 12.6$ cm; $\sigma_{\bar{X}} = 1.6/\sqrt{16} = 0.4$.
6. $z_{0.025} = \pm\, 1.96$.
7. $z = \dfrac{12.6 - 14.0}{0.4} = -3.5$
8. Since $-3.5 < -1.96$, the test statistic is in the rejection region and we reject H_0 and accept H_1.
9. The sample mean of 12.6 cm is *considerably different (less, since the test statistic is negative) than the assumed population mean of 14.0 cm.* Therefore, the population mean cannot be assumed equal to 14.0.

Generally, two-tailed tests with α level of significance are equivalent to $(1 - \alpha)\%$ confidence intervals described in Chapter 8. Comparing the results here to those in Example 8.2 (see Chapter 8), the values of the LCL and UCL (11.816 and 13.384) of the 95% confidence interval match the critical z-values of ±1.96 (from the above example). However, in a confidence interval, critical values are expressed in real terms or in the units of the mean. Since the assumed population mean of 14.0 cm is not included within the limits of 11.816 and 13.384, we can assume that the population mean is not equal to 14.0 cm. In other words, H_0 is rejected, verifying the result above.

Example 9.3. Suppose that in Example 9.2 above, the population standard deviation was not known, but was estimated to be 2.09 from the 16 observations (see also Example 8.2, Chapter 8). Can it be assumed that the unknown population mean is 14.0 cm if the sample mean based on 16 observations is 12.6 cm? Use $\alpha = 0.05$.

1. H_0: $\mu = 14.0$.
2. H_1: $\mu \neq 14.0$.
3. $\alpha = 0.05$.
4. Use Eqn 9.2, since σ^2 is not known.
5. $n = 16$; $\bar{x} = 12.6$ cm; $\sigma_{\bar{X}} = \dfrac{2.09}{\sqrt{16}} = 0.5225$.
6. $t_{0.025(15)}$ ± 2.13.
7. $t_{(15)} = \dfrac{12.6 - 14.0}{0.5225} \approx -2.68$.
8. Since $-2.68 < -2.13$, the test statistic is in the rejection region (Fig. 9.7) and we reject H_0 and accept H_1.
9. The sample mean of 12.6 cm is *considerably different (less) than the assumed population mean of 14.0 cm*, meaning that the population mean cannot be assumed to be equal to 14.0 cm based on our sample of 16 observations.

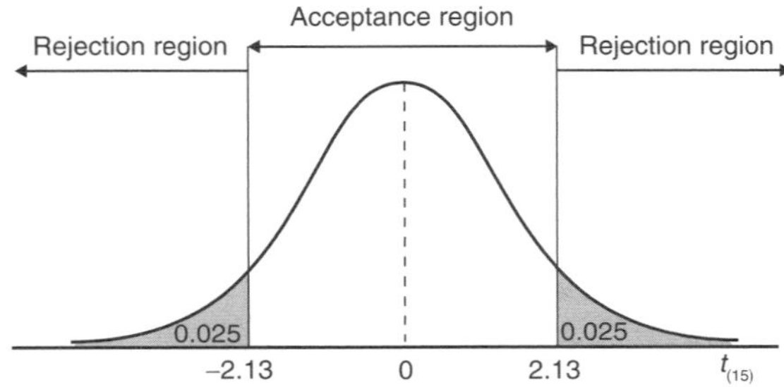

Fig. 9.7. Critical regions in *t* units, two-tailed test.

Example 9.4. We are told that a particleboard plant maintains a specific gravity of at least 0.6 for a certain brand of board. It is also known that the measurements of specific gravity of this type of particleboard are normally distributed. If a sample of 24 measurements has a mean of 0.578 and a standard deviation of 0.052, can we assume that the population specific gravity is at least 0.6? Use a $\alpha = 0.01$ level of significance.

This is a quality control problem and, since a higher specific gravity usually implies higher strength properties, it is important to maintain a certain specific gravity. This knowledge indicates that a one-tailed test is appropriate. To check the quality of production, H_0 should be rejected at the lower tail of the distribution only.

1. H_0: $\mu = 0.6$.
2. H_1: $\mu < 0.6$.
3. $\alpha = 0.01$.
4. Use Eqn 9.2, since σ^2 is not known.
5. $n = 24$; $\bar{x} = 0.578$; $s = 0.052$; $s_{\bar{X}} = \dfrac{0.052}{\sqrt{24}} = 0.0106$.
6. $t_{0.01(23)} = -2.50$.
7. $t_{(23)} = \dfrac{0.579 - 0.6}{0.0106} \approx -2.08$.
8. Since $-2.08 > -2.50$, the test statistic is in the acceptance region and we 'accept' H_0.
9. Based on our sample of 24 boards, the sample mean of 0.578 is *not significantly less than the assumed population mean of 0.6*. We can therefore assume that the population mean is, in fact, 0.6.

9.3 Tests Concerning Proportions

Tests of statistical hypotheses concerning proportions are very important in several fields of forestry and related sciences. In a manufacturing plant, a manager may be interested in the proportion of defective parts. A forester may be interested in the proportion of seedlings surviving a year after plantation. In a national park, a manager may be interested in the proportion of visitors from out of country during the summer months. And so on. As with testing of means, an assumed proportion, c, is declared in the null hypothesis as an equality and one of the three alternatives is stated in the alternative hypothesis:

H_0: $p = c$,

H_1: $p \neq c$; or $\quad p < c$; or $\quad p > c$.

For the test statistic presented here (proportions), it is required that c is not very close to 0 or 1 and n is greater than 30 (in other words, the binomial distribution approaches the normal distribution).

$$z = \frac{\hat{p} - c}{\sqrt{\frac{c(1-c)}{n}}} = \frac{\hat{p} - c}{\sigma_{\hat{p}}}, \tag{9.3}$$

or

$$z = \frac{\hat{p} - c}{\sqrt{\frac{\hat{p}\hat{q}}{n}}} = \frac{\hat{p} - c}{s_{\hat{p}}}. \tag{9.4}$$

The critical values are obtained from the z-table (see Table A.3, Appendix A) or, easier yet, from the last line of the t-table (see Table A.5, Appendix A). To compute the test statistic, one of two equations can be used, each varying in the way the standard error term is computed. If we have a good idea about what the assumed unknown proportion is (in other words, the value of $\hat{p}$ is close to c), Eqn 9.3 is recommended. Otherwise, Eqn 9.4 provides a more reliable test.

Example 9.5. It is assumed that no more than 40% of panels produced by a mill are grade B. Can this assumption be verified at the 0.05 level of significance if 27 out of 60 randomly selected panels examined were found to be grade B?

The question implies (with the statement 'no more than') that we should reject H_0 if our proportion is substantially greater than 40%. A one-tailed test should be used with the rejection region being in the upper tail of the distribution.

1. H_0: $p = 0.4$.
2. H_1: $p > 0.4$.
3. $\alpha = 0.05$.
4. Use either Eqn 9.3 or 9.4. We will use both to illustrate the difference.
5. $n = 60$; $\hat{p} = 27/60 = 0.45$; $\hat{q} = 1 - \hat{p} = 0.55$; $\sigma_{\hat{p}} = \sqrt{(0.4)(0.6)/60} \approx 0.0632$;

   $s_{\hat{p}} = \sqrt{(0.45)(0.55)/60} \approx 0.0642$.

6. $z_{0.05} = 1.645$.

7. Using the population standard error of the proportion:

$$z_1 = \frac{0.45 - 0.4}{0.0632} \approx 0.791.$$

Alternatively, using the sample standard error of the proportion

$$z_2 = \frac{0.45 - 0.4}{0.0642} = 0.778.$$

As the sample estimate of 0.45 is not very different from the assumed proportion of 0.4, the difference between the two test statistics is small and Eqn 9.3 using the population standard error term is recommended (we have shown both for demonstration purposes only).

8. Since 0.791 (or 0.778) < 1.645, the test statistic is in the acceptance region and we 'accept' H_0.

9. The sample proportion, 0.45, is *not significantly greater than the assumed population proportion of 0.4.* Thus, we can assume that the population proportion is 0.4.

Example 9.6. A politician claims that she will receive at least 45% of the votes in her riding in an upcoming election. Is it likely that her claim is correct at the 0.01 level of significance if 40 out of 110 randomly selected voters indicate that they will vote for her?

Again, this question clearly indicates that the politician's claim will be rejected at the lower tail of the distribution and a one-tailed test will be used.

1. H_0: $p = 0.45$.
2. H_1: $p < 0.45$.
3. $\alpha = 0.01$.
4. Use either Eqn 9.3 or 9.4.
5. $n = 110$; $\hat{p} = 40/110 = 0.364$; $\hat{q} = 0.636$; $\sigma_{\hat{p}} = \sqrt{(0.45)(0.55)/110} \approx 0.0474$;

   $s_{\hat{p}} = \sqrt{(0.364)(0.636)/110} \approx 0.0459$.
6. $z_{0.01} = -2.33$.
7. Using $\sigma_{\hat{p}}$ $\quad z_1 = \dfrac{0.364 - 0.45}{0.0474} \approx -1.81.$

   Using $s_{\hat{p}}$ $\quad z_2 = \dfrac{0.364 - 0.45}{0.0459} \approx -1.87.$

In this example, it can be seen that even with an almost 9% difference between the assumed proportion of 0.45 (45%) and the sample estimate of 0.364 (36.4%), the difference between the two test statistics is small. However, if the difference were large (say, 0.5), then we might reach a different conclusion.

8. Since −1.81 (or −1.87) > −2.33, the test statistic is within the acceptance region and we 'accept' H_0.

9. The sample proportion of 0.364 is *not significantly less than the declared population proportion of 0.45 (45%).* Therefore, we can assume that her claim is correct (a rarity for politicians!).

9.4 Tests Concerning Variances

Tests for variances are important in areas where knowledge of the uniformity of the population is important. For example, if the thickness of a certain kind of dimensional lumber is highly variable in a sawmill, it could indicate that the saws are not functioning according to specification. If the variation in the height of seedlings produced by a nursery exceeds a certain amount, the manager may have cause to be concerned.

In the case of testing variances, the null hypothesis states that the unknown population variance is equal to an assumed constant, c, against one of the three possible alternative hypotheses.

H_0: $\sigma^2 = c$,

H_1: $\sigma^2 \neq c$, or $\quad \sigma^2 < c$, or $\quad \sigma^2 > c$.

In order to use the following equation to calculate the test statistic, it is assumed that the population is approximately normally distributed. The critical values can then be obtained from the χ^2-table (see Table A.6, Appendix A), with $(n-1)$ degrees of freedom.

$$\chi^2_{(n-1)} = \frac{(n-1)s^2}{\sigma^2} \tag{9.5}$$

Example 9.7. The thicknesses of 20 randomly selected pieces of particleboard were measured in a mill. The standard deviation of these measurements was found to be 0.26 cm. Is there any problem with the quality of the production if we know that the population standard deviation should not exceed 0.2 cm? Use $\alpha = 0.05$.

This question indicates a one-tailed test with the rejection region at the upper tail of the χ^2 distribution. Since we do not have a test for standard deviation, we will conduct our testing in terms of variances:

1. H_0: $\sigma^2 = 0.2^2 = 0.04$.
2. H_1: $\sigma^2 > 0.04$.
3. $\alpha = 0.05$.
4. Use Eqn 9.5.
5. $n = 20$; $s = 0.26$; $s^2 = 0.0676$.
6. $\chi^2_{0.05(19)} = 30.1$.
7. $\chi^2_{(19)} = \dfrac{(20-1)0.0676}{0.04} \approx 32.11$.

8. Since 32.11 > 30.1, the test statistic is in rejection region (Fig. 9.8) and we reject H_0 and accept H_1.

9. The sample variance of 0.0676 is *significantly greater than the assumed variance of 0.04*. Thus, the product can be assumed to be more variable than the set standard and the mill manager should be concerned.

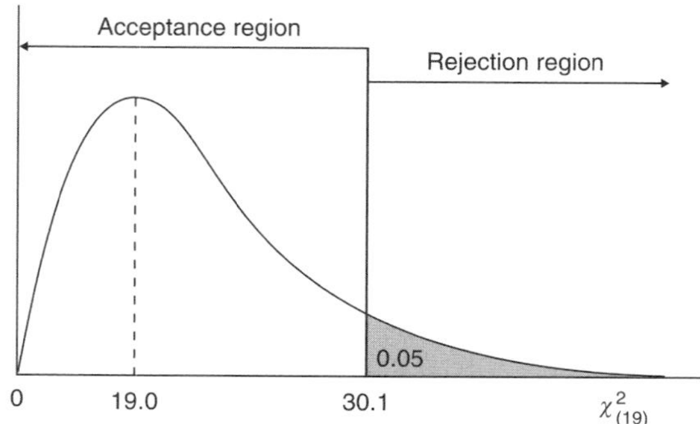

Fig. 9.8. Critical regions in χ^2 units, one-tailed test.

Example 9.8. It is assumed that the variance of the tree heights in a forest stand is 4.25 m^2. Using a 0.01 level of significance, does a variance of 5.37 m^2 obtained from measuring 17 randomly selected trees from the stand support this assumption?

From the statements above, it is not known that the rejection of H_0 is preferred at the lower or upper tail of the distribution. In cases like this, we use a two-tailed test.

1. H_0: $\sigma^2 = 4.25$.
2. H_1: $\sigma^2 \neq 4.25$.
3. $\alpha = 0.01$.
4. Use Eqn 9.5.
5. $n = 17$; $s^2 = 5.37$.
6. $\chi^2_{0.995(16)} = 34.30 \quad \chi^2_{0.005(16)} = 5.14$.
7. $\chi^2_{(16)} = \dfrac{(17-1)5.37}{4.25} \approx 20.22$.

8. Since 5.14 < 20.22 < 34.30, the test statistic is within the acceptance region and we 'accept' H_0.

9. The sample variance of 5.37 is *not significantly different from 4.25*. Therefore, the assumption that the variance equals 4.25 is reasonable based on this sample.

9.5 Tests Concerning the Difference between Two Means

Independent populations

In many situations, we are interested in comparing two population means, μ_1 and μ_2. For example, a forester would like to find out which of two fertilizers is the more effective in improving the height growth of young seedlings; a wood scientist would like to compare some strength properties in spruce and hemlock; an ecologist may want to compare the effect of two insecticides on creek water quality, to list just a few.

In testing the difference between two unknown population means, it is assumed that the difference is equal to a constant, c, against one of the usual three alternative hypotheses:

$$H_0: \mu_1 - \mu_2 = c,$$

$$H_1: \mu_1 - \mu_2 \neq c, \text{ or} \qquad \mu_1 - \mu_2 < c, \text{ or} \qquad \mu_1 - \mu_2 > c.$$

Frequently, in order to test whether the two population means are the same, the assumed constant is zero. Two different test statistics can be used to test the above hypotheses, depending on the information available about the two population variances. When the two population variances are known, the test statistic used is:

$$z = \frac{(\bar{x}_1 - \bar{x}_2) - c}{\sqrt{(\sigma_1^2/n_1) + (\sigma_2^2/n_2)}} = \frac{(\bar{x}_1 - \bar{x}_2) - c}{\sigma_{\bar{X}_1 - \bar{X}_2}} \tag{9.6}$$

The critical values are obtained from the z-table (see Table A.3, Appendix A) or from the last line of the t-table (see Table A.5, Appendix A). To use Eqn 9.6, it is assumed that either the two populations are normally distributed and/or the sample sizes used to calculate the sample means, n_1 and n_2, are both greater than 30.

Example 9.9. The actual (measured) nicotine content (in mg) of cigarettes is compared based on the following information.

	Brand A	Brand B
Mean	26.1	31.8
σ	5.2	4.7
n	24	31

Using a 0.05 level of significance, can we assume that the two unknown population means are equal? Since there is no indication whether one population mean is assumed to be greater than the other in the description of the problem, a two-tailed test is used.

1. H_0: $\mu_1 - \mu_2 = 0$.
2. H_1: $\mu_1 - \mu_2 \neq 0$.
3. $\alpha = 0.05$.
4. Use Eqn 9.6 as σ_1^2 and σ_2^2 are known.
5. $n_1 = 24$; $n_2 = 31$; $\bar{x}_1 - \bar{x}_2 = 26.1 - 31.8 = -5.7$; $\sigma_{\bar{X}_1 - \bar{X}_2} = \sqrt{\dfrac{5.2^2}{24} + \dfrac{4.7^2}{31}} \approx 1.356$
6. $z_{0.025} = \pm 1.96$.
7. $z = \dfrac{(26.1 - 31.8) - 0}{1.356} \approx -4.20$
8. Since $-4.20 < -1.96$, the test statistic is in the rejection region and we reject H_0 and accept H_1.
9. The difference between the two sample means, –5.7, is *significantly different (less because it is negative) than the assumed difference of zero.* Usually, the practical conclusion from tests like this is that the population means are assumed to be different. In other words, *26.1 mg is significantly different (less) than 31.8 mg* or *the nicotine content in Brand A is significantly less than that of Brand B.*

If the population variances are unknown, the following test statistic is used:

$$t_{(v)} = \frac{(\bar{x}_1 - \bar{x}_2) - c}{s_{\bar{X}_1 - \bar{X}_2}} \tag{9.7}$$

The critical values are obtained from the t-table (see Table A.5, Appendix A) with v degrees of freedom. When using Eqn 9.7, it is assumed that both populations are approximately normally distributed. However, this assumption can be violated slightly when both n_1 and n_2 are greater than 30.

As in finding confidence intervals for the difference between two unknown population means, we distinguish between three procedures when the population variances are unknown. For each case, the standard error of the difference between two means and the degrees of freedom are calculated differently.

Case 1. Both n_1 and $n_2 \geq 30$ (large samples).

$$s_{\bar{X}_1 - \bar{X}_2} = \sqrt{\frac{s_1^2}{n_1} + \frac{s_2^2}{n_2}} \text{ and } v = \infty.$$

Case 2. n_1 or n_2 or both < 30 (small samples) and it is assumed that σ_1^2 and σ_2^2

$$s_{\overline{X}_1-\overline{X}_2} = s_p\sqrt{\frac{1}{n_1}+\frac{1}{n_2}}$$

where $s_p = \sqrt{\dfrac{s_1^2(n_1-1)+s_2^2(n_2-1)}{n_1+n_2-2}}$ and $\nu = n_1+n_2-2$

Case 3. n_1 or n_2 or both < 30 (small samples) and it is assumed that $\sigma_1^2 \neq \sigma_2^2$

$$s_{\overline{X}_1-\overline{X}_2} = \sqrt{\frac{s_1^2}{n_1}+\frac{s_2^2}{n_2}}$$

and $\nu = \dfrac{\left[\dfrac{s_1^2}{n_1}+\dfrac{s_2^2}{n_2}\right]^2}{\dfrac{(s_1^2/n_1)^2}{n_1-1}+\dfrac{(s_2^2/n_2)^2}{n_2-1}}$ or ν = smaller of n_1-1 or n_2-1.

Example 9.10. In Example 8.7 (see Chapter 8), we presented the following data of samples taken from two lots of logs. These statistics were calculated from measurements of the top diameter of the logs.

	n	$\overline{x}$	s
Lot 1	42	15.9 cm	2.8 cm
Lot 2	36	18.6 cm	3.4 cm

Using α = 0.05, can we assume that the unknown population mean of Lot 2 exceeds the unknown population mean of Lot 1 by at least 2.0 cm?

In solving this problem, note that the constant stated is 2.0 and not zero. In problems involving differences in means (especially one-tailed problems), be very cautious of the order in which the means are presented. Here, we are interested in knowing if the mean of Lot 2 exceeds the mean of Lot 1 by 2 cm. This implies that we can set up the alternative hypotheses in two ways: (i) $\mu_2 - \mu_1 > 2.0$; or (ii) $\mu_1 - \mu_2 < -2.0$. Either will yield the correct answer, even though the rejection regions will lie on opposite ends of the distribution. What is more important is that the order of the means remains consistent throughout the solution. Here, for demonstration purposes, let us use the second alternative hypothesis: $\mu_1 - \mu_2 < -2.0$. This indicates a one-tailed test and that the rejection region will lie in the lower tail of the sampling distribution.

1. H_0: $\mu_1 - \mu_2 = -2.0$.
2. H_1: $\mu_1 - \mu_2 < -2.0$.
3. $\alpha = 0.05$.
4. Use Eqn 9.7, Case 1, since n_1 and $n_2 > 30$.
5. $n_1 = 42$; $n_2 = 36$; $(\overline{x}_1 - \overline{x}_2) = 15.9 - 18.6 = -2.7$; $s_{\overline{X}_1-\overline{X}_2} = \sqrt{\dfrac{2.8^2}{42}+\dfrac{3.4^2}{36}} \approx 0.7126$; $\nu = \infty$.

6. $z_{0.05} = t_{0.05(\infty)} = -1.645.$

7. $t_{(\infty)} = \dfrac{(15.9-18.6)-(-2.0)}{0.7126} \approx -0.98.$

8. Since –0.98 > –1.645, the test statistic is in the acceptance region; we 'accept' H_0.

9. The difference between the two sample means, –2.7, *is not significantly different from –2.0*, which means that *the unknown population mean of Lot 2 does not exceed the unknown population mean of Lot 1 by at least 2.0 cm.*

Example 9.11. From Example 8.8 (see Chapter 8), information on the volumes of the two lots of logs was summarized as follows:

	Lot 1	Lot 2
n	12	8
$\bar{x}$	1.72 m^3	1.93 m^3
s	0.56 m^3	0.61 m^3

Is it safe to assume that the two unknown population means are equal, using a 0.05 level of significance?

In this example, the assumed constant is zero and, from the description, we can conclude that there is no directional preference regarding the rejection of H_0. Therefore, we will use a two-tailed test.

1. H_0: $\mu_1 - \mu_2 = 0.$
2. H_1: $\mu_1 - \mu_2 \neq 0.$
3. $\alpha = 0.05.$
4. Use Eqn 9.7, Case 2, since it can be assumed that $\sigma_1^2 = \sigma_2^2$ (see Example 8.13, Chapter 8) and n_1 and $n_2 < 30$.
5. $n_1 = 12$; $n_2 = 8$; $\bar{x}_1 - \bar{x}_2 = 1.72 - 1.93 = -0.21$; $s_1 = 0.56$; $s_2 = 0.61$;

$$s_p = \sqrt{\frac{(12-1)0.56^2 + (8-1)0.61^2}{12+8-2}} \approx 0.58; \quad s_{\bar{x}_1 - \bar{x}_2} = 0.58\sqrt{\frac{1}{12} + \frac{1}{8}} \approx 0.265;$$

$\nu = 12 + 8 - 2 = 18.$

6. $t_{0.025(18)} = \pm 2.10.$

7. $t_{(18)} = \dfrac{(1.72-0.93)-0}{0.265} \approx -0.798.$

8. Since –2.10 < –0.798 < +2.10, the test statistic is in the acceptance region and we 'accept' H_0.

9. The difference between the two sample means, –0.21, *is not significantly different from the assumed zero*, meaning that *the two population means can be assumed the same.* Scientific literature will often state this type of conclusion as *the sample mean of 1.72 is not significantly different from 1.93.*

Example 9.12. From Example 8.9 (see Chapter 8), a summary of seedling biomass production measurements is given below:

	Lighting	
	Normal	Artificial
Mean	4.65	2.57
s	1.20	0.50
n	12	10

Using a probability level of 0.01, can we assume that the unknown population mean of biomass production under normal lighting exceeds that of artificial lighting?

Again, the constant here is zero. A one-tailed test will be used because, based on anecdotal evidence, the researcher assumes that the biomass production would be higher under normal lighting.

1. H_0: $\mu_1 - \mu_2 = 0$.
2. H_1: $\mu_1 - \mu_2 > 0$.
3. $\alpha = 0.01$.
4. Use Eqn 9.7, Case 3, since it can be assumed that $\sigma_1^2 \neq \sigma_2^2$ (see Example 8.14, Chapter 8) and n_1 and $n_2 < 30$.
5. $n_1 = 12; n_2 = 10; \bar{x}_1 - \bar{x}_2 = 4.65 - 2.57 = 2.08; s_1 = 1.2; s_2 = 0.50;$

$$s_{\bar{x}_1-\bar{x}_2} = \sqrt{\frac{1.20^2}{12} + \frac{0.50^2}{10}} \approx 0.381;$$

$$v = 10 - 1 = 9 \qquad \text{or} \qquad v = \frac{\left[\frac{1.2^2}{12} + \frac{0.5^2}{10}\right]^2}{\frac{\left(\frac{1.2^2}{12}\right)^2}{12-1} + \frac{\left(\frac{0.5^2}{10}\right)}{10-1}} \approx 15.25 \approx 15$$

6. $t_{0.01(15)} = 2.60$; $t_{0.01(9)} = 2.82$.

7. $t_{(v)} = \frac{(4.65 - 2.57) - 0}{0.381} \approx 5.46.$

8. Since 5.46 > 2.60 (or 2.82), the test statistic is in the rejection region; we reject H_0 and accept H_1.
9. The difference between the two sample means, 2.08, is *significantly greater than zero*, indicating that *the population mean of biomass under normal lighting is greater than that under artificial lighting.* Frequently, a loose statement like, *the sample mean of 4.65 is significantly greater than the sample mean of 2.57*, will be found in the scientific literature.

Dependent populations

In Chapters 7 and 8, we discussed a special case of the sampling distributions of the differences between two means when the samples from the two populations are not independent. For tests of the difference between two unknown dependent (paired) population means, H_0 and H_1 are the same as those of independent means. In testing hypotheses using dependent samples, there are two different cases for calculating the test statistic, and thus we use one of two different equations to calculate the test statistic: Eqn 9.8 is used when the population variance is known; and Eqn 9.9 is used when the population variance is not known.

$$z = \frac{\bar{d} - c}{\sigma_d / \sqrt{n}} = \frac{\bar{d} - c}{\sigma_{\bar{d}}}, \tag{9.8}$$

or

$$t_{(n-1)} = \frac{\bar{d} - c}{s_d / \sqrt{n}} = \frac{\bar{d} - c}{s_{\bar{d}}} \tag{9.9}$$

The critical values are obtained from the z-table (see Table A.3, Appendix A) or the last line of the t-table (see Table A.5, Appendix A) for Eqn 9.8, and from the t-table (see Table A.5, Appendix A) with $(n - 1)$ degrees of freedom for Eqn 9.9.

Example 9.13. In Example 8.10 (see Chapter 8), two scales were compared for accuracy. The logical question can now be raised, 'Are the unknown population means of the two scales equal?' Use $\alpha = 0.01$.

This should be a two-tailed test, because we have no reason to assume that one of the two scales will be more accurate than the other.

1. H_0: $\mu_1 - \mu_2 = 0$.
2. H_1: $\mu_1 - \mu_2 \neq 0$.
3. $\alpha = 0.01$.
4. Use Eqn 9.9, since σ is not known.
5. $n_1 = 8$; $\bar{d} = -0.035$; $s_d = 0.0316$; $s_{\bar{d}} = \dfrac{0.0316}{\sqrt{8}} \approx 0.0112$

(see Example 8.10 in Chapter 8).

6. $t_{0.005(7)} = \pm 3.36$.
7. $t_{(7)} = \dfrac{-0.035 - 0}{0.0112} \approx -3.13$.
8. Since $-3.36 < -3.13 < +3.36$, the test statistic is in the acceptance region and we 'accept' H_0.
9. The difference between the two sample means (or in this case, the mean of the paired differences), -0.035, is *not significantly different from zero*, indicating that *the accuracy of the two scales can be assumed to be the same*.

Since this was a two-tailed test, the 99% confidence limits (LCL and UCL) calculated in Chapter 8 are equivalent to the critical values seen here, except the confidence interval is expressed in the terms of the original units of the data (grams).

9.6 Tests Concerning the Difference between Two Proportions

Tests for differences between two proportions can also be very useful in forestry and natural resources applications. For instance, a manager of a nursery might want to test the difference between the germination rates in two seed lots. A sawmill manager may want to test proportions of No. 1 grade lumber recovered from harvested logs at two different sites. A manager of a national park may want to compare if the proportion of local visitors is different between the summer and winter months.

When testing the difference between two proportions, an assumed constant difference, c, is stated in the null hypothesis, against the usual three alternatives.

H_0: $p_1 - p_2 = c$,

H_1: $p_1 - p_2 \neq c$, or $p_1 - p_2 < c$, or $p_1 - p_2 > c$.

Any positive or negative value between –1 and +1 can be used for the value of c. However, the most frequently used value of c is zero, which assumes that the two unknown population proportions are equal. If the values of both p_1 and p_2 are not close to zero or one and the sample sizes, n_1 and n_2 (from which the sample proportions are calculated), are both greater than 30, the following equation can be used to calculate the test statistic:

$$z = \frac{(\hat{p}_1 - \hat{p}_2) - c}{s_{\hat{p}_1 - \hat{p}_2}}, \tag{9.10}$$

where

$$s_{\hat{p}_1-\hat{p}_2} = \sqrt{\hat{p}_c \hat{q}_c \left(\frac{1}{n_1} + \frac{1}{n_2} \right)},$$

and $\hat{p}_c = \dfrac{n_1\hat{p}_1 + n_2\hat{p}_2}{n_1 + n_2}$ and $\hat{q}_c = 1 - \hat{p}_c$; $\hat{p}_c =$ pooled or combined proportion;

or $s_{\hat{p}_1-\hat{p}_2} = \sqrt{\dfrac{\hat{p}_1\hat{q}_1}{n_1} + \dfrac{\hat{p}_2\hat{q}_2}{n_2}}$

Note that there are two equations given for the calculation of the standard error term. If $\hat{p}_1$ and $\hat{p}_2$ are numerically close, we use the equation with the pooled proportions, $\hat{p}_c$ and $\hat{q}_c$. When there is a considerable difference between $\hat{p}_1$ and $\hat{p}_2$, the second equation is recommended.

Critical values are obtained from the z-table (see Table A.3, Appendix A) or from the last line of the t-table (see Table A.5, Appendix A).

Example 9.14. Out of 50 trees examined in a managed forest stand, 6 were infested by bark beetles. In a nearby natural stand, 12 out of 60 were infested. With a 0.05 level of significance, can we assume that the unknown population proportion of infested trees is smaller in the managed stand than in the natural stand?

This test is clearly a one-tailed test, since it is implied by the question that managed stands are more resistant to bark beetle infestation than natural stands. Note that, like Example 9.10, the hypotheses can be set up or ordered in two different ways.

1. $H_0: p_1 - p_2 = 0$.
2. $H_1: p_1 - p_2 < 0$.
3. $\alpha = 0.05$.
4. Use Eqn 9.10.
5. $n_1 = 50$; $n_2 = 60$; $\hat{p}_1 = \frac{6}{50} = 0.12$; $\hat{p}_2 = \frac{12}{60} = 0.20$;

   $\hat{p}_1 - \hat{p}_2 = -0.08$; $\hat{p}_c = \dfrac{50 \times 0.12 + 60 \times 0.20}{50 + 60} = 0.164$; $\hat{q}_c = 0.836$;

   $s_{\hat{p}_1-\hat{p}_2} = \sqrt{(0.164)(0.836)\left(\frac{1}{50} + \frac{1}{60}\right)} \approx 0.071$ or

$$s_{\hat{p}_1-\hat{p}_2} = \sqrt{\frac{(0.12)(0.88)}{50} + \frac{(0.2)(0.8)}{60}} \approx 0.069.$$ Both equations for the standard error are presented here and they give almost identical results. Thus, either can be used.

6. $z_{0.05} = -1.645$.

7. $z = \frac{(0.12-0.20)-0}{0.071} \approx -1.13$ or $z = \frac{(0.12-0.20)-0}{0.069} \approx -1.16$.

8. Since −1.13 (or −1.16) > −1.645, the test statistic is in the acceptance region and we 'accept' H_0.

9. The difference between the two sample proportions, −0.08, is *not significantly different from zero*, which means that the two unknown population proportions are the same.

Example 9.15. A survey of home centre shoppers found that out of 100 women, 37% preferred certified lumber; out of 100 men, 22% preferred certified lumber. At a 0.01 level of significance, test the claim that the unknown population proportion of women's preference for certified lumber exceeds the unknown population proportion of men's preference for certified lumber by at least 10%.

This test is a one-tailed test because women's greater preference for certified lumber is assumed in the statement. If we set up the hypotheses such that p_2 is subtracted from p_1, the rejection region lies in the upper tail of the distribution. Also it should be noted that $c = 0.10$, instead of zero.

1. H_0: $p_1 - p_2 = 0.1$.
2. H_1: $p_1 - p_2 > 0.1$.
3. $\alpha = 0.01$.
4. Use Eqn 9.10.
5. $n_1 = 100$; $n_2 = 100$; $\hat{p}_1 = 0.37$; $\hat{p}_2 = 0.22$; $\hat{p}_1 - \hat{p}_2 = 0.15$;

$$\hat{p}_c = \frac{(0.37)100 + (0.22)100}{100+100} = 0.295;\ \hat{q}_c = 0.705;$$

$$s_{\hat{p}_1-\hat{p}_2} = \sqrt{(0.295)(0.705)\left(\frac{1}{100} + \frac{1}{100}\right)} \approx 0.0645.$$

6. $z_{0.01} = 2.33$.

7. $z = \frac{(0.37-0.22)-0.1}{0.0645} \approx 0.775$.

8. Since 0.775 < 2.33, the test statistic is in the acceptance region and we 'accept' H_0.

9. The difference between the two sample proportions, 0.15, is *not significantly greater than the claimed difference of 0.1 between the two population proportions*, meaning that women's preference for certified lumber *does not exceed men's preference* by more than 10%.

9.7 Tests Concerning the Ratio of Two Variances

Tests concerning the ratio of two variances have practical importance in studying the uniformity of two populations. For example, a sawmill manager may be interested in buying timber from two wood lots. All things being equal, the sawmill would prefer to have logs with less variation in diameter because production processes are easier to control. That being the case, they may test the variances of dbh from the two wood lots to see if there is a statistical difference in the way that the log sizes vary. However, the most important value of these tests will be seen in advanced statistical tests such as regression analysis, analysis of variance and covariance analysis. The first two of these will be discussed in Chapters 11 and 12. Tests concerning the ratio of two variances are also important in comparing the difference between two unknown population means when an assumption about the equality of the two population variances is required (see Independent Populations, Section 9.5.1).

It would make intuitive sense that in testing the difference between variances, we would use a difference (equal to a constant, usually zero) as in previous hypothesis tests. However, due to the skewed nature of variance distributions, there is no simple distribution that describes the differences between variances. That being the case, we test the ratio of two variances. In testing the ratio of two variances, it is assumed in the null hypothesis that the ratio of two unknown population variances, σ_1^2/σ_2^2, is equal to a constant, c, against one of the three usual alternatives:

$$H_0: \sigma_1^2/\sigma_2^2 = c,$$

$$H_1: \sigma_1^2/\sigma_2^2 \neq c, \quad \text{or} \quad \sigma_1^2/\sigma_2^2 < c, \quad \text{or} \quad \sigma_1^2/\sigma_2^2 > c.$$

Although any positive constant can be used as the value of c, in most cases it is set to one in order to test the equality of two unknown population variances, $\sigma_1^2 = \sigma_2^2$.

Assuming that the two populations from which the samples are taken are approximately normally distributed, the following test statistic can be used to test the above hypothesis:

$$F_{(n_1-1, n_2-1)} = \frac{s_1^2}{s_2^2}\frac{1}{c}, \tag{9.11}$$

and if $c = 1$, Eqn 9.11 becomes

$$F_{(n_1-1, n_2-1)} = \frac{s_1^2}{s_2^2} \tag{9.12}$$

Critical values are obtained from the F-table (see Table A.7, Appendix A) with $n_1 - 1$ and $n_2 - 1$, respectively describing the degrees of freedom in the numerator and denominator terms. Equation 9.12 is oftentimes referred to as the **variance ratio** test.

Example 9.16. In Example 8.13 (see Chapter 8), log volumes were compared from two populations, yielding $s_1^2 = 0.314$ with a sample size of 12 and $s_2^2 = 0.372$ with a sample size of 8. Can we assume with 0.05 level of significance that the two unknown population variances are equal?

From the information given, c is 1.0 and a two-tailed test is appropriate because no information is given about the comparative variability of the two populations.

1. $H_0: \dfrac{\sigma_1^2}{\sigma_2^2} = 1.0.$

2. $H_1: \dfrac{\sigma_1^2}{\sigma_2^2} \neq 1.0.$

3. $\alpha = 0.05.$

4. Use Eqn 9.12 since $c = 1.0$.

5. $n_1 = 12; \quad n_2 = 8; \quad s_1^2 = 0.314; \quad s_2^2 = 0.372.$

6. $F_{0.025(11,7)} = 4.72; \quad \dfrac{4.76 + 4.67}{2} \approx 4.72$ (by interpolation);

$$F_{0.975(11,7)} = \frac{1}{F_{0.025(7,11)}} = \frac{1}{3.76} \approx 0.266.$$

7. $F_{(11,7)} = \dfrac{0.314}{0.372} \approx 0.844.$

8. Since $0.266 < 0.844 < 4.72$, the test statistic is in the acceptance region (Fig. 9.9) and we 'accept' H_0.

9. It can be safely assumed that *the two unknown population variances are equal.*

Since the above test is a two-tailed test, the 95% confidence limits (LCL and UCL) calculated in Example 8.13 (see Chapter 8) are equivalent to the critical values used in this test, but expressed in terms of the actual ratio of two variances.

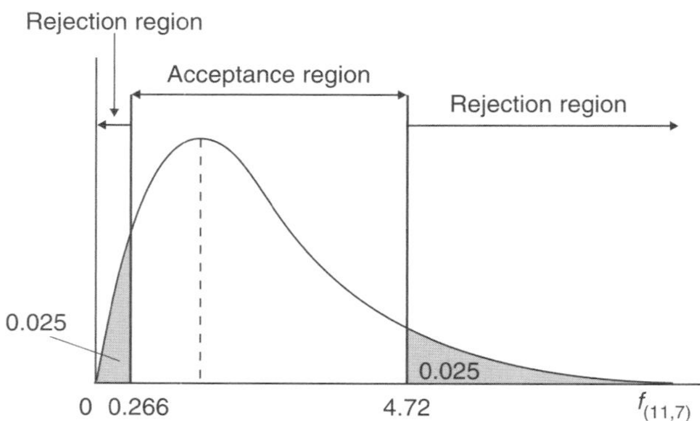

Fig. 9.9. Critical regions in *F* units, two-tailed test.

When the equality of two variances is tested using a two-tailed test, there is a short-cut procedure that is strongly recommended. We simply assign the larger of the two variances to be s_1^2. This means that when the s_1^2/s_2^2 ratio is set up, the larger variance is divided by the smaller one, which will always produce a ratio greater than 1.0. As a result, we are forcing the test to the upper side of the *F* distribution and a table look-up is required for the critical value on the right side of the distribution only. However, when using this short cut, it is important to remember that the critical value in the upper tail of the distribution is obtained for $\alpha/2$ probability (i.e. the significance value must still be halved, even though we are looking up one value only).

Applying this in Example 9.16, we have to exchange the positions of σ_1^2 and σ_2^2 in the hypotheses in order to divide the larger sample variance by the smaller one.

1. $H_0: \dfrac{\sigma_2^2}{\sigma_1^2} = 1.0.$

2. $H_1: \dfrac{\sigma_2^2}{\sigma_1^2} \neq 1.0.$

3. $\alpha = 0.05.$

4. Use Eqn 9.12 since $c = 1.0$.

5. $n_1 = 12;$ $n_2 = 8;$ $s_1^2 = 0.314;$ $s_2^2 = 0.372.$

6. $F_{0.025(7,11)} = 3.76$. Note that the degrees of freedom are reversed compared to the previous calculation and that we used $\alpha/2$ probability. However, we looked up the critical value at the upper end of the *F* distribution only.

7. $F_{(7,11)} = \dfrac{0.372}{0.314} \approx 1.185$. With the order of the two samples reversed, this ratio exceeds one.
8. Since $1.185 < 3.76$, the test statistic is in the acceptance region and we 'accept' H_0.
9. It can be safely assumed that *the two unknown population variances are equal.*

Example 9.17. In Examples 8.9 and 8.14 (see Chapter 8), we obtained the variance of the biomass of seedlings under two lighting scenarios. Based on 12 observations in natural lighting, the variance was 1.44. Based on 10 observations for artificial lighting, the variance was 0.25.

a. Can we assume with a 0.05 level of significance that the two unknown population variances are equal?
b. Can we assume with a 0.1 level of significance that the unknown population variance under natural lighting is at least three times as high as the unknown population variance (i.e. a ratio of variances exceeding three) under artificial lighting?

A two-tailed test is required for a, with $c = 1.0$. Since the question asks if the unknown population variance under natural lighting is 'at least three times as high' as the unknown population variance under artificial lighting, a one-tailed test is required for b, with the critical region in the upper tail of the distribution and $c = 3.0$.

a.

1. $H_0\text{:}\dfrac{\sigma_1^2}{\sigma_2^2} = 1.0.$

2. $H_1\text{:}\dfrac{\sigma_1^2}{\sigma_2^2} \neq 1.0.$

3. $\alpha = 0.05.$
4. Use Eqn 9.12 since $c = 1.0$.
5. $n_1 = 12$; $n_2 = 10$; $s_1^2 = 1.44$; $s_2^2 = 0.25$.

6. $F_{0.025(11,9)} = \dfrac{3.96 + 3.86}{2} \approx 3.92$ (by interpolation);

$F_{0.975(11,9)} = \dfrac{1}{F_{0.025(9,11)}} = \dfrac{1}{3.59} \approx 0.279$. Note that the second look-up is not entirely necessary because the larger variance will be divided by the smaller one.

7. $F_{(11,9)} = \dfrac{1.44}{0.24} \approx 5.76.$

8. Since $5.76 > 3.92$, the test statistic is in the rejection region; we reject H_0 and accept H_1.
9. The ratio of the two sample variances, 5.76, is *significantly different from the assumed ratio of 1.0*, meaning that *the population variance of the biomass under natural lighting is significantly different than the population variance under artificial lighting.*

Since the test is a two-tailed test, the 95% confidence limits (LCL and UCL) calculated in Example 8.14 (see Chapter 8) are equivalent to the critical values used in the test expressed in terms of ratios of two variances.

b.

1. $H_0\text{:}\dfrac{\sigma_1^2}{\sigma_2^2} = 3.0.$

2. $H_1\text{:}\dfrac{\sigma_1^2}{\sigma_2^2} > 3.0.$

3. $\alpha = 0.1$.
4. Use Eqn 9.11 since $c \neq 1.0$.
5. $n_1 = 12$; $n_2 = 10$; $s_1^2 = 1.44$; $s_2^2 = 0.25$.
6. $F_{0.1(11,9)} = 2.40$; (by interpolation $\frac{2.42 + 2.38}{2} = 2.40$).
7. $F_{(11,9)} = \frac{1.44}{0.24}\frac{1}{3.0} \approx 1.92$.
8. Since $1.92 < 2.40$, the test statistic is in the acceptance region and we 'accept' H_0.
9. It can be assumed that *the ratio of two unknown population variances is not significantly greater than 3.0.*

9.8 *p*-Values

When statistical packages are used to analyse data, a ***p*-value** is often provided as part of the output. In most statistical packages, these *p*-values indicate the smallest level of significance for rejecting H_0. For a left-tailed test, the *p*-value indicates the probability of obtaining a value in the sampling distribution of the test statistic less than the calculated test statistic. For a right-tailed test, the *p*-value indicates the probability of obtaining a value in the sampling distribution of the test statistic greater than the calculated test statistic. For a two-tailed test, the *p*-value indicates the probability of obtaining a value in the sampling distribution of the test statistic less than (to the left) or greater than (to the right) the calculated test statistic. Intuitively, the following rule is used to interpret *p*-values:

if *p*-value $\leq$ a predefined α, reject H_0.

A word of caution! It is a mistake to interpret the *p*-value as the level of significance, mainly because a low level of significance could result in a very high level type II error (β). To use statistical tests properly, the level of significance, α, must be decided prior to calculating the sample statistics and the test statistics (including their *p*-values). Some scientific journals require scientists to publish their *p*-values, instead of the level of significance. Unfortunately, though, this allows readers to select their own levels of significance, which can lead to conflicting and erroneous conclusions, especially when there is ambiguity on whether a one- or two-tailed test is appropriate.

Exercises

Section 9.1

9.1. It is assumed that the mean specific gravity of Douglas-fir is 0.45 with a standard deviation of 0.098. A random sample of 36 specimens was examined to test the hypothesis that $\mu = 0.45$ against the alternative that $\mu > 0.45$. Assume type I error is set at 0.04:

a. Find the critical value and show the critical region.
b. Find the type II error when the population mean is
   i. $\mu_1 = 0.49$.
   ii. $\mu_1 = 0.53$.
c. Repeat Exercises 9.1a and 9.1b using 60 observations.

9.2. A manufacturer claims that at least 20% of his consumers prefer his company's product. A random sample of 100 people is taken to check this claim.

a. If type I error is set at 0.05, what is the critical value below which the manufacturer's claim can be rejected?
b. Find the type II error if the 'real' (or 'known') preference is 10% and the sample size is 100.
c. Find the type II error if the 'real' (or 'known') preference is 10% and the sample size is 50. Compare your results to b.
d. Find the type II error if the 'real' (or 'known') preference is 5% and the sample size is 100. Compare your results to b.

9.3. State the null and alternative hypotheses for testing the following claims and indicate graphically where the critical region(s) is (are) located:

a. At least 80% of the readers will solve this exercise.
b. The average snowfall in Prince George, BC for the month of December is 25.2 cm.
c. No more than 10% of seedlings will die during the first year after plantation.
d. The proportion of reject parts coming off the assembly line is less than 0.07.
e. The average height of trees in a forest stand is 18.0 m.

Section 9.2

9.4. The hourly wages in the secondary wood products sector are normally distributed with a mean of US$19.20 and a standard deviation of US$2.80. If a company in this industry employs 20 summer students at an average of US$16.50 per hour, can this company be accused of paying inferior wages to students? Use $\alpha = 0.05$.

9.5. A coffee machine is set to dispense 150 ml of coffee with a standard deviation of 16 ml. The machine is regulated so that the amount dispensed is approximately normally distributed. If a mean of 144 ml was obtained by taking a sample of 10 cups, can we conclude that the machine is set up properly? Use $\alpha = 0.05$.

9.6. Repeat Exercise 9.5 assuming that the population standard deviation is not known, but estimated from the sample of 10 as 14 ml.

9.7. Use the data from Exercise 8.6 (see Chapter 8) to test the manufacturer's claim, with a 0.01 level of significance, that the price per thousand board feet of spruce–pine–fir lumber is at least US$600.

9.8. From a distance of 5 m, 10 students were asked to judge the dbh of a tree to the nearest 1 cm. The tree was measured to be exactly 25.0 cm. They gave the following estimates: 21, 18, 23, 23, 26, 25, 23, 25, 21, 25.

a. Do these observations indicate that the students have difficulty in accurately estimating the dbh of a tree? Use a 5% level of significance.
b. Calculate the 95% confidence interval for the population mean and compare your results to your answer in part a.

9.9. Consider the data given in Exercise 8.7. Can we assume that the population mean is 32.0 cm, with a 0.05 level of significance? Compare your results to those obtained in Exercise 8.7 (see Chapter 8).

Section 9.3

9.10. A manufacturer claims that at least 95% of the logging equipment they supply conforms to specifications. A study of a sample of 70 pieces of equipment revealed that 5 were faulty. Test the manufacturer's claim at a significance level of 0.01.

9.11. A manufacturer of kitchen countertops claims that no more than 20% of their product shows any noticeable defect. Out of 40 countertops examined, 10 show some defects. Test this statement using $\alpha = 0.05$.

9.12. It is assumed that the fire-retardant chemical tested in Exercise 8.8 (see Chapter 8) is effective 80% of the time. Given the data in Exercise 8.8 (see Chapter 8), is this assumption acceptable? Use $\alpha = 0.05$. Compare your results to the 95% confidence interval calculated in Exercise 8.8a.

Section 9.4

9.13. In Exercise 9.7, can we assume that the population variance is equal to 1400? Use $\alpha = 0.01$.

9.14. In Exercise 9.1, can we assume that the population standard deviation is 0.098? Is this assumption acceptable with a 0.05 level of significance if the standard deviation of a random sample of 36 is 0.085?

9.15. Scientists assume that the population standard deviation for weights of the rainbow trout, described in Exercise 8.19 (see Chapter 8), does not exceed 0.10 kg. Test this assumption using a 0.05 level of significance. Compare this to the original confidence interval calculated in Exercise 8.19 (see Chapter 8).

9.16. Finger-jointed stock coming into your plant is claimed to have a mean width of 4 in and a variance of at most 0.6 in^2. Would you accept this claim if the variance of a random sample of 10 boards was 1.2 in^2? Use $\alpha = 0.05$.

Section 9.5

9.17. In the logging truck tyre experiment described in Exercise 8.10 (see Chapter 8), can we assume that the two unknown population means are equal? Use 0.01 type I error in the test and compare your results to the 99% confidence interval calculated in Exercise 8.10 (see Chapter 8).

9.18. Two brands of chainsaws are available on the market. The manufacturers have provided the following specifications about the lifespan of these chainsaws: the standard deviation of Brand A is 1.5 years, while the standard deviation of Brand B is 2.0 years. Samples of 15 (Brand A) and 10 (Brand B) are tested and their mean lives are 6.1 years and 7.2 years, respectively. Can we assume that Brand B outperforms Brand A by at least 0.5 year? Use $\alpha = 0.05$.

9.19. Can it be assumed that the tensile strengths of two commercial fishing lines, as described in Exercise 8.12 (see Chapter 8), have the same population mean? Use $\alpha = 0.05$. Compare this test to the 95% confidence intervals calculated in Exercise 8.12 (see Chapter 8).

9.20. Two brands of laminated wood beams were tested for breaking load, with the following results:

Brand	*n*	Mean	$\Sigma(x_i - x)^2$
A	45	1565	48,000
B	34	1610	37,000

Can we conclude that Brand B beams outperform Brand A beams using a 1% level of significance?

9.21. Use the data given in Exercise 8.11 (see Chapter 8) to test the equality of the two unknown population means. Use $\alpha = 0.01$ and compare your results to the 99% confidence interval calculated in Exercise 8.11a.

9.22. At the start of the growing season, type I fertilizer was applied to 7 plots and type II was applied to 6 plots. The total biomass measurements (in g) per plot of 1-year-old ponderosa pine seedlings were recorded at the end of the growing season:

Type I	570	592	630	512	634	493	558
Type II	502	593	503	583	482	445	

Test the effect of the two types of fertilizers (*hint: is* $\mu_1 = \mu_2$?) with a 0.1 level of significance. Assume that $\sigma^2_1 = \sigma^2_2$.

9.23. Use the data in Exercise 8.13 (see Chapter 8) to test the equality of the two unknown population means, with $\alpha = 0.01$. Compare your results to the 99% confidence interval calculated in Exercise 8.13 (see Chapter 8).

9.24. Two diets were used in an experiment to study the gains in weight (in kg) of 12 steer.

Diet 1	45.9	38.7	44.1	49.0	41.4		
Diet 2	36.2	74.6	43.7	60.3	41.4	39.2	51.3

Is Diet 2 superior to Diet 1? Use $\alpha = 0.05$. Assume that $\sigma^2_1 = \sigma^2_2$.

9.25. Can you assume that the unknown population means are equal for the data described in Exercise 8.14 (see Chapter 8) using a 0.05 level of significance? Compare your results to the confidence interval calculated in Exercise 8.14 (see Chapter 8).

9.26. Volumes (m^3/ha) in a mature Douglas-fir stand on 10 plots were estimated independently by two separate crews. Assuming that the population is normally distributed, would these results lead you to conclude that the estimates of the two crews are the same? Use $\alpha = 0.05$.

Plot	1	2	3	4	5	6	7	8	9	10
Crew 1	480	878	910	980	1021	620	850	931	792	1002
Crew 2	436	925	878	1040	955	650	861	892	753	996

Section 9.6

9.27. In the example described in Exercise 8.16 (see Chapter 8), it is assumed that soil preparation on Area II is better than on Area I. Can we reach this conclusion by comparing the regeneration on Area II to that of Area I? Use $\alpha = 0.05$. Can these results be compared to either one of the confidence intervals calculated in Exercise 8.16 (see Chapter 8)?

9.28. A manufacturer of microprocessors buys its chips from two suppliers. If 22 out of 120 chips from Supplier I are defective and 20 out of 160 from Supplier II are defective, can we conclude that there is no difference between the real proportions of defective chips from the two suppliers? Use $\alpha = 0.01$.

Section 9.7

9.29. In the study described in Exercise 9.24, it is claimed that the variation of weight gains in Diet 2 is not more than five times the variation in Diet 1. Test this hypothesis with a 0.01 level of significance.

9.30. Test the equality of the 2 unknown population variances described in Exercise 8.13 (see Chapter 8) using $\alpha = 0.05$. Compare your results and conclusions to the 95% confidence intervals calculated in Exercise 8.21 (see Chapter 8).

9.31. In Exercises 9.22 and 9.24, we assumed that the population variances were equal. Test whether these claims are true using $\alpha = 0.1$ for the first claim and $\alpha = 0.05$ for the second.

10 Goodness-of-fit and Test for Independence

Testing Distributions

In Chapter 9, we discussed tests concerning single parameters such as μ, p, σ^2, and differences and ratios between two parameters. In this chapter, we introduce **tests concerning distributions** of one or more populations. The **goodness-of-fit test** is used to determine whether a population follows a specified theoretical distribution, and the **test for independence** (or a **contingency table**) is used to compare two or more distributions.

10.1 Goodness-of-fit Test

In some situations, we would like to know if a population follows a particular theoretical frequency distribution. For example, we may want to know if the diameter at breast height (dbh) measurements from a particular stand are normally distributed, so that we can use the Z distribution to construct confidence limits around the average observed diameter. Or we may want to know if observed germination rates for Douglas-fir seeds follow a binomial distribution. Numerous other examples in natural resources exist for the other distributions we have studied, such as the binomial, Poisson, hypergeometric, geometric, negative binomial, uniform, normal and exponential. To test if observed sample data follow a particular theoretical distribution, we perform a **goodness-of-fit test.** First, we construct a frequency distribution, which may be *categorical*, *ungrouped* or *grouped*. Frequency classes constructed this way are usually referred to as **cells**, and their class frequencies are called the **observed frequencies** (O_i). For each frequency class or cell, the **expected frequencies** (E_i) are calculated by using the equations for the specified theoretical distributions or the available tables of the specified distributions.

In the goodness-of-fit test, it is assumed in the null hypothesis that all the observed frequencies are equal to the expected frequencies, while the alternative hypothesis states that at least one is different. The hypotheses can be stated mathematically, or in words:

H_0: $O_i = E_i$, for all is (frequencies); or, the population distribution follows the specified distribution, and

H_1: $O_i \neq E_i$, for at least one i (frequency); or, the population distribution does not follow the specified distribution.

The calculation of the test statistic uses the number of frequency classes, c, and the expected and observed frequencies:

$$\chi^2_{(v)} = \sum_{i=1}^{c} \frac{(O_i - E_i)^2}{E_i} \qquad (10.1)$$

It can be shown that the test statistic, $\chi^2_{(v)}$ (which is an estimate of the random variable, χ^2), approximately follows a chi-square distribution with v degrees of freedom, where v is equal to the number of cells, c, minus *the number of independent quantities* obtained from sample data to calculate the expected cell frequencies. This will be discussed in more detail later.

From the H_1 statement, it may look as though the goodness-of-fit test is a two-tailed test. However, closer examination of Eqn 10.1 indicates that the value of $\chi^2_{(v)}$ is zero when the conditions stated in H_0 are perfectly met: that is, $O_i = E_i$ for all is. Since the extreme left tail of the chi-square distribution starts at zero (Fig. 10.1), we cannot reject H_0 at the lower tail of the distribution. Thus, the acceptance region here is always from zero to a specified critical χ^2-value, and the rejection region is in the upper tail of the distribution. We demonstrate the goodness-of-fit test with a very simple example and also offer another explanation for why is it a one-tailed test.

Example 10.1. A coin is flipped 90 times and the results are 50 tails and 40 heads. Is this a balanced coin or not? Use $\alpha = 0.05$.

In solving this problem, the observed frequencies are 50 tails and 40 heads. In the case of a balanced coin, we would expect (as our theoretical distribution) 45 heads and 45 tails:

	Tails	Heads
O_i	50	40
E_i	45	45

1. H_0: $O_i = E_i$ for all is; or, the distribution follows a uniform distribution.
2. H_1: $O_i \neq E_i$ for at least one i; or, the distribution does not follow a uniform distribution.
3. $\alpha = 0.05$.
4. Use Eqn 10.1.
5. $c = 2$; the expected and observed values are given above in two frequency classes; $v = 2 - 1$, because there are two cells of data (tails and heads) and we used one independent quantity, the total number of trials ($N = 90$), to calculate the expected values.
6. $\chi^2_{0.05(1)} = 3.84$
7. $\chi^2_{(1)} = \frac{(50-45)^2}{45} + \frac{(40-45)^2}{45} \approx 1.11$
8. Since 1.11 < 3.84, the test statistic is in the acceptance region (Fig. 10.1) and we 'accept' H_0.
9. It can be assumed that the coin is balanced: that is, the number of heads and tails are uniformly distributed and the proportion of tails (and heads) is not significantly different from 0.5.

We could also do a z-test with the proportion of tails (or heads) to test the same hypothesis:

1. H_0: $p = 0.5$.
2. H_1: $p \neq 0.5$.
3. $\alpha = 0.05$.
4. Use Eqn 9.3 (see Chapter 9).

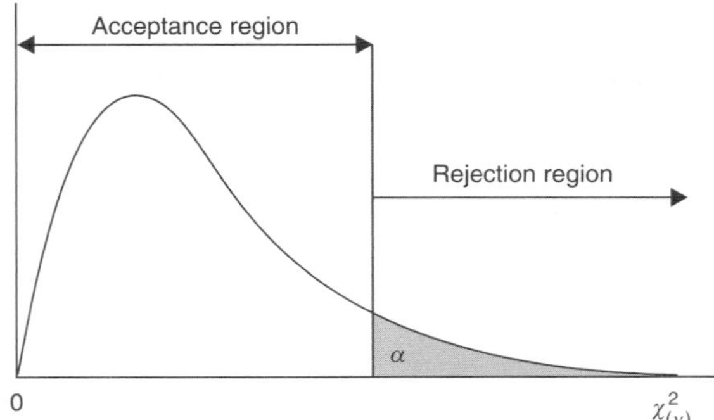

Fig. 10.1. Critical and acceptance regions in the χ^2-test.

5. $n = 90$; $p \approx 0.556$; $q = 0.444$; $\sigma_p = \sqrt{\frac{0.5 \times 0.5}{90}} \approx 0.0527$

6. $z_{0.025} = \pm 1.96$.

7. $z = \frac{0.556 - 0.5}{0.0527} \approx 1.05418$.

8. Since $-1.96 < 1.05418 < 1.96$, the test statistic is in the acceptance region and we 'accept' H_0.
9. Same conclusion as the chi-square test above.

It is interesting to note that the critical value of $\chi^2 = 3.84$ is equal to 1.96^2, the square of the z critical value. Similarly, the test statistic for the goodness-of-fit test is 1.11, which is approximately equal to 1.05418^2, the square of the test statistic for the proportion. Also note from above that the z-test is a two-tailed test, while the goodness-of-fit test is a one-tailed test. It can be shown mathematically that:

$$z^2_{\alpha/2} = \chi^2_{\alpha(1)}$$

This indicates that one can square either the negative or the positive z-values, and their probabilities sum to become the upper tail χ^2-values. This also shows why the goodness-of-fit test is always a one-tailed test.

Example 10.2. The frequencies of industrial accidents per month in a large sawmill over the last 4 years are summarized below (see Table 10.1). Is it reasonable to assume that the frequency of accidents in this sawmill follows a Poisson distribution with a 0.05 level of significance?

1. H_0: $O_i = E_i$ for all is; or the distribution follows a Poisson distribution.
2. H_1: $O_i \neq E_i$ for at least one i; or the distribution does not follow a Poisson distribution.
3. $\alpha = 0.05$.
4. Use Eqn 10.1.
5. We use the probability function of the Poisson distribution (see Eqn 5.8, Chapter 5) to calculate the probabilities for the various outcomes:

$$f(x; \mu) = \frac{e^{-\mu}\mu^x}{x!}$$

Table 10.1. Frequency distribution of industrial accidents per month in a sawmill over the last 4 years.

Number of accidents	Frequency (O_i)
0	8
1	15
2	13
3	8
4+	4
Total	48

Since μ is not known, we must estimate it from the frequency distribution (see Eqn 2.5, Chapter 2):

$$\bar{x} = \frac{\sum_{j=1}^{c} f_j m_j}{\sum_{j=1}^{c} f_j} = \frac{(0)(8)+(1)(15)+(2)(13)+(3)(8)+(4)(4)}{48} \approx 1.69$$

Substituting $\bar{x}$ for μ in the above probability function, the probabilities of x numbers of accidents occurring per month can be calculated:

$$P(x=0;1.69) = \frac{e^{1.69}1.690}{0!} = 0.183$$

$$P(x=1;1.69) = \frac{e^{1.69}1.69^{1}}{1!} = 0.310$$

and so on.

Based on the probabilities for the various frequency classes, the expected values are $E_i = 48 \times p_i$, from which the following table can be generated:

Number of accidents	p_i	E_i		O_i	
0	0.183	8.8		8	
1	0.310	14.9		15	
2	0.264	12.7		13	
3	0.149	7.1	11.6	8	12
4+	0.094	4.5		4	

There are two important notes to consider in generating the above table. First, in order to include all possible outcomes, the probability for the 4+ frequency class is calculated as $1 - P(X \le 3) = 1 - (P(X = 0) + P(X = 1) + P(X = 2) + P(X = 3))$. Second, the $\chi^2_{(v)}$ values (calculated using Eqn 10.1) do not give a good approximation to the chi-square distribution when the expected values in any class are < 5. It is therefore common practice to combine any classes where the expected values are < 5 with their neighbouring classes. This is why the last two classes in our example are combined into a single class and the number of classes, c, is four in the calculation of the degrees of freedom for the χ^2-test statistic.

6. Since the sample mean (1.69) and the total frequency (48) were used to calculate the expected frequencies, we lose two degrees of freedom, for a total of $4 - 2 = 2$ degrees of freedom for the χ^2 critical value and test statistic.

$\chi^2_{0.05(2)} = 5.99$

7. $\chi^2_{(4-2)} = \dfrac{(8-8.8)^2}{8.8} + \dfrac{(15-14.9)^2}{14.9} + \dfrac{(13-12.7)^2}{12.7} + \dfrac{(12-11.6)^2}{11.6} = 0.094$

8. Because 0.094 < 5.99, the test statistic is in the acceptance region and we 'accept' H_0.

9. It can be assumed that the population of monthly accident numbers in the sawmill studied follows a Poisson distribution.

Example 10.3. The frequency distribution of the 50 dbh measurements taken from the example forest stand used in Chapter 2 is shown again in Table 10.2, below. Is it reasonable to assume that the population of all dbh measurements in the stand follows a normal distribution? Use $\alpha = 0.01$.

Table 10.2. Frequency distribution of 50 dbh measurements.

Class boundaries	Class limits	Frequency O_i	Class mark
7.55–9.85	7.6–9.8	2	8.7
9.85–12.15	9.9–12.1	3	11.0
12.15–14.45	12.2–14.4	12	13.3
14.45–16.75	14.5–16.7	14	15.6
16.75–19.05	16.8–19.0	13	17.9
19.05–21.35	19.1–21.3	4	20.2
21.35–23.65	21.4–23.6	2	22.5

1. H_0: $O_i = E_i$ for all is; or the population follows a normal distribution.
2. H_1: $O_i \neq E_i$ for at least one i; or the population does not follow a normal distribution.
3. $\alpha = 0.01$.
4. Use Eqn 10.1.
5. We use the Z transformation and the z-table (see Table A.3, Appendix A) to find the probabilities between each class boundaries. We then multiply these probabilities by 50 (the total number of dbh measurements, or the sum of the frequencies) to get the expected values. The expected values will follow a normal distribution.

In generating the expected normal distribution values, the first frequency class probability is usually calculated from $-\infty$ to the upper class boundary of the first class, while the last frequency class probability is calculated between the lower class boundary of the last class and $+\infty$.

Since we do not know the population mean, μ, and the population standard deviation, σ, their estimates, $\bar{x}$ and s, can be used for the Z transformation. These statistics, $\bar{x}$ and s, can be calculated from the frequency distribution in Table 10.2 (we previously calculated these values as $\bar{x}$ = 15.74 cm and s = 3.10 cm in Examples 2.4 and 2.8, Chapter 2). The probabilities are calculated as follows:

$$P(x < 9.85) = P\left(z < \frac{9.85 - 15.74}{3.10}\right) = P(z < -1.90) = 0.0287$$

$$P(9.85 < x < 12.15) = P\left(-1.90 < z < \frac{12.15 - 15.74}{3.10}\right) = P(-1.90 < z < -1.16)$$
$$= 0.1230 - 0.0287 = 0.0943$$

$$P(12.15 < x < 14.45) = P\left(-1.16 < z < \frac{14.45 - 15.74}{3.10}\right) = P(-1.16 < z < -0.42)$$
$$= 0.3372 - 0.1230 = 0.2142$$

$$P(14.45 < x < 16.75) = P\left(-0.42 < z < \frac{16.75 - 15.74}{3.10}\right) = P(-0.42 < z < 0.33)$$

$$= 0.6255 - 0.3372 = 0.2911$$

$$P(16.75 < x < 19.05) = P\left(0.33 < z < \frac{19.05 - 15.74}{3.10}\right) = P(0.33 < z < 1.07)$$

$$= 0.8577 - 0.6293 = 0.2284$$

$$P(19.05 < x < 21.35) = P\left(1.07 < z < \frac{21.35 - 15.74}{3.10}\right) = P(1.07 < z < 1.81)$$

$$= 0.9649 - 0.8577 = 0.1072$$

$$P(x > 21.35) = P\left(z > \frac{21.35 - 15.74}{3.10}\right) = P(z > 1.81) = 1.0 - 0.9649 = 0.0351$$

From here, the expected values are calculated as $E_i = p_i N$, where $N = 50$.

Class boundaries	p_i		E_i		O_i	
7.55 – 9.85	0.0287		1.4	6.1	2	5
9.85 – 12.15	0.0943		4.7		3	
12.15 – 14.45	0.2142		10.7		12	
14.45 – 16.75	0.2911		14.6		14	
16.75 – 19.05	0.2284		11.4		13	
19.05 – 21.35	0.1072		5.4	7.2	4	6
21.35 – 23.65	0.0351		1.8		2	
Total	0.9990	(≈ 1)	50.0		50	

Again, the classes with expected frequency values that are < 5 are combined with their neighbouring classes. Consequently, the number of classes used in the degrees of freedom calculation is five. We lose three degrees of freedom because we used three independent quantities to calculate the expected values from the data: the sample mean, sample standard deviation and N, the sum of the frequencies. The degrees of freedom for the χ^2-value are then:

$$\nu = 5 - 3 = 2.$$

6. $\chi^2_{0.01(2)} = 9.21$

7. $\chi^2_{(2)} = \frac{(5-6.1)^2}{6.1} + \frac{(12-10.7)^2}{10.7} + \frac{(14-14.6)^2}{14.6} + \frac{(13-11.4)^2}{11.4} + \frac{(6-7.2)^2}{2.2}$

$$= 0.198 + 0.158 + 0.025 + 0.225 + 0.200 = 0.806$$

8. Since 0.806 < 9.21, the test statistic is in the acceptance region and we 'accept' H_0.

9. It can be assumed that the population of dbh measurements for this stand follows a normal distribution.

10.2 Test for Independence

When frequency observations are grouped by two or more classification criteria, it is often important to determine whether these various criteria are independent of one another. For example, it may be important to know for the purpose of marketing a product whether wood species preference is independent of consumer gender.

Similarly, a forester might be interested in studying whether different progenies of a particular tree species react differently to a certain insect infestation. Like the goodness-of-fit test, a χ^2-value can be calculated to test the independence of the two criteria. Although the test is available for more than two classifications, we restrict our discussion in this book to testing two classifications.

Let us use an example to introduce the process of the **test for independence**. Table 10.3 summarizes the classification of 221 pieces of 2 × 8 dimensional lumber by grade and work shift. Shift 1 is from 8:00 am to 4:00 pm, shift 2 is from 4:00 pm to 12 midnight, and shift 3 is from 12 midnight to 8:00 am. Since newer employees tend to be assigned to shifts 2 and 3, it is hypothesized that the quality of lumber being produced will be lower during these shifts, which would show in the distribution of lumber by grade. This set of observations is called a 3 × 4 **contingency table**, because it consists of 3 rows (r) and 4 columns (c), referring to shifts and grades, respectively, in this case. In general, the size of a contingency table for problems with two criteria is defined by $r \times c$.

Observations like those in Table 10.3 are considered independent when the relative frequencies of the row classification criteria are similar across the column classification criteria. In our example, grade is considered independent of shift when the various grades for the three shifts are similarly proportioned: that is, the distributions of grades are similar across all shifts. If the distributions vary from shift to shift, then they are shift-dependent. Conversely, the data are considered to be independent if the relative frequencies of the various shifts within each of the four grades are similarly related.

For general notation, we will use O_{ij} to denote the **observed frequencies**, where $i = 1, 2, \ldots, r$ (row number), $j = 1, 2, \ldots, c$ (column number). Also, we denote the row totals by $R_1, R_2, \ldots, R_c$, the column totals by $C_1, C_2, \ldots, C_r$, and the grand total (sum of all frequencies) by N. In order to calculate the **expected frequencies**, E_{ij}, where $i = 1, 2, \ldots, r$, and $j = 1, 2, \ldots, c$, the rule of *independence* from probability theory is used (see Eqn 3.16, Chapter 3):

$$P(A \cap B) = P(A)\,P(B).$$

Table 10.3. Distribution of lumber grades by shift.

	Shift	Lumber grade A	B	C	Reject	Total
(a)	1	10	25	28	5	68
	2	8	32	30	7	77
	3	6	15	25	10	76
	Total	24	72	103	22	221
	Row	Column 1	2	3	4	Total
(b)	1	O_{11}	O_{12}	O_{13}	O_{14}	R_1
	2	O_{21}	O_{22}	O_{23}	O_{24}	R_2
	3	O_{31}	O_{32}	O_{33}	O_{34}	R_3
	Total	C_1	C_2	C_3	C_4	N

This rule states that the probability of the intersection of two independent events is the product of the probabilities of the two events. Applying this to our example, the probability that a piece is produced in shift 1 is:

$P(\text{shift } 1) = 68/221.$

The probability that a piece is grade B is:

$P(\text{grade B}) = 72/221.$

The intersection of these two events, if they are independent is:

$P(\text{shift } 1 \cap \text{grade B}) = P(\text{shift } 1)\ P(\text{grade B}) = (68/221)\ (72/221) = p_{12}.$

p_{12} is the probability that we observe an outcome (observation) in the cell of row 1 and column 2. From here, the general equation to calculate the probability for any outcome, under the assumption that the rows (shift) and columns (grade) are independent, is:

$p_{ij} = (R_i/N)(C_j/N).$

The expected frequency for row i and column j is equal to the product of p_{ij} and N. This is mathematically equivalent to multiplying the sum of row i and the sum of column j and then dividing by the sum of all frequencies:

$$E_{ij} = N\, p_{ij} = N\,(R_i/N)\,(C_j/N) = (R_i)\,(C_j)/N. \tag{10.2}$$

These expected values are calculated so that they are row and column independent, meaning that both the relative distributions of each row and each column are the same.

In tests for independence, the null hypothesis states that for each cell, the expected and observed frequencies are the same (i.e. the rows and columns are independent). The alternative hypothesis is that at least one of the observed frequencies is not equal to its corresponding expected frequency:

H_0: $O_{ij} = E_{ij}$, for all possible *i*s and *j*s; or, the observed frequencies are independent of shift and grade,

and

H_1: $O_{ij} \neq E_{ij}$, for at least one set of i and j; or, the observed frequencies are dependent on shift and grade.

The test statistic is then computed as:

$$\chi^2_{(r-1)(c-1)} = \sum_{i=1}^{r}\sum_{j=1}^{c}\frac{\left(O_{ij} - E_{ij}\right)^2}{E_{ij}} \tag{10.3}$$

The critical values are obtained from the χ^2-table (see Table A.6, Appendix A) with $(r-1)\,(c-1)$ degrees of freedom. Recall the general rule of degrees of freedom, which stated that the degrees of freedom for the χ^2 distribution were equal to the total number of cells minus the number of 'independent quantities' obtained from the sample data to calculate the expected frequencies. It can be seen that:

$(r-1)\,(c-1) = rc - r - c + 1 = rc - (r + c - 1),$

where rc = total number of cells and $(r + c - 1)$ = number of independent quantities used to calculate the expected frequencies.

From Eqn 10.2, a casual observer might think that $(r + c + 1)$ quantities are used to calculate the expected frequencies: r row sums, c column sums, and 1 for N, the sum of all frequencies (see Eqn 10.2). However, these are not all independent. Since we know the r number of row sums, only $(c - 1)$ column sums are necessary to perform the calculations because the c column sums must add up to the same sum (N) as the row sums. For the same reason, N is known if we know all of the row sums and/or the column sums. In other words, to calculate the expected frequencies, we only need to know the values of $(r - 1)$ of the r rows or $(c - 1)$ of the c columns, which gives us degrees of freedom of $rc - (r + c - 1)$, algebraically equivalent to $(r - 1)(c - 1)$.

We now continue with our example and present the complete test of independence for grade distributions by shift:

1. H_0: $O_{ij} = E_{ij}$ for all is and js; the distribution of grades is independent of shift.
2. H_1: $O_i \neq E_{ij}$ for at least one pair of i and j; the distribution of grades is dependent on shift.
3. $\alpha = 0.05$.
4. Use Eqn 10.3.
5. The expected values are:

$E_{11} = (24)(68)/221 = 7.4$ $E_{21} = (24)(77)/221 = 8.4$ $E_{31} = (24)(76)/221 = 8.3$
$E_{12} = (72)(68)/221 = 22.2$ $E_{22} = (72)(77)/221 = 25.1$ $E_{32} = (72)(76)/221 = 24.8$
$E_{13} = (103)(68)/221 = 31.7$ $E_{23} = (103)(77)/221 = 35.9$ $E_{33} = (103)(76)/221 = 35.4$
$E_{14} = (22)(68)/221 = 6.8$ $E_{24} = (22)(77)/221 = 7.7$ $E_{34} = (22)(76)/221 = 7.6$

Thus, the expected values by lumber grade and shift are:

	Lumber grade			
Shift	A	B	C	Reject
1	7.4	22.2	31.7	6.8
2	8.4	25.1	35.9	7.7
3	8.3	24.8	35.4	7.6

6. $\chi^2_{0.05(6)} = 12.6$

7. $$\chi^2_{(6)} = \frac{(10-7.4)^2}{7.4} + \frac{(25-22.2)^2}{22.2} + \frac{(28-31.7)^2}{31.7} + \frac{(5-6.8)^2}{6.8} + \frac{(8-8.4)^2}{8.4} + \frac{(32-25.1)^2}{25.1} + \frac{(30-35.9)^2}{35.9} + \frac{(7-7.7)^2}{7.7} + \frac{(6-8.3)^2}{8.3} + \frac{(15-24.8)^2}{24.8} + \frac{(45-35.4)^2}{35.4} + \frac{(10-7.6)^2}{7.6} = 12.996$$

8. Since $12.996 > 12.6$, the test statistic is in the rejection region and we reject H_0 and accept H_1.
9. The *distribution of grades is dependent on shift*, meaning that the distributions of grades vary from shift to shift.

In the test for independence, the two classification criteria compared for independence do not have to be qualitative, as in our example above. They can also be quantitative, which makes it possible to compare two or more frequency distributions (grouped or ungrouped) obtained from two or more populations. In this case, one classification criteria would be the source of the population and the other would be the frequency classes. Note that when several frequency distributions are compared using the test for independence, the frequency class limits (boundaries) must be the same for each frequency distribution. For example, we could compare frequency distributions of dbh observations from three forest stands, or frequency distributions of accidents from two sawmills, like the one described in Example 10.2, using the chi-square test described in this section. If the test shows independence when comparing distributions, it can be assumed that the various populations being compared follow similar distributions.

A special case of the contingency table occurs when there are either two rows or two columns. In this case, the test for independence is equivalent to testing more than two proportions. This special case can be treated as an extension of the tests for the difference between two proportions (see Section 9.7, Chapter 9). In **testing several proportions**, the null hypothesis – that all proportions are equal – is tested against the alternative that at least one is different:

$H_0: p_1 = p_2 = \ldots = p_k$

H_1: at least one is different.

The test statistic for several proportions is seen in Eqn 10.3. This test again results in a one-tailed test for the same reasons discussed above.

Example 10.4. Three chemical treatments were applied to random samples of Douglas-fir seeds. An additional random sample was left as a control (i.e. not treated). After the treatment, the germination tests showed the results seen below in Table 10.4. Using a 0.01 level of significance, do these data indicate that the chemical treatments and control have different effects on germination?

Table 10.4. Distribution of Douglas-fir seed germination by chemical treatment.

		Treatment			
Observed frequencies	Control	1	2	3	Total
Germinated	84	72	88	62	306
Not germinated	10	8	9	10	37
Total	94	80	97	72	343

1. H_0: $O_{ij} = E_{ij}$ for all *i*s and *j*s; or $p_c = p_1 = p_2 = p_3$.
2. H_1: $O_{ij} \neq E_{ij}$ for at least one pair of *i* and *j*; or, at least one proportion is different.
3. $\alpha = 0.01$.
4. Use Eqn 10.3.
5. $E_{11} = (94)(306)/343 = 83.9$ $\quad E_{21} = (94)(37)/343 = 10.1$
   $E_{12} = (80)(306)/343 = 71.4$ $\quad E_{22} = (80)(37)/343 = 8.6$
   $E_{13} = (97)(306)/343 = 86.5$ $\quad E_{23} = (97)(37)/343 = 10.5$
   $E_{14} = (72)(306)/343 = 64.2$ $\quad E_{24} = (72)(37)/343 = 7.8$

Thus, the expected frequencies by treatment are:

	Control	1	2	3
Germinated	83.9	71.4	86.5	64.2
Not germinated	10.1	8.6	10.5	7.8

6. $\chi^2_{0.01(3)} = 11.3$.

7. $$\chi^2_{(3)} = \frac{(84-83.9)^2}{83.9} + \frac{(72-71.4)^2}{71.4} + \frac{(88-86.5)^2}{86.5} + \frac{(62-64.2)^2}{64.2}$$
$$+ \frac{(10-10.1)^2}{10.1} + \frac{(8-8.6)^2}{8.6} + \frac{(9-10.5)^2}{10.5} + \frac{(10-7.8)^2}{7.8} = 0.9842$$

8. Since 0.9842 < 11.3, the test statistic is in the acceptance region and we 'accept' H_0.
9. It can be assumed that *the four unknown population germination proportions (0.89, 0.90, 0.9 and 0.86) are the same.* In other words, the four unknown population proportions are not significantly different.

Another special case occurs when only two rows and two columns are used in a test for independence. These are called **2 × 2 contingency tables.** In this case, the test statistic (Eqn 10.3) used for testing independence is not exact. The sampling distribution for the 2 × 2 case is only *approximated* by a chi-square distribution, and this approximation is often poor. To correct for this, the English statistician, Frank Yates, introduced a **correction for continuity** for use in testing 2 × 2 contingency tables:

$$\chi^2_{(2-1)} = \sum_{i=1}^{2}\sum_{j=1}^{2} \frac{\left\{\left|O_{ij}\right| - 0.5\right\}^2}{E_{ij}} \tag{10.4}$$

The correction is strongly recommended when some or all of the frequencies are ≤ 10. When all of the cell frequencies are large (i.e. > 10), the values of the χ^2-test statistic are almost identical whether the correction is used or not: in these cases, the correction is not recommended.

As a final note, the test for the 2 × 2 contingency tables is equivalent to the test of comparing the differences between two proportions (see Section 9.7, Chapter 9). However, the results of these tests are only mathematically approximate, as the latter test relies on the normal approximation to the binomial distribution.

Example 10.5. A manufacturer of furniture components qualitatively evaluated the surface roughness of pieces produced by two operators (Table 10.5). Can we assume that the proportion of acceptable pieces are the same (independent) for the two operators, using $\alpha = 0.05$?

Table 10.5. Acceptable and unacceptable surface roughness frequencies for two operators.

	Roughness		
Operator	Unacceptable	Acceptable	Total
1	8	84	92
2	23	52	75
Total	31	136	167

1. H_0: $O_{ij} = E_{ij}$ for all *i*s and *j*s; or $p_1 = p_2$ (where p_i is the proportion of acceptable pieces from operator *i*).
2. H_1: $O_{ij} \neq E_{ij}$ for at least one pair of *i* and *j*; or $p_1 \neq p_2$.

3. $\alpha = 0.05$.
4. Use Eqn 10.4, since one of the frequencies is < 10.
5. $E_{11} = (31)(92)/167 = 17.1 \quad E_{21} = (31)(75)/167 = 13.9$
   $E_{12} = (136)(92)/167 = 74.9 \quad E_{22} = (136)(75)/167 = 61.1$

Thus, the expected frequencies by operator are:

Operator	Unacceptable	Acceptable
1	17.1	74.9
2	13.9	61.1

6. $\chi^2_{0.05(1)} = 3.84$

7. $$\chi^2_{(1)} = \frac{\{|8-17.1|-0.5\}^2}{17.1} + \frac{\{|84-74.9|-0.5\}^2}{74.9} + \frac{\{|23-13.9|-0.5\}^2}{13.9} + \frac{\{|52-61.1|-0.5\}^2}{61.1} = 11.85$$

8. Since 11.85 > 3.84, the test statistic is in the critical region and we reject H_0 and accept H_1.
9. *The unknown population proportions of unacceptable pieces for each operator are not equal.*

We can also solve this problem by using a *z*-test for the difference between the two proportions of unacceptable pieces:

1. H_0: $p_1 - p_2 = 0$.
2. H_1: $p_1 - p_2 \neq 0$.
3. $\alpha = 0.05$.
4. Use Eqn 9.10.
5. $n_1 = 92$; $n_2 = 75$; $$s_{\hat{p}_1-\hat{p}_2} = \sqrt{\frac{(0.087)(0.913)}{92} + \frac{(0.307)(0.693)}{75}} \approx 0.061$$

Note that we did not use the combined proportion (p_c) here because the proportions are considerably different.

6. $Z_{0.025} = \pm 1.96$.
7. $$Z = \frac{(0.087 - 0.307) - 0}{0.061} \approx -3.62$$
8. Since –3.62 < –1.96, the test statistic is in the critical region; we reject H_0 and accept H_1.
9. Same as for the χ^2-test above.

Again, the *z*-test performed here is a two-tailed test and the χ^2-test is a one-tailed test. Also, $(-3.62)^2 \approx 13.10$ is approximately equal to the χ^2 statistic, 13.27, that we would get by using Eqn 10.3 instead of Eqn 10.4 (with Yates' correction), for the same reasons discussed under the goodness-of-fit test.

Before closing this chapter, we give a word of caution. If any of the expected class frequencies are < 5, for any size of contingency table, the sampling distribution of the test statistic discussed in this section is very poorly approximated by the chi-square distribution. A special procedure, the *Fisher–Irwin exact test*, is available for cases when some of the expected frequencies are < 5. Details of this test can be found in more advanced statistical texts.

Exercises

Section 10.1

10.1. A certain software manufacturer advertises that its random number generator produces numbers that follow a uniform distribution. Based on the following sample of 500 digits generated using this software, can we support their claim? Use $\alpha = 0.01$.

Digit	0	1	2	3	4	5	6	7	8	9
Frequency	43	52	55	46	48	57	49	60	41	49

10.2. In a regeneration survey, the number of established seedlings on 150 survey plots was recorded. The following table summarizes this data:

No. of seedlings	Frequency
0	21
1	41
2	40
3	25
4	11
5	7
6+	5

a. Identify the distribution.
b. Test whether your identification in 10.2a is correct. Use $\alpha = 0.05$.

10.3. One-hundred-and-five independent germination tests were carried out on a batch of Douglas-fir seeds. Ten seeds were used in each test and the resulting frequency distribution is shown below:

Number of seeds germinated	Frequency
0	0
1	0
2	0
3	0
4	2
5	0
6	12
7	18
8	33
9	30
10	10

a. Identify the distribution.
b. Test whether your identification in 10.3a is correct. Use $\alpha = 0.01$.

10.4. Consider the following frequency distribution of tree crown lengths (in metres) collected in a young stand.

Class limits	Frequency
< 4.4	4
4.5–6.4	14
6.5–8.4	22
8.5–10.4	27
10.5–12.4	16
12.5–14.4	7

Test the hypothesis that the distribution is normal using a 0.01 level of significance.

Section 10.2

10.5. Four forest types were sampled to examine their tree species distributions. The results are summarized as follows:

	Forest type			
Species	1	2	3	4
Douglas-fir	35	28	37	42
Hemlock	20	22	34	26
Other	12	8	10	6

Test the hypothesis that species and forest type are independent. Use $\alpha = 0.05$.

10.6. A park manager wanted to investigate whether a visitor's preference to stay at 1 of 3 available campsites was related to their income. The manager conducted a survey and the results are summarized below.

	Campsite		
Income (US$)	1	2	3
< 15,000	20	9	12
15,000–50,000	38	39	20
50,000 <	17	13	19

Is level of income independent of campsite preference? Use $\alpha = 0.01$.

10.7. The distribution of numbers of accidents per month in 2 large secondary wood manufacturing plants are as follows:

	Accidents per month				
Plant	0	1	2	3	4+
A	27	40	18	12	7
B	14	18	8	7	6

Is the distribution of accidents independent of plant? Use a = 0.05.

10.8. A study was conducted to find out whether germination rates from 4 different seed sources were the same or different.

	Seed source			
Germinated	A	B	C	D
Yes	85	76	92	87
No	10	8	9	10

a. Are the proportions of germinated seeds the same from the 4 sources? Use α = 0.01.
b. Repeat the test above for sources A and B only, using two different tests.

10.9. A company harvested 3 areas and the forester tabulated the distribution of grades by area for all 10 m logs brought to market.

	Peeler 1	Peeler 2	Saw-log	Pulp
Area A	90	180	210	120
Area B	20	40	42	21
Area C	54	90	111	50

Are the distributions of grade area dependent? Use α = 0.01.

10.10. In early May 1999, a late frost killed a significant number of seedlings from a certain nursery. To study the effect of frost on the 2 species grown in the nursery, the following data were collected:

	Killed	Survived
White spruce	97	223
Lodgepole pine	167	253

Is there a relationship between survival rate and species? Use a 0.01 level of significance and two different procedures to test the hypothesis.

This page intentionally left blank

11 Regression and Correlation

Relationships between Variables

Regression and correlation analyses are perhaps two of the most important statistical tools in forestry, as well as in most other physical and social science fields. Regression and correlation analyses provide an understanding of the mathematical relationship between two or more variables and are particularly important in the area of forest mensuration and measurements. The knowledge of the relationship between variables enables us to estimate or predict variables that are difficult or expensive to measure by using other variables that are easier or more cost-effective to measure. For example, when timber cruising, the volume of a tree is both very difficult and costly to measure. A good estimate would require cutting down the tree and submerging it in a large tank of water to measure the amount of water that is displaced. Obviously, this is not only untenable, but is also destructive (if we want to keep the tree where it is!). However, the diameter at breast height (dbh), or the total tree height, or the basal area at breast height, are all much easier and relatively cheaper variables to measure. We can use these variables to predict tree volume based on the relationships (equations) that we observe between the variables. In this chapter, we will cover statistical procedures to derive mathematical relationships between sampled tree volume and sampled dbh, tree height and/or basal area. The tools that we will use to derive these relationships are regression and correlation analyses.

Regression is often used in forestry to predict variables that are difficult to measure. Some examples include estimations of tree height from dbh, amount of decay from tree size or age, tree height from ages (also called site index equations), dbh from stump diameter in stump cruising and merchantable tree volume from tree size (usually dbh).

In many cases, we may not need to predict difficult or expensive variables from more affordable variables; instead, the objective is simply to study whether or not there is a significant relationship between two or more variables. Here, we would use **correlation**. Examples include relationships between: stand density and stand age; the strength properties of particleboard and the amount of resin used in its manufacture; the level of mountain pine beetle infestation and temperature; and the price of Canadian lumber and interest rates in the USA.

There are not many differences between the statistical procedures used when the main objective is to find the relationship between two or more variables for prediction purposes versus simply investigating the relationship between two or more variables. The former procedure is called **regression analysis**, while the latter is called **correlation analysis**.

When the relationship between two variables is a straight line, we use what is called a **simple linear regression**. When the relationship between two variables is characterized by a curve, it is called a **curvilinear regression**. When these relationships are based on more than two variables, they are called either **multiple linear regressions** or **multiple curvilinear regressions**, depending on their shapes. In this chapter, simple linear regression will be discussed in detail and some introductory comments will be made on both multiple linear and curvilinear regression. For further information on these more advanced forms of regression analysis, the student is directed to any number of texts devoted entirely to this subject.

11.1 Simple Linear Regression

Simple linear regression involves the straight-line relationship between two variables: the **dependent variable** (or the ***y*** **variable**) and the **independent variable** (or the ***x*** **variable**). In many cases, it is easy to distinguish between these two variables. For example, the height of a tree depends on its age. In this case, age is the independent variable and height is the dependent variable. However, in a relationship such as the one between height and dbh, it is very difficult – if not impossible – to decide which variable depends on the other. In these cases, the variable which is more difficult or more expensive to measure is typically called the dependent variable. In other words, height would be the dependent variable because dbh is the more easily measured variable. In timber inventories, we are also generally trying to predict height, not dbh. The independent variable is therefore frequently referred to as the **predictor variable** and the dependent variable is referred to as the **response variable**. In general, the dependent variable is the one that we wish to predict and it usually cannot be controlled or manipulated. In contrast, the independent variable can easily be measured and can usually be controlled and/or manipulated.

Determination of the regression equation

The recommended first step in both regression and correlation analyses is to create a **scatter plot** or **scatter diagram**. Here, pairs of data are plotted with the independent variable on the horizontal (x) axis and the dependent variable on the vertical (y) axis (see Fig. 11.1). These scatter plots serve two very important roles. First, they graphically show the nature of the relationship between the independent and the dependent variables. For example, Fig. 11.1a shows a straight-line relationship, Fig. 11.1b shows a curvilinear relationship and Fig. 11.1c shows no relationship at all between the dependent and independent variables. Scatter plots are also useful for indicating **outliers** in the data (see Fig. 11.2). Outliers are extreme observations, oftentimes a result of some source of measurement or experimental error. Outliers, like the one shown in Fig. 11.2, can seriously influence a regression analysis. If the source of error can be identified, we recommend that it be corrected. If the source of error cannot be located, and the researcher is convinced the observation is erroneous, it should be deleted. The reader is referred to more advanced textbooks for more information on outliers (e.g. Draper and Smith, 1998).

To introduce the procedures used in simple linear regression, we will refer to a practical example throughout this chapter. Assume that we have ten pairs of observations of dbh and crown radius taken from a sample of 10 young white pine trees. The dbh measurements (in cm) will be the independent variable (x), while the crown radius measurements (in m) will be the dependent variable (y). Note that dbh

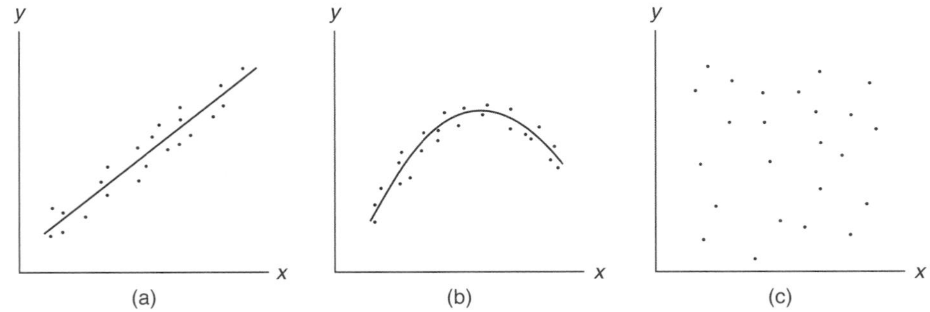

Fig. 11.1. Scatter diagrams indicating (a) a linear relationship, (b) a curvilinear relationship and (c) no relationship.

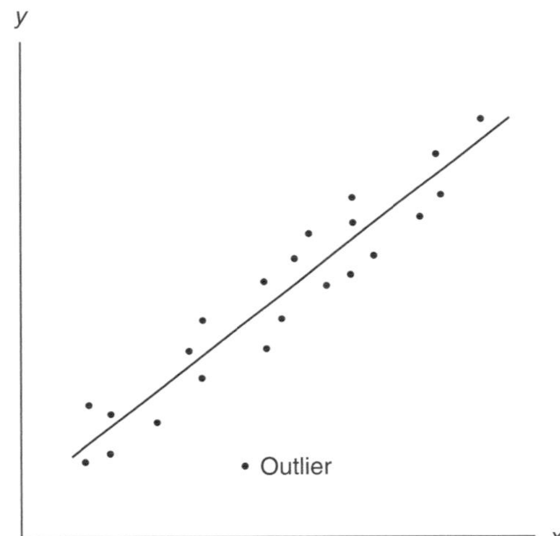

Fig. 11.2. The effect of an outlier.

is very easy to measure (with a diameter tape, known as a D-tape), while a measurement for crown radius is more difficult to obtain. The data are shown below and the scatter diagram of our paired measurements presented in Fig. 11.3 reveals a reasonably straight-line (linear) relationship.

x	5.0	12.7	7.6	17.8	5.1	15.2	10.2	22.9	20.3	10.1
y	0.91	1.83	1.22	2.18	1.22	2.30	1.70	2.74	2.65	1.52

If the scatter diagram indicates an outlier-free straight-line relationship, the next task is to find the 'best' equation to describe the relationship of data points. In mathematics, the equation for a straight-line relationship between x and y is typically described by:

$$y = mx + b,$$

where m is the slope for the equation and b is the y-intercept. In regression analysis, the same general equation is used, although the terminology is different:

$y = b_0 + b_1x$ (sometimes this is expressed as $y = a + bx$),

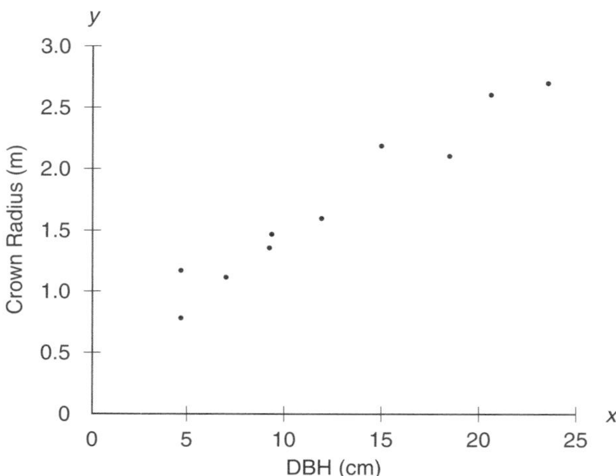

Fig. 11.3. Scatter diagram of the dbh and crown radius data.

where b_0 is the y-intercept and b_1 is the slope for the equation. This is called a *deterministic relationship* between x and y, because any value of y is completely determined by a value of the independent variable, x. To describe the actual relationship between variables *in a sample*, we use yet another form (which is mathematically the same as the above). More specifically, two slightly different equations are introduced, the first describing the straight regression line and the second describing the location of each individual observation:

$$\hat{y}_i = b_0 + b_1x_i \tag{11.1}$$

and

$$y_i = b_0 + b_1x_i + e_i \tag{11.2}$$

where

$\hat{y}_i$ = a point on the regression line for a given x_i (a prediction of y_i for a given x_i);
x_i = an observation of the independent variable;
$e_i = y_i - \hat{y}_i$, the residual error for the ith observation (the vertical difference between the ith plotted point and the point corresponding to x_i on the line);
$i = 1, 2, 3, \ldots, n$ (an index or count of the observations in a sample);
n = number of observations (number of pairs of xs and ys);
b_0 = y-intercept of the regression line, and
b_1 = the slope of the regression line.

Equation 11.1 describes the straight-line relationship between the points (Fig. 11.4), while Eqn 11.2 incorporates a **residual** term to describe the location of every single point around the straight regression line. Equation 11.1 is often called the **regression equation.** Equation 11.2 is called the **regression model** and more generally referred to as an **additive probabilistic model** (in this book, we use the terms 'equation' and 'model' interchangeably). The notations b_0 and b_1 are used to describe the relationship between values of x and y of a sample taken from a population, and are called **regression coefficients.**

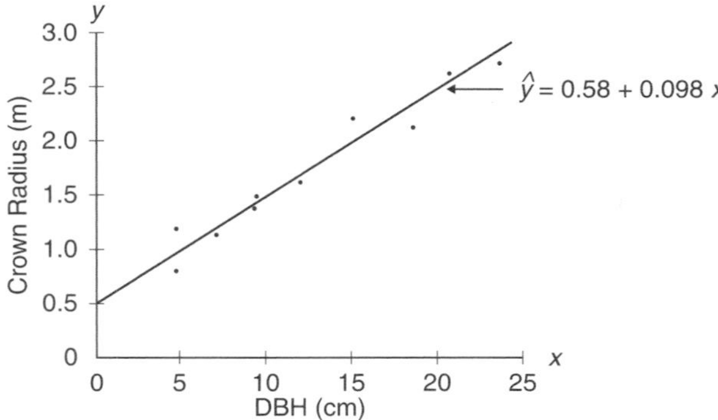

Fig. 11.4. The regression line for the dbh and crown radius data.

The following equations are used to describe the relationship between the x and y values *for the entire population*:

$$\mu_{y|x} = \beta_0 + \beta_1 x_i \tag{11.3}$$

and

$$y_i = \beta_0 + \beta_1 x_i + \varepsilon_i \tag{11.4}$$

where

y_i and x_i = as above, and
i = 1, 2, 3, ..., N (an index or count of all of the observations in a population);
N = number of all possible observations (pairs of x and y values) in the population;
$\varepsilon_i = y_i - \mu_{y|x_i}$, the ith residual;
β_0 = y-intercept for the population, and
β_1 = slope of the straight-line relationship for the population.

β_0 and β_1 are called **parameters** of the regression equations or models because relationships within a population are being described.

The **method of least squares**, which determines the line in which the **sum of squares for error** is minimized, is used to find the numerical value of the regression coefficients, b_0 and b_1, for a sample. In other words, our goal is to find the line of 'best fit' for the data at hand. To illustrate this point, Fig. 11.5a shows the line of best fit:

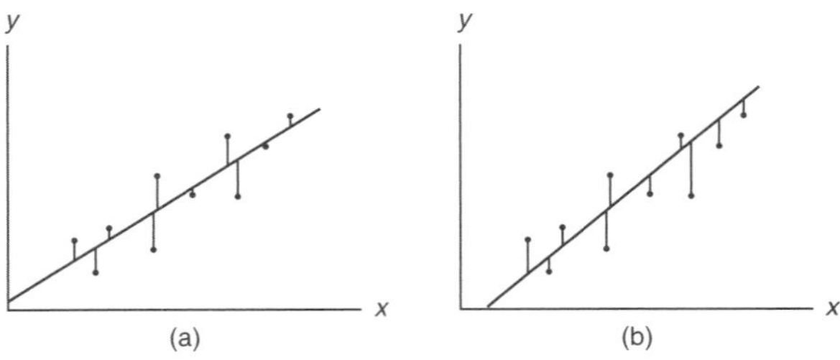

Fig. 11.5. Two hypothetical regression lines showing (a) the line of best fit and (b) a line that does not fit the data well.

this is equivalent to the line in which the sum of squares for error is minimized. Conversely, Fig. 11.5b shows a line that does not fit the data well; here, the sum of square errors is not minimized.

Frequently, the process of finding the numerical values of the regression coefficients is referred to as *fitting the regression*. The complete derivation of the method of least squares requires differential calculus, which is beyond the scope of this text and will only be outlined as follows.

If we let $\hat{y}_i$ denote the predicted straight-line values of the dependent variable, the residual sum of squares is:

$$SS_{Res} = \sum_{i=1}^{n}\left(y_i - \hat{y}_i\right)^2$$

Substituting for $\hat{y}_i$ (Eqn 11.1)

$$SS_{Res} = \sum_{i=1}^{n}\left[y_i - \left(b_0 + b_1 x_i\right)\right]^2$$

Using the method of least squares, we want to minimize SS_{Res} to obtain the line of best fit. Therefore, we take the partial derivatives of the above function with respect to b_0 and b_1 and set these equal to zero, which produces two equations:

$$b_0 n + b_1 \sum x_i = \sum y_i$$

$$b_0 \sum x_i + b_1 \sum x_i^2 = \sum x_i y_i \, ,$$

where the index, i, for each Σ goes from 1 to n.

For the two equations above, the solution for the two unknowns, b_0 and b_1, which minimizes SS_{Res}, is:

$$b_1 = \frac{\sum_{i=1}^{n} x_i y_i - \left(\sum_{i=1}^{n} x_i\right)\left(\sum_{i=1}^{n} y_i\right) \Big/ n}{\sum_{i=1}^{n} x_i^2 - \left(\sum_{i=1}^{n} x_i\right)^2 \Big/ n} = \frac{SP_{xy}}{SS_x} \qquad (11.5)$$

and

$$b_0 = \frac{\sum_{i=1}^{n} y_i - b_1 \sum_{i=1}^{n} x_i}{n} = \bar{y} - b_1 \bar{x}, \qquad (11.6)$$

where

$\bar{y}$ = mean of the dependent variable;
$\bar{x}$ = mean of the independent variable;
SS_x = corrected sum of squares of x; and
SP_{xy} = corrected sum of products x and y.

See Chapter 2 for the equations of $\bar{y}$, $\bar{x}$, and SS_x. Note that SS_x divided by $(n - 1)$ would result in the variance of x, and that SP_{xy} divided by $(n - 1)$ would result in the covariance (see Chapter 4) of x and y.

We recommend the following steps to find the values of b_0 and b_1. Find:

1. n = number of observations;

2. $\bar{x}$ = mean of x;
3. $\bar{y}$ = mean of y;
4. SS_x = corrected sum of squares of x;
5. SS_y = corrected sum of squares of y; and
6. SP_{xy} = corrected sum of products of x and y.

Going back to the dbh and crown radius example, these are:

$n = 10$
$\bar{x} = 126.9/10 = 12.69$
$\bar{y} = 18.27/10 = 1.827$
$SS_x = 1960.49 - 126.9^2/10 \approx 350.13$
$SS_y = 36.9267 - 18.27^2/10 \approx 3.5474$
$SP_{xy} = 266.282 - (126.9)(18.27)/10 \approx 34.4357$
$b_1 = 34.4357/350.13 \approx 0.0984 \approx 0.098$
$b_0 = 1.827 - (0.0984)(12.69) \approx 0.5783 \approx 0.58$

Note that SP_{xy} can be negative, which indicates a negative relationship between x and y. In other words, as x increases, y decreases.

If the numerical values of b_0 and b_1 are calculated using the least squares method, the regression line will have the following properties:

1. It passes through the point of $(\bar{x}, \bar{y})$ (see Fig. 11.5).
2. The sum of the residuals around the regression line is equal to zero $\sum_{i=1}^{n}(y_i - \hat{y}_i)$, which indicates a well-balanced line between the points.
3. The sum of squares of the residuals is at its minimum.

This last point does not need any mathematical proof, as this is how the least squares line was derived. The first and second can be verified mathematically:

1. Substituting $\bar{x}$ for x_i in the regression model of Eqn 11.1, we have $\hat{y}_i = b_0 + b_1\bar{x}$.

Substituting $\bar{y} - b_1\bar{x}$ for b_0 (see Eqn 11.6), we get $\hat{y}_{\bar{x}} = \bar{y} - b_1\bar{x} + b_1\bar{x} = \bar{y}$.

Therefore, $(\bar{x}, \bar{y})$ is a point on the regression line.

2. The sum of residuals is $\sum_{i=1}^{n}(y_i - \hat{y}_i)$. Substituting the value of $\hat{y}_i$ from Eqn 11.1 in the summation, we have $\sum_{i=1}^{n}\left[y_i - (b_0 + b_1x_i)\right] = \sum_{i=1}^{n} y_i - \sum_{i=1}^{n} b_0 - \sum_{i=1}^{n} b_1x_i = \sum_{i=1}^{n} y_i - nb_0 - b_1\sum_{i=1}^{n} x_i$.

Then, substituting the value of b_0 from Eqn 11.6, we have $\sum_{i=1}^{n} y_i - n\bar{y} + nb_1\bar{x} - b_1\sum_{i=1}^{b} x_i$, since $n\bar{y} = \sum_{i=1}^{n} y_i$ and $n\bar{x} = \sum_{i=1}^{n} x_i$, the above expression reduces to zero.

Figure 11.6 shows the algebraic and trigonometric meaning of the y-intercept and the slope. In general, the slope indicates the amount of change in the dependent variable, y, for every unit change of the independent variable, x. The inverse tangent of the slope gives the angle that the regression line makes with the x-axis. The intercept indicates the value of the dependent variable, y, when $x = 0$. Numerically, for the dbh and crown radius example, the slope of 0.098 means that the crown radius increases (on average) by 0.098 m (9.8 cm) increase for every 1 cm increase in dbh.

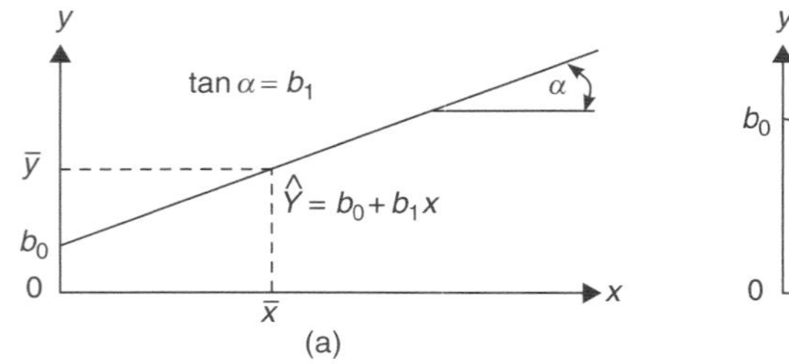

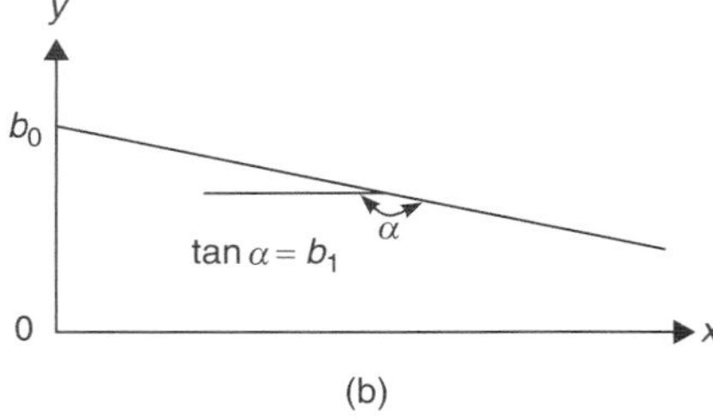

Fig. 11.6. The *y*-intercept and the slope: (a) positive slope and (b) negative slope.

The intercept of 0.58 indicates that the crown radius is 0.58 m when the dbh is zero. However, a word of caution is in order here. If the observations for the independent variable are much > 0, the interpretation of the intercept can be misleading. Unless there are good reasons to believe that the model fitted (in this case, the straight line) is valid beyond the range of observed data, predictions of the dependent variable outside the range of the sampled independent variable should only be made with extreme caution. In our example, the smallest dbh is 5.0 cm and we do not know whether the relationship between $x = 0$ and $x = 5$ is a straight line or a curve, so the interpretation of a crown radius of 0.58 m may be misleading. In fact, it is fairly safe to assume that if a tree has a dbh of 0 cm, it also lacks a crown radius (in other words, it does not exist). If, however, there are observations close to $x = 0$, the interpretation of the intercept is fairly reliable.

Regression analysis

Since the method of least squares is a mathematical procedure, no assumptions about the observations are necessary to find the regression coefficients, b_0 and b_1. However, when statistical inferences are required about the regression parameters, predictions made from the model, or the quality (goodness) of the relationship, the following assumptions must be met for the case of simple linear regression:

1. A simple linear regression model, $y_i = \beta_0 + \beta_1 x_i + \varepsilon_i$, is appropriate for the population (i.e. the relationship within the population between x and y is linear).
2. The x_i values in the data pairs of $\{(x_i, y_i); i = 1, 2, 3, \ldots, n\}$ are *fixed* (or measured without error), and are not values of a random variable.
3. The y_i values in the data pairs of (x_i, y_i) for any given x_i are values of a random variable, Y, and are *normally* distributed.
4. The y_i values for any given x_i have a *uniform variance*; that is, the unknown population variances from one given x_i to another are the same.
5. A y_i value for a given x_i value is independent from any other y_i values in the sample.
6. The samples of y_i values for any given x_i value are randomly selected from the population of the random variable of Y.

Assumption 1 is a hypothetical statement about the relationship between the x and y values for the entire population. Assumption 2 indicates that values of the independent variable need not be randomly selected, but can be selected arbitrarily.

For our example, this means that we should select trees with certain dbh values (perhaps encompassing a wide range of values) and then measure their crown radius. Also, as stated, the dbh should be measured without error, something that is virtually impossible. The only thing we can do is to measure the values of dbh with as much accuracy as possible. In most textbooks, Assumptions 3–5 are combined into one and stated as the values of ε_i are 'iid $N(0,\sigma^2)$', meaning that the residuals are independent, identically distributed normal (N) variates, with a zero mean and a constant variance of σ^2. Lastly, Assumption 6 indicates that for a selected dbh, there are many trees with various radius measurements and we take a random sample of these trees.

If the above assumptions are met, various inferences can be made about the quality of the relationship, about β_0 and β_1 and about the unknown population mean of y at a given x value.

After calculating b_0 and b_1, the logical questions are: '*How good is the relationship between* x *and* y*?*', or '*How is* y *affected by* x*?*' Several diagnostic tests are available to answer these questions and, in all of them, we test the assumption that there is no relationship between the independent and the dependent variables. H_0 is set up to claim that the unknown population slope is equal to zero (a flat line would indicate no relationship because, as x changes, nothing happens to y). The alternative hypothesis claims that the population slope is non-zero:

$$H_0: \beta_1 = 0,$$

$$H_1: \beta_1 \neq 0.$$

These hypotheses can be tested either by using an F-test (analysis of variance), r-test (test for the correlation coefficient), or t-test. It should be noted that these three tests are mathematically equivalent and lead to the same decisions and conclusions. For this reason, they are discussed together, but in turn.

However, before discussing these tests, we introduce the important concept of partitioning the total variation in a regression. For a data set consisting of independent and dependent variables, the variation that we are concerned with occurs in the dependent variable, because this is what we are trying to predict. The total variation (in terms of sum of squares) of the dependent variable, y, is called SS_{Total}, or SS_T, or SS_y. This total sum of squares can be partitioned into two components: the variation caused by the regression, or accounted for by the regression (SS_{Reg}), and the residual variation, or the variation around the regression line (SS_{Res}).

Figure 11.7 graphically illustrates the way that the total sum of squares is partitioned into the two components: (a) shows the variation (SS_T) of the y_i values (raw data) around their mean, $\bar{y}$ (which would be our best estimate of the population mean for y, given no information regarding its relationship with x); (b) is the variation (SS_{Reg}) of the predicted values around the mean, $\bar{y}$; and (c) is the variation (SS_{Res}) of y_i values around the regression line.

Mathematically stated, the way in which the total sum of squares (SS_T) is partitioned is as follows:

$$SS_T = SS_{Reg} + SS_{Res} \tag{11.7}$$

or

$$\sum_{i=1}^{n}\left(y_i - \bar{y}\right)^2 = \sum_{i=1}^{n}\left(\hat{y}_i - \bar{y}\right)^2 + \sum_{i=1}^{n}\left(y_i - \hat{y}_i\right)^2 \tag{11.8}$$

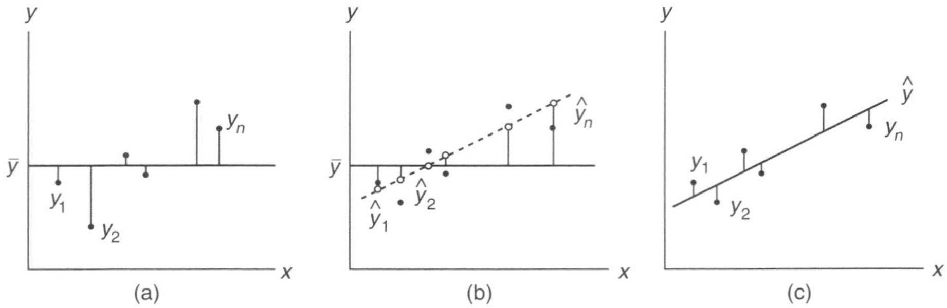

Fig. 11.7. The concept of partitioning the sum of squares of the dependent variable (a) SS_T, (b) SS_{Reg} and (c) SS_{Res}.

with the degrees of freedom of the three respective sum of squares terms being:

$$(n - 1) = 1 + (n - 2). \tag{11.9}$$

As discussed in Chapter 2, SS_T has $(n - 1)$ degrees of freedom, because only one statistic is used from the data to calculate this sum of squares. The degrees of freedom of SS_{Res} is $(n - 2)$, which can be explained in two different ways. Since we use n statistics (Eqn 11.8) to calculate the residual sum of squares, it looks as though it should have zero $(n - n)$ degrees of freedom. However, since it is known from basic algebra that two points define a straight line, and all of the values occur on a straight line, only two out of the n values are independent. That is, if we know any two of the values, all the others can be calculated. Thus, the degrees of freedom of SS_{Res} is $(n - 2)$. Another explanation of this is that the values are calculated using two statistics: the regression coefficients, b_0 and b_1. Since these are calculated from the data, SS_{Res} has $(n - 2)$ degrees of freedom. Similarly, from Eqn 11.8, it looks as though the sum of squares caused by the regression (SS_{Reg}) should have $(n - 1)$ degrees of freedom, but since only two of the n values are independent, the degrees of freedom are $(2 - 1) = 1$.

The first diagnostic test, the **F-test** or **analysis of variance** test, is based on this partitioning of variation, with the following logic. When the SS_{Reg} accounts for a very small portion of the total sum of squares (SS_T), we have a poor relationship between x and y; in fact, when it is zero, it means that $\beta_1 = 0$ exactly and there is no relationship between the independent and dependent variables at all. On the other hand, when SS_{Res} is zero, this means that we have a mathematically perfect relationship between x and y; that is, each data point is on the regression line. The difference between the sum of squares regression and the sum of squares residual is tested in such a way that we assume that the **variance due to regression** ($SS_{Reg}/1$) is equal to the **residual variance** ($SS_{Res}/(n - 2)$). The alternative hypothesis states that the variance due to regression is greater than the residual variance, meaning that we have a good relationship between x and y:

$$H_0: \sigma^2_{Reg}/\sigma^2_{Res} = 1.0$$

$$H_1: \sigma^2_{Reg}/\sigma^2_{Res} > 1.0.$$

The above null and the alternative hypotheses can be restated in terms of β_1, as shown earlier. The two statements are equivalent, since:

$$E(MS_{Res}) = \sigma^2_{Res}$$

and

$$E(MS_{Reg}) = \sigma^2_{Reg} = \sigma^2_{Res} + \beta_1^2\,(x_i - \bar{x})^2,$$

therefore if $\beta_1 = 0$,

$$\sigma^2_{Reg} = \sigma^2_{Res}.$$

For the above null hypothesis, the correct test statistic is:

$$F_{(1,n-2)} = \frac{SS_{Reg}/1}{SS_{Res}/(n-2)} = \frac{MS_{Reg}}{MS_{Res}} \tag{11.10}$$

for which the critical values are obtained from the *F*-table (see Table A.7, Appendix A), with 1 and $(n - 2)$ degrees of freedom. In Eqn 11.10, MS_{Reg} and MS_{Res} are point estimators of σ^2_{Reg} and σ^2_{Res}, respectively. Fisher (see Chapter 7, Section 7.7) coined the term **mean square (MS)** in place of variance, which is short for *mean sum of squares*. Thus, variance and mean square are analogous – a sum of squares term divided by its corresponding degrees of freedom. However, the mean square terminology is generally used in regressions and analyses of variance (see Chapter 12).

In most cases, the above test can be summarized in an **analysis of variance** table (discussed in more detail in Chapter 12), as in Table 11.1. When the sums of squares are calculated using pocket calculators, 'working' or 'machine' equations should be used for ease of calculation. These working equations are algebraically the same as those presented above:

$$SS_T = SS_Y = \sum_{i=1}^{n} y_i^2 - \left(\sum_{i=1}^{n} y_i\right)^2 \Big/ n,$$

$$SS_{Reg} = b_1 SP_{xy}$$

and

$$SS_{Res} = SS_T - SS_{Reg} \text{(from Eqn 11.7).}$$

The numerical values for our example are:

$$SS_T = 3.5474 \text{ (from above)}$$

$$SS_{Reg} = (0.0984)(34.4357) \approx 3.3885$$

$$SS_{Res} \approx 3.5474 - 3.3885 \approx 0.1589.$$

The test to see whether the relationship between dbh and crown radius (based on our sample) is 'well defined' or 'significant' is set up as:

$$H_0 : \sigma^2_{Reg} / \sigma^2_{Res} = 1.0$$

$$H_1 : \sigma^2_{Reg} / \sigma^2_{Res} > 1.0$$

$\alpha = 0.01$ F-critical $= F_{0.01(1,8)} = 11.26.$

Table 11.1. Analysis of variance table.

Source of variation	Degrees of freedom	Sum of squares	Mean squares	F
Regression	1	SS_{Reg}	MS_{Reg}	MS_{Reg}/MS_{Res}
Residual	$n-2$	SS_{Res}	MS_{Res}	
Total	$n-1$	SS_T		

Since 170.27 > 11.26, we reject H_0: the regression is well defined (significant). Table 11.2 summarizes the analysis of variance for the dbh and crown radius data.

Two rather important descriptive statistics can be calculated from the sums of squares presented above. These are the **standard error of estimate:**

$$S_{y.x} = \sqrt{MS_{Res}} \tag{11.11}$$

and the **coefficient of determination** (also known as r^2):

$$r^2 = \frac{SS_{Reg}}{SS_T} = 1 - \frac{SS_{Res}}{SS_T} \tag{11.12}$$

The standard error of estimate describes the variation of the observations around the regression line. The coefficient of determination shows the proportion of the sum of squares of the dependent variable 'accounted for' or 'explained by' the independent variable, or by the regression. For our example, the calculations are as follows:

$$S_{y.x} = \sqrt{0.0199} \approx 0.1411 \approx 0.14$$

$$r^2 = \frac{3.3885}{3.5474} \approx 0.955$$

If the assumptions of regression analysis are met, especially Assumptions 3 and 4, the meaning of the standard error of estimate of 0.14 m is that about 68% of the observations (from which the regression equation was derived) can be found within ±0.14 m around the regression line, and that about 95% of the observations can be found within 2 × 0.14 = ±0.28 m around the regression line (Fig. 11.8). This is the *empirical rule* (see Chapter 2) applied to each one of the x values.

Since the standard error of estimate is a descriptive statistic, describing the spread of 'observations' around the regression line, it behaves like a 'standard deviation' and, technically, should not be called 'standard error': this term is generally reserved for describing the variation of a 'statistic'. Some books do not use the term 'standard error of estimate' at all, using either 'square root residual variance' or 'RMSE' (root mean square error) instead.

The meaning of 0.955 for the coefficient of determination is that about 96% (95.5% to be exact) of the sum of squares of the crown radius is accounted for or explained by the change in dbh. While this indicates a very good relationship between dbh and crown radius, it is difficult to provide general guidance on what constitutes a good r^2 value, since there are so many different relationships that we try to predict in forestry. The more important result is whether or not the r^2 value is significant. Due to the manner in which the coefficient of determination is interpreted, it is frequently expressed in percentage rather than as a proportion. That said, it can be shown that the range of r^2 values must lie between:

$$0.0 \leq r^2 \leq 1.0 \text{ (or } 0\% \leq 100r^2 \leq 100\%\text{)},$$

Table 11.2. Analysis of variance for the dbh and crown radius data.

Source of variation	*DF*	*SS*	*MS*	*F*
Regression	1	3.3885	3.3885	170.27
Residual	8	0.1589	0.0199	
Total	9	3.5474		

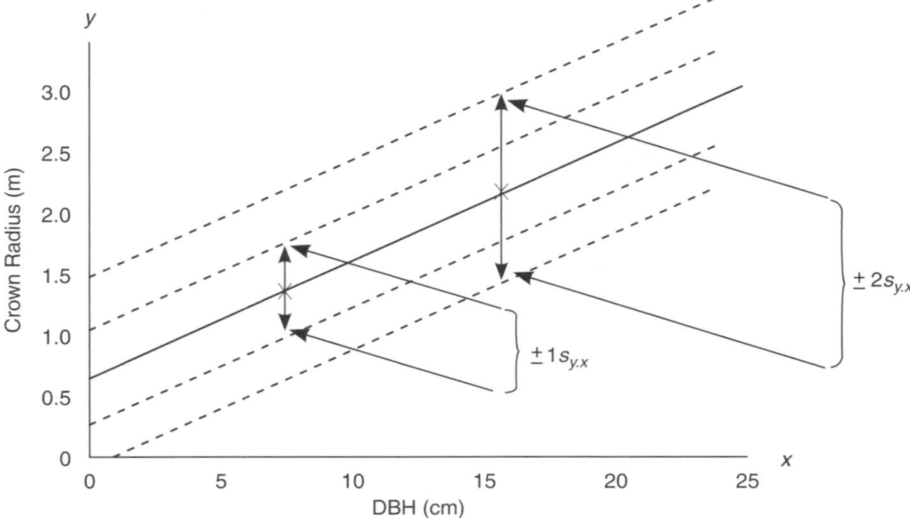

Fig. 11.8. The concept of the standard error of estimate for the dbh and crown radius data.

where zero means no relationship at all (a scattering of points with no discernable trend), while 1.0 means a perfect relationship between the two variables tested (each of the data points lies exactly on the regression line).

The **correlation coefficient** is the square root of the coefficient of determination and can be calculated as:

$$r = \sqrt{\frac{SS_{Reg}}{SS_T}} = \frac{SP_{xy}}{\sqrt{SS_x\, SS_y}} \tag{11.13}$$

The correlation coefficient measures the direction and strength of the linear association between two random variables. Frequently, it is called the *Pearson product moment correlation coefficient*, named after Karl Pearson (1857–1936), the statistician who introduced correlation analysis. Its value is in the range:

$$-1.0 \leq r \leq 1.0.$$

A negative value indicates a negative or downward sloping relationship between the dependent and independent variables (Fig. 11.6); that is, as the independent variable increases, the dependent variable decreases (for example, the number of frosty days in April *versus* the survival rate of seedlings). In other words, a negative correlation coefficient is associated with a regression line that has a negative slope (b_1). Conversely, a positive value indicates an upward slope, with the dependent variable increasing as the independent variable increases (our dbh and crown radius example, for instance).

The absolute value of the correlation coefficient is another useful diagnostic tool that can be used to test whether the regression is significant (or well defined). In this case, the absolute value is compared to a critical value obtained from Table A.8 (see Appendix A), with (n – 2) degrees of freedom for one independent variable. If the absolute value of the calculated r (test statistic) exceeds the critical value, H_0: $\beta_1 = 0$ is rejected. For our example, the critical value from Table A.8 (see Appendix A) is:

$$r_{0.05(8)} = 0.632.$$

Since the calculated r is 0.977, H_0 is rejected, again showing that we have a significant relationship. This test is equivalent to the F-test for testing whether the relationship between two variables is significant (see Eqn 11.10); each can be derived from the other. Like the F-test, this test is considered a one-tailed test in that we reject the null hypothesis only if the unknown population variance due to regression is greater than the unknown population residual variance.

For the final diagnostic test, the correlation coefficient can also be transformed into a t-statistic and a ***t*-test** can be performed with the following hypotheses regarding the unknown population correlation coefficient, ρ

$$H_0: \rho = 0,$$

$$H_1: \rho \neq 0, \qquad \text{or} \qquad \rho < 0, \qquad \text{or} \qquad \rho > 0.$$

The test statistic for testing this hypothesis is calculated as:

$$t_{(n-2)} = r\sqrt{\frac{n-2}{1-r^2}} \tag{11.14}$$

The results are not shown here, but the t-value in this two-tailed test would be equal to the square root of the F-value obtained in the analysis of variance testing the significance of the regression, and the conclusions are identical.

The critical value is obtained from the t-table (see Table A.5, Appendix A) with $(n - 2)$ degrees of freedom. For this test, it is assumed that both the dependent and independent variables are randomly selected from a bivariate (see Section 4.2, Chapter 4) normal distribution.

Sampling distributions and tests concerning the regression coefficients and predictions

If all possible samples of size n are taken from a population which has the simple linear relationship of Eqn 11.4 between the dependent and independent variables, it can be shown that the *Central Limit Theorem* (see Section 7.2, Chapter 7) can be applied to the statistics, b_0, b_1 and $\hat{y}_{x_k}$ (see Fig. 11.9).

These three statistics possess the following properties:

1. Since the values of all three statistics change from sample to sample, they are random variables.
2. They are unbiased:

$$E(b_0) = \beta_0$$

$$E(b_1) = \beta_1$$

$$E\left(\hat{y}_{x_k}\right) = \mu_{y|x_k}$$

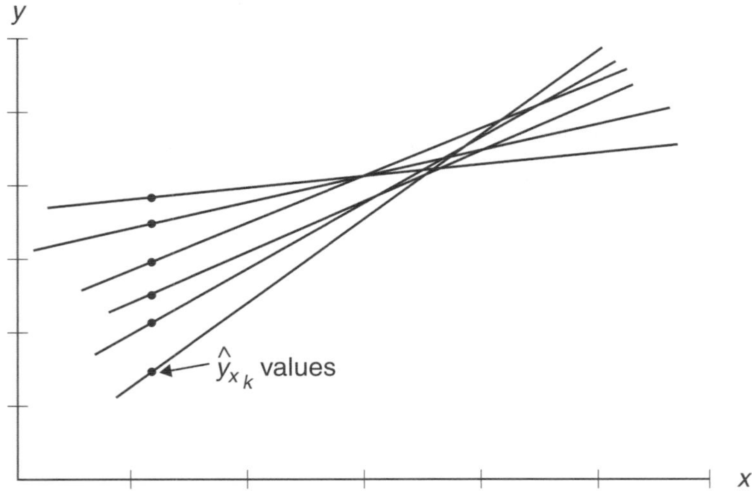

Fig. 11.9. Regression lines obtained from repeated samples of size *n* from a population.

3. Their standard errors are:

$$S_{b_0} = \sqrt{\frac{MS_{Res}}{n} \frac{\sum_{i=1}^{n} x_i^2}{SS_x}} \tag{11.15}$$

$$S_{b_1} = \sqrt{\frac{MS_{Res}}{SS_x}} \tag{11.16}$$

$$S_{\hat{y}x_k} = \sqrt{MS_{Res}\left[\frac{1}{n} + \frac{(x_k - \bar{x})^2}{SS_x}\right]} \tag{11.17}$$

where S_{b_0} = standard error of the intercept; S_{b_1} = standard error of the slope; $S_{\hat{y}x_k}$ = standard error of the predicted *y* value at x_k.

4. If the sample size, *n*, is greater than 30, then their sampling distributions are approximately normal (Fig. 11.10).

Because of the above properties, probabilities regarding these random variables (b_0, b_1 and $\hat{y}_{x_k}$) can be calculated using the *Z* or *t* transformations. More importantly, we can derive equations for confidence intervals and equations to calculate test statistics for statistical hypothesis testing concerning these statistics. Since the population residual variance in regression analysis is almost never readily available, the equations given below are for cases when the residual variance is estimated from a sample. The equations used to calculate confidence intervals are:

Confidence interval for the intercept:

$$P\left(b_0 - t_{\alpha/2(n-2)} S_{b_0} < \beta_0 < b_0 + t_{\alpha/2(n-2)} S_{b_0}\right) = 1 - \alpha \tag{11.18}$$

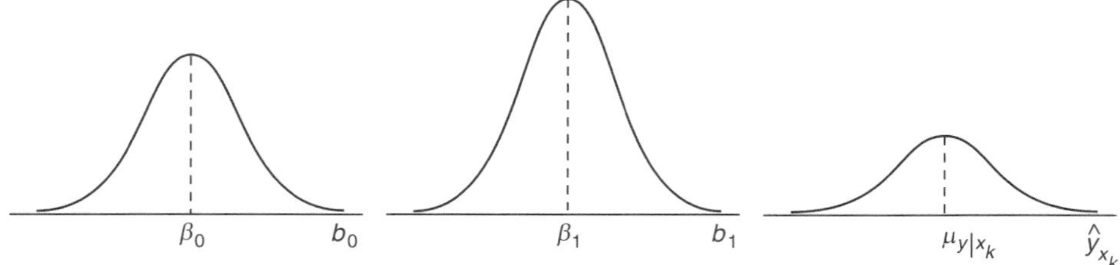

Fig. 11.10. Sampling distributions of b_{0_1}, b_1 and $\mu_{y|x_k}$.

Confidence interval for the slope:

$$P\left(b_1 - t_{\alpha/2(n-2)}S_{b_1} < \beta_1 < b_1 + t_{\alpha/2(n-2)}S_{b_1}\right) = 1-\alpha \tag{11.19}$$

Confidence interval for the population mean of y at a particular x_k:

$$P\left(\hat{y}_{x_k} - t_{\alpha/2(n-2)}S_{\hat{y}_{x_k}} < \mu_{y|x_k} < \hat{y}_{x_k} + t_{\alpha/2(n-2)}S_{\hat{y}_{x_k}}\right) = 1-\alpha \tag{11.20}$$

It is also possible to estimate the variance and the confidence interval for a single y value (an observation) at a particular x_k. The equations are:

$$S_{y_{x_k}} = \sqrt{MS_{\text{Res}}\left[1+\frac{1}{n}+\frac{\left(x_k-\bar{x}\right)^2}{SS_x}\right]} \tag{11.21}$$

$$P\left(\hat{y}_{x_k} - t_{\alpha/2(n-2)}S_{y_{x_k}} < y_{x_k} < \hat{y}_{x_k} + t_{\alpha/2(n-2)}S_{y_{x_k}}\right) = 1-\alpha \tag{11.22}$$

Example 11.1. Using the dbh and crown radius data, find the 95% confidence interval for the intercept, the 99% confidence interval for the slope and the 95% confidence interval for the mean predicted crown radius at 10.5 cm dbh.

The 95% confidence interval for the intercept:

$$S_{b_0} = \sqrt{\frac{(0.0199)(1960.49)}{(10)(350.13)}} = 0.106; \qquad t_{0.025(8)} = 2.31$$

$$P\left[0.58-(2.31)(0.106) < \beta_0 < 0.58+(2.31)(0.106)\right] = 0.95$$

$$P(0.335 < \beta_0 < 0.825) = 0.95$$

The probability is 0.95 that the unknown population intercept is between 0.335 and 0.825.

The 99% confidence interval for the slope:

$$S_{b_1} = \sqrt{\frac{0.0199}{350.13}} \approx 0.0075; \qquad t_{0.005(8)} = 3.36$$

$$P\left[0.098-(3.36)(0.0075) < \beta_1 < 0.098 + (3.36)(0.0075)\right] = 0.99$$

$$P(0.073 < \beta_1 < 0.123) = 0.99$$

The probability is 0.99 that the unknown population slope is between 0.073 and 0.123. Since zero is not included in the interval, it can be concluded that the unknown population slope is significantly different from zero and the regression is well defined (significant), verifying the results of our diagnostic tests above.

The 95% confidence interval for the unknown population mean of crown radius at 10.5 cm dbh:

$$\hat{y}_{10.5} = 0.58 + (0.098)(10.5) \approx 1.61$$

$$S_{\hat{y}_{10.5}} = \sqrt{0.0199\left[\frac{1}{10} + \frac{(10.5 - 12.69)^2}{350.13}\right]} \approx 0.048; \qquad t_{0.025(8)} = 2.31$$

$$P\left[1.61 - (2.31)(0.048) < \mu_{y|x_k} < 1.61 + (2.31)(0.048)\right] = 0.95$$

$$P\left(1.499 < \mu_{y|x_k} < 1.721\right) = 0.95$$

The probability is 0.95 that the unknown population mean of the crown radius for 10.5 cm dbh trees is between 1.499 and 1.721 m.

An interesting property of the equation to construct a confidence interval for a predicted y at a given x value (Eqn 11.20) is that the interval for the predicted y value is at a minimum at $\bar{x}$. This results in a bow-shaped confidence interval, or so-called 'confidence belt', which is graphically conveyed in Fig. 11.11.

Sometimes, we are interested in testing whether the unknown population intercept is equal to an assumed constant, c. We do so by testing a null hypothesis against one of the three usual alternative hypotheses:

H_0: $\beta_0 = c$,
H_1: $\beta_0 \neq c$, or $\beta_0 < c$, or $\beta_0 > c$,

for which the test statistic is:

$$t_{(n-2)} = \frac{b_0 - c}{S_{b_0}} \tag{11.23}$$

and the critical value is obtained from the t-table (see Table A.5, Appendix A) with $(n - 2)$ degrees of freedom.

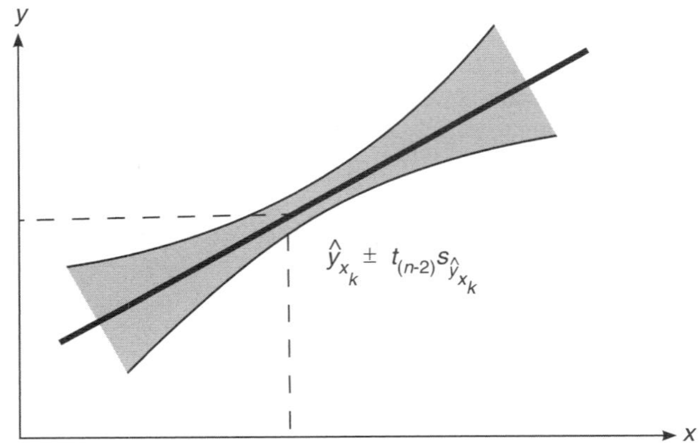

Fig. 11.11. Confidence intervals for the predicted y values at given x values.

Example 11.2. Is it reasonable to believe that the unknown population intercept of the dbh versus crown radius regression model is zero (i.e. the line runs through the origin)? Use $\alpha = 0.05$.

Since there is no prior knowledge available about the directional preference for the rejection of H_0, a two-tailed test will be used.

1. H_0: $\beta_0 = 0$.
2. H_1: $\beta_0 \neq 0$.
3. $\alpha = 0.05$.
4. Use Eqn 11.23.
5. $b_0 \approx 0.58$; $S_{b_0} \approx 0.106$.
6. $t_{0.025(8)} = \pm 2.31$.
7. $t_{(8)} = \dfrac{0.58 - 0}{0.106} \approx 5.47$
8. Since 5.47 < 2.31, we reject H_0.
9. Our data indicate that the population intercept of the dbh versus crown radius regression model is significantly greater than zero and that the regression line does not run through the origin.

We can also test the assumption that the unknown population slope of a regression line is equal to a constant, *c*, against one of three usual alternative hypotheses:

$$H_0: \beta_1 = c,$$

$$H_1: \beta_1 \neq c, \quad \text{or} \quad \beta_1 < c, \quad \text{or} \quad \beta_1 > c,$$

for which the test statistic is:

$$t_{(n-2)} = \frac{b_1 - c}{s_{b_1}} \tag{11.24}$$

and the critical value is obtained from the *t*-table (see Table A.5, Appendix A) with $(n - 2)$ degrees of freedom (see Example 11.3). If *c* in the above test is assumed to be zero and a two-tailed test is used, the above test is equivalent to the *F*-test (Eqn 11.10) discussed above. In fact, it can be shown that:

$$F_{\alpha(1,\nu)} = t^2_{\alpha/2(\nu)}.$$

Example 11.3. Use the dbh–crown radius data to answer the following:

a. Is it reasonable to believe that the unknown population slope is zero (i.e. it is a flat, horizontal line, meaning that there is no relationship between dbh and crown radius)? Use a 0.01 level of significance.
b. Is it reasonable to assume that the unknown population slope is at least 0.1? In other words, for every cm increase in dbh, does the crown radius increase by at least 0.1 m (10 cm)? Use a 0.05 level of significance.

In Example 11.3a, we are essentially testing whether the regression is well defined once again. Therefore, we will reject a negative or a positive slope and use a two-tailed test. In Example 11.3b, a one-tailed test will be used because it asks about assuming that there is *at least* a 0.1 m increase in crown radius for every cm increase of dbh. The rejection region will be at the lower tail of the sampling distribution.

a.

1. H_0: $\beta_1 = 0$.
2. H_1: $\beta_1 \neq 0$.

3. $\alpha = 0.01$.
4. Use Eqn 11.24.
5. $b_1 \approx 0.098$; $S_{b_1} \approx 0.0075$.
6. $t_{0.005(8)} = \pm 3.36$.

7. $t_{(8)} = \dfrac{0.098 - 0}{0.0075} \approx 13.07$ (note that $13.07^2 \approx 170.82$, which is within rounding error to the *F*-value in Table 11.2).
8. Since 13.07 > 3.36, we reject H_0 and accept H_1.
9. The unknown population slope is significantly different from zero and the regression is well defined (the regression is significant).

b.
1. H_0: $\beta_1 = 0.1$.
2. H_1: $\beta_1 < 0.1$.
3. $\alpha = 0.05$.
4. Use Eqn 11.24.
5. See Example 11.3a.
6. $t_{0.05(8)} = -1.86$.

7. $t_{(8)} = \dfrac{0.098 - 0.1}{0.0075} \approx -0.27$

8. Since −0.27 > −1.86, the test statistic is in the acceptance region and we 'accept' H_0.
9. The unknown population slope is either equal to 0.1 or greater than 0.1, so we can safely assume that it is at least 0.1.

Because a two-tailed test at $\alpha = 0.01$ was used in Example 11.3a, it is equivalent to the 99% confidence interval calculated in Example 11.1. The two critical values of ±3.36 can be converted to LCL and UCL values.

$$0.098 \pm (3.36)\,(0.0075)$$

$$P(0.073 < \beta_1 < 0.123) = 0.99.$$

Since the interval does not include zero, this matches the results in Example 11.1.

To test whether or not an unknown population mean of the dependent variable is equal to some constant for a given value of the independent variable (x_k), the following hypotheses can be constructed:

$$H_0: \mu_{y|x_k} = c$$

$$H_1: \mu_{y|x_k} \neq c; \quad \text{or} \quad \mu_{y|x_k} < c; \quad \text{or} \quad \mu_{y|x_k} > c$$

The appropriate test statistic is:

$$t_{(n-2)} = \frac{\hat{y}_{x_k} - c}{S_{\hat{y}_{x_k}}}, \tag{11.25}$$

for which the critical values are obtained from the *t*-table (see Table A.5, Appendix A) with $(n - 2)$ degrees of freedom.

Example 11.4. Is it reasonable to believe, with a 0.05 level of significance, that the unknown population mean of the crown radius for trees measuring 10.5 cm in dbh is 1.75 m?

Here, a two-tailed test will be used because there is no information indicating that the crown radius should be more or less than 1.75 m.

1. H_0: $\mu_{y|10.5} = 1.75$
2. H_1: $\mu_{y|10.5} \neq 1.75$
3. $\alpha = 0.05$.
4. Use Eqn 11.25.
5. See Example 11.1.
6. $t_{0.025(8)} = \pm 2.31$.
7. $t_{(8)} = \dfrac{1.61 - 1.75}{0.048} \approx -2.92$
8. Since –2.92 < –2.31, the test statistic is in the rejection region and we reject H_0.
9. The unknown population mean of the crown radius is significantly different from 1.75 m for trees with a 10.5 cm dbh.

Again, the critical values of ± 2.31 could be restated in crown radius units (m) and they would be the same as the LCL and UCL in Example 1.

Lack of fit

A well-defined or significant relationship in regression only indicates a significant linear dependency. It does not necessarily mean that the fitted model is adequate for practical applications like prediction (see Fig. 11.12). Even when the r^2 value is very high, a model's adequacy should be tested with a **lack of fit** test or by **plotting the residuals** before it can be used for predictive purposes. To test for lack of fit, the residual sum of squares of the linear regression, SS_{Res}, can be partitioned into two components: the variation due to lack of fit, SS_{Lf}, and the variation due to pure error, SS_{Pe}. These components are then compared using an *F*-test. If the variation due to lack

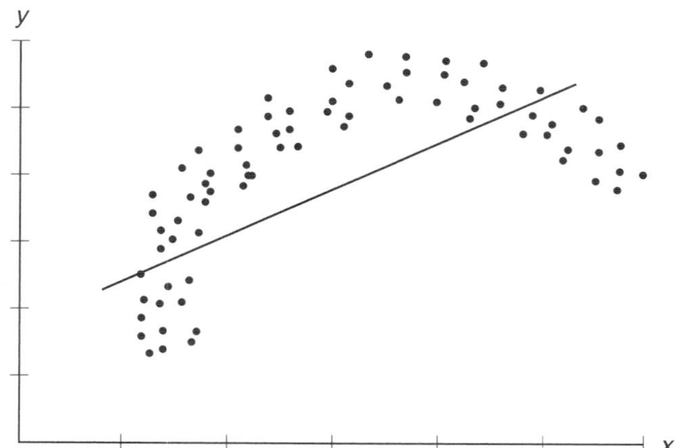

Fig. 11.12. Lack of fit.

of fit is significantly greater than the variation due to pure error, the model tested is inadequate due to a lack of fit. Note that this test can be carried out only if we have repeated y_{lj} observations for several x_j values (see Example 11.5). In general, we have:

x_j, where $j = 1, 2, 3, \ldots, k$ and y_{lj}, where $l = 1, 2, 3, \ldots, n_j$; and $\sum_{j=1}^{k} n_j = n$

From here we can state that:

$SS_{Res} = SS_{Lf} + SS_{Pe}$,

which is equivalent to:

$$\sum_{j=1}^{k}\sum_{l=1}^{n_j}\left(y_{lj}-\hat{y}_j\right)^2 = n_j\sum_{j=1}^{k}\left(\bar{y}_{\cdot j}-\hat{y}_j\right)^2 + \sum_{j=1}^{k}\sum_{l=1}^{n_j}\left(y_{lj}-\bar{y}_{\cdot j}\right)^2$$

where $\bar{y}_{\cdot j}$ = mean value of y_{ij} at x_j; $\hat{y}_j$ = predicted value of y at x_j; y_{ij} = observed value of y at x_j.

Since $\sum_{j=1}^{k} n_j = n$, the total number of observations for the regression, SS_{Res}, can be written as:

$$SS_{Res} = \sum_{i=1}^{n}\left(y_i-\hat{y}_i\right)^2 = \sum_{j=1}^{k}\sum_{l=1}^{n_j}\left(y_{lj}-\hat{y}_j\right)^2$$

The degrees of freedom of the three respective sums of squares are:

$$\left(\sum_{j=1}^{k} n_j\right) - 2 = (k-2) + \sum_{j=1}^{k}(n_j - 1), \qquad \text{or} \qquad n - 2 = (k-2) + (n-k).$$

The partitioned sums of squares divided by their corresponding degrees of freedom give estimates of the variance of pure error and lack of fit, respectively. To test the equality of the unknown population 'pure error' variance and 'lack of fit' variance, hypotheses are stated in the following manner:

H_0: $\sigma^2_{Lf}/\sigma^2_{Pe} = 1.0$,

H_1: $\sigma^2_{Lf}/\sigma^2_{Pe} > 1.0$.

The test statistic is calculated as follows:

$$F_{[k-2,(\sum_{j=1}^{k} n_j)-k]} = \frac{MS_{Lf}}{MS_{Pe}} \qquad (11.26)$$

where $MS_{Lf} = \dfrac{SS_{Lf}}{k-2}$, and $MS_{Pe} = \dfrac{SS_{Pe}}{\left(\sum_{j=1}^{k} n_j\right) - k}$

The critical values are obtained from the F-table (Table A.7) with $(k - 2)$ and $\left(\sum_{j=1}^{k} n_j\right) - k$ degrees of freedom.

Example 11.5. The following data show nail sizes and the corresponding ultimate loads (strengths) of nailed joints:

Nail size	30	40	50	60
Loads	11.6	12.5	11.8	14.2
	10.2	11.6	12.7	13.7
	10.5	11.1	12.8	14.0
		11.4	12.6	

Find the linear regression model. Test for the significance of the regression and lack of fit ($\alpha = 0.05$).

First, we determine the regression model:

$$n = 14;\ \bar{x} = 45.0;\ \bar{y} = 12.19;\ SS_x = 1550;\ SS_y = 19.969;\ SP_{xy} = 160.5$$

$$b_1 = 160.5/1550 = 0.1035$$

$$b_0 = 12.19 - (0.1035)(45.0) = 7.53$$

Hence

$$\hat{y}_i = 7.53 + 0.1035x_i.$$

Next, we see if the relationship between nail size and ultimate load is significant:

1. H_0: $\beta_1 = 0$.
2. H_1: $\beta_1 \neq 0$.
3. $\alpha = 0.05$.
4. Use Eqn 11.10.
5. $SS_{Reg} = (0.1035)(160.5) = 16.612;\ SS_{Res} = 19.969 - 16.612 = 3.357$

$MS_{Reg} = 16.612/1 = 16.612;\ MS_{Res} = 3.357/12 = 0.280$

6. $F_{0.05(1,12)} = 4.75$.
7. $F_{(1,12)} = 16.612/0.280 = 59.33$.
8. Since 59.33 > 4.75, the test statistic is in the critical region; we reject H_0 and accept H_1.
9. The relationship is well defined and the slope is significant.

Then, we perform a lack of fit test:

1. H_0: $\sigma^2_{Lf}/\sigma^2_{Pe} = 1.0$
2. H_1: $\sigma^2_{Lf}/\sigma^2_{Pe} > 1.0$
3. $\alpha = 0.05$.
4. Use Eqn 11.26.
5. The means and predicted values for the various nail sizes are calculated as:

Nail size	30	40	50	60
Mean	10.77	11.65	12.48	13.97
Predicted value	10.64	11.67	12.71	13.74

from here:

$$SS_{Lf} = 3(10.77 - 10.64)^2 + 4(11.65 - 11.67)^2 + 4(12.48 - 12.71)^2$$
$$+3(13.97 + 13.74)^2 \approx 0.423$$

$$SS_{Pe} = SS_{Res} - SS_{Lf} = 3.357 - 0.423 = 2.934$$

$$MS_{Lf} = 0.423/(4-2) \approx 0.2115 \qquad SS_{Pe} = 2.934/(14-4) \approx 0.2934$$

6. $F_{0.05(2,\ 10)} = 4.96.$

7. $F_{(2,10)} = \dfrac{0.2115}{0.2934} \approx 0.72$

8. Since 0.72 < 4.96, the test statistic is in the acceptance region and we 'accept' H_0.

9. The model is acceptable; there is no significant lack of fit.

When there are no repeated measurements of the dependent variable at various values of the independent variable, the adequacy of the model can be evaluated by plotting the residuals, defined as the differences between points and the regression line ($\hat{y}_j - y_i$). Residuals can be plotted over the independent variable and/or over the predicted dependent variable. Figure 11.13 shows three possible residual plots or scatter graphs: Fig. 11.13a indicates an adequate model (no lack of fit), while Fig. 11.13b shows a typical lack of fit case. Figure 11.13c shows a non-constant variance problem (the variance is increasing with the independent variable), where Assumption 4 of the regression analysis is not met.

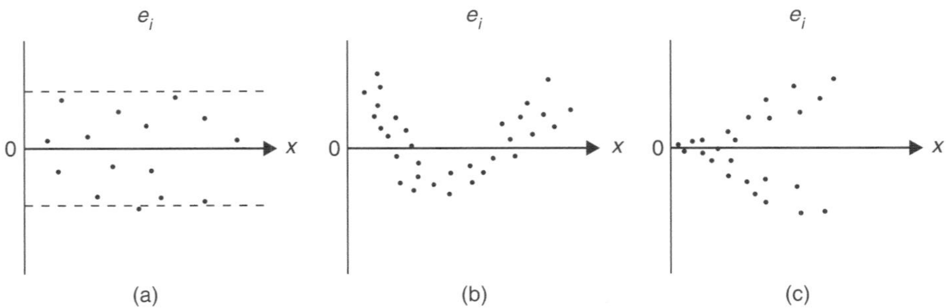

Fig. 11.13. Scatter graphs of the residuals: (a) no lack of fit, (b) lack of fit and (c) violation of the assumption of uniform variance.

11.2 Correlation Analysis

Mathematically, correlation analysis is the same as regression analysis. There are two major differences between regression and correlation analyses. The first is that the main objective in regression analysis is to find an equation to predict values of dependent variables for certain fixed values of independent variables, while the main objective in correlation analysis is to evaluate the linear association between two or more (multiple correlation) variables. Secondly, in regression analysis, values of the independent variables are known and can be selected and controlled by the experimenter. In correlation analysis, however, both variables are assumed to be random variables with a bivariate normal distribution (see Section 4.2, Chapter 4).

That said, there are also many similarities between the two analyses. Equation 11.13 can be used to calculate the correlation coefficient and Eqn 11.14 can be used to test its significance. The significance of a correlation coefficient can also be tested against the values in Table A.8 (see Appendix A), as discussed earlier.

A word of caution! It is easy to draw incorrect conclusions about the relationship between two random variables. For example, the number of students registered at the University of British Columbia yearly between 1956 and 1975 is significantly positively correlated with the number of crimes in Vancouver, BC. It is certainly hoped that there is no real cause-and-effect relationship between these two random variables; rather, that they were both affected by the rapid growth of the population of Vancouver during this time.

11.3 Multiple Regression

It is possible, oftentimes desirable, to predict the values of a dependent variable from the measurements of several independent variables. For example, the volume of standing trees can be predicted from their height, dbh, age and site index. The breaking load of 2 × 4s can be predicted from their specific gravity, number of annual rings per cm and the width of late wood within annual rings. As in simple linear regressions, the dependent variable (there is still only one) in multiple regressions is a random variable, while the independent variables can be selected or fixed by the experimenter. The various forms of multiple regression models for samples and populations are as follows:

Model describing the sample regression surface:

$$\hat{y}_i = b_0 + b_1 x_{1i} + b_2 x_{2i} + \cdots + b_m x_{mi} \tag{11.27}$$

Model describing the points around the sample regression surface:

$$y_i = b_0 + b_1 x_{1i} + b_2 x_{2i} + \cdots + b_m x_{mi} + e_i \tag{11.28}$$

Model describing the population regression surface:

$$\mu_{y|x_{1i}, x_{2i}, \ldots, x_{mi}} = \beta_0 + \beta_1 x_{1i} + \beta_2 x_{2i} + \cdots + \beta_m x_{mi} \tag{11.29}$$

Model describing the points around the population regression surface:

$$y_i = \beta_0 + \beta_1 x_{1i} + \beta_2 x_{2i} + \cdots + \beta_m x_{mi} + \varepsilon_i \tag{11.30}$$

The least squares estimates of b_0, b_1, b_2, ..., b_m in Eqn 11.28 are obtained by using differential calculus and solving the following systems of linear equations. These equations are also known as the normal equations and are based on uncorrected sums of squares and products. For two independent variables, the normal equations are as follows:

$$b_0 n + b_1 \sum x_{1i} + b_2 \sum x_{2i} = \sum y_i$$

$$b_0 \sum x_{1j} + b_1 \sum x_{1i}^2 + b_2 \sum x_{1i} x_{2i} = \sum x_{1i} y_i$$

$$b_0 \sum x_{2i} + b_1 \sum x_{1i} x_{2i} + b_2 \sum x_{2i}^2 = \sum x_{2i} y_i$$

Expressing b_0 from the first equation and substituting it into the other two, we get normal equations based on corrected sums of squares and products:

$$b_1 SS_{x_1} + b_2 SP_{x_1 x_2} = SP_{x_1 y}$$
$$b_1 SP_{x_1 x_2} + b_2 SS_{x_2} = SP_{x_2 y}\,.$$

The corrected sum of squares for x_1 is (to simplify notation, $\Sigma = \sum_{i=1}^{n}$, in all the following equations):

$$SS_{x_1} = \Sigma(x_{1i} - \bar{x}_1)^2 = \Sigma x_{1i}^2 - (\Sigma x_{1i})^2 / n$$

The corrected sum of products for x_1 and x_2 is:

$$SP_{x_1 x_2} = \Sigma\,(x_{1i} - \bar{x}_1)(x_{2i} - \bar{x}_2) = \Sigma x_{1i} x_{2i} - (\Sigma x_{1i})(\Sigma x_{2i})/n$$

Dividing through by n, the first normal equation becomes:

$$b_0 = \bar{y} - b_1 \bar{x}_1 - b_2 \bar{x}_2$$

These equations can be solved either by substitutions or by matrix algebra (the latter being outside the scope of this text).

In order to test the quality of a multiple regression model, we hypothesize that:

H_0: $\beta_1 = \beta_2 = \ldots = \beta_m = 0$

(in other words, the independent variables have no linear relationship with the dependent variable), against the alternative of:

H_1: at least one is not equal to zero.

These hypotheses are tested in terms of variances in the same way as simple linear regressions by partitioning the sum of squares of the dependent variable into sum of squares regression and sum of squares residual:

$$SS_T = SS_{Reg} + SS_{Res},$$

$$\sum_{i=1}^{n} (y_i - \bar{y}_i)^2 = \sum_{i=1}^{n} (\hat{y}_i - \bar{y})^2 + \sum_{i=1}^{n} (y_i - \hat{y}_i)^2$$

with the following degrees of freedom:

$$(n - 1) = m + (n - m - 1).$$

The test statistic is:

$$F_{(m,n-m-1)} = \frac{SS_{Reg}/m}{SS_{Res}/(n-m-1)} = \frac{MS_{Reg}}{MS_{Res}} \qquad (11.31)$$

The critical value is obtained from the F-table (see Table A.7, Appendix A) with m and $(n - m - 1)$ degrees of freedom. The surface of a multiple linear regression equation with m independent variables is defined by $(m + 1)$ points, since a predicted value of the dependent variable is determined by $(m + 1)$ statistics, namely: $b_0, b_1, \ldots, b_m$. Therefore, SS_{Res} has $\{n - (m + 1)\} = (n - m - 1)$ degrees of freedom. For the same reasons, the SS_{Reg} has $\{(m + 1) - 1\} = m$ degrees of freedom. Figure 11.14 shows the surface of a multiple linear regression model with two independent variables, which is not a line but a *plane*.

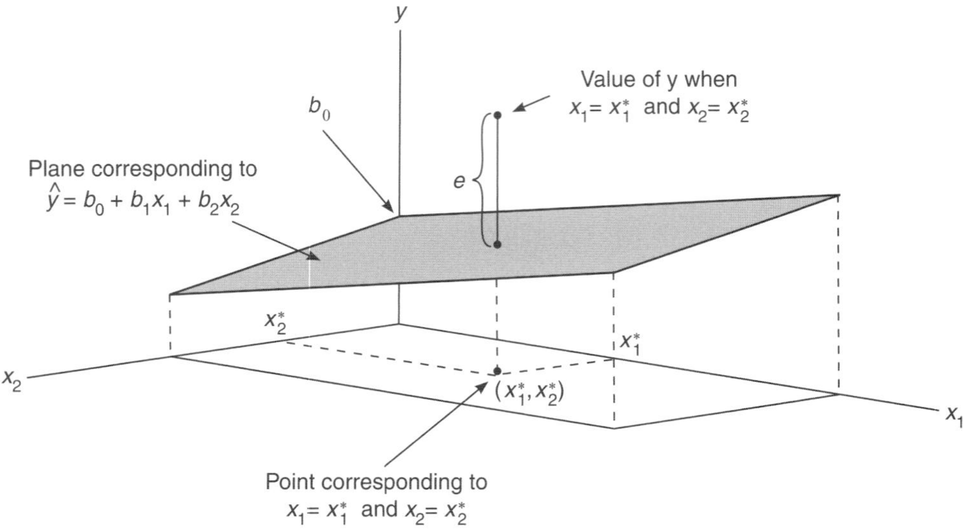

Fig. 11.14. Graph of the multiple regression surface with two independent variables.

In multiple regression, the coefficient of determination defined earlier is called the **multiple coefficient of determination.** This statistic and the so-called 'standard error of estimate' for multiple linear regression equations are calculated as:

$$R^2 = \frac{SS_{Reg}}{SS_T} \tag{11.32}$$

$$S_{y \cdot x_1, x_2, \ldots x_m} = \sqrt{MS_{Res}} \tag{11.33}$$

Like simple linear regression, the coefficient of determination in a multiple regression is the proportion of the sum of squares of the dependent variable explained by or attributable to all of the independent variables (or to the regression model). It does not show, however, which independent variable is more important than the others. Likewise, the standard error of estimate is again a measure of spread of the observations (from which the regression model is derived) around the regression surface. However, here the spread of observations is around a surface or a plane, not a line, as in simple linear regression. That said, if all of the assumptions are met, the empirical rule applies.

Example 11.6. The following data show modulus of rupture (kp), specific gravity (gm/cm^3) and moisture content (%) of wood specimens randomly taken from 15 trees. Find the multiple linear regression model for modulus of rupture (the dependent variable) with specific gravity and moisture content as the independent variables.

Modulus of rupture, Y	Specific gravity, X_1	Moisture content, X_2
80,492	0.588	62
87,293	0.608	68
89,868	0.579	59
86,830	0.566	55
85,169	0.556	57
29,191	0.412	42
26,226	0.386	44
28,116	0.373	40
30,102	0.402	42
29,221	0.397	43
43,236	0.472	48
39,271	0.451	47
48,238	0.444	50
45,281	0.452	52
50,618	0.500	55

For the above data, the normal equations based on corrected sums of squares and products are:

$$0.09227b_1 + 8.9961b_2 = 27874.4$$
$$8.9961b_1 + 964.93b_2 = 2651940.0.$$

Solving this system of equations, we have:

$$b_1 = 374833.0;\ b_2 = -746.25.$$

The intercept is found with the equation:

$$b_0 = 52810.1 - b_1(0.4791) - b_2(50.93) = -88750.9.$$

The test for the quality of the regression model is:

1. H_0: $\beta_1 = \beta_2 = 0$.
2. H_1: at least one is different.
3. $\alpha = 0.01$.
4. Use Eqn 11.31.
5. $SS_{Reg} = b_1SP_{x_1y} + b_2SP_{x_2y} = 0.846922 \times 10^{10}$

$SS_{Res} = SS_T - SS_{Reg} = 0.886409 \times 10^{10} - 0.846922 \times 10^{10} = 32,905,800.$
6. $F_{(2,12)} = 6.93.$
7. $F_{(2,12)} = \dfrac{SS_{Reg}/2}{SS_{Res}/12} = 128.69$
8. Since 128.69 > 6.93, the test statistic is in the rejection region; we reject H_0 and accept H_1.
9. There is a well-defined, significant regression for modulus of rupture with specific gravity and moisture content.

$$R^2 = \frac{0.846922 \times 10^{10}}{0.886409 \times 10^{10}} \approx 0.9554$$

Specific gravity and moisture content explain about 95.5% of the variation (in terms of sum of squares) of modulus of rupture.

$$S_{y \cdot x_1x_2} = \sqrt{32,905,800} \approx 5736.4.$$

This number shows how the observations vary around the regression surface. About 68% of the observed modulus of rapture values are within 5736.4 kilopascals of the regression surface (a plane).

11.4 Non-linear Models

When the relationship between an independent and a dependent variable is not a straight line (e.g. the dbh and height example in Fig. 11.15), we can oftentimes describe the relationship as being curvilinear. Statistically speaking, some curvilinear relationships are said to be *linear in the parameters*. **Polynomial regression** models, for example, can be *transformed* into a form that is linear (see description following). Some **exponential** and **hyperbolic regressions** can also be transformed in a similar manner. For example, curvilinear models of second and third degree polynomials appear as follows:

$$\hat{y}_i = b_0 + b_1 x_i + b_2 x_i^2, \tag{11.34}$$

$$\hat{y}_i = b_0 + b_1 x_i + b_2 x_i^2 + b_3 x_i^3. \tag{11.35}$$

To obtain a model that is linear *in parameters*, x_i, x_i^2 and x_i^3 are simply treated as x_1, x_2 and x_3 in a multiple linear regression model, and the procedures discussed in Section 11.3 can be used.

One of the most frequently used exponential regression models occurs in the following form:

$$\hat{y}_i = a x_i^b c^{x_i} \tag{11.36}$$

After taking the logarithm of both sides of the equation, it becomes:

$$\ln(\hat{y}_i) = \ln(a) + b(\ln(x_i) + \ln(c)x_i$$

from here

$$\ln(\hat{y}_i) = b_0 + b_1 \ln(x_i) + b_2 x_i \tag{11.37}$$

where $b_0 = \ln(a)$; $b_1 = b$; $b_2 = \ln(c)$.

In this case, $\ln(y_i)$ is treated as the dependent variable and $\ln(x_i)$ as x_1 and x_i as x_2. We can then proceed with a multiple linear regression analysis with two independent variables.

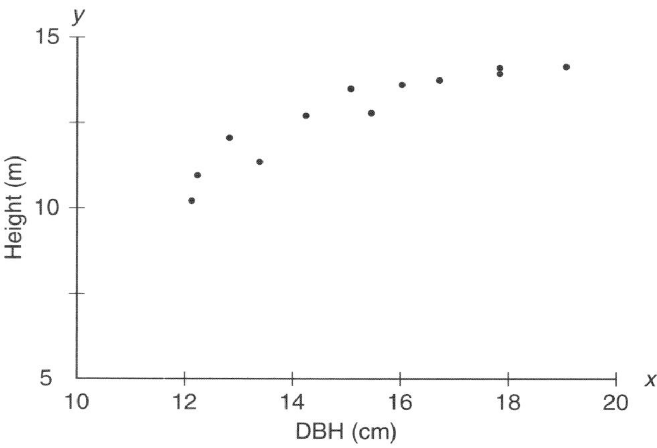

Fig. 11.15. Scatter diagram for the dbh and height data (see Example 11.7).

Consider a hyperbola shape:

$$\hat{y}_i = b_0 + b_1(1/x_i) + b_2 x_i \tag{11.38}$$

In order to linearize the model, $(1/x_i)$ is considered as x_1, x_i is considered as x_2, and we can again proceed with a multiple linear regression analysis with two independent variables (see Example 11.7). Both Eqns 11.37 and 11.38 can be reduced to a simple linear regression case if β_2 (estimated by b_2) can be assumed to be zero (the test for this is beyond the scope of this text). In this case, Eqn 11.36 simplifies to:

$$\hat{y}_i = a x_i^b, \tag{11.39}$$

and Eqn 11.38 becomes:

$$\hat{y}_i = b_0 + b_1(1/x_i). \tag{11.40}$$

Example 11.7. Total tree height (m) and dbh (cm) were measured on 12 randomly selected trees.

dbh:	12.9	14.0	13.4	14.8	15.1	15.7	12.5	16.6	17.5	18.4	18.4	19.5
height:	11.1	11.1	12.7	12.8	13.9	13.3	9.8	14.1	14.2	14.3	14.0	14.2

Find the second-degree polynomial for the above data set (see Fig. 11.15).

The normal equations based on corrected sums of squares and products for the second-degree polynomial are:

$$60.48b_1 + 1925.32b_2 = 33.05$$
$$1925.32b_1 + 61497.0b_2 = 1024.79.$$

Solving the systems of equations, we have:

$$b_1 = 4.64; \ b_2 = -0.1286$$

and substituting these values, we have:

$$b_0 = 12.96 - b_1(15.73) - b_2(252.58) = -27.56.$$

Model:

$$\hat{y}_i = -27.56 + 4.64x - 0.1286x^2$$

The test for the significance of the regression model is:

1. H_0: $\beta_1 = \beta_2 = 0$.
2. H_1: at least one is different.
3. $\alpha = 0.05$.
4. Use Eqn 11.31.
5. $SS_{Reg} = b_1 SP_{x_1y} + b_2 SP_{x_2y} = 21.59$

$SS_{Res} = SS_T - SS_{Reg} = 25.25 - 21.59 = 3.66$

6. $F_{0.05(2,9)} = 4.26$
7. $F_{(2,9)} = \dfrac{21.59/2}{3.66/9} \approx 26.54.$
8. Since 26.54 > 4.26, the test statistic is in the rejection region; we reject H_0 and accept H_1.
9. There is a well-defined or significant curvilinear relationship between tree height and dbh.

$$R^2 = \frac{21.59}{25.25} \approx 0.843$$

Therefore, dbh explains 84.3% of the variation (in terms of sum of squares) in tree height.

$$S_{y \cdot x} = \sqrt{\frac{3.66}{9}} = 0.637$$

Therefore, about 68% of the observations of height are within 0.637 m of the regression curve.

An example of a very commonly used multiple curvilinear model in forestry is the volume equation for standing trees developed by Schumacher and Hall (1933). The dependent variable is total tree volume (V) and the independent variables are dbh (D) and total tree height (H):

$$V = aD^bH^c. \tag{11.41}$$

After logarithmic transformation, this equation becomes:

$$\ln(V) = b_0 + b_1\ln(D) + b_2\ln(H). \tag{11.42}$$

Equation 11.42 can be fitted as a multiple linear regression model with two independent variables ($\ln(D)$ and $\ln(H)$) and $\ln(V)$ as the dependent variable.

Note that this has been a very brief and cursory introduction of advanced regression methods such as multiple regression and non-linear models. The interested reader is referred to one of many advanced textbooks on these subjects.

Exercises

Section 11.1

11.1. The following observations are from a morphometric study of cottonwood trees. The widths of 12 leaves from a single tree were measured (in mm) while fresh and after drying.

Fresh leaf width – X	Dry leaf width – Y
90	88
115	109
55	52
110	105
76	71
100	95
84	78
95	90
84	77
95	91
100	96
90	86

One of the study objectives is to use fresh leaf width to predict dry leaf width.

a. Plot the observations and determine the linear relationship between dry leaf width and fresh leaf width.
b. Is the relationship significant ($\alpha = 0.01$)?
c. Calculate the coefficient of determination and discuss its meaning.
d. Calculate the standard error of estimate and discuss its meaning.
e. Is the intercept significantly different from zero ($\alpha = 0.05$)?

f. Is the slope significantly different from 1.0 ($\alpha = 0.05$)? Define the meaning of the slope.
g. Find the 95% confidence interval for the unknown population mean of dry leaf widths at 100 mm fresh leaf width.

11.2. The amounts of fertilizer dissolved in 100 g of water at various water temperatures were recorded as follows. At each temperature, the test was repeated three times.

	Fertilizer (grams) – *Y*		
Water temperature (°C) – *X*	Test 1	Test 2	Test 3
0	8	6	9
5	10	12	13
10	20	23	19
15	28	31	33
20	44	40	47
25	46	48	50

a. Plot the observations and determine the linear relationship between temperature and amount of dissolved fertilizer.
b. Is the relationship significant ($\alpha = 0.05$)?
c. Calculate the coefficient of determination and discuss its meaning.
d. Calculate the standard error of estimate and discuss its meaning.
e. Is the intercept significantly different from 10.0 ($\alpha = 0.01$)?
f. Is the slope significantly different from 2.0 ($\alpha = 0.05$)? Define the meaning of the slope.
g. Find the 95% confidence interval for the unknown population mean of the amount of dissolved fertilizer at a temperature of 15°C.
h. Test the lack of fit with a 0.05 level of significance.

11.3. According to a secondary wood products machinery supplier, the cost of maintaining a CNC moulder appears to increase with age.

Age (years) – *X*	6-month cost (US$) – *Y*
4.5	619
4.5	1040
4.5	1030
4.0	495
4.0	723
4.0	681
5.0	890
5.0	1522
5.5	987
5.0	1194
0.5	163
0.5	182
6.0	764
6.0	1373
1.0	978
1.0	466

a. Plot the observations and determine the linear relationship between age and maintenance cost.
b. Is the relationship significant ($\alpha = 0.01$)?
c. Calculate the coefficient of determination and discuss its meaning.
d. Calculate the standard error of estimate and discuss its meaning.
e. Is the intercept significantly greater than zero ($\alpha = 0.05$)? Discuss the meaning of the intercept in this exercise.
f. Is the slope significantly different from 240 ($\alpha = 0.05$)? Discuss the meaning of the slope.
g. Find the 99% confidence intervals for the unknown population means of cost at:
   i. 2.5 years, and
   ii. 5.5 years.
   Compare the two intervals.
h. Test the lack of fit with a 0.05 level of significance.

11.4. To find whether or not a relationship exists between the size of the boulders in a stream and their distance from the source, samples of boulders were measured every 0.5 km downstream, beginning at 1.0 km from the source.

Distance (km) – X	Boulder size (cm) – Y
1.0	104.9
1.5	86.4
2.0	82.6
2.5	89.4
3.0	73.2
3.5	65.2
4.0	76.5
4.5	65.3
5.0	57.4
5.5	51.6
6.0	47.8
6.5	43.4

a. Plot the observations and determine the linear relationship between distance from the source and boulder size.
b. Is the relationship significant ($\alpha = 0.01$)?
c. Calculate the coefficient of determination and discuss its meaning.
d. Calculate the standard error of estimate and discuss its meaning.
e. Is the intercept significantly different from 100 cm ($\alpha = 0.05$)? Discuss the meaning of the intercept in this exercise.
f. Is the slope significantly different from –8.0 ($\alpha = 0.05$)? Discuss the meaning of the slope.
g. Find the 95% confidence interval for the unknown population mean of boulder size at 4.0 km downstream of the source.
h. Can the test for lack of fit be carried out for this data set?

Section 11.2

11.5. Values for specific gravity and modulus of rupture (in 1000 lb/inch) were obtained for 12 randomly selected air-dried spruce specimens:

Specific gravity – X	Modulus of rupture – Y
0.680	15.72
0.348	6.14
0.413	8.21
0.698	17.07
0.252	5.18
0.456	10.43
0.550	14.46
0.576	12.39
0.262	4.67
0.612	13.73
0.375	8.93
0.387	9.36

a. Find the correlation coefficient between the two variables.
b. Is the correlation between specific gravity and modulus of rupture significant? Use $\alpha = 0.05$.

11.6. A study was carried out to find the correlation between dbh (cm) and crown radius (m) for ponderosa pine.

dbh	30	39	40	72	66	66	45	50	55	57
radius	3.5	3.6	4.8	7.1	6.4	7.9	5.6	4.8	5.7	7.0

a. Find the correlation coefficient between the two variables.
b. Is the correlation between dbh and crown radius significant? Use $\alpha = 0.05$.

Section 11.3

11.7. Crown diameter (m) and tree height (m) can easily be measured from aerial photographs, whereas total tree volume (m^3) cannot. A multiple regression equation between crown diameter (x_1), tree height (x_2) and total tree volume (y) is proposed so that total tree volume can be better estimated. The following data are available.

X_1	2.92	3.35	3.41	3.23	2.44	2.47	2.68	3.66	3.02	2.99
X_2	28.7	48.0	36.9	37.8	25.1	28.4	26.1	37.8	28.7	30.6
Y	1.03	2.87	1.96	2.21	0.98	1.86	1.02	2.15	1.06	1.81

a. Estimate the multiple regression model $\hat{y} = b_0 + b_1 x_1 + b_2 x_2$.
b. Is the multiple regression significant (use $\alpha = 0.05$)?
c. Find the multiple coefficient of determination and explain its meaning.
d. Calculate the standard error of estimate and explain its meaning.

Section 11.4

11.8. Stand age (X) and number of trees per hectare (Y) are observed in 15 low-site Douglas-fir stands.

X	20	25	30	45	45	60	60	70	70	80	90	95	95	100	120
Y	210	205	172	139	127	102	96	90	70	60	49	51	48	47	46

Find the second-degree polynomial and the hyperbolic relationship between age and number of trees. Are these equations significant (use $\alpha = 0.05$)? Which equation explains more of the sum of squares of the dependent variable?

12 Analysis of Variance

Testing Differences between Several Means

Several procedures for comparing two unknown population means were presented in Chapter 9. In many practical problems, however, we may be interested in comparing more than two population means. For instance, to study the effects of three different fertilizers on the height growth of Douglas-fir seedlings, a forest practitioner would need to test a hypothesis involving three population means. Similarly, to investigate the effects of these three fertilizers on the water quality in nearby creeks, a hydrologist would also need to compare three population means. In this chapter, we introduce a technique called **analysis of variance**, which enables us to compare the equality of two or more population means.

Analysis of variance, often referred to by the acronym ANOVA, is one of the most powerful and frequently used techniques in statistics. It is used to analyse data obtained through both **experimental designs** and **sampling designs**, some of which will be introduced in Chapter 13.

We offer two definitions of analysis of variance:

1. A statistical tool that compares two or more unknown population means.
2. A method for partitioning the total variation of the data into meaningful components, and comparing these different sources of variation.

In regression analysis (see Chapter 11), we saw two examples of partitioning variation. First, we partitioned the total sum of squares of the dependent variable (SS_T) into sum of squares due to regression (SS_{Reg}) and sum of squares residual (SS_{Res}). Second, we partitioned the sum of squares residual into sum of squares pure error (SS_{Pe}) and sum of squares due to lack of fit (SS_{Lf}). In the above study into the effects of the three fertilizers on the height growth of Douglas-fir seedlings, the total variation can be partitioned into sum of squares due to or caused by the fertilizers (the treatment) and sum of squares due to seedling-to-seedling variation within each fertilizer treatment (the experimental error). We could further complicate this experiment and apply the three fertilizers to three different species – say, Douglas-fir, western hemlock and western red cedar – and then partition the total variation of the data into: sum of squares due to fertilizers, sum of squares due to species, sum of squares due to the interaction (discussed later in this chapter) between fertilizer and species, and sum of squares due to individual seedling's variation within each fertilizer and species (experimental error).

In the first example, there is only *one* meaningful component; that is, we are interested in the treatment variation only. The variation due to experimental error is found only so that we may quantify the variation associated with the use of different fertilizers. Consequently, it is called a **one-way analysis of variance** or a **one-way**

classification. In the second example, there are two meaningful components of variation that need to be separated from experimental error: variation due to fertilizers and variation due to species. This, therefore, is called a **two-way analysis of variance**, or **two-way classification.** This can also be extended to more complex problems of **three-way**, or ***m*-way classifications.** However, we will discuss only the one- and two-way classifications in this book.

12.1 One-way Analysis of Variance

For a one-way analysis of variance, it is assumed that *random* samples of size n are selected from k *independent* populations that are *normally* distributed with population means $\mu_1, \mu_2, \ldots, \mu_k$ and *common variance*, σ^2. The hypothesis tested in a one-way analysis of variance is:

$$H_0: \mu_1 = \mu_2 = \ldots = \mu_k$$

and

H_1: at least two of the means are not equal (*note that this does not tell us which of the two means are unequal*).

Table 12.1 shows the general notation used to describe sample data collected for a one-way analysis of variance. Here, y_{ij} is the jth observation selected from the ith population, and $T_{i.}$ is the sum of the observations from the ith population. We use a special symbol to denote means and totals, using a 'dot' as a placeholder for the subscript i or j: $\bar{y}_{i.}$ is the sample mean of the ith population, $\bar{y}_{..}$ is the overall or grand mean (calculated from the nk total number of observations), $T_{..}$ is the grand total or sum of all the observations and $T_{i.}$ is the total of all the observations in the ith population.

Each observation in the k populations can be described by the following linear model:

$$y_{ij} = \mu_i + \varepsilon_{ij}.$$

The ijth error, ε_{ij}, is the deviation between the jth observations in the ith population from the ith population mean: μ_i. By using the ith mean instead of the overall mean, we remove the effect of groups and isolate the variation that is inherent in the experiment. In the above linear model, the mean of the ith population, μ_i, can be related to the overall population mean using its *group effect*, τ_i:

Table 12.1. General notation used in one-way classification analysis of variance.

	Population				
	1	2	……	k	
	y_{11}	y_{21}	……	y_{k1}	
	y_{12}	y_{22}	……	y_{k2}	
	⋮	⋮		⋮	
	y_{1n}	y_{2n}	……	y_{kn}	
Total	$T_{1\cdot}$	$T_{2\cdot}$	……	$T_{k\cdot}$	$T_{\cdot\cdot}$
Mean	$\bar{y}_{1\cdot}$	$\bar{y}_{2\cdot}$	……	$\bar{y}_{k\cdot}$	$\bar{y}_{\cdot\cdot}$

$\mu_{i.} = \mu + \tau_i.$

Hence, we have:

$y_{ij} = \mu + \tau_i + \varepsilon_{ij}.$

Since μ is the overall mean of the k population means (or treatment effect), it is easy to show that:

$$\sum_{i=1}^{k} \tau_i = \sum_{i=1}^{k} (\mu_i - \mu) = 0$$

Consequently, the above-stated H_0 and H_1 can also be found in a modified form, such that:

H_0: $\tau_1 = \tau_2 = \ldots = \tau_k = 0$

and

H_1: at least one τ_i is not equal to zero.

Note that the two sets of null and alternative hypotheses presented here are equivalent.

For the discussion to follow, we introduce a simplified practical example.

Example 12.1. Three treatments, consisting of two types of fertilizer (organic and inorganic) and a control (i.e. no fertilizer), were tested to investigate their effects on the height growth (in cm) of 1-year-old Douglas-fir seedlings. The original data, totals and means are summarized in Table 12.2. Our task is to compare the equality of three unknown population means (height growth) using one-way analysis of variance with a level of significance of 0.05.

The total variation of the 12 observations, in terms of sums of squares (SS_T or SS_{Total} or SS_y), is due to two sources. First, there is some variation between the observations because they come from three distinct fertilizer treatments, or groups. This is called **group-to-group variation** (SS_G). We can isolate this group-to-group variation by replacing every observation in Table 12.2 with its group mean (see Table 12.3) – which is to say that the sample data would contain only group-to-group variation. This would be an ideal, or at least simplified, experiment with only one source of variation due to fertilizer treatments. However, since the observations also vary from seedling to seedling, we must address another source of variation in this data set: experimental error, or the tree-to-tree variation that occurs within groups. This natural variation in the height growth among seedlings that received exactly the same treatment (in our case, fertilizer type) – which could be due to genetic variation between seedlings or microclimatic effects, or some combination of these and other factors – is called **within-group variation** (SS_W). As shown in Table 12.4, we can isolate this variation in the data by replacing the observations with their deviation from their respective group means.

It follows from the discussion above that we can partition the total variation in the data (SS_T) into two parts: variation between groups (SS_G) and variation within groups (SS_W). We can therefore write the following identity:

$$SS_T = SS_G + SS_W \qquad (12.1)$$

and restate this identity algebraically as:

$$\sum_{i=1}^{k}\sum_{j=1}^{n} (y_{ij} - \bar{y}_{..})^2 = n\sum_{i=1}^{k} (\bar{y}_{i.} - \bar{y}_{..})^2 + \sum_{i=1}^{k}\sum_{j=1}^{n} (y_{ij} - \bar{y}_{i.})^2.$$

Table 12.2. Height growth (in cm) of 1-year-old Douglas-fir seedlings.

	Fertilizer			
	Control	Organic	Inorganic	
	3.0	4.0	4.1	
	3.3	4.3	4.0	
	3.5	4.0	4.2	
	3.0	4.1	3.7	
Total	12.8	16.4	16.0	45.2
Mean	3.2	4.1	4.0	3.8

Table 12.3. Hypothetical observations showing group-to-group effect only.

	Fertilizer			
	Control	Organic	Inorganic	
	3.2	4.1	4.0	
	3.2	4.1	4.0	
	3.2	4.1	4.0	
	3.2	4.1	4.0	
Total	12.8	16.4	16.0	45.2
Mean	3.2	4.1	4.0	3.8

Table 12.4. Hypothetical observations showing within-group effect only.

	Fertilizer			
	Control	Organic	Inorganic	
	–0.2	–0.1	0.1	
	0.1	0.2	0.0	
	0.3	–0.1	0.2	
	–0.2	0.0	–0.3	
Total	0.0	0.0	0.0	0.0
Mean	0.0	0.0	0.0	0.0

To prove this identity, we start with the definition of the total corrected sum of squares on the left-hand side of the equation. We both add and subtract $\bar{y}_i$ inside the parentheses and rearrange the terms. Next, we square the terms and rearrange once more to get:

$$\begin{aligned}\sum_{i=1}^{k}\sum_{j=1}^{n}\left(y_{ij}-\bar{y}_{..}\right)^2 &= \sum_{i=1}^{k}\sum_{j=1}^{n}\left[\left(\bar{y}_{i.}-\bar{y}_{..}\right)+\left(y_{ij}-\bar{y}_{i.}\right)\right]^2 \\ &= \sum_{i=1}^{k}\sum_{j=1}^{n}\left[\left(\bar{y}_{i.}-\bar{y}_{..}\right)^2+2\left(\bar{y}_{i.}-\bar{y}_{..}\right)\left(y_{ij}-\bar{y}_{i.}\right)+\left(y_{ij}-\bar{y}_{i.}\right)^2\right] \\ &= \sum_{i=1}^{k}\sum_{j=1}^{n}\left(\bar{y}_{i.}-\bar{y}_{..}\right)^2+2\sum_{i=1}^{k}\sum_{j=1}^{n}\left(\bar{y}_{i.}-\bar{y}_{..}\right)\left(y_{ij}-\bar{y}_{i.}\right)+\sum_{i=1}^{k}\sum_{j=1}^{n}\left(y_{ij}-\bar{y}_{i.}\right)^2.\end{aligned}$$

Since the first sum on the right-hand side does not need to incorporate the subscript, j, we can rewrite this as:

$$\sum_{i=1}^{k}\sum_{j=1}^{n}(y_{i\cdot}-\bar{y}_{\cdot\cdot})^2 = n\sum_{i=1}^{k}(\bar{y}_{i\cdot}-\bar{y}_{\cdot\cdot})^2$$

The second term then sums to zero, because:

$$\sum_{j=1}^{n}\left(y_{ij}-\bar{y}_{i\cdot}\right) = \sum_{j=1}^{n} y_{ij} - n\bar{y}_{i\cdot} = \sum_{j=1}^{n} y_{ij} - n\left(\frac{\sum_{j=1}^{n} y_{ij}}{n}\right) = 0$$

which leaves us with:

$$\sum_{i=1}^{k}\sum_{j=1}^{n}\left(y_{ij}-\bar{y}_{\cdot\cdot}\right)^2 = n\sum_{i=1}^{k}(\bar{y}_{i\cdot}-\bar{y}_{\cdot\cdot})^2 + \sum_{i=1}^{k}\sum_{j=1}^{n}\left(y_{ij}-\bar{y}_{i\cdot}\right)^2 \tag{12.2}$$

Equation 12.2 shows that the degrees of freedom for the total sum of squares is the total number of observations minus one ($nk - 1$), because one statistic is used in the calculation of the sum of squares. Since the group-to-group sum of squares is calculated from k independent observations (k sample means) and one statistic is used in the equation, its degrees of freedom is ($k - 1$).

There are two ways to derive the within-group sum of squares degrees of freedom. One way is to consider that nk independent observations and k statistics (k sample means) are used to calculate the sum of squares. The degrees of freedom is therefore ($nk - k$), which equals $k(n - 1)$. The within-group sum of squares can also be interpreted as the sum of the k sums of squares of the observations calculated for each of the k groups – meaning that the degrees of freedom equals k times ($n - 1$).

To summarize, the degrees of freedom for the three sums of squares (SS_T, SS_G and SS_W) are:

$$nk - 1 = k - 1 + k(n - 1) \tag{12.3}$$

Many books refer to the group-to-group sum of squares as **treatment sum of squares** and the within-group sum of squares as **experimental error sum of squares** (or simply, **error sum of squares**) because analysis of variance is mainly used to analyse data generated by designed experiments, where the groups are typically treatments and the within-group variation is unexplained experimental error.

If Eqn 12.1 shows that the group-to-group sum of squares is higher than the within-group sum of squares, this means that the sample means are considerably different from one another, and it is very likely that the unknown group population means are also considerably different. The question is: how much higher should the group-to-group sum of squares be, compared to the within-group sum of squares, in order to reject the above stated H_0 (in terms of the equality of μs or τs)? To answer this question, we need to create a test statistic. We start by asserting that *if H_0 is true*, the constant population variance stated in the assumptions, σ^2, can be estimated by:

$$nS^2_{\bar{y}_{i.}} = MS_G = \frac{n\sum_{i=1}^{k}(\bar{y}_{i.}-\bar{y}_{\cdot\cdot})^2}{k-1} \tag{12.4}$$

The variance of the sample means is $\sigma^2_{\bar{y}} = \frac{\sigma^2}{n}$, which can be rearranged as $\sigma^2 = n\sigma^2_{\bar{y}}$.

Regardless of the truth or falsity of H_0, a second independent estimate of σ^2 can be obtained by calculating the mean squares within group:

$$S_2^2 = MS_W = \frac{\sum_{i=1}^{k}\left[\sum_{j=1}^{n}\left(y_{ij} - \bar{y}_{..}\right)^2\right]}{k(n-1)} \tag{12.5}$$

The notation, *MS*, refers to the **mean squares** (essentially, these are variance terms with sums of squares being divided by degrees of freedom). MS_G is the group-to-group mean square (variance) and MS_W is the within-groups mean square (variance). These are estimators of the variances, σ_G^2 and σ_W^2.

From here, H_0 and H_1 can be restated in terms of variances, rather than means or group effects, as:

H_0: $\sigma_G^2/\sigma_W^2 = 1.0$,

H_1: $\sigma_G^2/\sigma_W^2 > 1.0$.

Since this hypothesis is stated as a ratio of variances, it must be tested with an *F*-test statistic:

$$F_{[(k-1),k(n-1)]} = \frac{MS_G}{MS_W} \tag{12.6}$$

The critical value is obtained from the *F*-table (see Table A.7, Appendix A) with $(k-1)$ and $k(n-1)$ degrees of freedom. For the group and within-group sums of squares, the computational equations are:

$$SS_T = \sum_{i=1}^{k}\sum_{j=1}^{n} y_{ij}^2 - \frac{T_{..}^2}{nk}, \tag{12.7}$$

$$SS_G = \frac{\sum_{i=1}^{n} T_{i.}^2}{n} - \frac{T_{..}^2}{nk} \tag{12.8}$$

and

$$SS_W = SS_T - SS_G \text{ (from Eqn 12.1).} \tag{12.9}$$

Equations 12.7–12.9 are algebraically equivalent to the sum of squares computations summarized in Eqn 12.2. While these may be less intuitive, they are much easier to use and are therefore recommended for all practical purposes.

As with regression analysis, the results of the above *F*-test are generally presented in the form of an analysis of variance table, as seen in Tables 12.5 (notation) and 12.6 (results).

Table 12.5. One-way analysis of variance table.

Source of variation	Degrees of freedom	Sum of squares	Mean squares	Computed *F*
Group-to-group	$k-1$	SS_G	$MS_G = \frac{SS_G}{k-1}$	$\frac{MS_G}{MS_W}$
Within-group	$k(n-1)$	SS_W	$MS_W = \frac{SS_W}{k(n-1)}$	
Total	$kn-1$	SS_T		

Table 12.6. Analysis of variance table for the Douglas-fir fertilizer test.

Source of variation	*DF*	*SS*	*MS*	Computed *F*	Critical *F*
Group-to-group	2	1.947	0.9735	22.91	4.26
Within-group	9	0.383	0.0425		
Total	11	2.330			

DF, degrees of freedom; *SS*, sum of squares; *MS*, mean squares.

We can now complete Example 12.1:

1. H_0: $\mu_1 = \mu_2 = \mu_3$.
2. H_1: at least one is different.
3. $\alpha = 0.05$.
4. Use Eqn 12.6.
5. $SS_T = 3.0^2 + 3.3^2 + \dots 4.2^2 + 3.7^2 - 45.2^2/12 \approx 2.330$

   $SS_G = \dfrac{12.8^2 + 16.4^2 + 16.0^2}{4} - 45.2^2/12 \approx 1.947$

   $SS_W = 2.330 - 1.947 \approx 0.383$.
6. $F_{0.05(2,\,9)} = 4.26$.
7. $F_{(2,9)} = \dfrac{0.9735}{0.0425} \approx 22.91$.
8. Since 22.91 > 4.25, the test statistic is in the critical region: we reject H_0 and accept H_1.
9. The data indicate that at least one of the three unknown population means is significantly different from another one (i.e. at least two of the means are significantly different).

While analysis of variance tells us whether unknown population means are statistically similar or different, it does not indicate which of the means is different from the others. A *post hoc* test, known as a **multiple comparison** test or a **mean separation** test, is required for this (we will introduce two multiple comparison tests to compare several population means in the next section). However, if the null hypothesis is accepted in the analysis of variance, all of the means are assumed to be equal and no further tests are required.

When performing an analysis of variance, one common problem is that we may not have equal numbers of observations in all of the groups under investigation. Observations may be lost during experimentation, or an experiment may simply not have been planned around equal numbers within groups. In either case, we have an *unbalanced* analysis of variance. Although this imbalance does not affect the validity of the analysis discussed above, we must make slight modifications to the equations in order to compute the sums of squares and degrees of freedom.

Consider a general case where $n_1, n_2, \dots, n_k$ samples are selected from k groups. To simplify some of the equations, let $\sum_{i=1}^{k} n_i = N$. The modified working equations for SS_T, SS_G, SS_W and the generalized analysis of variance table (Table 12.7) are as follows:

$$SS_T = \sum_{i=1}^{k}\sum_{j=1}^{n_i} y_{ij}^2 - \frac{T_{..}^2}{N}, \tag{12.10}$$

Table 12.7. One-way analysis of variance table for an unequal number of observations.

Source of variation	*DF*	*SS*	*MS*	Computed *F*
Group-to-group	$k-1$	SS_G	$MS_G = \frac{SS_G}{k-1}$	$\frac{MS_G}{MS_W}$
Within-group	$N-k$	SS_W	$MS_W = \frac{SS_W}{N-k}$	
Total	$N-1$	SS_T		

$$SS_G = \sum_{i=1}^{k} \frac{T_{i.}^2}{n_i} - \frac{T_{..}^2}{N}, \tag{12.11}$$

$$SS_W = SS_T - SS_G. \tag{12.12}$$

The corresponding degrees of freedom for SS_T, SS_G and SS_W are then:
$(N-1) = (k-1) + (N-k)$.

Example 12.2. A fibreboard manufacturer is interested in studying the modulus of rupture of four types of fibreboard (A, B, C and D). Although the experimenters planned to measure 4 boards per type, the number of boards actually available for the study was limited to 3 in type A and 2 in type C (see Table 12.8). Can it be assumed, with a 0.05 level of significance, that the four unknown population means are equal?

Table 12.8. Modulus of rupture of 4 fibreboard types (measured in megapascals).

	Fibreboard types				
	A	B	C	D	
	61.9	42.4	39.2	46.2	
	67.4	52.5	42.4	55.9	
	63.3	54.9		55.6	
		60.1		58.2	
Total	192.6	209.9	81.6	215.9	700.0
Mean	64.200	52.475	40.800	53.975	53.846
n_i	3	4	2	4	13

1. H_0: $\mu_1 = \mu_2 = \mu_3 = \mu_4$.
2. H_1: at least one is different.
3. $\alpha = 0.05$.
4. Use Eqn 12.6
5. $SS_T = 61.9^2 + 67.4^2 + \ldots\ 55.6^2 + 58.2^2 - 700.0^2/13 \approx 941.23$

   $SS_G = \frac{192.6^2}{3} + \frac{219.4^2}{4} + \frac{81.6^2}{2} + \frac{215.9^2}{4} - 700.0^2/13 = 669.6$

   $SS_W = 941.23 - 669.60 = 271.63.$

The sums of squares and mean sums of squares are summarized in the analysis of variance table (Table 12.9).

Table 12.9. Analysis of variance for the modulus of rupture by fibreboard type data.

Source of variation	*DF*	*SS*	*MS*	Computed *F*	Critical *F*
Board type	3	669.60	223.20	7.40	3.86
Within-board type	9	271.63	30.18		
Total	12	941.23			

6. $F_{0.05(3,\ 9)} = 3.86$.

7. $F_{(3,9)} = \frac{223.20}{30.18} \approx 7.40$.

8. Since 7.40 > 3.86, the test statistic is in the critical region: we reject H_0 and accept H_1.

9. The data indicate that at least one of the four unknown population means is significantly different from another one.

Again, since the unknown population means are found to be different, a post hoc multiple comparison technique is needed to decide which of the means are different (this is introduced in the next section).

At this point, it is important to list the assumptions of analysis of variance. Like regression analysis, certain assumptions must be met in order to test the equality of k population means:

1. Each population is normally distributed.
2. Each population has the same variance, σ^2.
3. The observations in a given population are independent from the observations in the other populations.
4. The observations are randomly selected from each population.

If these assumptions are met, the within-group variance (MS_W) is an unbiased estimate of σ^2. Consequently, the standard error of estimate for *any* of the k means is:

$$S_{\bar{y}_i.} = \sqrt{\frac{MS_W}{n_i}} \tag{12.13}$$

When there are equal numbers of observations per group, this simplifies to:

$$S_{\bar{y}_i.} = \sqrt{\frac{MS_W}{n}} \tag{12.14}$$

Hence, the confidence interval for any of the k population means is:

$$P\left(\bar{y}_{i.} - t_{\alpha/2(N-k)} S_{\bar{y}_i.} < \mu_i < \bar{y}_{i.} + t_{\alpha/2(N-k)} S_{\bar{y}_i.}\right) = 1 - \alpha \tag{12.15}$$

The degrees of freedom of the t-value in the confidence interval follows that of the within-groups mean squares: for equal numbers of observations, the degrees of freedom are $k(n - 1) = N - k$. When the analysis of variance is balanced, the width of the confidence interval is the same around all of the means. When the analysis of variance is not balanced, the confidence intervals are all constructed with the same t-value, but the widths of these intervals vary based on n_i.

Example 12.3. Calculate the 95% confidence interval for the unknown population mean of height growth for the seedlings treated with inorganic fertilizer in Example 12.1.

$$\bar{y}_{3.} = 4.0 \qquad t_{0.025(9)} = 2.26 \qquad S_{\bar{y}_{i.}} = \sqrt{\frac{0.0425}{4}} = 0.1031$$

$$P[4.0 - (2.26)(0.1031) < \mu_3 < 4.0 + (2.26)(0.1031)] = 0.95$$

$$P(3.77 < \mu_3 < 4.23) = 0.95$$

There is a 0.95 probability that the unknown population mean height growth of seedlings treated with inorganic fertilizer is between 3.77 and 4.23 cm.

Example 12.4. Calculate the 99% confidence interval for the unknown population mean modulus of rupture for fibreboard type B in Example 12.2.

$$\bar{y}_{2.} = 524.75 \qquad t_{0.005(9)} = 3.25 \qquad S_{\bar{y}_{i.}} = \sqrt{\frac{3018.17}{4}} = 27.49$$

$$P[524.75 - (3.25)(27.49) < \mu_2 < 524.75 + (3.25)(27.49)] = 0.99$$

$$P(435.41 < \mu_2 < 614.09) = 0.99.$$

There is a 0.99 probability that the unknown population mean modulus of rupture for fibreboard type B is between 435.41 and 614.09 MPa.

If desired, we could also calculate confidence intervals based on the data from a single group, independent of the other groups (using equations from Chapter 8, Section 8.3). For instance, using Examples 12.3 and 12.4, we could calculate:

$$\bar{y}_{3.} = 4.0 \qquad t_{0.025(3)} = 3.18 \qquad S_3 = 0.2160 \qquad S_{\bar{y}_{3.}} = \frac{0.2160}{\sqrt{4}} = 0.1080$$

$$P[4.0 - (3.18)(0.1080) < \mu_2 < 4.0 + (3.18)(0.1080)] = 0.95$$

$$P(3.66 < \mu_3 < 4.44) = 0.95$$

and

$$\bar{y}_{2.} = 524.75 \qquad t_{0.005(9)} = 3.25 \qquad S_2 = 74.24 \qquad S_{y_{2.}} = \frac{74.24}{\sqrt{4}} = 37.14$$

$$P[524.75 - (5.84)(37.14) < \mu_2 < 524.75 + (5.84)(37.14)] = 0.99$$

$$P(307.85 < \mu_2 < 741.65) = 0.99.$$

In both cases, the width of the confidence intervals increases when calculated independently of the other groups. The reason for this is simply that the variance estimated from the analysis of variance is based on many more independent observations (i.e. the pooled variance for all groups) than the variance calculated from an isolated group. The pooled data therefore have higher degrees of freedom. Since *t*-values decrease with increasing degrees of freedom, higher degrees of freedom produce a narrower confidence interval. This is a very important advantage of analysis of variance.

The standard error of the difference between any two of the *k* means can be calculated as:

$$S_{\bar{y}_p.-\bar{y}_r.} = \sqrt{MS_W\left(\frac{1}{n_p}+\frac{1}{n_r}\right)}, \tag{12.16}$$

where $\bar{y}_{p.}$ and $\bar{y}_{r.}$ are two out of the k means; n_p = number of observations from the pth treatment; and n_r = number of observations from the rth treatment.

If the sample sizes are equal, Eqn 12.16 reduces to:

$$S_{\bar{y}_p.-\bar{y}_r.} = \sqrt{\frac{2MS_W}{n}}. \tag{12.17}$$

From this, confidence intervals can be obtained for the difference between any two unknown population means:

$$P\begin{bmatrix}(\bar{y}_{p.}-\bar{y}_{r.})-t_{\alpha/2(N-k)}S_{\bar{y}_p.-\bar{y}_r.} < \mu_p-\mu_r < (\bar{y}_{p.}-\bar{y}_{r.}) \\ +t_{\alpha/2(N-k)}S_{\bar{y}_p.-\bar{y}_r.}\end{bmatrix} = 1-\alpha, \tag{12.18}$$

where the degrees of freedom of the t-value for an equal number of observations is $k(n-1) = N-k$.

12.2 Multiple Comparisons

In analysis of variance, rejection of the null hypothesis does provide some limited information concerning the k population means (i.e. at least one population mean is different from the others), but it does not indicate which of the means differ. Several multiple comparison procedures have been proposed for simultaneous comparison of all the k population means when the null hypothesis is rejected: *Bonferroni's Procedure*, *Duncan's New Multiple Range Test*, *Scheffé's Method* and *Tukey's Method*, to name a few. Other procedures exist for conducting predetermined comparisons on a subset of all possible population means: *Fisher's Least Significant Difference* and *Orthogonal Contrasts* are two examples.

If there are m possible comparisons, we have m opportunities for committing a *type I error*: this is why we perform a multiple comparison test instead of testing each of the means against each other. For instance, if we are testing three population means (see Example 12.1), then there are three possible comparisons: 1 *versus* 2, 1 *versus* 3 and 2 *versus* 3. The number of comparisons increases, at an increasing rate, with the number of means: there are six comparisons with four means and ten comparisons with five. Using the rules for combinations, if k groups are to be compared, there are:

$$m = {}_kC_2 = \frac{k!}{2!(k-2)!} = \frac{k(k-1)}{2}$$

pairwise comparisons of means.

The fact that we do not have to perform m t-tests individually is another clear advantage of the analysis of variance procedure because individual tests will result in lower degrees of freedom for each comparison and a probability of committing *type I* error that is higher than the stated significance level. On this latter point, the probability of committing *type I* error with three means and $\alpha = 0.05$, would be:

$$P(\text{type I error}) = 1 - P(\text{no type I error}) = 1 - 0.95^3 \approx 1 - 0.86 \approx 0.14.$$

In general, if there are m comparisons, type I error is computed as:

$$P(\text{type I error}) = 1 - (1 - \alpha)^m.$$

By using *post hoc* multiple comparisons, the critical values are adjusted to compensate for the 'inflation' of type I error demonstrated above. However, multiple comparison tests should be used with some caution. As with other statistical tests, if there is more than one procedure available, we can expect that none of them are perfect. This is also the case for multiple comparison tests: if two or three procedures are used to compare the k means, it is possible to obtain different results. Some tests are therefore referred to as more 'sensitive', or more 'liberal', than others because they are more likely to indicate significant differences between means.

There are many available methods for multiple comparisons. We will discuss **Bonferroni's Procedure** and **Scheffé's Method**, both of which can be readily adjusted to counter the inflation of *type I* error that comes with multiple comparison of means.

Bonferroni's Procedure

In Bonferroni's Procedure the level of significance is set to α/m in a one-tailed test and $\alpha/2m$ in a two-tailed test, where α is the desired rate of significance and m is the number of comparisons to be performed. The following steps are recommended to carry out the comparisons:

1. Rank the means in ascending or descending order.
2. Calculate the **critical difference (CD)**.
3. Calculate the pairwise absolute differences between ranked means.
4. Draw a line under any subset of adjacent means that are not significantly different from each other or use a symbol (see Example 12.6).

Since the pairwise comparisons are almost always two-tailed tests, Bonferroni's critical difference is:

$$CD = t_{\alpha/(2m),\ N-k}\sqrt{MS_{\mathrm{W}}\left(\frac{1}{n_p} + \frac{1}{n_r}\right)} \qquad (12.19)$$

If we have a balanced design (equal sample sizes in each group), this simplifies to:

$$CD = t_{\alpha/(2m),k(n-1)}\sqrt{\frac{2MS_{\mathrm{W}}}{n}} \qquad (12.20)$$

If the pairwise difference of $\left|\bar{y}_{p.} - \bar{y}_{r.}\right| > CD$, then $\mu_{p.}$ and $\mu_{r.}$ are different.

If we have $m = 6$ comparisons and $\alpha = 0.05$, we will need to find $t_{(N-k)}$ at $0.05/12 = 0.0042$. Because most of the available t-tables (see Table A.5, Appendix A) do not contain the critical t-values for this and other small probability levels, they must be obtained either from computer packages or from specially created tables. Table A.9

(see Appendix A), for example, contains the critical t-values as a function of the within-group degrees of freedom (first column) and the number of desired comparisons, m (first row), such that a type I error of 0.05 is maintained.

Example 12.5. Use Bonferroni's Procedure to compare the unknown population means described in Example 12.1, with a 0.05 level of significance.

1.

Group:	Control	Inorganic	Organic
Mean:	3.2	4.0	4.1

$m = 3$ and $t_{\alpha/2m,\,9} = t_{0.05/6,\,9} = t_{0.0083,\,9} = 2.93$ (see Table A.9, Appendix A).

2. $CD = 2.93\sqrt{\dfrac{(2)(0.0425)}{4}} = 0.1458 = 0.15.$

3. Since $|3.2 - 4.0| = 0.8 > 0.15$, the control mean (3.2) is significantly different from the inorganic mean (4.0). Note that we do not need to compare the control (3.2) against the organic fertilizer (4.1), since the mean of the inorganic fertilizer (4.0) is lower than the mean of the organic fertilizer and we know that the distance between the control and the organic means must be greater than the distance between the inorganic and the control means. In other words, since there is a significant difference between the inorganic and the control, there must also be a significant difference between the organic and the control. To test the inorganic versus the organic fertilizers, we compute $|4.0 - 4.1| = 0.1 < 0.15$. Thus, the inorganic fertilizer (4.0) is not significantly different from the organic fertilizer (4.1). It is customary to make a diagram of the means in ascending order, where means that are not significantly different are underlined (although it is possible for one mean to be underlined more than once, as we shall see in the next example).

Control	Inorganic	Organic
3.2	<u>4.0</u>	<u>4.1</u>

Note that if equal numbers of observations are used in the analysis of variance, the same critical difference applies for all comparisons. This is one of the advantages of using equal numbers of observations.

Scheffé's Method

In Scheffé's Method, a test statistic (which is a modified F-value) is calculated for every possible comparison of the pth and rth group means:

$$F_S = \frac{\left(\bar{y}_{p.} - \bar{y}_{r.}\right)^2}{MS_W\left(1/n_p + 1/n_r\right)} \tag{12.21}$$

For equal numbers of observations, Eqn 12.21 simplifies to:

$$F_S = \frac{\left(\bar{y}_{p.} - \bar{y}_{r.}\right)^2}{2MS_W / n} \tag{12.22}$$

F_S is compared with a critical value of:

$$F_{SC,\alpha(k-1,N-k)} = (k-1)F_{\alpha(k-1,N-k)} \tag{12.23}$$

No special table is required: $F_{\alpha(k-1,N-k)}$ comes from the F-table (see Table A.7, Appendix A). If $F_S > F_{SC,\alpha(k-1,\,N-k)}$, $\bar{y}_{p.}$ is said to be significantly different from $\bar{y}_{r.}$.

The process of comparing a series of means is similar to Bonferroni's Procedure, with some modifications to the second step. In Scheffé's Method, we first calculate the modified critical F-value, F_{SC}. Next, instead of calculating the critical difference, we calculate a test statistic, F_S, for each comparison. These are then compared against our critical F-value, F_{SC}.

Example 12.6. Use Scheffé's Method to compare the four unknown population means described in Example 12.2.

1.

Group:	C	B	D	A
Mean:	408.00	524.75	539.75	642.00
n_i	2	4	4	3

2. $F_{SC,0.05(3,\,9)} = (3.86)\,(4-1) = 11.58$

C versus B $\quad F_S = \dfrac{(408.00 - 524.75)^2}{3018.17(1/2 + 1/4)} = 6.02 < 11.58$

C versus D $\quad F_S = \dfrac{(408.00 - 539.75)^2}{3018.17(1/2 + 1/4)} = 7.67 < 11.58$

C versus A $\quad F_S = \dfrac{(408.00 - 642.00)^2}{3018.17(1/2 + 1/3)} = 21.77 > 11.58$

B versus D $\quad F_S = \dfrac{(524.75 - 539.75)^2}{3018.17(1/4 + 1/4)} = 0.15 < 11.58$

B versus A $\quad F_S = \dfrac{(524.75 - 642.00)^2}{3018.17(1/4 + 1/3)} = 7.81 < 11.58$

D versus A $\quad F_S = \dfrac{(539.75 - 642.00)^2}{3018.17(1/4 + 1/3)} = 5.94 < 11.58$

The F_S-test statistics were calculated for all possible comparisons because there are unequal numbers of observations in each treatment. The only significant difference observed occurs between fibreboard types C and A.

The significant differences can also be presented in a diagram and/or with symbols (subscripts). In the first diagram, any two means underlined with a continuous line are not significantly different from each other. In the second diagram, any two means bearing the same letter subscript are not significantly different. When results indicating significant differences between means are presented in this manner, they are generally accompanied with a statement such as, 'any two means underlined by the same line (or labelled with the same subscript) are not significantly different', to ensure clarity.

Note here that two means (B and D) are underlined twice, indicating that they (B and D) are not significantly different from C (taken alone) or A (taken alone), while means C and A are significantly different from one another. This result is simply a function of the distributional nature of the means: think of a series of sampling distributions or normal curves where C, B and D overlap; B, D and A overlap; but C and A do not overlap.

3. C B D A

C	B	D	A
408.00	524.75	539.75	642.00

or

C	B	D	A
408.00_a	524.75_{ab}	539.75_{ab}	5642.00_b

For both multiple comparison procedures presented here, the number of required comparisons can be considerably reduced in most cases *when the number of observations in each group is equal.* Simply start with the mean at the extreme left of the ordered means and compare one-by-one, in order, to all of the other means occurring to its right. Once the first significant difference is found, the remaining (right-hand) means are also significantly different and so do not have to be compared. This process is then repeated, starting with the second-to-left of the ordered means, and so on. This is another advantage of using equal numbers of samples per group in analysis of variance.

12.3 Test for Equality of Variances

One of the most important assumptions of both analysis of variance and *post hoc* pairwise comparisons of means is that the *variances* of the k populations are equal. It is advisable to test this assumption before any analysis of variance is carried out. If the assumption of equality of variances is not met, a transformation of the data prior to analysis may alleviate this problem (this is covered in more advanced statistical texts). **Bartlett's test,** which produces a test statistic that approximately follows a χ^2 distribution with $(k - 1)$ degrees of freedom, is recommended for comparing the equality of k unknown population variances. The null and alternative hypotheses are formulated as follows:

H_0: $\sigma_1^2 = \sigma_2^2 = \ldots = \sigma_k^2$, and

H_1: at least one variance is different.

The following steps are suggested to calculate Bartlett's test statistic:

1. Compute the variance for each group, $S_1^2, S_2^2, \ldots, S_k^2$
2. Compute the pooled variance for the k groups as:

$$S_p^2 = \frac{\sum_{i=1}^{k} (n_i - 1)S_i^2}{N - k} \qquad (12.24)$$

3. Compute the test statistic:

$$\chi^2_{(k-1)} = \frac{q}{h} \qquad (12.25)$$

where

$$q = (N - k) \ln S_p^2 - \sum_{i=1}^{k} (n_i - 1) \ln S_i^2$$

ln = natural logarithm, and

$$h = 1 + \frac{1}{3(k-1)}\left[\sum_{i=1}^{k} \frac{1}{n_i - 1} - \frac{1}{N - k}\right]$$

4. Compare the test statistic to the $\chi^2_{(k-1)\alpha}$ critical value obtained from the χ^2-table (see Table A.6, Appendix A) with $(k-1)$ degrees of freedom.

You may recognize that the pooled variance, S_p^2, is equal to the within-group mean squares, MS_W. In other words: if $k = 2$, Eqn 12.24 is the same equation used to calculate the pooled variance for two groups of data, when it is assumed that $\sigma_1^2 = \sigma_2^2$ (see Case 2 under Section 7.5, Chapter 7).

Example 12.7. Compare the four unknown population variances for the fibreboard data described in Example 12.2, using a 0.05 significance level.

1. H_0: $\sigma_1^2 = \sigma_2^2 = \sigma_3^2 = \sigma_4^2$.
2. H_1: at least one is different.
3. $\alpha = 0.05$.
4. Use Eqn 12.25.
5. $S_1^2 = 817.0$; $S_2^2 = 5517.6$; $S_3^2 = 512.0$; $S_4^2 = 2821.6$

$$S_p^2 = \frac{2(817.0) + 3(5517.6) + 1(512.0) + 3(2821.6)}{9} = 27{,}163.6$$

$\ln S_p^2 = 8.013$; $\ln S_1^2 = 6.705$; $\ln S_2^2 = 8.616$

$\ln S_3^2 = 6.238$; $\ln S_4^2 = 7.944$;

$$q = 9(8.013)\ [2(6.705) + 3(8.616) + 1(6.238) + 3(7.944)] = 2.789$$

$$h = 1 + \frac{1}{3(4-1)}\left[\frac{1}{2} + \frac{1}{3} + \frac{1}{1} + \frac{1}{3} - \frac{1}{9}\right] = 1.228$$

6. $\chi^2_{0.05(3)} = 7.81$.
7. $\chi^2_{(3)} = \frac{2.789}{1.228} \approx 2.2789$.
8. Since $2.27 < 7.81$, the test statistic is in the acceptance region and we 'accept' H_0.
9. The data indicates that the assumption of constant variance has been met, meaning that an analysis of variance can be performed.

12.4 Two-way Analysis of Variance

In a one-way classification analysis of variance, we study a single *factor*. For instance, in Example 12.1 we were interested in the three types of fertilizer (control, inorganic and organic) and in Example 12.2 we were interested in the four types of fibreboard (A, B, C and D). In general, we refer to the three fertilizer types and the four fibreboard types as *levels* within the factor of fertilizer or fibreboard type, respectively. Therefore, Example 12.1 describes a study of one factor with three levels and Example 12.2 describes a study of one factor with four levels.

When the effects of two *factors* are investigated at the same time, we have a **two-way classification analysis of variance.** For example, if the three fertilizers from Example 12.1 were applied to three species (e.g. Douglas-fir, western red cedar and western hemlock), it would become a *two-factor* experiment with three levels in each factor. Similarly, if all four fibreboard types from Example 12.2 were produced in three different manufacturing plants, it would become a *two-factor* experiment with four levels in the first factor (fibreboard type) and three levels in the second (manufacturing plant).

Let us consider k levels in factor A, t levels in factor B and n *replications* (an equal number of observations) within each factor combination, or 'cell' (see Table 12.10). Consequently, the total number of observations in the data set is ktn, and each observation is identified with three subscripts: y_{ijl}, where $i = 1, 2, \ldots, k$; $j = 1, 2, \ldots, t$; and $l = 1, 2, \ldots, n$. As in the one-way analysis of variance, we first present the linear model (equation). For the two-way analysis of variance, we have:

$$y_{ijl} = \mu + \tau_i + \beta_j + \omega_{ij} + \varepsilon_{ijl}.$$

As in the one-way case, μ is the overall population mean, τ_i is the ith-level effect of factor A and ε_{ijl} is the within-group effect (or experimental error) for the lth observation from the ith level of factor A and jth level of factor B. For the two-way case, we add β_j, the jth-level effect of factor B, and an interaction effect, ω_{ij}, which describes the unique effects (if any) of the ith level of factor A and jth level of factor B acting together on an experimental unit (this notion of interactions is explained more fully later).

For the two-way analysis of variance, the total sum of squares (SS_T) is partitioned into four sources:

$$SS_T = SS_A + SS_B + SS_{AB} + SS_E \tag{12.26}$$

$$\sum_{i=1}^{k}\sum_{j=1}^{t}\sum_{l=1}^{n}\left(y_{ijl}-\bar{y}_{...}\right)^2 = tn\sum_{i=1}^{k}\left(\bar{y}_{i..}-\bar{y}_{...}\right)^2 + kn\sum_{j=1}^{t}\left(\bar{y}_{.j.}-\bar{y}_{...}\right)^2 + n\sum_{i=1}^{k}\sum_{j=1}^{t}\left(\bar{y}_{ij\cdot}-\bar{y}_{i..}-\bar{y}_{.j.}+\bar{y}_{...}\right)^2 + \sum_{i=1}^{k}\sum_{j=1}^{t}\left[\sum_{l=1}^{n}\left(y_{ijl}-\bar{y}_{ij.}\right)^2\right] \tag{12.27}$$

Table 12.10. General notation used for observations in two-way classification analysis of variance.

	Factor A					
Factor B	1	2	....	k	Total	Mean
1	y_{111} y_{112} ...... y_{11n} $T_{11\cdot}$ $\bar{y}_{11\cdot}$	y_{211} y_{212} ...... y_{21n} $T_{21\cdot}$ $\bar{y}_{21\cdot}$	....	y_{k11} y_{k12} ...... y_{11n} $T_{11\cdot}$ $\bar{y}_{k1\cdot}$	$T_{.1.}$	$\bar{y}_{.1.}$
2	y_{121} y_{122} ...... y_{12n} $T_{12\cdot}$ $\bar{y}_{12\cdot}$	y_{221} y_{222} ...... y_{22n} $T_{22\cdot}$ $\bar{y}_{22\cdot}$	....	y_{k21} y_{k22} ...... y_{k2n} $T_{k2\cdot}$ $y_{k\cdot}$	$T_{.2.}$	$\bar{y}_{.2.}$
⋮	⋮	⋮		⋮	⋮	⋮
t	y_{1t1} y_{1t2} ...... y_{1tn}	y_{2t1} y_{2t2} ...... y_{2tn}	....	y_{kt1} y_{kt2} ...... y_{ktn}	$T_{.t.}$	$\bar{y}_{.t.}$
Total	$T_{1\cdot\cdot}$	$T_{2\cdot\cdot}$	....	$T_{k\cdot\cdot}$	$T_{...}$	$\bar{y}_{...}$
Mean	$\bar{y}_{1\cdot\cdot}$	$\bar{y}_{2\cdot\cdot}$		$\bar{y}_{k\cdot\cdot}$		

The top line in each cell shows the notation for observations and the bottom row shows the notation for both the totals and the means computed within each cell.

Their respective degrees of freedom are:

$$(ktn - 1) = (k - 1) + (t - 1) + (k - 1)(t - 1) + kt(n - 1) \quad (12.28)$$

where

SS_T = sum of squares total;
SS_A = sum of squares due to factor A;
SS_B = sum of squares due to factor B;
SS_{AB} = sum of squares due to the interaction between factors A and B; and
SS_E (or SS_W) = sum of squares error (or sum of squares within-group).

All sums of squares and degrees of freedom – except for the interaction, SS_{AB} – can be interpreted in the same manner as a one-way analysis of variance. Interpreting the sum of squares due to the interaction between A and B is a more complicated matter and will be discussed later with an example. For now, SS_{AB} measures the variation of the means of one factor (A) within the various levels of the other factor (B). For instance, if the means of factor A within the various levels of B are similar, SS_{AB} is low. If the means of factor A are very different within the various levels of B, then SS_{AB} is high. The degrees of freedom for the interaction can also be interpreted in relation to the two factors. Notice that:

$$(k - 1)(t - 1) = kt - k - t + 1 = kt - (k + t - 1).$$

These degrees of freedom here are very similar to those seen in contingency tables (see Section 10.2, Chapter 10). Note that the theoretical equation of the sum of squares of the interaction (Eqn 12.27) indicates that the degrees of freedom should be $kt - (k + t + 1)$, since kt observations and $(k + t + 1)$ means are used in the calculation. However, this is not correct because the $(k + t + 1)$ means are not all independent: if we know the k means for factor A, only $(t - 1)$ of the t means are independent for factor B. Furthermore, the k means of factor A average to the grand mean, as do the t means of factor B. This reduces the degrees of freedom by $(k + (t - 1))$.

The analysis of variance table for the two-way classification is summarized in Table 12.11. The recommended computational equations for the sums of squares are:

$$SS_T = \sum_{i=1}^{k}\sum_{j=1}^{t}\sum_{l=1}^{n} y_{ijl}^2 - \frac{T_{\cdots}^2}{ktn} \quad (12.29)$$

$$SS_A = \frac{1}{tn}\sum_{i=1}^{k} T_{i\cdot\cdot}^2 - \frac{T_{\cdots}^2}{ktn} \quad (12.30)$$

Table 12.11. Two-way analysis of variance table.

Source of variation	*DF*	*SS*	*MS*	Computed *F*
Factor A	$k-1$	SS_A	$MS_A = \frac{SS_A}{k-1}$	MS_A/MS_E
Factor B	$t-1$	SS_B	$MS_B = \frac{SS_B}{t-1}$	MS_B/MS_E
A × B	$(k-1)(t-1)$	SS_{AB}	$MS_{AB} = \frac{SS_{AB}}{(k-1)(t-1)}$	MS_{AB}/MS_E
Error	$kt(n-1)$	SS_E	$MS_E = \frac{SS_E}{kt(n-1)}$	
Total	$ktn-1$	SS_T		

$$SS_B = \frac{1}{kn}\sum_{i=1}^{t} T_{\cdot j \cdot}^2 - \frac{T_{\cdots}^2}{ktn} \tag{12.31}$$

$$SS_{AB} = \frac{1}{n}\sum_{i=1}^{k}\sum_{j=1}^{t} T_{ij\cdot}^2 - \frac{T_{\cdots}^2}{ktn} - SS_A - SS_B \tag{12.32}$$

$$SS_E = SS_T - SS_A - SS_B - SS_{AB} \tag{12.33}$$

Example 12.8. To investigate how different tree species respond to fertilizers, the fertilizer trial from Example 12.1 was expanded to include two more tree species: western hemlock and western red cedar. Table 12.12 expands on Table 12.3, showing height growth data for 1-year-old seedlings using the three fertilizer levels (organic, inorganic and control) and three species (Douglas-fir, western hemlock and western red cedar). For each factor (treatment) combination, four observations were made. Analyse the data using a 0.01 level of significance.

Table 12.12. The effect of fertilizer treatment and tree species on the height growth (in cm) of 1-year-old seedlings.

	Douglas-fir				Western hemlock				Western red cedar			
	C	F_1	F_2		C	F_1	F_2		C	F_1	F_2	
	3.0	4.0	4.1		2.9	3.1	4.0		2.8	3.2	4.0	
	3.3	4.3	4.0		3.1	3.0	4.2		3.0	3.0	3.9	
	3.5	4.0	4.2		3.4	3.2	4.4		3.1	3.1	4.1	
	3.0	4.1	3.7		3.0	3.1	4.2		3.1	2.9	3.2	
Sum:	12.8	16.4	16.0	45.2	12.4	12.4	16.8	41.6	12.0	12.2	15.8	40.0
Ave:	3.2	4.1	4.0	3.77	3.1	3.1	4.2	3.47	3.0	3.05	3.95	3.33

C, control; F_1, inorganic fertilizer; F_2, organic fertilizer.

$$\bar{y}_{1..} = \frac{12.8 + 12.4 + 12.0}{12} = 3.10 \qquad \bar{y}_{2..} = \frac{16.4 + 12.4 + 12.2}{12} = 3.42$$

$$\bar{y}_{3..} = \frac{16.0 + 12.8 + 15.8}{12} = 4.05$$

Since we have three effects (factor A, factor B and their interaction), we now have three hypotheses to test:

1. H_0:

a. Treatment $\mu_{1..} = \mu_{2..} = \mu_{3..}$

b. Species $\mu_{.1.} = \mu_{.2.} = \mu_{.3.}$

c. Interaction $\mu_{11.} - \mu_{12.} = \mu_{21.} - \mu_{22.} = \mu_{31.} - \mu_{32.}$

$\mu_{12.} - \mu_{13.} = \mu_{22.} - \mu_{23.} = \mu_{32.} - \mu_{33.}$

$\mu_{11.} - \mu_{13.} = \mu_{21.} - \mu_{23.} = \mu_{31.} - \mu_{33.}$

(Note that the hypothesis for the interaction of the two factors is set up in such a way that the differences of two means at any two levels of a single factor are compared at all levels of the other factor. This hypothesis is often simply stated in words as: 'there is no interaction'.)

H_1:

a. At least two of the means are not equal.

b. At least two of the means are not equal.

c. At least two of the differences are not equal.

2. $\alpha = 0.01$.
3. See Table 12.11 for equations.

4. $SS_T = 3.0^2 + 3.3^2 + \cdots\cdots 4.1^2 + 3.2^2 - \dfrac{126.8^2}{36} \approx 9.4222$

$$SS_A = \frac{1}{12}\left(37.2^2 + 41.0^2 + 48.6^2\right) - \frac{126.8^2}{36} \approx 5.6156$$

$$SS_B = \frac{1}{12}\left(45.2^2 + 41.6^2 + 40.0^2\right) - \frac{126.8^2}{36} \approx 1.1822$$

$$SS_{AB} = \frac{1}{4}\left(12.8^2 + 16.4^2 + 16.0^2 + 12.4^2 + 12.4^2 + 16.8^2 + 12.0^2 + 12.2^2 + 15.8^2\right)$$

$$- \frac{126.8^2}{36} - 5.6156 - 1.822 \approx 1.8444$$

$$SS_E = 9.4222 - 5.6156 - 1.1822 - 1.8444 = 0.7800$$

5. $F_{(2,27)0.05} = 5.49$, $F_{(4,27)0.05} = 4.11$.

6. See Table 12.13 (computed F).

Table 12.13. Two-way analysis of variance for height growth data.

Source of variation	*DF*	*SS*	*MS*	Computed *F*	Critical *F*
Treatment	2	5.6156	2.8078	97.19	5.49
Species	2	1.1822	0.5911	20.46	5.49
T × S	4	1.8444	0.4611	15.96	4.11
Error	27	0.7800	0.0289		
Total	35	9.4222			

7. Since 15.96 > 4.11, the test statistic for the interaction is in the critical region: we reject H_0 (3) and accept H_1 (3).
8. The interpretation of the results of a two-way analysis of variance is generally more difficult than those of a one-way analysis (see the following discussion).

To interpret the results of a two-way analysis of variance, one must first look at the *F*-value of the interaction. Example 12.8 (above) resulted in a significant *F*-value for the interaction. Had it not been significant, we would have assumed there was no interaction between the two factors. In this case, a non-significant interaction *F*-value would mean that the fertilizer treatments affected the height increments of all three tree species in the same way. Figure 12.1 uses hypothetical means to illustrate the absence of an interaction. When the means for a particular level of the first factor (i.e. organic fertilizer) are plotted over the levels of the second factor (i.e. tree species), a non-significant interaction will result in nearly parallel lines. Conversely, plotting two factors with a significant interaction will produce lines that have different slopes.

If the *F*-test for the interaction is not significant, the *F*-values for each of the two factors (treatment and tree species, in this example) can be interpreted independently, as in a one-way analysis of variance. In addition, if the *F*-value is significant for either one or both factors, any of the multiple comparison procedures we have described can be used to compare the means of each factor. The equations for calculating the standard error of the means and finding the confidence intervals for the unknown population means are as follows:

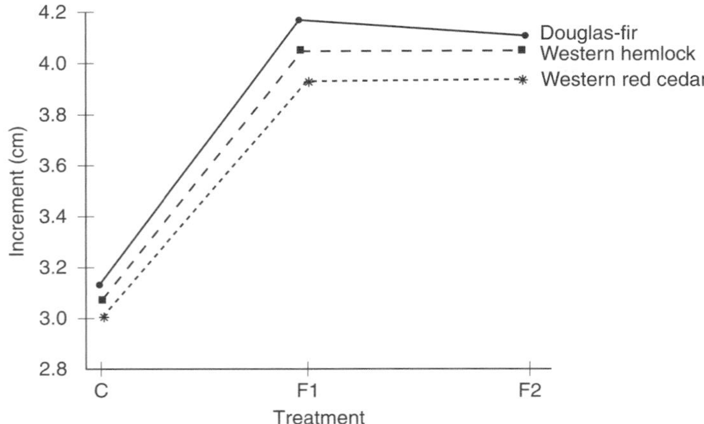

Fig. 12.1. Hypothetical plot of tree species by fertilizer treatment means, with lack of (no) statistically significant interaction between tree species and fertilizer treatments.

For factor A (fertilizer treatment in our example):

$$S_{\bar{y}_{i..}} = \sqrt{\frac{MS_E}{tn}} \tag{12.34}$$

$$P\left(\bar{y}_{i..} - t_{\alpha/2,kt(n-1)} S_{\bar{y}_{i..}} < \mu_{i\cdot} < \bar{y}_{i..} + t_{\alpha/2,kt(n-1)} S_{\bar{y}_{i..}}\right) = 1 - \alpha \tag{12.35}$$

For factor B (tree species, in our example):

$$S_{\bar{y}_{\cdot j\cdot}} = \sqrt{\frac{MS_E}{kn}} \tag{12.36}$$

$$P\left(\bar{y}_{\cdot j\cdot} - t_{\alpha/2,kt(n-1)} S_{\bar{y}_{\cdot j\cdot}} < \mu_{.j} < \bar{y}_{\cdot j\cdot} + t_{\alpha/2,kt(n-1)} S_{\bar{y}_{\cdot j\cdot}}\right) = 1 - \alpha \tag{12.37}$$

If the interaction is significant, such as in our example, interpretation of the *F*-values for the individual factors (fertilizer treatment and tree species) can lead to erroneous conclusions. In cases like this, it is strongly recommended that the interpretation begins with plotting the interaction means ($\bar{y}_{ij}$). The plot of the actual mean values for this example (Fig. 12.2a) indicates that fertilizer effects vary by tree species: both fertilizers were effective for Douglas-fir, while only the inorganic fertilizer was effective for western hemlock and western red cedar. Figure 12.2a can also be constructed such that species are on the horizontal axis and the different lines represent the three fertilizers (Fig. 12.2b).

If multiple comparisons are required, they can be conducted separately for each of the three species and/or separately for each of the three fertilizers. In multiple comparisons for two-way analyses of variance, we take advantage of a higher within-group degrees of freedom of $kt(n - 1)$ versus analysing the three species separately, where the within-group degrees of freedom are only $k(n - 1)$. The higher degrees of freedom make the multiple comparisons more sensitive to detecting significant differences between means.

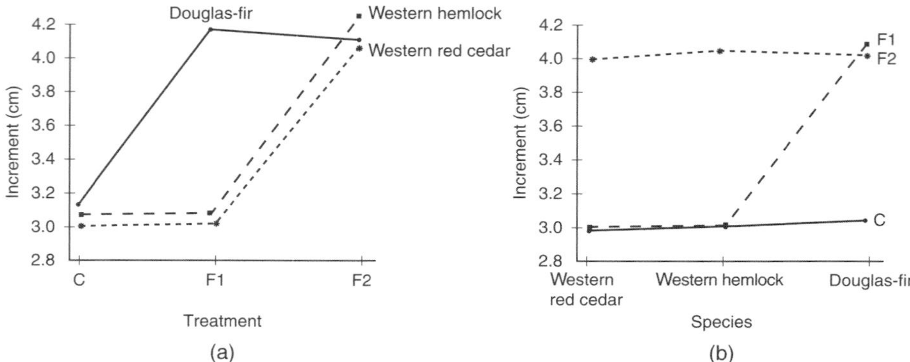

Fig. 12.2. Statistically significant interaction of (a) species by treatment and (b) treatment by species.

If confidence intervals are required for interaction means, equations for the standard error of the mean and confidence intervals are as follows:

$$S_{\bar{y}_{ij.}} = \sqrt{\frac{MS_E}{n}} \tag{12.38}$$

$$P\left(\bar{y}_{ij.} - t_{\alpha/2,kt(n-1)} S_{\bar{y}_{ij.}} < \mu_{ij} < \bar{y}_{ij.} + t_{\alpha/2,kt(n-1)} S_{\bar{y}_{ij.}}\right) = 1 - \alpha \tag{12.39}$$

Figure 12.3 illustrates the possible risks of interpreting the significance of individual factors independently when the interaction is significant. Here, we have plotted only two (versus three) levels for each factor: control and organic fertilizer, and Douglas-fir and western hemlock. The resulting figure suggests that the organic fertilizer *did not affect* the height growth of western hemlock seedlings and that it *did affect* the height growth of Douglas-fir. The line between the two species shows the average effect of the organic fertilizer, which indicates a considerable change in height growth between the control and the organic fertilizer. This increased height growth could very well

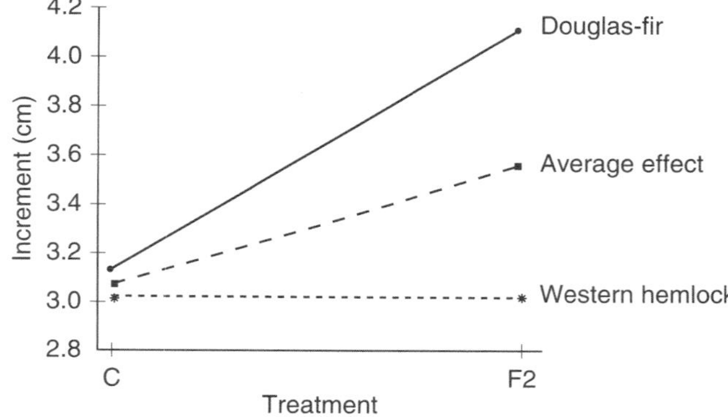

Fig. 12.3. Simplified representation of statistically significant interaction between fertilizer treatment and tree species (showing only two levels of each factor).

result in a significant F-value for the effect of fertilizer treatment (regardless of tree species). Our general conclusion would then be that the organic fertilizer was effective in increasing the height growth for both tree species: a statement that is obviously incorrect for western hemlock.

Many people may ask the question: why not carry out three separate one-way analyses of variances for the above data? We could have done this. However, we chose not to because a two-way analysis of variance offers several advantages:

- It has higher within-group degrees of freedom, which will result in narrower confidence intervals and more sensitive multiple comparison tests.
- If the data for the three tree species in Example 12.8 had been analysed separately, we could not have performed any statistical tests for tree species or for the treatment–species interaction (the same would hold true for fertilizer treatments, if we were to perform three separate analyses for each level of fertilizer).
- The standard error of the mean for the main factors in a two-way analysis of variance is smaller than those calculated from independent one-way analyses of variance (compare Eqns 12.34 and 12.35 to Eqn 12.14).

In Example 12.8, we had four *replicates*. That is, we were able to observe the effect of each combination of treatment levels on four different trees. In practice, an experiment of this sort may result in only one observation for each combination of the two factors. In our example, this would produce a single observation for each fertilizer level within each tree species (the first line of the data set – see Table 12.12). An analysis of such data is referred to as a **two-way classification analysis of variance without interaction**, or a **randomized complete block design**. All of the equations and procedures discussed above can be applied to this simplified version of the two-way analysis of variance, assuming that $n = 1$. Consequently, the equations to calculate the various sums of squares and the general analysis of variance table are simplified to:

$$SS_T = \sum_{i=1}^{k}\sum_{j=1}^{t} y_{ij}^2 - \frac{T_{..}^2}{kt} \qquad (12.40)\text{k}$$

Here, the grand total, $T_{..}$, and observations have only two subscripts and the observations are denoted: y_{ij}, $i = 1, 2, \ldots, k$ and $j = 1, 2, \ldots, t$. The working equations and analysis of variance table (Table 12.14) are as follows:

$$SS_A = \frac{1}{t}\sum_{i=1}^{k} T_{i.}^2 - \frac{T_{..}^2}{kt} \qquad (12.41)$$

Table 12.14. Two-way analysis of variance table, without interaction.

Source of variation	*DF*	*SS*	*MS*	Computed *F*
Factor A	$k - 1$	SS_A	$MS_A = \frac{SS_A}{k-1}$	MS_A/MS_E
Factor B	$t - 1$	SS_B	$MS_B = \frac{SS_B}{t-1}$	MS_B/MS_E
Error	$(k - 1)(t - 1)$	SS_E	$MS_E = \frac{SS_E}{(k-1)(t-1)}$	
Total	$kt - 1$			

$$SS_B = \frac{1}{k}\sum_{j=1}^{t} T_{\cdot j}^2 - \frac{T_{\cdot\cdot}^2}{kt} \quad (12.42)$$

$$SS_E = SS_T - SS_A - SS_B = SS_{AB}$$

In any two-way analysis of variance of this type, interactions cannot be easily tested (for more information on available tests, see references at the end of this book). However, interactions will be present and are commonly referred to as the 'error term'. The term, 'without interaction', used to describe this procedure, is therefore somewhat misleading. Two-way classification analyses of variance without interaction are usually used to analyse what are known as *randomized complete block designs*. In these cases, one of the factors is the experimental treatment effect and the other is the so-called block effect. This commonly used experimental design, as well as others, will be described in further detail in Chapter 13.

Exercises

Section 12.1

12.1. The specific gravity of wood was measured on four random samples taken from each of 3 tree species.

Western hemlock	0.42	0.37	0.40	0.38
Douglas-fir	0.56	0.51	0.44	0.46
Western red cedar	0.32	0.27	0.29	0.35

a. Are the means of the 3 species significantly different ($\alpha = 0.05$)? Assume $\sigma^2_1 = \sigma^2_2 = \sigma^2_3$.
b. Calculate and explain the meaning of the standard error of the mean for the specific gravity of Douglas-fir.
c. Calculate and explain the meaning of the standard error of the difference between the specific gravities of Douglas-fir and western red cedar.

12.2. Water samples were taken at 4 locations in a river to determine whether the quality of dissolved oxygen, a measure of water pollution, varied from location to location. Five water specimens were randomly selected at each location. Locations 1 (close to shore) and 2 (midstream), upstream from a pulp and paper plant; location 3, adjacent to the industrial water discharge for the plant; and location 4 (midstream), slightly downstream. The data are shown below (the lower the dissolved oxygen readings, the greater the pollution levels).

Location 1	5.9	6.1	6.3	6.1	6.0
Location 2	6.3	6.6	6.4	6.4	6.5
Location 3	4.8	4.3	5.0	4.7	5.1
Location 4	6.0	6.2	6.1	5.8	6.1

a. Is the level of water pollution significantly different from location to location ($\alpha = 0.01$)? Assume $\sigma^2_1 = \sigma^2_2 = \sigma^2_3 = \sigma^2_4$.
b. Find the 95% confidence interval for the unknown population mean of the oxygen readings at location 3.

c. Find the 95% confidence interval for the difference of two unknown population means of the oxygen readings at locations 1 and 3.

12.3. An experiment was conducted to determine the effect of 3 methods of soil preparation treatments on the first year of height growth (in cm) of slash pine seedlings. Three seedlings were available for measurement from soil preparation A; 5 from B; and 4 from C.

Soil preparation	Height growth (in cm)				
A	11.5	13.2	12.4		
B	15.3	14.5	17.6	20.4	16.4
C	10.0	12.3	14.2	13.2	

a. Compare the means of the height growth of the seedlings for the 3 soil preparations with a 0.05 level of significance. Assume $\sigma^2_1 = \sigma^2_2 = \sigma^2_3$.
b. Calculate the 99% confidence interval for the unknown population mean of height growth using soil preparation B.
c. Calculate the 95% confidence interval for the difference between the two unknown population means of height growths using soil preparations A and B.

12.4. An experiment was carried out to study the effect of various fertilizers on tree diameter increment growth. The measurements taken at the end of the growing season (in mm) are given below:

Fertilizer 1	11	16	15	
Fertilizer 2	13	14		
Fertilizer 3	18	17	23	20
Control	8	10	6	

a. Compare the 4 treatment means with 0.01 level of significance. Assume $\sigma^2_1 = \sigma^2_2 = \sigma^2_3 = \sigma^2_4$.
b. Calculate and compare the standard errors of the mean for increment growth in the Fertilizer 3 treatment group and the control.

Section 12.2

12.5. Compare the four means obtained in Exercise 12.2 using Bonferroni's Procedure. Use $\alpha = 0.05$.

12.6. Compare the four means obtained in Exercise 12.4 using Scheffé's Method with a 0.01 level of significance.

Section 12.3

12.7. Test for the equality of variances (use $\alpha = 0.05$) for the observations given in Exercise 12.13. Was the assumption of $\sigma^2_1 = \sigma^2_2 = \sigma^2_3$ justified?

12.8. Test for the equality of variances for the observations given in Exercise 12.4 with 0.01 level of significance. Was the assumption of $\sigma^2_1 = \sigma^2_2 = \sigma^2_3 = \sigma^2_4$ correct?

Section 12.4

12.9. Four levels of nitrogen fertilizers, N_0, N_3, N_6 and N_9, were applied to 1-year-old seedlings of 2 tree species. At the end of the growing season, each fertilizer treatment was measured for the diameter increment growth at root collar (in mm) of 2 randomly selected seedlings from each tree species.

	N_0	N_3	N_6	N_9
Douglas-fir	4.4	*5.6*	6.8	*6.5*
	3.8	5.0	6.5	6.2
Western hemlock	3.8	4.9	5.8	5.9
	4.1	4.7	5.3	5.6

a. Analyse this experiment using $\alpha = 0.05$.
b. Calculate the 95% confidence interval for the unknown population mean of diameter increment at root collar when treated with N_6.
c. Calculate the 95% confidence interval for the unknown population mean of diameter increment at root collar of western hemlock.
d. Compare the results of b and c.

12.10. A manufacturer of furniture components is studying the surface roughness (in micrometres) of pieces produced at 3 different machine centres during the morning and afternoon shifts. For each machine centre and shift, one randomly selected component was measured.

	Shift	
Machine centre	Morning	Afternoon
1	12	14
2	18	16
3	14	12
4	19	20

a. Analyse these data using $\alpha = 0.05$.
b. Compare the means of the 4 machine centres using Bonferroni's Procedure ($\alpha =$ 0.05).

13 Sampling Methods and Design of Experiments

Collecting Data

There are countless forms, sources and types of data in forestry and its related disciplines. In addition, there are a number of ways in which these data can be collected. Commonly in forestry, data arise from the implementation of what are known as **sampling designs** or **experimental designs** (sometimes referred to as **observational** and **experimental** studies, respectively). Both comprise a number of specific methodologies that allow for the collection of different types of data in systematic, manageable, orderly and logical ways – a necessity in a field like forestry in which there exists so much (and so many ways of obtaining) data.

In sampling designs (observational studies), the researcher or practitioner does not attempt to control or influence the variables of interest, but merely observes and measures them. The main purpose of sampling is to collect information from a subset of the population leading to prediction, or inferences, about the entire population. We have already discussed **simple random sampling** in Chapter 7 and we will introduce **stratified random sampling**, **two-stage sampling** and **systematic sampling** in this chapter. We will also briefly discuss **survey design**, a means of collecting information from and about people.

Experimental designs (experimental studies) involve controlling some of the factors affecting the variables of interest. The objectives of these studies are to investigate how these controlled factors affect the variables of interest. We will briefly discuss **completely randomized**, **randomized complete block** and **latin square designs** in this chapter. We will also discuss **factorial experiments**, which use these designs in the allocation of treatments.

The sampling designs and experimental designs discussed here are a few of the more commonly used methods in forestry applications. The interested reader is directed to advanced texts on the subjects, of which there are many, for more comprehensive overviews.

13.1 Sampling Methods

Most statistical inferences (estimation or hypothesis testing) about population parameters are made using incomplete knowledge (see Chapters 7 and 9). In sampling, we are usually concerned with estimating one or more population parameters, bearing in mind that the estimation should be as precise as possible given the constraints of time, practicality and, especially, costs. In statistics, a higher precision means a narrower confidence interval for an unknown population parameter. Various sampling designs have been developed that aim to improve precision, while keeping the cost of sampling at a reasonable level.

Simple random sampling

All of the sampling methods described in this book thus far have been related to simple random sampling. Simple random sampling is considered effective only if the population to be sampled is homogeneous, meaning that the measurements taken have a uniform variation throughout. This implies that the cost of obtaining any observation should be the same for each sampling unit.

Stratified random sampling

Often, some prior knowledge about a population can be used to increase the precision of the estimate of a parameter of interest. In stratified random sampling, the sampling units (individual measurements) in the population can be grouped together to form a stratum on the basis of similarity of some characteristic or characteristics. Once the groups are formed, each group or stratum is treated as an individual population and samples are collected from them in the same way as for simple random sampling. Individual group estimates (means and variances) are combined to obtain the estimates for the population. Equations to calculate the combined estimates, confidence intervals and sample sizes are not provided here. For further details, interested readers can refer to specialized books on sampling methods.

Stratification is effective only if the variation of the variable of interest within each stratum is less than the variation of the whole population without stratification. If stratification is done wisely, it can provide many advantages. Most notably, the estimation of the population mean will be more precise than simple random sampling (based on equal sample sizes). This is because part of the variation of the entire data set has been accounted for or explained by the fact that we have stratified the population into logical segments. In other words, the variation in the data will be smaller. In addition, stratified random sampling provides separate estimates for each stratum or subpopulation. In timber cruising, for example, strata are commonly set up corresponding to major forest types and the volume per hectare estimates are available not only for the total area of interest, but each forest type as well. Stratification is usually effective in timber cruising, because the plot-to-plot variation of the measurements (dbh, height or volume) within most of the forest types is less than that of the whole forested area. Lastly, stratified random sampling can be a more cost effective means of collecting data. For instance, in studying strength properties of certain types of dimensional lumber, observations can be stratified by lumber grade. Since strength properties differ from grade to grade, this results in increased precision of estimates and requires fewer observations than a simple random sampling scheme.

Two-stage sampling

In some situations, locating or getting to a sampling unit can be more expensive than collecting or measuring an observation once there. For example, if a large (say

100,000 ha) forested area is to be sampled for total volume, we may need to put in 200 sampling units consisting of 10 m × 10 m plots. Locating and visiting all of these 200 randomly scattered plots would be very time-consuming, difficult and expensive – on average, they would be 2.2 km apart! In cases like this, two-stage sampling can be more effective. The area is first subdivided into primary units (e.g. 200 square blocks of 500 ha each), from which we randomly select 20 of these 200 primary units. We then randomly select 10 plots (secondary units) within each of these 20 blocks to use for measurement. Allocating the required 200 plots in this way is much more cost effective than a completely random sample: our average distance between plots within primary units would be reduced to about 700 m. Again, the interested reader is referred to specialized books on sampling methods for more details on this topic.

Systematic sampling

Systematic sampling methods are different from simple random sampling in that the sampling units are numbered from 1 to N and n units are selected using a regular interval. For example, when samples are taken from a production line, every 10th item (from a starting point) that passes by may be pulled off the conveyor belt for a quality control check. Alternatively, samples can be taken at constant time intervals until the desired sample size, n, is obtained. In timber cruising and forest inventory activities, systematic samples are often set up on a grid, with samples taken at equally spaced intervals along equally spaced rows. In order to introduce some randomization into a systematic sample, the starting point may be randomly selected in space (or in time).

In comparison to simple random sampling, there are two distinct advantages to systematic sampling:

1. Locating sample units is much easier and it is therefore cheaper to collect samples and train employees.
2. Systematic sampling covers the population more uniformly (in time or space) than simple random sampling. Generally, this means that systematic samples are more representative of populations under study.

The equations used to estimate the population mean, population variance and sampling error from samples selected by systematic sampling are the same as those used for simple random sampling. While the estimation of population mean is generally unbiased, some statisticians argue that the estimation of population variance – and consequently the estimation of sampling error – can be biased if the population is not randomly ordered. Our practical experience in forestry has shown few instances where systematic sampling gave misleading results. That said, a bias might be introduced when the sampling units within a population show a definite pattern in their values. For example, consider the case of a forester systematically sampling a plot of 10 × 10 row-planted trees. If this plot is adjacent to a road, and every 10th tree is selected, there is a chance that only trees growing next to the road will be selected. This will obviously lead to misleading results because trees growing next to roads are exposed to more light and, thus, have more favourable growing conditions. When patterns like this are present, bias can be avoided by sampling along trend lines, such as establishing a grid pattern along the direction of topographic change in a mountainous area.

Survey design

While not strictly a 'sampling method', survey design is becoming an increasingly important component of conservation, sustainable forest management and wood products manufacturing as we seek to understand the opinions, attitudes and behaviour of the public at large, forestry stakeholders, consumer segments and so on.

In social sciences, there are many ways of interacting with and obtaining information from people. **Qualitative methods** refer to exploratory means of collecting information from people and are generally used to gain insight into a research problem or for theory development. **Quantitative methods** employ rigorous sampling methods (like the ones described above) and make it possible to draw inferences about the population in question using many of the statistical tools described in this book. In either case, information is generally collected either by means of personal interviews or with mail, telephone and/or Internet surveys.

Survey design is complicated, for a number of reasons. Unlike the measurement of tree diameters, for instance, the way that people think cannot be measured directly and scales must be used as proxies for obtaining this information. Most of us have seen the commonly used Likert scale, a five-point scale that measures the degree to which one agrees or disagrees with certain statements. The idea behind using such a scale is that human thought processes are not simple, but rather are best captured on a continuum of categorical responses. Many such scales exist for different purposes. They are readily available and are considered to be valid measures of attitudes. In addition, many are in the form of interval scales, which is to say that the statistical tools described in this book can be used in their interpretation. Other commonly used scales, such as rank scales, are ordinal in nature and their analyses and interpretation are somewhat less straightforward.

Another problem in survey design is that we must understand and account for the additional error associated with the fact that we are sampling people and measuring their attitudes. For example, one of the most frequently encountered types of bias in implementing surveys is the so-called 'non-response error'. Imagine a situation where we are mailing out 1000 surveys. Perhaps only 300 people who receive the survey will respond. If these 300 people are different in any way (related to the research objectives) from the 700 who did not respond, then non-response bias is present and it is difficult, if not impossible, to make inferences onto the population.

In the final analysis, a good survey should look like it was fairly simple to devise. However, designing and implementing a valid and reliable survey instrument is long and painstaking work. There are many rules and tricks to good survey design, most of them revolving around the twin goals of making the survey clear and easy to understand and maximizing response rates. If presented with the opportunity to conduct a survey, we strongly recommend consulting a social scientist who is an expert in survey design and/or one of the many excellent texts on survey design and implementation.

13.2 Experimental Designs

In experimental designs, one or more factors (which are assumed to affect the variable of interest) are controlled or kept at fixed levels in order to estimate their effect. These controlled factors are generally referred to as *treatments*. There are many types of treatments in forestry applications, such as different types of fertilizers, different doses of a given fertilizer, different tree species, different varieties of a single tree species, different irrigation levels, different soil preparations, different seed sources, different kinds of resins, different temperatures, different pressures and so on.

The variable of interest in experimental designs is called the *response variable* or the *dependent variable*. An *experimental unit* is anything (a tree, a forest, a piece of wood, a group of people, etc.) that receives the same treatment. If the same treatment is applied to more than one experimental unit within an experiment, we call these *replications*. *Experimental error* is the pooled variation among experimental units receiving the same treatment, which is to say that experimental error cannot be estimated without replications.

The main objectives of an experimental design are to evaluate whether the variation due to treatments (between various treatment) is *significantly greater* than the naturally occurring experimental error variation (within each treatment). These sources of variation were introduced in Chapter 12 as group-to-group and within-group variation.

It is always in our best interest to minimize experimental error. There are two practical means of doing so:

1. Increase the number of replications; and/or
2. Select the most appropriate experimental design.

In this chapter, we will briefly discuss the three *basic experimental designs*: **completely randomized designs, randomized complete block designs** and **latin square designs**. Following this, we will discuss **factorial experiments** (which are not designs, but make extensive use of them). Several other experimental designs exist and the interested reader is directed to one of many texts on this subject.

Completely randomized design

The completely randomized design is the simplest of the experimental designs: treatments are randomly assigned to each experimental unit (in time or space). This design is most useful when the material available for experimentation is uniform in nature. As such, completely randomized designs are frequently used in laboratory experiments, greenhouse studies, or in feeding experiments on animals of the same age/cohort.

In experimental design terminology, the term *layout* refers to the placement of the treatments on the experimental units. A layout provides a map of the allocation of the treatments over space, type of material, or time. A possible layout of a completely randomized design with four treatments (A, B, C and D), each replicated three times, is shown in Fig. 13.1, where each cell represents a separate experimental unit.

For example, the four treatments could be four soil preparations in an experimental plantation, where we have planted 50 seedlings per experimental unit. The response variable could be the number of seedlings surviving a year after planting (out of the 50 planted). In this case, the layout indicates the map of the location of the treatments in the field.

Alternatively, A, B, C and D could be four different types of resins used to produce laminated veneer lumber (LVL) and the experimental units could refer to different days and times. The observations could be shear strength measurements and the layout would indicate the day and time that the LVL was to be sampled.

A	B	A
A	C	B
D	B	D
C	D	C

Fig. 13.1. Layout of a completely randomized design with four treatments and three replications.

A completely randomized design is analysed by a one-way classification analysis of variance (see Table 13.1). For equations, procedures and interpretations of the one-way analysis of variance, refer to Chapter 12. Note that in Table 13.1, SS_{TR} and SS_E are the same, in terms of calculation and interpretation, as SS_G and SS_W, from one-way analysis of variance (see Chapter 12).

Table 13.1. Analysis of variance for a completely randomized design.

Source of variation	*DF*	*SS*	*MS*	Computed *F*
Treatment	$k-1$	SS_{TR}	$MS_{TR} = \frac{SS_{TR}}{k-1}$	$\frac{MS_{TR}}{MS_E}$
Experimental error	$k(n-1)$	SS_E	$MS_E = \frac{SS_E}{k(n-1)}$	
Total	$kn-1$	SS_T		

Randomized complete block design

The main disadvantage of the completely randomized design described above is that it is really only suited to homogeneous materials, which are rarely found in field experimentation or industrial settings. If the experimental units, area, time or material are not homogeneous, it is possible (in most cases) to subdivide the experimental units into smaller, more uniform groups. These groups are called blocks. This process is very similar to stratification in stratified random sampling, with a block equivalent to a stratum. When each treatment is applied to one experimental unit within each block, and treatments are randomly allotted to the experimental units independently within each block, we have what is known as a randomized complete block design. In this design, we have as many replications as we have blocks. Using the same example as in the completely randomized design, the four treatments (A, B, C and D) would be randomly placed in each one of three blocks, as shown in Fig. 13.2.

Relating this layout to the four soil preparation treatments, the blocks could be placed such that they correspond to three different levels of erosion, or three different drainage rates.

In the case of the LVL example, the experimenter may have reason to assume that the quality of production changes from day to day, in which case the blocks could represent the days.

Data collected from a randomized complete block design are analysed by a two-way classification analysis of variance without interaction (see Table 13.2). For equations and procedures, see Chapter 12.

Block 1	Block 2	Block 3
C	B	C
A	D	B
D	C	A
B	A	D

Fig. 13.2. Layout of a randomized complete block design with four treatments and three blocks.

Table 13.2. Analysis of variance for a randomized complete block design.

Source of variation	*DF*	*SS*	*MS*	Computed *F*
Treatment	$k-1$	SS_{TR}	$MS_{TR} = \dfrac{SS_{TR}}{k-1}$	$\dfrac{MS_{TR}}{MS_E}$
Block	$n-1$	SS_B	$MS_B = \dfrac{SS_B}{n-1}$	$\dfrac{MS_B}{MS_E}$
Experimental error	$(k-1)(n-1)$	SS_E	$MS_E = \dfrac{SS_E}{(k-1)(n-1)}$	
Total	$kn-1$	SS_T		

Note that SS_{TR}, SS_B and SS_E are the same, in terms of calculation and interpretation, as SS_A, SS_B and SS_E, respectively, from two-way analysis of variance (see Chapter 12).

Comparing Tables 13.1 and 13.2, we can see that the experimental error (i.e. within-group) variation and its degrees of freedom in the completely randomized design is partitioned into variation due to blocks and a new experimental error variation term in the randomized block design. This implies that the technique of blocking 'removes' a certain amount of variation from the experimental error. The effectiveness of randomized complete block designs is measured by testing this 'removed variation'. Although there is usually no practical reason for testing the significance of the blocks, the size of the block mean square relative to the error mean square does indicate the precision gained by blocking. Some practitioners assert that if the block mean square is at least two to three times the size of the experimental error mean square, there is enough evidence to say that blocking is effective.

One disadvantage of the randomized complete block design – in comparison to the completely randomized design – is that for experiments of the same size (i.e. same number of treatments and replications), the experimental error degrees of freedom is smaller. This would seem to indicate that estimating the treatment means or differences has less precision in a randomized block design than in a completely randomized design, which is true, unless the blocking removes a sizeable variation from the experimental error. For this reason, blocking in experimental designs is only recommended when the experimental units are non-homogeneous.

Latin square design

Latin square designs are used in experiments when blocking alone cannot reduce the natural variation between experimental units. The major difference between a randomized complete block design and a Latin square design is that the former removes the natural variation of the experimental units in only one direction, while Latin square designs remove the variation of the experimental units in two directions. For example, a field experiment may have soil fertility gradients running both parallel and perpendicular to the slope. Blocking would eliminate only one of those differences, but a Latin square design would eliminate both. Latin square designs are usually laid out in rows and columns in space (or time) and the treatments are randomly allocated to the experimental units, with the restriction that each treatment must appear once in every row and every column. This restriction requires that the number of replications, rows and columns must all be equal to the number of treatments used. Figure 13.3 represents a layout of a Latin square design with four treatments, A, B, C and D. Since the total sum of squares is partitioned into treatment, row, column and experimental error sums of squares, the Latin square design is analysed with a three-way classification analysis of variance (see Table 13.3). Interested readers are referred to higher-level experimental design textbooks for the equations used to calculate the sums of squares used in the analysis below.

Our example of the four soil preparation treatments could be laid out in a Latin square design in the field, but to do so would require four replications. The LVL example could be laid out in a Latin square design if the production was repeated for 4 days (columns), with panels selected in, say, 2 h time intervals (rows) during each day.

For the analysis of the Latin square design (see Table 13.3), we assume that rows are equivalent to blocks. Compared to the randomized complete block design (see

	Column 1	Column 2	Column 3	Column 4
Row 1	B	C	A	D
Row 2	A	D	B	C
Row 3	D	A	C	B
Row 4	C	B	D	A

Fig. 13.3. Layout of a Latin square design with four treatments and four replications.

Table 13.3. Analysis of variance for a Latin square design.

Source of variation	*DF*	*SS*	*MS*	Computed *F*
Treatment	$k-1$	SS_{TR}	$MS_{TR} = \frac{SS_{TR}}{k-1}$	$\frac{MS_{TR}}{MS_E}$
Row	$k-1$[a]	SS_B	$MS_R = \frac{SS_R}{k-1}$	$\frac{MS_R}{MS_E}$
Column	$k-1$[a]	SS_C	$MS_C = \frac{SS_C}{k-1}$	$\frac{MS_C}{MS_E}$
Experimental error	$(k-1)(k-2)$	SS_E	$MS_E = \frac{SS_E}{(k-1)(n-2)}$	
Total	k^2-1	SS_T		

[a] Since $k = n$.

Table 13.2), the experimental error sum of squares in Latin square design is further partitioned into a column sum of squares and a new experimental error sum of squares. Similar partitioning also occurs for the degrees of freedom. A rule of thumb: there is enough evidence that the Latin square design is effective if both the row and column mean squares are at least two to three times the size of the experimental error mean squares. Due to the restriction requiring an equal number of replications, rows, columns and treatments, the Latin square design is seldom used for more than ten treatments.

Factorial experiments

Factorial experiments are sometimes referred to as 'factorial designs'. Strictly speaking, this is not correct, since *design* refers to the physical layout of an experiment. For example, *randomized complete block* design refers to an experiment in which each of the treatments is randomly allocated in each block. The term, *factorial*, on the other hand, refers to a special arrangement of treatments. Factorial arrangements of treatments can actually be used in any one of the basic designs described above: completely randomized, randomized complete block and Latin square designs. Factorial experiments are aimed at investigating the effects of *all possible combinations* of two or more *factors* and their *interactions* on the response variable, without restricting the randomization of possible combinations. For example, if the combined effect of four levels of nitrogen content in fertilizer (0, 50, 100 and 150 kg per unit area) and three levels of water in irrigation (0, 3 and 6 cm per week) are studied, it is a factorial experiment if the treatment combinations are randomly allocated to the experimental units (see Table 13.4).

If the above treatments are laid out in one of the basic designs, it is called a 4 × 3 factorial experiment. The product of the levels of the factors studied defines the size of the factorial experiment and the total number of treatments to be studied. A possible layout of this experiment in a randomized complete block design using three blocks is seen in Fig. 13.4.

Table 13.4. Treatment combinations of 4 levels of nitrogen and 3 levels of irrigation.

	Nitrogen fertilizer			
Irrigation	n_0	n_1	n_2	n_3
i_0	i_0n_0	i_0n_1	i_0n_2	i_0n_3
i_1	i_1n_0	i_1n_1	i_1n_2	i_1n_3
i_2	i_2n_0	i_2n_1	i_2n_2	i_2n_3

Block 1	i_2n_0	i_0n_3	i_2n_1	i_0n_2	i_2n_2	i_1n_1	i_1n_2	i_0n_0	i_2n_3	i_1n_3	i_1n_0	i_0n_1
Block 2	i_1n_0	i_1n_1	i_0n_0	i_0n_1	i_1n_2	i_0n_3	i_2n_0	i_1n_3	i_2n_2	i_0n_2	i_2n_1	i_2n_3
Block 3	i_1n_3	i_2n_3	i_0n_3	i_0n_1	i_1n_2	i_1n_0	i_2n_2	i_2n_0	i_0n_2	i_0n_0	i_1n_1	i_2n_1

Fig. 13.4. Layout of a randomized complete block design with 12 treatments (factorial arrangement) and 3 blocks.

Table 13.5 shows the analysis of variance for a factorial experiment. Note that the analysis of this experiment could be carried out in the same manner as would be used for any randomized complete block design (see Table 13.2). However, if the treatments are factorial in nature, the treatment sum of squares and degrees of freedom can be partitioned into the effects of the factors and their interaction. For our example, the treatment sum of squares can be partitioned into the effects of nitrogen, irrigation and their interaction. Partitioning in this way makes the interpretation of the treatment effects easier and more meaningful. For equations to calculate the sums of squares and complete discussion of factorial experiments, the reader is referred to advanced experimental design textbooks.

Table 13.5. Analysis of variance for a factorial experiment in a randomized complete block design.

Source of variation	*DF*	*SS*	*MS*	Computed *F*
Block	$n-1$	SS_B	$MS_B = \frac{SS_B}{n-1}$	$\frac{MS_B}{MS_E}$
Treatment	$kt-1$	SS_{TR}	$MS_{TR} = \frac{SS_{TR}}{kt-1}$	$\frac{MS_{TR}}{MS_E}$
Nitrogen (N)	$k-1$	SS_N	$MS_N = \frac{SS_N}{k-1}$	$\frac{MS_N}{MS_E}$
Irrigation (I)	$t-1$	SS_I	$MS_I = \frac{SS_I}{t-1}$	$\frac{MS_I}{MS_E}$
N × I	$(k-1)(t-1)$	SS_{NxI}	$MS_{NxI} = \frac{SS_{NI}}{(k-1)(t-1)}$	$\frac{MS_{NxI}}{MS_E}$
Experimental error	$(kt-1)(n-1)$	SS_E	$MS_E = \frac{SS_E}{(kt-1)(n-1)}$	
Total	$ktn-1$	SS_T		

Where: *kt*, number of treatments; *n*, number of replications; *k*, number of levels in *N*; *t*, number of levels in *I*.

14 Non-parametric Tests

Testing when Distributions are Unknown

Before introducing *non-parametric* tests, we should look back at the tests covered in the preceding chapters and consider why those methods are called *parametric tests*. Parameters are values which uniquely define a probability distribution. For example, the heights of Douglas-fir seedlings may follow a normal distribution. There are an infinite number of normal distributions, but we can uniquely specify the distribution associated with these seedling heights by specifying the distribution parameters (e.g. a mean of 25 cm and a standard deviation of 5 cm). We can then use the methods described in Chapters 9 and 10 to perform hypothesis tests on these heights. These would be called parametric tests, because they involve testing estimates of the parameter values (i.e. the mean or the standard deviation), or comparing the distribution of heights to a distribution with assumed parameter values. Non-parametric tests, on the other hand, do not require knowledge or estimates of these parameter values. They can be performed without uniquely identifying the distribution, or its parameters.

Probably the most common assumptions we have made thus far are that **random samples** are taken from **normal** populations and/or that the population **standard deviations** are either known, or known to be equal. This may not seem to be a serious constraint, but in practice there are many situations where these assumptions cannot be met. In such situations, **non-parametric tests** or **distribution-free** tests are appropriate.

In addition to not depending upon a particular distribution (e.g. the normal distribution), non-parametric tests have several other advantages over traditional parametric tests:

- Most non-parametric tests are quick and relatively easy to conduct.
- The data used in non-parametric tests need not be quantitative (i.e. interval or ratio data). These tests can be used to analyse ranks and categorical data.
- Non-parametric tests require fewer restrictive assumptions than parametric tests.

However, non-parametric tests are not without their disadvantages:

- Many non-parametric tests are less efficient than parametric tests – particularly when sampling *is* from normal populations – mainly because they do not utilize all the information provided by the data. Consequently, non-parametric tests are not as sensitive as parametric tests. They require a larger sample and/or they result in higher probabilities for *type II* errors.
- Since these procedures do not rely upon or test parameters, the usual methods of describing test results are no longer appropriate. For example, we cannot use the

mean value to describe a group of data because we are not calculating a mean. This makes it difficult to formulate quantitative statements about differences between populations.

Non-parametric procedures are very useful when the assumption of normality cannot be justified, or the data that we wish to analyse are qualitative, rather than quantitative. However, if both parametric and non-parametric tests are applicable for a certain set of observations, it is advisable to use the more efficient parametric procedure.

You may be surprised to learn that we have already studied one non-parametric test: the chi-square test for independence of categorical data (see Section 10.2, Chapter 10). We will use this chapter to introduce several other commonly used non-parametric tests: the **sign test**, the **Wilcoxon signed rank test**, the **Wilcoxon rank sum test**, the **Kruskal–Wallis test**, the **runs test**, and **Spearman's rank correlation test.**

14.1 Sign Test

In Section 9.2 of Chapter 9, we considered an example in which 16 trees were randomly sampled from a large plantation and measured for diameter at breast height (dbh). In order to test the hypothesis that the mean dbh of trees in this plantation was 14 cm, we had to assume that the population of dbh measurements was normally distributed. This is because the null hypothesis for testing, $\mu = c$, is only valid when the sampled population is approximately normal, or the sample size is large ($n > 30$). In reality, however, we may not know the shape of the distribution of a population and, with small samples, it is rather difficult to test for normality.

In cases where we have a small sample from an unfamiliar population and we are interested in making statements about the 'centre' of a distribution, the **sign test** may be most appropriate. The sign test is a one-sample test for the **median,** $\tilde{\mu}$, of a continuous population (see Section 2.3, Chapter 2 for more information on the median). In symmetric populations, this test can also be applied for the mean; values in these populations are symmetrically distributed around the median and, therefore, the population mean and population median are approximately equal.

For the sign test, our null hypothesis is very similar to that discussed in Section 9.1 (see Chapter 9). To test that the median is equal to some hypothetical constant, c, we have:

$$H_0: \tilde{\mu} = c,$$

$$H_1: \tilde{\mu} \neq c; \quad \text{or} \quad \tilde{\mu} < c; \quad \text{or} \quad \tilde{\mu} > c.$$

To test H_0, a random sample is taken from the population. We assign a *plus* sign to those values that are greater than c and a *minus* sign to those less than c. If an observation is equal to c, it is discarded and the sample size, n, is reduced by one. H_0 is rejected if the number of plus signs is much higher than the number of minus signs, or the number of minus signs is much higher than the number of plus signs. The test statistic for the sign test is the random variable, X, which represents the number of plus signs. X follows a binomial probability distribution (see Eqn 5.2, Chapter 5):

$$f(x;n,p) = \binom{n}{x} p^x q^{n-x} = \frac{n!}{x!\,(n-x)!} p^x q^{n-x}, \qquad \text{for } x_1 = \quad 0, 1, 2, \ldots n.$$

When H_0 is true ($\tilde{\mu} = c$), the parameter, p, in this binomial distribution is equal to 0.5. The critical values used in the test when $n < 10$ are found using either the binomial equation or Table A.1 (see Appendix A). When $n \geq 10$, the normal approximation to the binomial can be used (since $np \geq 5$, see Section 6.4 in Chapter 6). The test statistic for this large sample case is:

$$z = \frac{x - n(0.5)}{\sqrt{n(0.5)(0.5)}}. \tag{14.1}$$

The one-sample sign test can also be used for analysis of paired data (see Section 9.5b, Chapter 9). In this case, the null and alternative hypotheses are:

H_0: $\mu_1 - \mu_2 = 0$,

and

H_1: $\mu_1 - \mu_2 \neq 0$; or $\mu_1 - \mu_2 < 0$; or $\mu_1 - \mu_2 > 0$.

In this test, we replace each difference, d_i, of the paired observations with a plus or minus sign, depending on the sign of d_i, and then continue as per the one-sample sign test. As above, observations with zero difference are omitted, with a consequent reduction in sample size.

The following examples demonstrate the use of the sign test.

Example 14.1. The number of seeds in 9 Douglas-fir cones were recorded as: 21, 18, 23, 25, 25, 26, 24, 19 and 27. Is it reasonable to assume that the median number of seeds per cone in the population is 23? Use a 0.05 level of significance.

1. H_0: $\tilde{\mu} = 23$.
2. H_1: $\tilde{\mu} \neq 23$.
3. $\alpha = 0.05$.
4. Use Table A.1 (see Appendix A).
5.

21	18	23	25	25	26	24	19	27
−	−		+	+	+	+	−	+

Number of minus signs = 3
Number of plus signs = 5.

6. Since $n < 10$, we look up the following probabilities in the binomial table with $n = 8$ (the original sample size was reduced by 1, because one of the values was equal to 23) and $p = 0.5$:

$P(x \leq 3) = 0.363$
$P(x \geq 5) = 0.363$

(Note that only one of the two test statistics is required, because the binomial distribution is symmetric when $p = 0.5$.).

7. Since we have a two-tailed test, we use a critical probability of $\alpha/2 = 0.025$.
8. Since both probabilities are greater than 0.025, we 'accept' H_0.
9. It is reasonable to assume that the median of this population is 23.

Example 14.2. The following data are wait times (in minutes) for 20 logging trucks at a particular weigh-scale: 12, 20, 21, 30, 27, 15, 19, 4, 20, 7, 13, 22, 10, 20, 18, 16, 23, 15, 24 and 11. Test the weigh-scale manager's claim that the median wait is not longer than 15 min. Use a 0.05 level of significance.

We can use the 20 observations to construct a histogram and get a rough idea of the shape of the distribution of the wait time values (see Fig. 14.1). Since this distribution appears to be approximately symmetric, we will assume that the means can be tested.

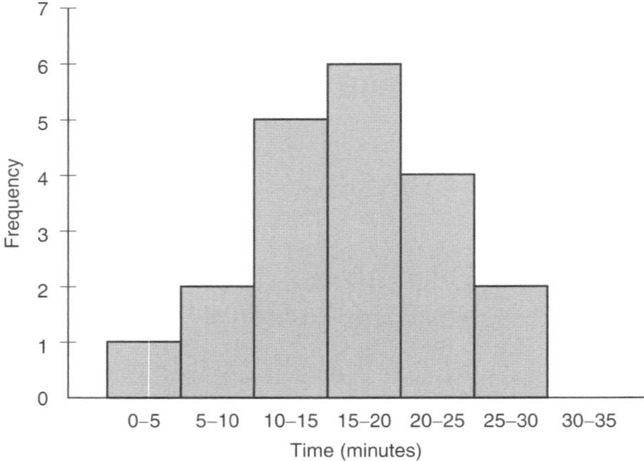

Fig. 14.1. Distribution of logging truck wait times.

1. H_0: $\mu = 15$.
2. H_1: $\mu > 15$.
3. $\alpha = 0.05$.
4. Use Eqn 14.1, since $n \geq 10$.
5.

12	20	21	20	27	15	19	4	20	7	13	22	10
−	+	+	+	+		+	−	+	−	−	+	−
20	18	16	23	15	24	11						
+	+	+	+		+	−						

$n = 18$ (note: 2 observations are discarded)
x = number of plus signs = 12.

6. $z_{0.05} = 1.645$.

7. $z = \dfrac{12 - 18(0.5)}{\sqrt{18(0.5)(0.5)}} \approx 1.41$.

8. Since $1.41 < 1.645$, we 'accept' H_0.
9. The data indicate that the median or mean wait time is not significantly longer than 15 min.

Example 14.3. Two crews of summer students independently estimated the volumes (in m^3/ha) of 12 plots located in a ponderosa pine stand. Using a significance level of 0.1, do the data support a conclusion that the estimates from the two crews are equal?

Plot	1	2	3	4	5	6	7	8	9	10	11	12
Crew 1	320	421	450	280	160	250	231	321	436	521	182	254
Crew 2	305	423	452	276	155	261	235	321	465	503	180	260

We will assume that the data are from a continuous symmetric population.

1. H_0: $\mu_1 - \mu_2 = 0$.
2. H_1: $\mu_1 - \mu_2 \neq 0$.
3. $\alpha = 0.1$.
4. Since $n > 10$, use Eqn 14.1.

5. Calculate the differences:

Plot	1	2	3	4	5	6	7	8	9	10	11	12
	15	–2	–2	4	5	–11	–4	0	–29	18	2	–6

$n = 11$ (there are 12 observations, but one zero);
$x = 5$ (number of positive signs).

6. $z_{0.05} = \pm 1.645$.

7. $z = \dfrac{5 - 11(0.5)}{\sqrt{11(0.5)(0.5)}} \approx -0.30.$

8. Since $-1.645 < -0.30 < 1.645$, we 'accept' H_0.

9. The data indicates that the two crews estimated equivalent volumes.

14.2 Wilcoxon Signed Rank Test

The sign test, discussed in Section 14.1, uses only plus and minus signs to identify differences between the observations and their median. In the 1940s, Frank Wilcoxon created a similar but more sophisticated test that uses both the direction and the magnitude of the differences between the observations and their median.

The null and alternative hypotheses for the so-called **Wilcoxon signed rank test** are the same as those in Section 14.1. As with the sign test, the Wilcoxon signed rank test can be used to test the null hypothesis of $\tilde{\mu} = c$ in a one-sample test and the null hypothesis of $\tilde{\mu}_1 - \tilde{\mu}_2 = 0$ in a paired difference two-sample test. Also, if the samples (either one-sample or paired) are taken from a continuous symmetric population, the signed rank test is applicable for testing unknown population means, as well as medians. In the one-sample case, the absolute values of the differences between the observations and the unknown hypothetical population median (or mean) are ranked. In the paired sample cases, the absolute values of the paired differences (d_i) are ranked. In both cases, zero differences are discarded in the process of ranking. If there are ties, we assign the average of the ranks that would have been assigned if the differences were distinguishable. This concept of 'ties' is perhaps best illustrated with an example. In the following problem, the lowest ranked differences are –2, –2 and 2. Since we are looking at absolute values only, this is considered a three-way tie. If the differences between these values were distinguishable, they would be ranked 1, 2 and 3. However, since they are tied, each receives an average rank of (1 + 2 + 3)/3 = 2.

If the null hypothesis is true, the total of the ranks corresponding to the positive differences (w_+) should be approximately equal to the total of the ranks of the negative differences (w_-). When repeated samples are taken from a population, the w_+ and w_- are considered individual values of W_+ and W_-, the random variables denoting positive and negative ranks, respectively.

If the number of non-zero differences, n, is ≥ 15 (large samples), the sampling distribution of W_+ (or W_-) approaches the normal distribution with mean and variance of:

$$\mu_{W_+} = \frac{n(n+1)}{4} \tag{14.2}$$

and

$$\sigma^2_{W_+} = \frac{n(n+1)(2n+1)}{24}. \tag{14.3}$$

Hence, the test statistic for Wilcoxon signed rank test is computed as:

$$z = \frac{w_+ - \mu_{W_+}}{\sigma_{W_+}}. \tag{14.4}$$

For $n < 15$, the minimum of the computed values of w_+ and w_- is compared to a critical value in Table A.10 (see Appendix A). The following examples demonstrate the use of the signed rank test for the one-sample case and paired observation cases, for small and large samples, respectively.

Example 14.4. Using the data given in Example 14.3 (see Section 14.1), apply the Wilcoxon signed rank test ($\alpha = 0.1$) to evaluate whether the two crews estimated equal volumes.

Again, we assume that the observations were taken from a continuous symmetric distribution, so that the means can be tested.

1. H_0: $\mu_1 - \mu_2 = 0$.
2. H_1: $\mu_1 - \mu_2 \neq 0$.
3. $\alpha = 0.1$.
4. Use Table A.10 (see Appendix A), since $n < 15$.
5. We assign ranks to the differences:

d_i	15	−2	−2	4	5	−11	−4	−29	18	2	−6
Rank	9	2	2	4.5	6	8	4.5	11	10	2	7

$n = 11$ (note that the sample size has reduced by 1, because there is one zero difference)
$w_+ = 31.5$
$w_- = 34.5$
Note that $(w_+ + w_-)$ must be equal to $\frac{n(n+1)}{2} = \frac{11 \times 12}{2} = 66$, and, therefore, one of the two of w_+ and w_- can be calculated by subtraction: $w_- = 66 - 31.5 = 34.5$.
6. For a two-sided test with $\alpha = 0.1$, the critical value from Table A.10 (see Appendix A) for $n = 11$ is $W \leq 14$.
7. The value of w to be used in Eqn 14.4 is the $\min(w_+, w_-) = \min(31.5, 34.5) = 31.5$.
8. Since $31.5 > 14$, we 'accept' H_0.
9. The estimates by the two crews are not significantly different.

Example 14.5. Repeat Example 14.2 (see Section 14.1) by applying the Wilcoxon signed rank test.

1. H_0: $\tilde{\mu} = 15$.
2. H_1: $\tilde{\mu} > 15$.
3. $\alpha = 0.05$.
4. Use Eqn 14.4.
5. Compute the differences between the observations and the hypothesized median of 15:

Differences	−3	5	6	15	12	0	4	−11	5	−8
Ranks	3.5	8.5	11	18	17		5.5	16	8.5	13.5
Differences	−2	7	−5	5	3	1	8	0	9	−4
Ranks	2	12	8.5	8.5	3.5	1	13.5		15	5.5

$n = 18$, since there are two zeros.

$$w_+ = 122 \qquad w_- = \frac{18 \times 19}{2} - 122$$

$$\mu_{W_+} = \frac{18 \times 19}{4} = 85.5 \qquad \sigma^2_{W_+} = \frac{18 \times 19 \times 37}{24} = 527.25$$

$$\sigma_{W_+} \approx 22.96$$

6. $z_{0.05} = 1.645$.
7. $z = \dfrac{122 - 85.5}{22.96} \approx 1.59$.
8. Since $1.59 < 1.645$, we 'accept' H_0.
9. The median waiting time is not significantly longer than 15 min.

Notice that this test comes much closer to rejecting H_0 than the sign test from Example 14.2. This is because it is a more sensitive test that uses more of the information contained in the data.

14.3 Wilcoxon Rank Sum Test

In this section, we discuss a non-parametric alternative for comparing two unknown population means when independent samples are selected from two continuous populations. The **Wilcoxon rank sum test** (also known as the **Mann–Whitney U-test**) is the alternative to the two-sample *t*-test discussed in Section 9.5 (see Chapter 9). The advantage of this test is that two unknown population means can be compared for equality without having to assume that the two populations sampled have normal distributions.

For the Wilcoxon rank sum test, the null and the alternative hypotheses are the same as those stated in Section 9.5 (see Chapter 9):

$$H_0: \mu_1 - \mu_2 = 0$$

and

$$H_1: \mu_1 - \mu_2 \neq 0; \qquad \text{or} \qquad \mu_1 - \mu_2 < 0; \qquad \text{or} \qquad \mu_1 - \mu_2 > 0.$$

In the process of testing the above hypotheses, the two samples of size n_1 and n_2 are combined and arranged in increasing order. Ranks are then assigned to each observation. As with the methods discussed in Section 14.2, tied observations are assigned the average of the ranks that would have been assigned if the differences were found to be distinguishable.

In computing the test statistic for this test, we assign subscript '1' for the smaller sample and '2' for the larger sample. Thus, n_1 is the size of the smaller sample and n_2 is the size of the larger sample; w_1 is the sum of the ranks in the smaller sample and w_2 is the sum of the ranks in the larger sample. In cases of equal sample sizes, the designations of *smaller* and *larger* sample are assigned arbitrarily. As was the case for $(w_+ + w_-)$ in the one-sample test, the total of $w_1 + w_2$ depends only on the number of observations in the two independent samples. It is unaffected by either the nature or magnitude of the individual observations. This total is calculated as:

$$w_1 + w_2 = \frac{(n_1 + n_2)(n_1 + n_2 + 1)}{2}. \tag{14.5}$$

Once w_1 is calculated, w_2 can be obtained as:

$$w_2 = \frac{(n_1 + n_2)(n_1 + n_2 + 1)}{2} - w_1$$

or

$$w_1 = \frac{(n_1+n_2)(n_1+n_2+1)}{2} - w_2.$$

Where repeated samples of size n_1 and n_2 are taken from two populations, the sums of the ranks, w_1 and w_2, will change from sample to sample. Therefore, w_1 and w_2 are observations of the random variables W_1 and W_2. To simplify the construction of tables of critical values, the rank sums (w_1 and w_2) are transformed into u_1 and u_2 – hence the name **U-test**:

$$u_1 = w_1 - \frac{n_1(n_1+1)}{2} \tag{14.6}$$

or

$$u_2 = w_2 - \frac{n_2(n_2+1)}{2}. \tag{14.7}$$

Critical values of the random variables U_1 and U_2 for small sample tests are listed in Table A.11 (see Appendix A). When both n_2 and n_1 contain at least ten observations, it is considered a large sample case and the sampling distribution of u_1 and u_2 approaches the normal distribution with a mean and variance of:

$$\mu_{U_1} = \frac{n_1 n_2}{2} \tag{14.8}$$

$$\sigma^2_{U_1} = \frac{n_1 n_2(n_1+n_2+1)}{12}. \tag{14.9}$$

The large sample case test statistic is therefore calculated as:

$$z = \frac{u_1 - \mu_{U_1}}{\sigma_{U_1}}. \tag{14.10}$$

The critical values for the large sample case are obtained from the z-table. We demonstrate both the small and large sample cases in the following examples.

Example 14.6. The following data are the total biomass per plot (in g) of 1-year-old ponderosa pine seedlings, following the application of two kinds of fertilizers:

Fertilizer I	570	592	630	512	634	493	558
Fertilizer II	502	593	503	582	482	445	

Which fertilizer was more effective? Use $\alpha = 0.05$.
This is considered a small sample case, since $n_1 = 6$ and $n_2 = 7$.

1. H_0: $\mu_1 - \mu_2 = 0$.
2. H_1: $\mu_1 - \mu_2 \neq 0$.
3. $\alpha = 0.05$.
4. Use Table A.11 (see Appendix A).
5.

Fertilizer I	570	592	630	512	634	493	558
Ranks	8	10	12	6	13	3	7
Fertilizer II	502	593	503	582	482	445	
Ranks	4	11	5	9	2	1	

$$w_1 + w_2 = \frac{(6+7)(6+7+1)}{2} = 91$$

$w_2 = 32$, then
$w_1 = 91 - 32 = 59$.

6. Critical value (6, 7) = 6, from Table A.11 (see Appendix A).

7. $u_1 = 32 - \frac{6(6+1)}{2} = 11$

8. Since $u_1 = 11 > 6$, we 'accept' H_0.

9. Our evidence suggests that there is no difference between the effectiveness of the two fertilizers.

Example 14.7. Two groups of employees from a large forestry firm were given a questionnaire to determine their job satisfaction. The total score ranged from 0 to 10. Members of Group I graduated from a college or university and members of Group II did not.

Group I 6, 8, 4, 9, 7, 10, 7, 9, 5, 10, 9
Group II 4, 8, 9, 5, 7, 5, 7, 6, 4, 2, 9, 7, 6, 8, 9, 10, 4, 3, 5, 5, 7

Can we conclude, at $\alpha = 0.05$, that those with a college or university education were more satisfied with their job than those without higher education?
Since $n_1 = 11$ and $n_2 = 21$, this is considered a large sample case.

1. H_0: $\mu_1 - \mu_2 = 0$.
2. H_1: $\mu_1 - \mu_2 > 0$.
3. $\alpha = 0.05$.
4. Equation 14.10.
5.

Group I	6	8	4	9	7	10	7	9	5	10	9
Ranks	13	22	4.5	26.5	17.5	31	17.5	26.5	9	31	26.5
Group II	4	8	9	5	7	5	7	6	4	2	9
Ranks	4.5	22	26.5	9	17.5	9	17.5	13	4.5	1	26.5
	7	6	8	9	10	4	3	5	5	7	
Ranks	17.5	13	22	26.5	31	4.5	2	9	9	17.5	

$w_1 = 225$ $\quad w_2 = \frac{(11+21)(11+21+1)}{2} - 225 = 303$

$$u_1 = 225 - \frac{11(11+1)}{2} = 159$$

$$u_2 = 303 - \frac{21(21+1)}{2} = 72$$

$$\mu_{U_1} = \frac{(11)(21)}{2} = 115.5$$

$$\sigma^2_{U_1} = \frac{(11)(21)(11+21+1)}{12} = 635.25 \qquad \sigma_{U_1} \approx 25.20.$$

6. $z_{0.05} = 1.645$.

7. $z = \frac{159 - 115.5}{25.20} \approx 1.73$.

8. Since $1.73 > 1.645$, we reject H_0 and accept H_1.

9. The data indicate that employees with college or university education are significantly more satisfied with their job than those without higher education.

14.4 Kruskal–Wallis Test

When the assumptions of normality and equal population variance cannot be met in a one-way analysis of variance (see Section 12.1, Chapter 12), the **Kruskal–Wallis** test (also known as the ***H*-test**) can be used to compare three or more (k) unknown population means. The null and alternative hypotheses for this test are:

H_0: $\mu_1 = \mu_2 = \ldots = \mu_k$

and

H_1: at least two of the means are not equal.

As in the Wilcoxon rank sum test, the data for all the k groups are combined and ordered from low to high and ranks are assigned to each observation. Let $n_1, n_2, \ldots, n_k$ represent the number of observations in each of the k groups, and $R_1, R_2, \ldots, R_k$ represent the sum of the ranks in each of the k groups. The test statistic, h, can then be calculated as:

$$h = \frac{12}{n(n+1)} \sum_{i=1}^{k} \frac{R_i^2}{n_i} - 3(n+1) \qquad (14.11)$$

where $n = n_1 + n_2 + \ldots + n_k$.

If repeated samples are taken from the k populations, h is an observation of the random variable, H. If each of the k groups has at least five observations, H approximately follows a χ^2-square distribution with $k - 1$ degrees of freedom. Consequently, the null hypothesis stated above is rejected if $h > \chi^2_{\alpha(\nu)}$, with $\nu = k - 1$ degrees of freedom.

Example 14.8. A manufacturer of furniture components is studying the number of items produced during three shifts. Six days were randomly selected and observations were made during the morning and afternoon shifts. The night shift, however, could only be observed during 5 days.

Shift	Production					
Morning	105	128	109	120	115	122
Afternoon	101	97	96	102	108	90
Night	87	91	94	97	100	

Can we assume the same level of production for the three shifts? Use $\alpha = 0.05$.

1. H_0: $\mu_1 = \mu_2 = \mu_3$.
2. H_1: at least one of the means is different.
3. $\alpha = 0.05$.
4. Use Eqn 14.11.
5.

Morning	105	128	109	120	115	122
Rank	11	17	13	15	14	16
Afternoon	101	97	96	102	108	90
Rank	9	6.5	5	10	12	2
Night	87	91	94	97	100	
Rank	1	3	4	6.5	8	

$n = 17$

$R_1 = 86$

$R_2 = 44.5$
$R_3 = 22.5$
$n_1 = 6$
$n_2 = 6$
$n_3 = 5$.

6. $\chi^2_{0.05(2)} = 5.99$.

7. $h = \dfrac{12}{17(17+1)}\left(\dfrac{86^2}{6} + \dfrac{44.5^2}{6} + \dfrac{22.5^2}{5}\right) - 3(17+1) \approx 11.25.$

8. Since 11.25 > 5.99, we reject H_0 and accept H_1.
9. At least one of the three unknown population means is different (note that we do not know which one is different).

14.5 Runs Test

Most, if not all, of the statistical tests (including the non-parametric tests) discussed in this book are based on the assumption that samples are chosen randomly. But there are many instances where it is difficult to judge whether this assumption is justified. Based upon the order in which the observations are obtained, the **runs test** is a non-parametric procedure for testing the null hypothesis that observations are drawn in a random order.

To demonstrate this test, consider data collected on 13 trees in a bark beetle infested forest using line transect sampling. Let the symbol '*I*' denote an infested tree and '*H*' denote a healthy tree:

$$\overbrace{I\,I}\;\overbrace{I\,H\,H\,H\,H}\;\overbrace{I\,I}\;\overbrace{H\,H\,H}\;\overbrace{I}.$$

Here, consecutive groups of identical letters constitute what are known as 'runs'. In this set of observations, for example, we have $v = 5$ runs. Whether our data are qualitative or quantitative, the runs test requires that observations can be classified into two mutually exclusive groups (i.e. infested or healthy, Douglas-fir or not Douglas-fir, male or female, heads or tails, defective or non-defective, above or below the median and so on). In this test, we let n_1 denote the number of symbols associated with one of the two categories, n_2 denote the number of symbols associated with the other and $n_1 + n_2 = n$. In our example, therefore, $n_1 = 6$, $n_2 = 7$ and $n = 13$. The null hypothesis of randomness will be rejected if the number of runs compared to the number of observations is either smaller or larger than expected. In our example, the least number of runs would be 2:

$$I\,I\,I\,I\,I\,I\,H\,H\,H\,H\,H\,H\,H,$$

while the most number of runs would be 13:

$$H\,I\,H\,I\,H\,I\,H\,I\,H\,I\,H\,I\,H.$$

In the runs test, the null and alternative hypotheses are set ups as follows:

H_0: the sequence is random

H_1: the sequence is not random.

The number of runs is a random variable, V, and we observe one realization of this random variable, v, for each test. If n_1 and n_2 are less than or equal to 10, Table A.12 (see Appendix A) gives critical values for V and their corresponding probabilities as a function of n_1 and n_2. The rejection criteria is based on the following probability statements:

$$P(v \leq a) \leq \alpha/2,$$

and

$$P(v \geq b) \leq \alpha/2.$$

This means that if the number of runs is outside of the interval from a to b, the null hypothesis is rejected. For our case, if $\alpha = 0.05$, Table A.12 (see Appendix A) shows:

$$P(v \leq 3) = 0.008 \leq \alpha/2$$

and

$$P(v \geq 11) = (1 - 0.992) = 0.008 \leq \alpha/2.$$

We then have $a = 3$ and $b = 11$, which gives us:

$$P(a < v < b) \geq 1 - \alpha = 1.0 - \{0.008 + (1.0 - 0.992)\} = 1 - 0.016.$$

This results in an actual *type I* error of 0.016, instead of 0.05. Since the number of runs is 5 (which is within the interval of 3 and 11), H_0 cannot be rejected and the sequence can be assumed to be random.

The runs test can also be used to test the departure from randomness in a sequence of quantitative observations. In this case, we replace each measurement (in the order it was collected) by a *minus* sign, if it falls below the median, or by a *plus* sign, if it falls above the median. Observations that are equal to the median are discarded. This generates a sequence of plus signs and minus signs that can then be tested in the same way as the sequence of infested and healthy trees above.

Example 14.9. Assume that the data given in Example 14.2 (see Section 14.1) are listed in the order they were taken. Can we claim, with $\alpha = 0.05$ level of significance, that the length of wait time is random?

1. H_0: the sequence is random.
2. H_1: the sequence is not random.
3. $\alpha = 0.05$.
4. Not applicable.
5.

12	20	21	30	27	15	19	4	20	7
–	+	+	+	+	–	+	–	+	–
13	22	10	20	18	16	23	15	24	11
–	+	–	+	–	–	+	–	+	–

   Median = 18.5
   $n_1 = 10$
   $n_2 = 10.$
6. Using Table A.12 (see Appendix A), the critical values are $a = 6$ and $b = 15$.
7. Number of runs = $v = 15$.
8. Since $b = v$, we reject H_0 and accept H_1.
9. The sequence of waiting times is not random.

For large samples (n_1 and $n_2 \geq 10$), the sampling distribution of V approaches the normal distribution with a mean and a variance of:

$$\mu_V = \frac{2n_1 n_2}{n_1 + n_2} + 1 \tag{14.12}$$

and

$$\sigma_V^2 = \frac{2n_1 n_2 (2n_1 n_2 - n_1 - n_2)}{(n_1 + n_2)^2 (n_1 + n_2 - 1)}. \tag{14.13}$$

Therefore, if both n_1 and n_2 are ≥ 10, the following test statistic can be used to test the null hypothesis:

$$z = \frac{v - \mu_V}{\sigma_V}. \tag{14.14}$$

In this case, critical values are found in the z-table.

Since $n_1 = n_2 = 10$ in Example 14.9, the runs test can be carried out with a z-test as well.

1. H_0: the sequence is random.
2. H_1: the sequence is not random.
3. $\alpha = 0.05$.
4. Equation 14.14.
5. $\mu_V = 11$

   $\sigma_V^2 \approx 4.734$

   $\sigma_V \approx 2.18$
6. $z_{0.025} = \pm 1.96$.
7. $z = \dfrac{15 - 11}{2.18} \approx 1.83$.
8. Since $-1.96 < 1.83 < 1.96$, we cannot reject H_0 (a different result from above).
9. The sequence of waiting time is not significantly different from random.

Note that the two versions of this test give different conclusions. When the test was carried out using the small sample critical values, H_0 was rejected; however, with the large-sample z-test, it was not. These apparently contradictory results can occur when we have 'borderline' cases. In cases such as these – where the samples are just barely large enough – the test based on the critical values from Table A.12 (see Appendix A) is more reliable and should therefore be used.

Example 14.10. The sequence of digits for the value of π is considered random. The first 90 digits of the 8,000,000,000 known decimals in π are as follows:

3.14159	16353	89793	23846	26433	83279
50288	41971	69399	37510	58209	74944
59230	78164	06286	20899	86280	34825

Test the first 90 digits for randomness, using E to represent the even digits and O to represent the odd digits. Use $\alpha = 0.05$.

1. H_0: the sequence is random.

2. H_1: the sequence is not random.
3. $\alpha = 0.05$.
4. Equation 14.14.
5. OEOOO OEOOO EOOOO EOEEE EEEOO EOEOO
OEEEEE EOOOO EOOOO OOOOE OEEEO OEOEE
OOEOE OEOEE EEEEE EEEOO EEEEE OEEEO

$n_1 = 44$ (odd) $\quad n_2 = 46$ (even) $\quad V = 39$

$\mu_V = 45.98 \quad \sigma_V^2 \approx 22.225 \quad \sigma_V \approx 4.71.$

6. $z_{0.025} = \pm 1.96$.
7. $z = \dfrac{39 - 45.98}{4.71} \approx 1.48$.
8. Since $-1.96 < 1.48 < 1.96$, we 'accept' H_0.
9. Based on the first 90 digits, the sequence of odd and even numbers in π appears to be random.

14.6 Spearman's Rank Correlation Test

In Chapter 11, we discussed the significance test for the sample correlation coefficient, known as the *Pearson product moment correlation coefficient*. We also identified that testing the significance of the correlation coefficients requires an assumption that both variables, X and Y, are randomly selected from a *bivariate normal distribution*. When this assumption cannot be met, a non-parametric equivalent known as **Spearman's rank correlation coefficient** (r_s) can be used to conduct **Spearman's rank correlation test.**

To calculate Spearman's rank correlation coefficient, we first independently arrange the values of the two random variables, X and Y, and assign ranks. These ranks are then substituted for the actual numeric values of the two variables. If there is a tie, we assign the mean of the ranks that they jointly occupy to each of the tied observations. If no ties exist, then the ranks can be substituted into Eqn 11.13 (see Chapter 11) to obtain the rank correlation coefficient. The same value can also be obtained more efficiently by calculating the differences between the each pair of ranks, d_i, and using the following equation:

$$r_s = 1 - \frac{6\sum_{i=1}^{n} d_i^2}{n\left(n^2 - 1\right)}. \tag{14.15}$$

When there are ties, Eqns 11.13 (see Chapter 11) and 14.15 will result in slightly different values. Equation 14.15 is considered to be more reliable and has the further advantage of being simpler to compute.

We present two procedures to test the significance of Spearman's rank correlation coefficient, for which the null and alternative hypotheses are stated as:

H_0: $\rho = 0$,

H_1: $\rho \neq 0$, or $\rho < 0$, or $\rho > 0$.

When X and Y are independent, the r_s values approach a normal distribution with a mean of zero and a standard deviation of:

$$\sigma = \frac{1}{\sqrt{n-1}}. \tag{14.16}$$

Hence, the significance of the rank correlation coefficient can be tested with:

$$z = \frac{r_s - 0}{1/\sqrt{n-1}} = r_s\sqrt{n-1}. \tag{14.17}$$

Critical values for $\alpha = 0.05$, 0.025, 0.01 and 0.005 are tabulated as a function of n in Table A.13 (see Appendix A) or, alternatively, can be obtained from the z-table. If the absolute value of the calculated r_s is greater than the tabulated value for a given n, the null hypothesis is rejected.

Example 14.11. An interior designer and a randomly selected customer rated 7 different kitchen cabinets for aesthetics. They used a scale from 1 (lowest) to 100 (highest).

Designer	48	76	30	88	61	93	55
Customer	35	44	28	50	75	85	77

At $\alpha = 0.05$, is there a significant correlation between the customer's and the designer's ratings?

1. H_0: $\rho_s = 0$.
2. H_1: $\rho_s \neq 0$.
3. $\alpha = 0.05$.
4. Equation 14.17.
5.

Designer	48	76	30	88	61	93	55
Rank	2	5	1	6	4	7	3
Customer	35	44	28	50	75	85	77
Rank	2	3	1	4	5	7	6
Difference	0	2	0	2	–1	0	–3
d_i^2	0	4	0	4	1	0	9

6. $z_{0.005} = \pm 1.96$.
7. $z = 0.679\sqrt{7-1} \approx 1.66$.
8. Since $-1.96 < 1.66 < 1.96$, we 'accept' H_0.
9. There is no correlation between the two sets of ratings.

Alternatively, we could have used Table A.13 (see Appendix A) to obtain the critical value for this test. The tabulated value for $n = 7$ with $\alpha = 0.025$ is 0.786. Since $0.676 < 0.786$, we arrive at the same conclusion: we cannot 'reject' H_0.

Exercises

Section 14.1

14.1. A meteorologist claims that the median May temperature in Vancouver, British Columbia is 14°C. To investigate this, the following data (in degrees Celsius) were collected from the Vancouver airport weather station at noon during 8 randomly selected days in May 2001:

12, 16, 13, 14, 18, 15, 12, 17

Without making any assumptions about the distribution of this data, is the meteorologist's claim supported at $\alpha = 0.05$?

14.2. The following are observations of daily sulphur oxide emission (in tonnes) from a certain pulp mill:

17	15	20	29	19	18	22	17	21	13
14	24	22	13	14	25	21	18	9	11

According to government regulations, the median emissions should not exceed 15 t/day. Using the sign test, evaluate whether this mill operates within the required regulations ($\alpha = 0.01$)?

14.3. In a cabinet plant, 9 machines were selected to test whether routine daily maintenance was more effective than weekly maintenance for decreasing the number of defective parts produced. The data show the number of defective parts observed in a 24-h period.

	Machine number								
Maintenance	1	2	3	4	5	6	7	8	9
Once per week	6	12	5	12	10	6	9	10	7
Once per day	5	9	5	6	8	5	4	11	5

Use the sign test ($\alpha = 0.05$) to evaluate whether daily maintenance improves the quality of production.

14.4. Leader growth (in cm) of 21 randomly selected 5-year-old Douglas-fir seedlings was measured for 2 consecutive years:

	Seedling number																				
Year	1	2	3	4	5	6	7	8	9	10	11	12	13	14	15	16	17	18	19	20	21
1	28	36	25	36	45	16	29	42	44	36	18	26	39	15	29	22	24	42	33	26	35
2	26	38	26	24	43	16	26	40	39	39	16	25	39	16	21	26	29	40	35	27	39

Assuming that these observations are symmetrically distributed about their means, test whether the unknown population means of leader growth from each of the 2 years are equal using the sign test ($\alpha = 0.01$).

Section 14.2

14.5. Repeat the test for the data described in Exercise 14.1 using the Wilcoxon signed rank test. Compare and discuss the results obtained by the two tests.

14.6. Is the unknown population mean of leader growth in Year 1 for the seedlings listed in Exercise 14.4 significantly different from 30 cm? Use the Wilcoxon signed rank test with $\alpha = 0.05$.

14.7. Using the data given in Exercise 14.4, compare the two unknown population means using the Wilcoxon signed rank test. Compare and discuss the two sets of results.

Section 14.3

14.8. The numbers of trees per prism plot measured in two forest types (I and II) were recorded as follows:

Forest Type I	12	9	13	14	10
Forest Type II	7	8	10		

Using the Wilcoxon rank sum test, can we assume that the unknown population mean of the number of measured trees in Type I is significantly greater than that of Type II ($\alpha = 0.05$)?

14.9. Two diets were used in an experiment to study the weight gain (in kg) of 12 steer (see Exercise 9.24, Chapter 9):

Diet 1	45.9	38.7	44.1	49.0	41.4		
Diet 2	36.2	74.6	43.7	60.3	41.4	39.2	51.3

Using the Wilcoxon rank sum test, can we assume that Diet 2 is superior to Diet 1 in terms of weight gain ($\alpha = 0.05$)? Compare your results with those obtained in Exercise 9.24 (see Chapter 9).

14.10. A large forest products company requires potential employees to complete a screening test prior to interviewing. The following observations are times (in minutes) required for a random sample of 15 women and 13 men to complete this test:

Women	8.5	12.5	17.3	9.0	15.2	14.8	15.4	9.1	10.5	11.9	12.7	9.6	10.4	13.1	14.4
Men	10.6	10.7	7.8	10.3	12.6	10.8	9.6	8.9	14.2	11.0	16.6	10.4	14.9		

Using the Wilcoxon rank sum test, test the hypothesis that the two unknown population means of time to complete the test are equal, with a 0.05 level of significance.

Section 14.4

14.11. The specific gravity of wood was measured from six random samples of 3 different tree species:

Douglas-fir	0.55	0.51	0.44	0.46	0.56	0.48
Western hemlock	0.42	0.37	0.40	0.38	0.42	0.43
Western red cedar	0.32	0.28	0.29	0.35	0.29	0.30

Use the Kruskal–Wallis test with a significance level of 0.05 to evaluate the hypothesis that the population means of the 3 species are the same.

14.12. The following observations are random samples from three normally distributed populations. It is known that the variances of the three populations are equal.

A:	8.7	11.0	9.7	6.7	9.9	7.8	10.9	9.5	7.5	9.6	
B:	12.7	9.0	10.5	10.4	10.4	11.4	12.2	7.8	10.5	10.8	12.9
C:	17.3	14.0	13.8	17.2	16.9	16.0	13.6	10.5	14.9	14.1	

Test the hypothesis that the population means are equal ($\alpha = 0.05$) by using:

a. One-way analysis of variance (ANOVA).
b. The Kruskal–Wallis test.

Compare the results of parts a and b. Which test would you expect to be more sensitive? Explain.

Section 14.5

14.13. Consider the first 90 digits of π listed in Example 14.10 that we tested for randomness using even and odd digits. Re-test these digits using the median (α = 0.05).

14.14. Assume that the data listed in Exercise 14.1 were taken in order of occurrence. Test these data for randomness using a 0.10 level of significance:

a. Using the median.
b. Letting 'E' represent an even and 'O' represent an odd number.
c. Compare and contrast the two results.

14.15. On 20 successive trips, a logging truck carried the following numbers of logs:

24	19	32	28	21	23	26	17	20	28
30	24	13	35	26	21	19	29	27	18

Test these data for randomness ($\alpha = 0.05$) by using the median.

Section 14.6

14.16. The following data are the number of hours that 11 students studied for a statistics test and their resulting test scores (as %):

Hours	8	6	15	5	21	15	8	12	10	9	14
Score	80	62	93	92	95	90	79	88	91	77	89

Calculate Spearman's rank correlation coefficient and test whether there is a *positive* correlation between the two random variables (α = 0.05). Why would you not recommend testing this data with Pearson's correlation coefficient?

14.17. Find Spearman's rank correlation coefficient for the leader growth data given in Exercise 14.4 and test whether the correlation between the 2 years is significant ($\alpha = 0.01$).

15 Quality Control

Statistics for Production and Processing

With the advent of assembly lines in the early 1900s, the industrial process was characterized by many workers contributing to the completion of a product. Because of this, process control became necessary to ensure that the product performed to expectations and factories began to train their employees to inspect completed and in-process products. When poor performance was detected, its cause had to be identified. In 1924, Walter A. Shewhart, working for Bell Laboratories, documented a new statistical approach for tracking process quality levels over time. Each part of the manufacturing process was checked in order to identify the need for corrections before the product was assembled. Known today as **statistical quality control**, or **statistical process control**, this procedure involves measuring certain production-related metrics, such as the thickness of lumber exiting a planer, and monitoring these metrics on **control charts**. Small samples are taken and charted, displaying a running record for such quantities as the *median*, *mean*, *range* and *standard deviation*. Their purpose is to detect whether or not a process is 'in control'. In other words: are products being manufactured correctly? Modern quality control methods are commonplace in the forest products industry today, being used in the production of a wide range of goods, from dimension lumber to pulp and paper to value-added products.

In order to construct a simple control chart for a product, several (k) samples of a certain size (n) are taken and the attribute being controlled is measured. For each of the k *samples*, measures of central tendency (median or mean) and variation (range or standard deviation) are calculated for the n measurements. To introduce and demonstrate the construction of various control charts, we will use k = 10 samples (see Table 15.1) of n = 5 2 × 4s taken from a sawmill. The attribute we are interested in controlling is board thickness (in inches). Assume that we are estimating a parameter, θ, a measure of the mean board thickness. Using all of the k samples, we can obtain parameter estimates of the mean, $\hat{\theta}$, and the standard deviation, $\hat{\sigma}$. The **lower control limit** and the **upper control limit** can then be calculated as:

$UCL = \hat{\theta} + 3\hat{\sigma}/\sqrt{n}$ upper control limit

$LCL = \hat{\theta} - 3\hat{\sigma}/\sqrt{n}$ lower control limit

We can also compute a **lower warning limit** and an **upper warning limit** such that:

$UWL = \hat{\theta} + 2\hat{\sigma}/\sqrt{n}$ upper warning limit

$LWL = \hat{\theta} - 2\hat{\sigma}/\sqrt{n}$ lower warning limit

Table 15.1. Two-by-four thicknesses[a] (in inches) of 10 samples from 5 randomly selected boards.

Sample	1	2	3	4	5	6	7	8	9	10
	1.647	1.671	1.672	1.672	1.666	1.683	1.647	1.640	1.663	1.667
	1.690	1.658	1.649	1.661	1.661	1.677	1.649	1.648	1.662	1.661
	1.678	1.673	1.661	1.653	1.671	1.669	1.644	1.678	1.655	1.683
	1.667	1.659	1.671	1.655	1.667	1.651	1.680	1.667	1.646	1.677
	1.666	1.662	1.667	1.655	1.669	1.678	1.666	1.665	1.656	1.683
Mean	1.670	1.665	1.664	1.659	1.667	1.672	1.657	1.660	1.656	1.674
Standard deviation	0.016	0.007	0.009	0.008	0.004	0.013	0.015	0.015	0.007	0.010
Range	0.043	0.015	0.023	0.019	0.010	0.032	0.036	0.038	0.017	0.022

[a] Two-by-four refers to a board's nominal dimensions. In practice, lumber dimensions are smaller than their nominal dimensions and are measured by calipers or lasers.

If the observations of the attribute of interest are normally distributed, the probability of obtaining a statistic from a sample of size n outside the 3-sigma limits is about 0.0027, while the probability of obtaining a statistic outside the 2-sigma limits is about 0.0456 (think of the normal or empirical rule from Section 2.4, Chapter 2). Using the control limits and the warning limits, control charts can be constructed, as in Figs 15.1 and 15.2.

Once the limits for the control charts are constructed, consecutive samples of size n are taken from production at different times and the appropriate statistics (medians, means, ranges or standard deviations) are calculated from each sample and plotted on the chart. Control charts show us when action should be taken to adjust the process because it is out of control (see Fig. 15.1), or when to leave the process alone (see Fig. 15.2) because it is in control. When one or several of the statistics fall outside of the upper or lower control limits, the manufacturing process is said to be 'out of control' and should be stopped in order to locate and correct the source, or sources, of the problem. When one or several means fall outside of the upper or lower warning limits, closer attention is paid to the manufacturing process to identify the problem, but

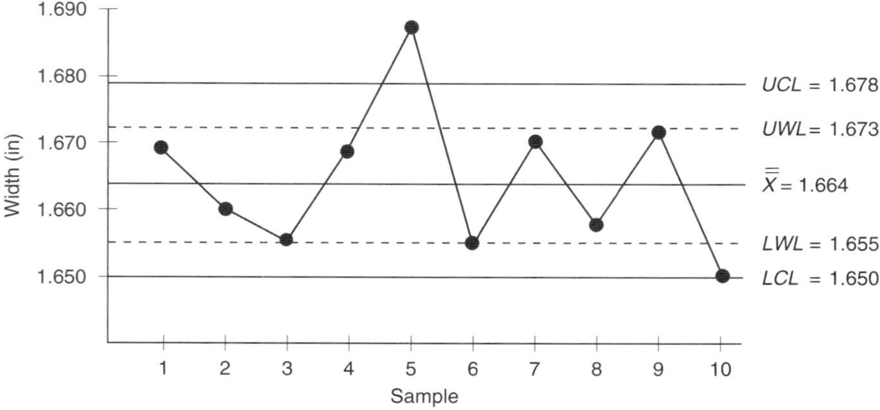

Fig. 15.1. Control chart for an 'out of control' process.

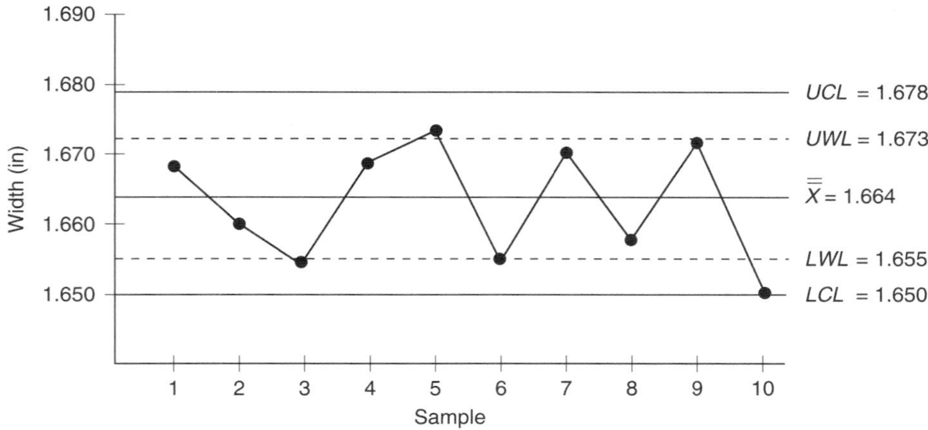

Fig. 15.2. Control chart for an 'in control' process.

production is generally not halted (this is often costly). Note that the process displayed in Fig. 15.2 is said to be 'in control' because none of the points fall outside of the lower or upper control limits. That said, since two points do fall outside of the warning limits, closer attention should be paid to the production process.

Additional 'Runs Rules' can also often be applied to determine whether a process is out of control, such as:

1. A 'run' of seven or more points in a row occurring below or above the centre line (see Fig. 15.3);
2. Six consecutive points showing an increasing or decreasing trend (see Fig. 15.4); or
3. Two of three points in a row occurring in the region between the warning and control limits (see Fig. 15.5).

Control charts can be constructed either for **variable data** (data that can be measured: i.e. lengths, diameters, weights or strengths) or **attribute data** (data which require an

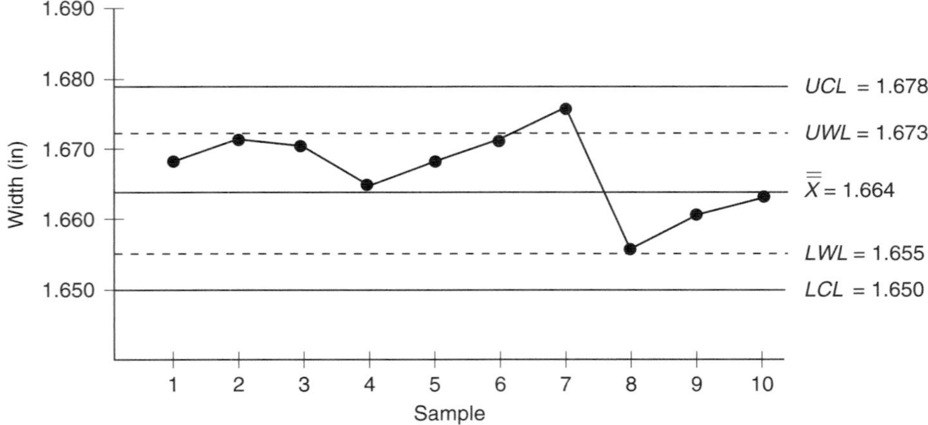

Fig. 15.3. Control chart for an 'out of control' process, using Runs Rule No 1.

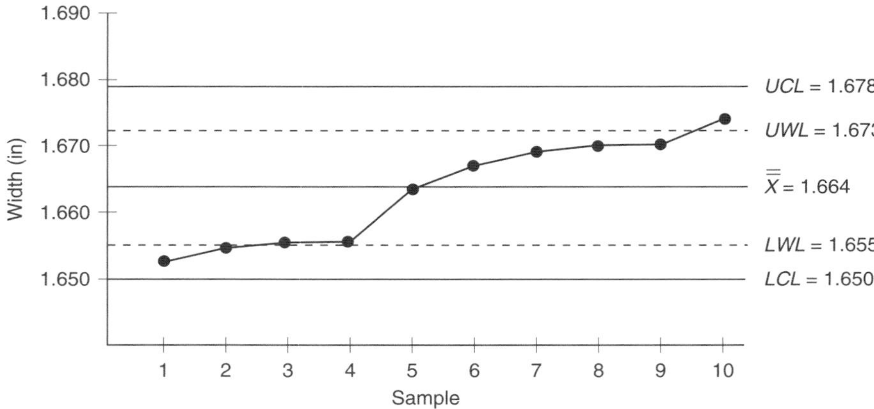

Fig. 15.4. Control chart for an 'out of control' process, using Runs Rule No 2.

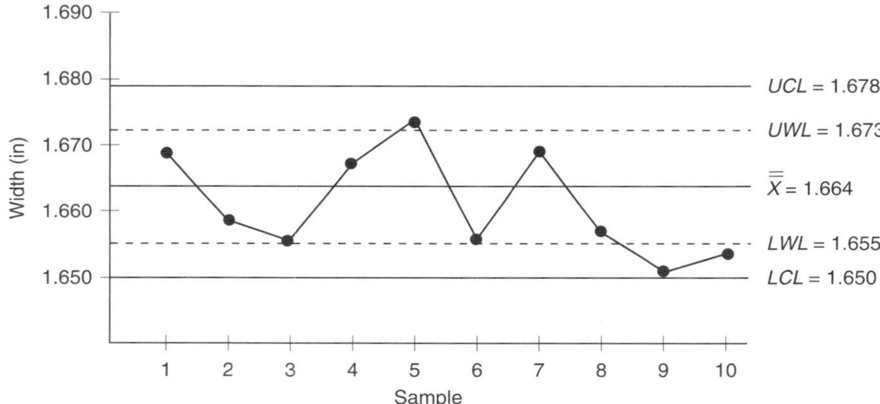

Fig. 15.5. Control chart for an 'out of control' process, using Runs Rule No 3.

operational definition of acceptable and defective products). Of the several variable charts used in practice, we will discuss the most common variable charts: the **$\bar{X}$ chart, R chart** and **S chart.** We will also discuss an appropriate and commonly used chart for attribute data: the **p chart**. All of these charts play a very important role in the quality control of forest products processes.

15.1 Variable Charts

To illustrate the use of variable charts, we will use one of the most commonly encountered quality control applications in forest products: the dimensional control of lumber. In Table 15.1, we show lumber thickness data collected from 10 samples of 5 randomly selected 2×4 boards. The means, standard deviations and ranges for each of the 10 samples are also given, from which most variable charts can be constructed.

The **$\bar{X}$ chart** is constructed after k samples have been taken from a process. Using the data from Table 15.1, the grand mean of all the samples is calculated first and is used as the centre line of the chart. Then the standard error of the mean is estimated from either the mean of the standard deviations or the mean of the ranges. These are used to calculate the upper and lower control limits and the upper and lower warning limits.

1. Calculate the grand mean (or mean of the means):

$$\overline{\overline{X}} = \frac{1}{k}\sum_{j=1}^{k} \overline{x}_j \approx 1.664$$

where $\overline{x}_j$ is the mean of the jth sample.

2. Calculate the mean of the standard deviations:

$$\overline{s} = \frac{1}{k}\sum_{j=1}^{k} s_j \approx 0.010,$$

then

$$\hat{\sigma} = \overline{s}/c_4(n) = 0.010/0.9400 \approx 0.0106,$$

where s_j is the standard deviation from the jth sample, and $c_4(n)$ is a correction factor commonly called a *control chart constant* which depends on n (see Table 15.2). Its use guarantees an unbiased estimate of σ.

3. If the standard deviations are not available, σ can be estimated from the ranges:

$$\overline{R} = \frac{1}{k}\sum_{j=1}^{k} R_j = 0.026,$$

then

$$\hat{\hat{\sigma}} = \overline{R}/d_2(n) = 0.026/2.3259 \approx 0.0112,$$

where R_j is the range of the jth sample, and $d_2(n)$ is a *control chart constant* which depends on n (see Table 15.2). Its use also guarantees an unbiased estimate of σ.

From here, the $\overline{X}$ chart control and warning limits can be constructed as:

$$UCL = 1.664 + (3)(0.0106)/\sqrt{5} \approx 1.678$$
$$LCL = 1.664 - (3)(0.0106)/\sqrt{5} \approx 1.650$$

and

$$UWL = 1.664 + (2)(0.0106)/\sqrt{5} \approx 1.673$$
$$LWL = 1.664 - (2)(0.0106)/\sqrt{5} \approx 1.655$$

Table 15.2. Control chart constants $c_4(n)$, $d_2(n)$ and $d_3(n)$ for estimating σ.

n	$c_4(n)$	$d_2(n)$	$d_3(n)$
2	0.7979	1.1283	0.8525
3	0.8862	1.6926	0.8884
4	0.9213	2.0587	0.8798
5	0.9400	2.3259	0.8641
6	0.9515	2.5343	0.8480
7	0.9594	2.7044	0.8332
8	0.9650	2.8471	0.8198
9	0.9693	2.9699	0.8078
10	0.9727	3.0774	0.7971

Similar values could also be obtained by using the estimated variance from the range, $\hat{\hat{\sigma}}$.

Once the control chart is constructed, production can easily be monitored by plotting the means of ten consecutive samples of size $n = 5$ on the chart (see Fig. 15.6). Since each mean occurs within the control limits for our example, the process is said to be 'in control' and no adjustments to the process are required.

Frequently, control of process variability is as important to control as the process mean; after all, it is a high degree of variability that leads to an inconsistent product. Either the **S chart** or the **R chart** is used for this purpose. Both are constructed much like the $\bar{X}$ chart, using k samples taken from the process. Although the S chart is known to be more efficient than the R chart, the R chart is more popular because sample ranges are easier to compute than sample standard deviations. When sample sizes are small, the R chart performs as well as the S chart. However, as sample size increases, the efficiency of the R chart decreases to the point where it is not recommended when $n > 5$.

In the process of calculating control limits for the S chart, it is assumed that the sampling distribution of the standard deviations is symmetric. Using the data from Table 15.1, the control limits are calculated as:

$\bar{s} = 0.010$ (the centre line)

$UCL = \bar{s} + 3\hat{\sigma}_{\bar{s}}$ upper control limit

$LCL = \bar{s} - 3\hat{\sigma}_{\bar{s}}$ lower control limit

where

$\sigma_{\bar{s}} = \sigma\sqrt{1 - c_4^2}$ (see Table 15.2 for the constants).

The value of σ is not usually known and, thus, the standard error of the standard deviations, $\sigma_{\bar{s}}$, is estimated as:

$$\hat{\sigma}_{\bar{s}} = \bar{s}/c_4\sqrt{1 - c_4^2}$$

In our example:

$$\hat{\sigma}_{\bar{s}} = 0.010/0.9400\sqrt{1 - 0.9400^2} \approx 0.00276$$

$$UCL = 0.010 + 3(0.00276) = 0.0183$$

$$LCL = 0.010 - 3(0.00276) = 0.0017$$

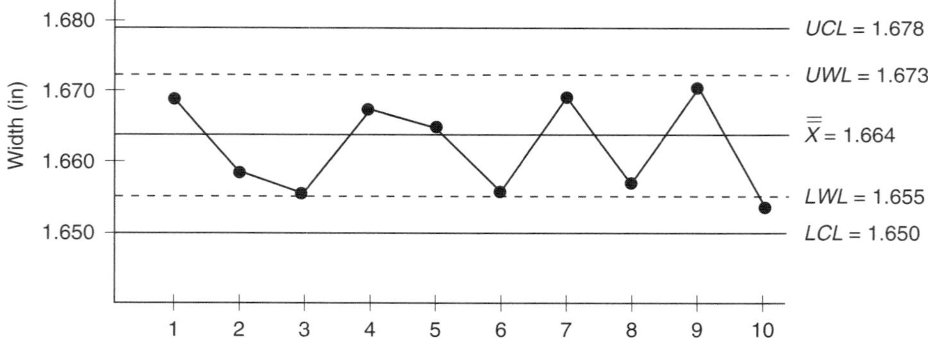

Fig. 15.6. $\bar{X}$ chart for the 2 × 4 example.

$$UWL = 0.010 + 2(0.00276) = 0.0155$$

$$LWL = 0.010 - 2(0.00276) = 0.0045.$$

In the case of a negative control limit for an S chart or R chart, zero is used in place of the negative limit. Once the control limits are constructed, variation in production can be monitored by plotting the standard deviations of consecutive samples of size n (see Fig. 15.7). Since each standard deviation in our example is within the control limits, this process is said to be 'in control'.

Using the data from Table 15.1, the control limits for the R chart (see Fig. 15.8) are calculated and analysed like the S chart.

The centre line is $\bar{R} = 0.026$ and the standard error of $\bar{R}$ is:

$$\hat{\sigma}_{\bar{R}} = \bar{R} d_3 / d_2$$

where d_3 is a *control chart constant* which depends on n (see Table 15.2). Its use guarantees an unbiased estimate of $\hat{\sigma}_{\bar{R}}$.

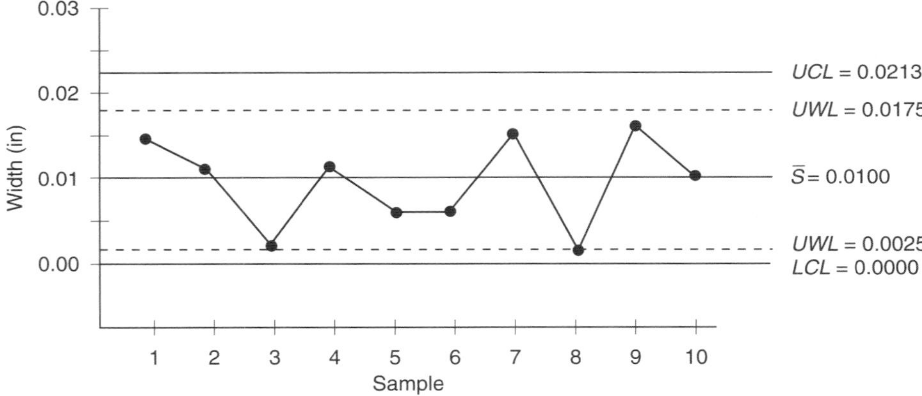

Fig. 15.7. S chart for the 2 × 4 example.

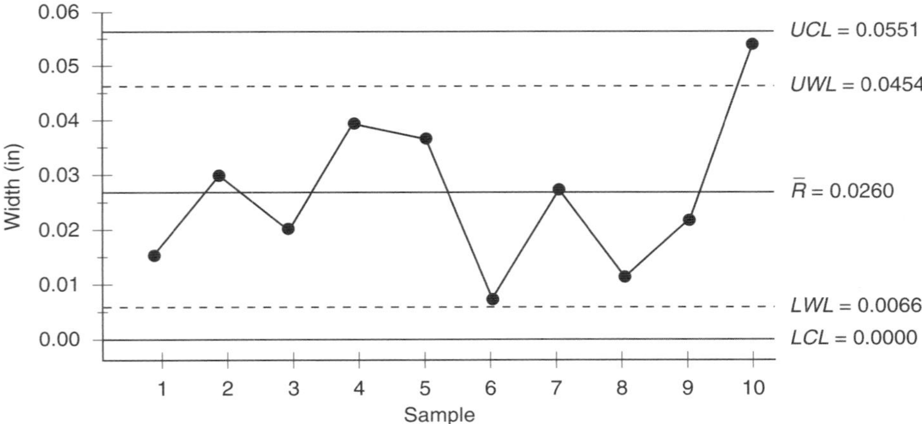

Fig. 15.8. R chart for the 2 × 4 example.

In our example, $\hat{\sigma}_{\overline{R}} = (0.026)(0.8641)/2.3259 \approx 0.0097$ and the limits are as follows:

$$UCL = \overline{R} + 3\hat{\sigma}_{\overline{R}} = 0.026 + 3(0.0097) \approx 0.0551$$

$$LCL = \overline{R} - 3\hat{\sigma}_{\overline{R}} = 0.026 - 3(0.0097) \approx -0.0031, \text{ use } 0.0$$

$$UWL = \overline{R} + 2\hat{\sigma}_{\overline{R}} = 0.026 + 2(0.0097) \approx 0.0454$$

$$LWL = \overline{R} - 2\hat{\sigma}_{\overline{R}} = 0.026 - 2(0.0097) \approx 0.0066$$

Figure 15.8 shows the resulting R chart. Once again, this process is in control, which is in agreement with the S chart (see Fig. 15.7).

15.2 Attribute Charts

We use **attribute charts** when the items in a manufacturing process can be classified into two groups: *acceptable* and *defective*. Thus, it can be assumed that the observations follow a binomial distribution. For example, stained kitchen cabinets might contain streaks or bubbles, or dimensional lumber products might contain visible defects such as knots or holes. To ensure that the proportion of defective items produced is within certain limits, a **p chart** is commonly used. The p chart is constructed in the same manner as an $\overline{X}$ chart. Based on k samples of size n taken from the production process, the centre line and the lower and upper control limits are calculated as:

$$\overline{p} = \sum_{j=1}^{k} \text{defectives} / \sum_{j=1}^{k} n$$

$$UCL = \overline{p} + 3\sqrt{\frac{\overline{p}(1-\overline{p})}{n}}$$

$$LCL = \overline{p} - 3\sqrt{\frac{\overline{p}(1-\overline{p})}{n}}$$

If desired, the lower and upper warning limits can also be calculated as:

$$UWL = \overline{p} + 2\sqrt{\frac{\overline{p}(1-\overline{p})}{n}}$$

$$LWL = \overline{p} - 2\sqrt{\frac{\overline{p}(1-\overline{p})}{n}}$$

In Table 15.3, the number of defective pieces out of 60 kitchen cabinets is given for six samples.

In this example, the p chart is constructed as follows:

$$\overline{p} = \sum_{j=1}^{k} \text{defectives} / \sum_{j=1}^{k} n = 22/360 \approx 0.061$$

$$UCL = 0.061 + 0.093 \approx 0.154$$

$$LCL = 0.061 - 0.093 \approx -0.031, \text{ use } 0.0$$

Table 15.3. The proportion of defectives of six samples from 60 2 × 4 boards.

Sample	Sample size (n)	Number of defectives	$\hat{p}$
1	60	4	0.067
2	60	2	0.033
3	60	5	0.083
4	60	3	0.050
5	60	6	0.100
6	60	2	0.033

$UWL = 0.061 + 0.062 \approx 0.123$

$LWL = 0.061 - 0.062 \approx -0.001$, use 0.00.

Once the control chart is constructed (Fig. 15.9), the process can be analysed by taking several samples of 60 kitchen cabinets and plotting the proportion of defective pieces. Since each proportion is within the control limits, this process is said to be 'in control'.

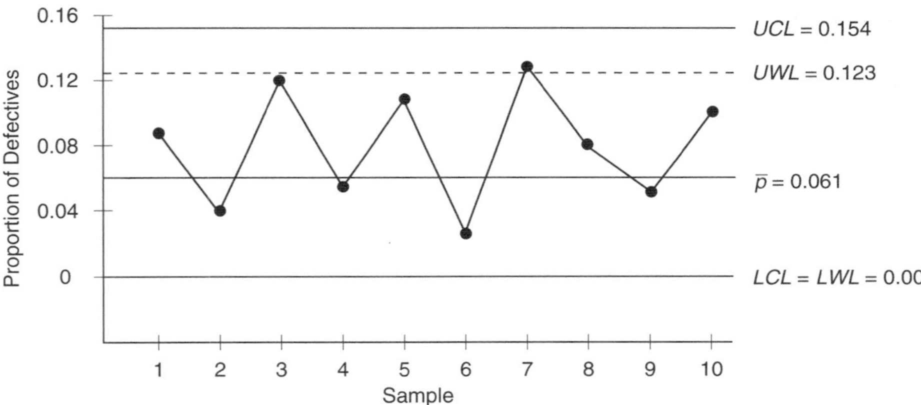

Fig. 15.9. p chart for the 2 × 4 example.

Exercises

Section 15.1

15.1. Water quality can be assessed by measuring the amount of dissolved oxygen contained within it. Six samples of size 5 were taken from a river when the water quality was known to be acceptable. The data are shown below.

Sample					
1	6.0	6.2	6.1	5.8	6.1
2	6.3	6.6	6.4	6.2	6.0
3	6.1	6.3	6.4	6.2	5.9
4	5.9	5.8	6.2	6.1	6.0
5	6.2	5.9	5.7	6.2	6.1
6	5.7	5.9	5.9	6.0	6.1

Construct $\bar{X}$, S and R charts. Calculate the control limits, as well as the warning limits for each chart.

15.2. Five samples of 4 rechargeable batteries were charged for 40 min and the length of time that they held the charge was measured. The observations (in hours) are given below.

Sample				
1	14.2	14.0	13.9	13.8
2	13.6	13.9	14.3	13.8
3	14.2	14.3	14.1	14.0
4	14.0	14.8	14.9	13.2
5	14.2	14.0	14.1	14.3

Construct $\bar{X}$, S and R charts for this data. Calculate the control limits, as well as the warning limits for each chart.

15.3. Four samples of 6 boards were taken from the production line of a certain fibreboard manufacturing process and tested for strength. The moduli of rupture values are listed below:

Sample						
1	492	621	521	562	584	596
2	596	585	593	601	574	596
3	502	563	598	536	604	610
4	582	589	521	603	597	571

Construct $\bar{X}$, S and R charts. Calculate the control limits, as well as the warning limits for each chart.

Section 15.2

15.4. Six samples of 25 wooden chairs were inspected to assess the quality of their joints. Chairs that could not withstand a certain load were called 'defectives':

Sample	Size	Number of defectives
1	25	3
2	25	1
3	25	3
4	25	0
5	25	2
6	25	4

Construct a p chart for the data. Calculate the control limits, as well as the warning limits.

15.5. Five samples of 50 seedlings lifted from a nursery were inspected to assess their quality. Seedlings that were damaged or smaller than a certain size were called 'rejects':

Sample	Size	Number of rejects
1	50	4
2	50	7
3	50	3
4	50	4
5	50	3

Construct a p chart for the data. Calculate the control limits, as well as the warning limits.

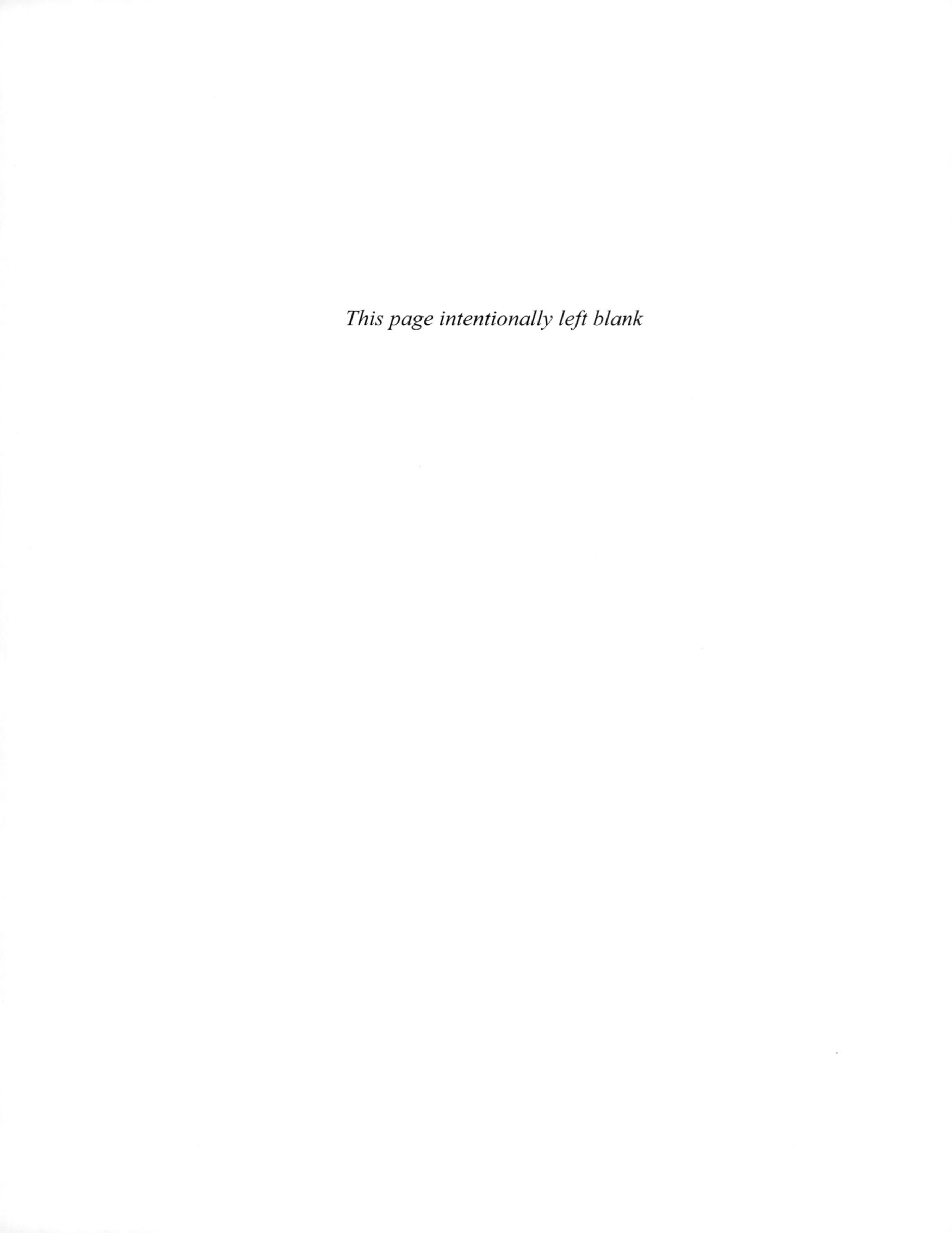

This page intentionally left blank

Bibliography

For those interested in further reading on the topic of statistics, we recommend the following texts. Many were instrumental in the development of this book, and we gratefully acknowledge the contribution of these authors.

Elementary Statistics

Bliss, C.I. 1964. *Statistics in Biology.* McGraw-Hill, New York.

Bluman, A.G. 2007. *Elementary Statistics: A Step by Step Approach.* 6th Edition. McGraw-Hill, New York.

Freese, F. 1974. *Elementary Statistical Methods for Foresters.* Agricultural Handbook 317. US Department of Agriculture, Forest Service.

Freund, J.E. and M.P. Benjamin. 1999. *Statistics. A First Course.* 7th Edition. Prentice-Hall, Upper Saddle River, New Jersey.

Huntsberger, D.V. and Billingsley, P. 1973. *Elements of Statistical Inference.* 3rd Edition. Allyn and Bacon, Inc, Boston, Massachusetts.

Menenhall, W. 1987. *Introduction to Probability and Statistics.* 7th Edition. Duxbury Press, Boston, Massachusetts.

Ott, L. and W. Mendenhall. 1990. *Understanding Statistics.* 5th Edition. PWS-Kent Publishing Co., Boston, Massachusetts.

Prodan, M. 1961. *Forest Biometrics.* (Translated by: Gardiner, S.H. 1968.) Program Press, Oxford, UK.

Snedecor, G.W. 1956. *Statistical Methods.* Iowa State University, Press, Ames, Iowa.

Steel, R.G.D. and J.H. Torrie. 1960. *Principles and Procedures of Statistics.* McGraw-Hill, New York.

Walpole, R.E. 1982. *Introduction to Statistics.* 3rd Edition. Macmillan Publishing Co. Inc., New York.

Weimer, R.C. 1993. *Statistics.* 2nd Edition. Wm. C. Brown Publishers, Dubuque, Iowa.

Statistical Software

Dilorio, F.C. 1991. *SAS Applications Programming: A Gentle Introduction.* Duxbury Press, Boston, Massachusetts.

Dixon, W.J. 1981. *BMDP Statistical Software.* University of California Press, Berkeley, California.

Field, A. 2000. *Discovering Statistics: Using SPSS for Windows.* Sage Publications Ltd, London, UK.

Littell, R.C., G.A. Milliken, W.W. Stroup, R.D. Wolfinger and O. Schabenberger. 2006. *SAS System for Mixed Models,* 2nd Edition. SAS Institute, Cary, North Carolina.

Schaefer, R.L. and B.A. Anderson. 1989. *The Student Edition of MINITAB. Statistical Software ... Adapted for Education.* Addison-Wesley Publishing Co. Inc., New York.

SPSS, Inc. Staff. 1984. *SPSS-X User's Guide.* 2nd Edition. McGraw-Hill, New York.

Wilkinson, L. 1990, *SYSTAT: The System for Statistics.* SYSTAT Inc., Evantson, Illinois.

Advanced Statistics

Everitt, B.S. 1992. *The Analysis of Contingency Tables.* 2nd Edition. Chapman and Hall, New York.

Neter, J., W. Wasserman, M.H. Kutner and C. Nachtsheim. 1996. *Applied Linear Statistical Model.* 4th Edition. Irwin McGraw-Hill, Chicago, Illinois.

Schabenberger, O. and F. J. Pierce. 2001. *Contemporary Statistical Models for the Plant and Soil Sciences.* CRC Press, Boca Raton, Florida.

Experimental Design

Federer, W.T. 1963. *Experimental Design – Theory and Application.* The Macmillan Company, New York.

Hicks, R.C. 1993. *Fundamental Concept in the Design of Experiments.* 4th Edition. Saunders College Publishing, New York.

Jeffers, J.N.R. 1959. *Experimental Design and Analysis in Forest Research.* Almqvist & Wiksell, Stockholm.

Montgomery, D.C. 2004. *Design and Analysis of Experiments.* 5th Edition. John Wiley & Sons, New York.

Multivariate Statistics

Dillon, W.R. and M. Goldstein. 1988. *Multivariate Analysis: Methods and Applications.* John Wiley & Sons, New York.

Manly, B.F.J. 2004. *Multivariate Statistical Methods: A Primer,* 3rd Edition. CRC Press, Boca Raton, Florida.

Tabachnick, B.G. and L.S. Fidell. 2007. *Using Multivariate Statistics,* 5th Edition. Allyn and Bacon Inc., Boston, MA.

Quality Control

Brown, T.D. 1982. *Quality Control in Lumber Manufacturing.* Miller Freeman, San Francisco, California.

Montgomery, D.C. 2005. *Introduction to Statistical Quality Control.* 5th Edition. John Wiley & Sons, New York.

Regression and Correlation

Draper, N.R. and H. Smith. 1998. *Applied Regression Analysis*. 3rd Edition. John Wiley & Sons, New York.

Ezekial, M. and K.A. Fox. 1959. *Methods of Correlation and Regression Analysis: Linear and Curvilinear.* John Wiley & Sons, New York.

Freese, F. 1964. *Linear Regression Methods for Forest Research.* FPL 17. U.S. Department of Agriculture, Forest Service.

Kleinbaum, D.G., L.L. Kupper, K.E. Muller and A. Nizam. 1998. *Applied Regression Analysis and Multivariable Methods,* 3rd Edition. Duxbury Press, Boston, Massachusetts.

Myers, R.H. 1986. *Classical and Modern Regression with Applications.* Duxbury Press, Boston, Massachusetts.

Sampling

Cochrane, W.G. 1963. *Sampling Techniques*, 2nd Edition. John Wiley & Sons, New York.

Freese, F. 1962. *Elementary Forest Sampling.* Agricultural Handbook 232. U.S. Department of Agriculture, Forest Service.

Williams, W.H. 1978. *A Sampler on Sampling.* John Wiley & Sons, New York.

Survey Design

Babbie, E. 2001. *Practice of Social Research.* 9th Edition. Wadsworth Thompson Learning, Belmont, California.

Dillman, D.A. 2000. *Mail and Internet Surveys: The Tailored Design Method.* 2nd Edition. John Wiley & Sons, New York.

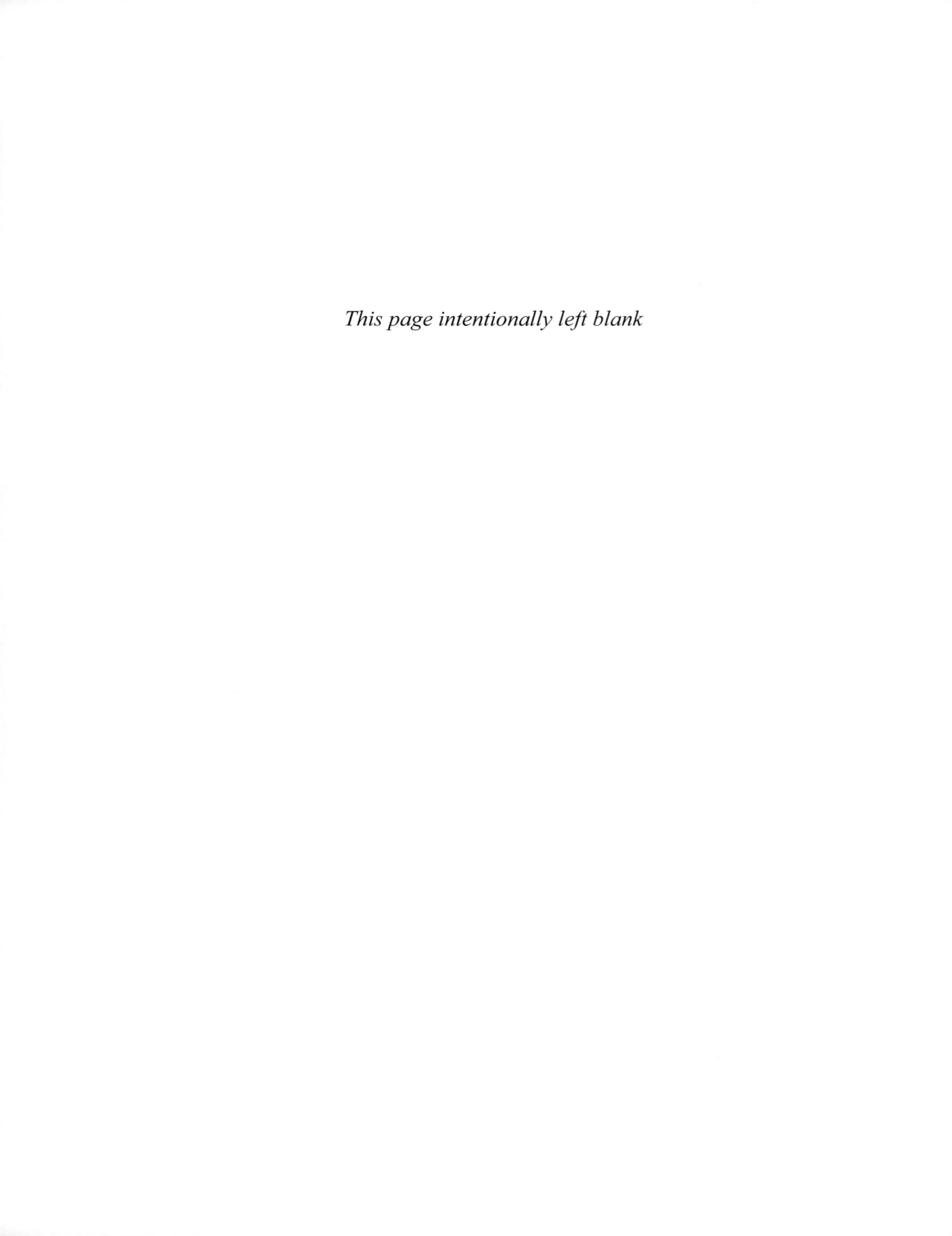
This page intentionally left blank

Solutions to Odd-Numbered Questions

Chapter 1

1.1. 1. A calculated value (mean, for example) from some observations.

2. Statistics recorded during a tennis match (number of double faults, number of aces, number of unforced errors, etc.).

3. Heights and weights of the students in a class.

1.3. By calculating the mean from 25 randomly selected trees, estimate the average height of all trees in a plantation.

1.5.
- a. The thickness of all boards produced during the shift.
- b. As 4 samples are taken during every one of the 8 h, the result is a sample of 32 observations of the board thicknesses.
- c. She can use both, as she can describe all of the measurements and/or she can estimate the unknown population mean of all possible board thicknesses.

1.7.
- a. Discrete.
- b. Not quantitative.
- c. Discrete.
- d. Continuous.
- e. Not quantitative.
- f. Continuous.
- g. Not quantitative.
- h. Continuous.
- i. Discrete.
- j. Not quantitative.
- k. Discrete (because these are given to the nearest degree).
- l. Discrete (because annual wage is given in whole dollars).
- m. Discrete (because they are whole numbers).
- n. Not quantitative (unless the date is given as 1 March 2006, for example, in which case it is discrete).

1.9. Experimental design measuring the effect of chemicals on board strength – two chemicals (A and B) are being tested and 'no chemical' is the controlled factor.

1.11. If we measured only 5 randomly selected boards out of the 10 treated ones from each of the three treatments.

Chapter 2

2.1.

Number of accidents	Frequency	Relative frequency
0	4	0.20
1	6	0.30
2	3	0.15
3	3	0.15
4	3	0.15
5	1	0.05
Total	20	1.00

2.3.

Number of trees	Frequency	Relative frequency
2	4	0.133
3	4	0.133
4	5	0.167
5	9	0.300
6	5	0.167
7	2	0.067
8	1	0.033
Total	30	1.000

2.5. $C = 3.3$ $\log(60) + 1 \approx 6.87 \approx 7$ classes Range = 6.1 − 2.5 = 3.6.

Class boundaries	Class limits	Frequency	Class mark	Relative frequency	Cumulative frequency	Relative cumulative frequency	Inverse cumulative frequency	Relative inverse cumulative frequency
≤ 3.05	≤ 3.0	9	2.8	0.150	9	0.150	60	1.000
3.05–3.55	3.1–3.5	15	3.3	0.250	24	0.400	51	0.850
3.55–4.05	3.6–4.0	13	3.8	0.217	37	0.617	36	0.600
4.05–4.55	4.1–4.5	11	4.3	0.183	48	0.800	23	0.383
4.55–5.05	4.6–5.0	5	4.8	0.083	53	0.883	12	0.200
5.05–5.55	5.1–5.5	5	5.3	0.083	58	0.967	7	0.117
≥ 5.55	≥ 5.6	2	5.8	0.033	60	1.000	2	0.033

Note: The class width was determined to be 0.53. Because of its proximity, we chose to use a class width of 0.5 rather than rounding up to the nearest odd number (0.7). This will generally result in the need for open classes to house the largest and smallest observations (6.1 and 2.5). If we used a class width of 0.7, an alternative for the first class would be 2.35–3.05.

2.7.

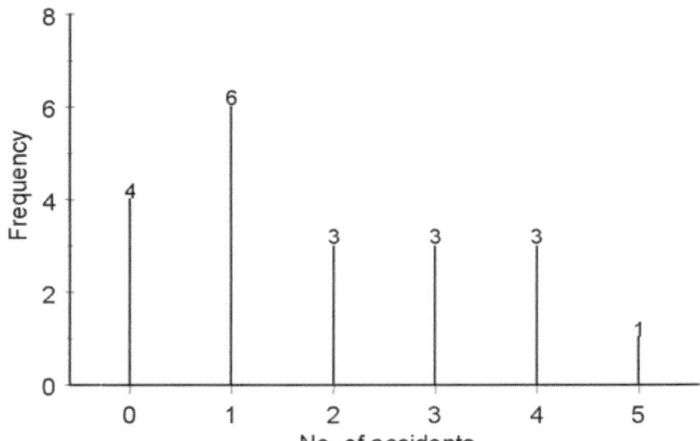

2.9.

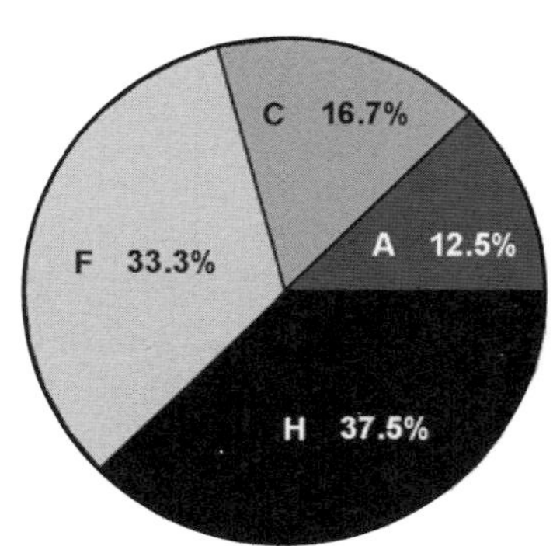

2.11. Histogram:

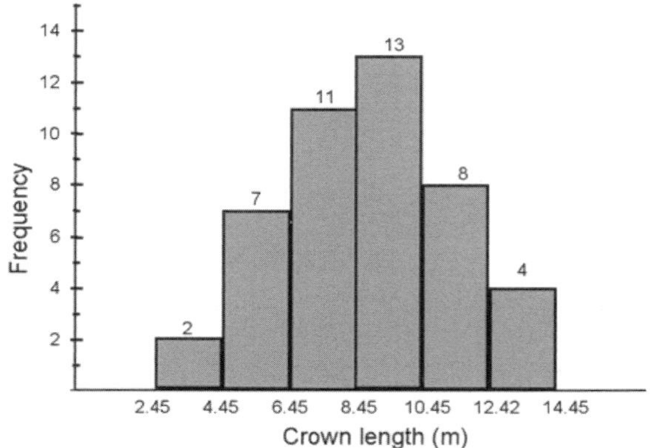

Frequency polygon:

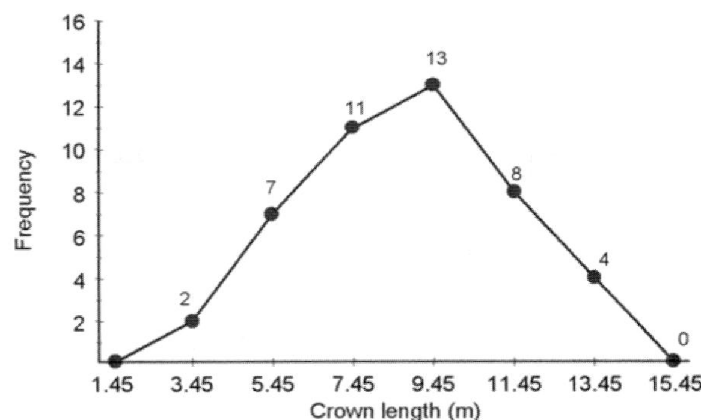

2.13.

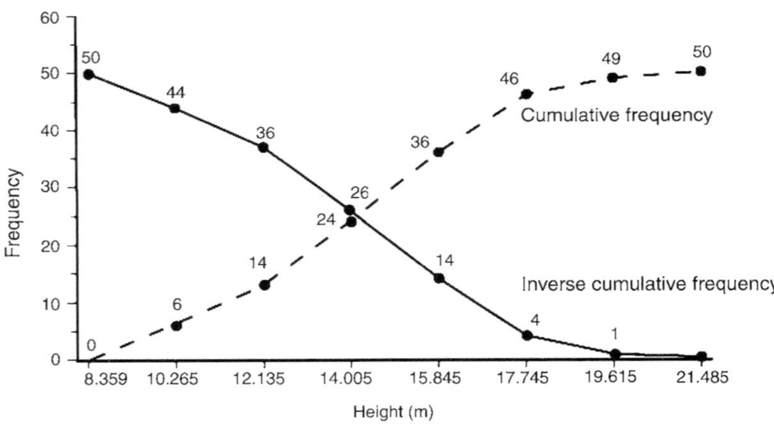

2.15. $\bar{x} = 0.4893$; $\tilde{x} = 0.484$; mode = does not exist.

2.17. $\bar{x} = 1.9$; $\tilde{x} = 1.5$; mode = 1.

2.19. a. $\tilde{x} = 14.95$; mode = 14.1.
b. $\bar{x}_w = 15.80$.

2.21. $\bar{x}$ (from raw data) = 3.857; $\bar{x}$ (from grouped data) = 3.892.
The means are different because the mean from the raw data uses all of the data points, while the mean from the grouped data uses the midpoints of each class as a representation of the points within the classes.

2.23. $s^2 = 2.4105$; $s = 1.5526$.

2.25. $s^2 = 11.3636$; $s = 3.3710$.

2.27. 2.15. CV = 25.34;

2.16. CV = −361.68.
The CVs indicate the relative measure of the spread. One is positive and the other is negative because they carry the sign of the mean.

2.29. a. $z(4.5) = 0.756$; $z(3.7) = -0.185$.
Standard scores indicate how many standard deviations an observation is above or below the mean.
b. 4.45.
c. $P(4.5) = 77.5\%$; $P(3.7) = 50.8\%$.

Chapter 3

3.1. a. S = {1H, 1T, 2H, 2T, 3H, 3T, 4H, 4T, 5H, 5T, 6H, 6T}
number of outcomes = 12.
b. A = {2H, 2T, 4H, 4T, 6H, 6T}.
c. B = {1H, 2H, 3H, 4H, 5H, 6H}.
d.

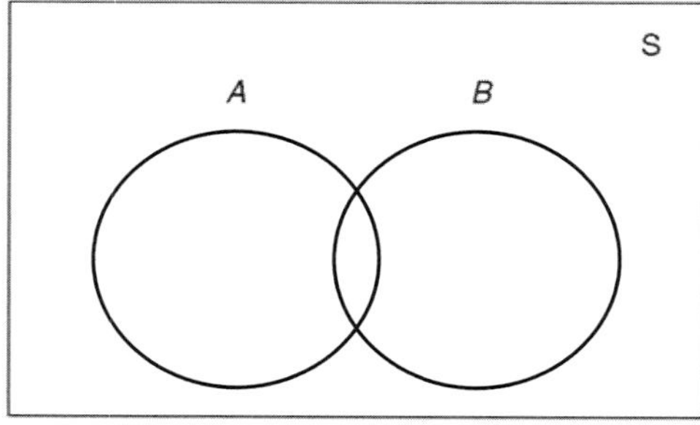

e. $A \cap B$ = {2H, 4H, 6H}.
f. $A \cup B$ = {1H, 2H, 2T, 3H, 4H, 4T, 5H, 6H, 6T}.
g. A' = {1H, 1T, 3H, 3T, 5H, 5T}.
h. B' = {1T, 2T, 3T, 4T, 5T, 6T}.
i. Less than 2 on the die and T on the coin, C = {1T}.
j. Less than 2 on the die, D = {1H, 1T}.

3.3. a. {cedar}.
b. {Douglas-fir, hemlock, cedar, spruce}.
c. {spruce}.
d. {cedar}.
e. {spruce}.

3.5. a.

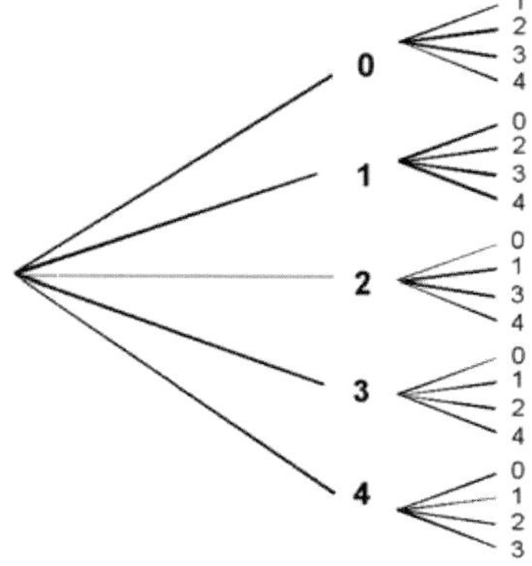

$n = 20$

b. $n = 5 \times 4 = 20.$

c. $n = {}_5P_2 = \dfrac{5!}{(5-2)!} = 20.$

3.7. a.

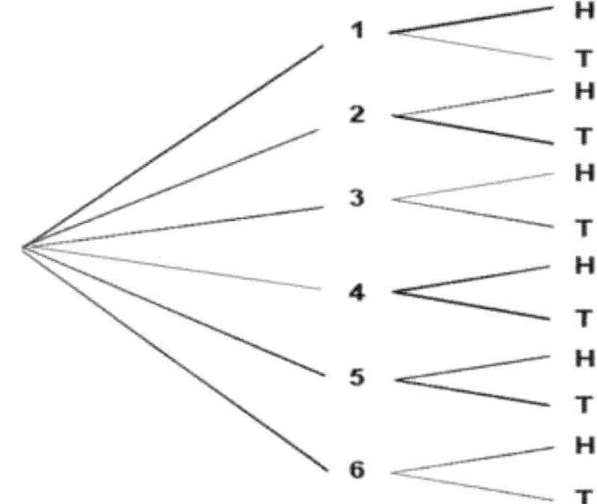

$n = 12$

b. $n = 6 \times 2 = 12.$

3.9. 216.

3.11. 4,000,000.

3.13. a. 100.
b. 52.
c. 32.

3.15. 35.

3.17. 277,200.

3.19. a. 2,598,960.
b. 1584.
c. 3744.

3.21. $P(A) = 0.5$ $P(B) = 0.5$

$P(A') = 0.5$ $P(B') = 0.5.$

3.23. a. 0.0526.
b. 0.4737.
c. 0.4737.
d. 0.4737.
e. 20:18 (or 10:9).

3.25. a. 3:2.
b. 2:3.

3.27. a. 0.6429.
b. 0.6429.

3.29. 0.015.

3.31. a. 0.20.
b. 0.80.
c. 0.4375.

3.33. a. $A \cap B = \{HH\}$ $P(A \cap B) = \frac{1}{4}$

Since $P(A)P(B) = \frac{1}{2} \times \frac{1}{2} = \frac{1}{4}$, A and B are independent.

b. $A \cap C = \{HH\}$ $P(A \cap C) = \frac{1}{4}$

Since $P(A)P(C) = \frac{1}{2} \times \frac{1}{2} = \frac{1}{4}$, A and C are independent.

c. $A \cap B \cap C = \{HH\}$ $P(A \cap B \cap C) = \frac{1}{4}$

Since $P(A)P(B)P(C) = \frac{1}{2} \times \frac{1}{2} \times \frac{1}{2} = \frac{1}{8} \neq \frac{1}{4}$, the three events are not independent.

3.35 a. $P(P) = \frac{80}{220}$ $P(I) = \frac{40}{220}$ $P(P \cap I) = \frac{10}{220}$.

Since $P(P)\,P(I) = \frac{80}{220} \times \frac{40}{220} \neq P(P \cap I)$, P and I are independent.

b. $P(F) = \frac{70}{220}$ $P(N) = \frac{180}{220}$ $P(F \cap N) = \frac{52}{220.}$

Since $P(F)\,P(N) = \frac{70}{220} \times \frac{180}{220} \neq P(F \cap N)$, F and N are independent.

3.37. a. 0.000125.
b. 0.85375.
c. 0.007125.

3.39. a. 0.14.
b. 0.20.
c. 0.060.
d. 0.38.
e. 0.4286.
f. 0.0893.
g. 0.4821.

Chapter 4

4.1.
1. Roll a die and record the number of dots.
2. Flip a coin and record the number of tails.
3. Deal 5 cards and record the number of clubs.
4. Measure one of the strength properties of a piece of lumber.
5. Volume of trees.

4.3.
1. Height of trees.
2. Specific gravity of Douglas-fir specimens.
3. Weight of students.

4.5 a.

x	2	3	4	5	6	7	8	9	10	11	12
$P(X=x)$	$\frac{1}{36}$	$\frac{2}{36}$	$\frac{3}{36}$	$\frac{4}{36}$	$\frac{5}{36}$	$\frac{6}{36}$	$\frac{5}{36}$	$\frac{4}{36}$	$\frac{3}{36}$	$\frac{2}{36}$	$\frac{1}{36}$

b.

x	−5	−4	−3	−2	−1	0	1	2	3	4	5
$P(X=x)$	$\frac{1}{36}$	$\frac{2}{36}$	$\frac{3}{36}$	$\frac{4}{36}$	$\frac{5}{36}$	$\frac{6}{36}$	$\frac{5}{36}$	$\frac{4}{36}$	$\frac{3}{36}$	$\frac{2}{36}$	$\frac{1}{36}$

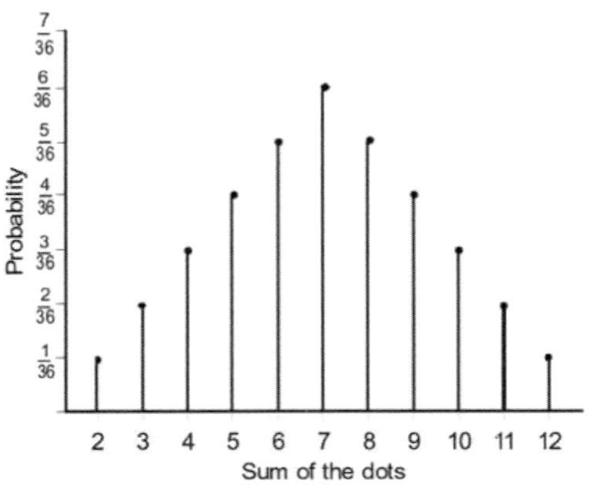

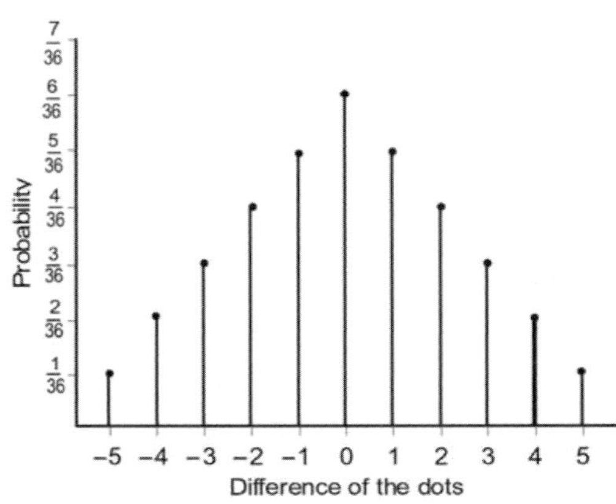

4.7 a.

x	0	1	2	3	4
$P(X=x)$	0.6588	0.2995	0.0399	0.0017	0.00002

a. 0.9983.
b. 0.6588.

4.9 a.

x	1	2	3	4	5	6	7	8	9	10
$P(X=x)$	0.5	0.25	0.125	0.0625	0.0313	0.0156	0.0078	0.0039	0.0020	0.0010

a. 0.5.
b. 0.9375.

4.11 a.

x	0	1	2
$P(X=x)$	0.10	0.60	0.30

$$P(X=x)=\frac{{}_2C_x\,{}_3C_{3-x}}{{}_5C_3}$$

a. 0.3.
b. 1.0.

4.13. a.

		X = Douglas-fir			
		0	1	2	3	$h(y)$
Y = pine	0	×	$\frac{3}{70}$	$\frac{9}{70}$	$\frac{3}{70}$	$\frac{15}{70}$
	1	$\frac{2}{70}$	$\frac{18}{70}$	$\frac{18}{70}$	$\frac{2}{70}$	$\frac{40}{70}$
	2	$\frac{3}{70}$	$\frac{9}{70}$	$\frac{3}{70}$	×	$\frac{15}{70}$
	$g(x)$	$\frac{5}{70}$	$\frac{30}{70}$	$\frac{30}{70}$	$\frac{5}{70}$	$\frac{70}{70}$

$$P(X = x; Y = y) = \frac{{}_3C_x\ {}_2C_y\ {}_3C_{(4-x-y)}}{{}_8C_4}$$

b. 0.5429.
c. 0.0714.
d. 0.
e.

x	0	1	2	3
$g(x)$	$\frac{5}{70}$	$\frac{30}{70}$	$\frac{30}{70}$	$\frac{5}{70}$

f.

y	0	1	2
$g(y)$	$\frac{15}{70}$	$\frac{40}{70}$	$\frac{15}{70}$

4.15. $\mu_x = 1.5$
$\mu_y = 1.0.$

4.17. US$880 (gain).

4.19. US$10,200.

4.21. a. US$2.00.
b. US$2.30.

4.23. 4.16. $\sigma^2 \approx 8.75$; $\sigma \approx 2.96$.
4.17. $\sigma^2 = 135{,}705{,}600$; $\sigma \approx 11{,}649.3$.
4.18. $\sigma^2 \approx 167{,}729{,}600$; $\sigma \approx 12{,}951.0$.

4.25. 4.20. $\sigma^2 \approx 424$; $\sigma \approx 20.59$.
4.21. $\sigma^2 \approx 27.78$; $\sigma \approx 5.27$.

Chapter 5

5.1. $f(X; 20) = \frac{1}{20}, \qquad X = 1,2\ldots 20$

a. 0.15.
b. 0.25.

5.3. a. 0.5033.
b. 0.2936.
c. 0.9896.

5.5. $\mu = 1.6$; $\sigma^2 = 1.28$; $\sigma = 1.13$.
For $k = 1.5$, at least 56% of the observations are contained within the interval, –0.095 – 3.295.
For $k = 2$, at least 75% of the observations are contained within the interval, –0.66 – 3.86.

5.7. ≈ 0.0732.

5.9. 0.1133.

5.11. a. 0.6.
b. $\mu = 2.0$; $\sigma = 0.632$.
c. For $k = 1.5$, at least 56% of the observations are contained within the interval, 1.052 – 2.948.
For $k = 2$, at least 75% of the observations are contained within the interval, 0.736 – 3.264.

5.13. 0.1779.

5.15. a. 0.0656.
b. 0.02916.
c. 0.00486.

5.17. a. 0.0155.
b. 0.0456.
c. 0.0557.

5.19. a. 0.1743.
b. 0.6204.
c. 0.2053.
d. $\sigma = 1.766$.

Chapter 6

6.1. a. 0.3333.
b. 0.6667.
c. 0.3333.

6.3. 6.1. $\mu = 550$; $\sigma^2 = 7500.00$; $\sigma = 86.60$.
6.2. $\mu = 18.5$; $\sigma^2 = 4.083$; $\sigma = 2.021$.

6.5. a. i. 0.5507.
ii. 0.3012.
iii. 0.1418.
b. $\sigma = 2.5$.
It is preferable to use Chebyshev's Theorem because this distribution is not normal.

For $k = 1.5$, at least 56% of the observations are contained within the interval, –1.25 – 6.25.
For $k = 2$, at least 75% of the observations are contained within the interval, –2.50 – 7.50.

6.7. $P(X > 10) = 0.2494 \approx 0.25$.
Since 2/8 = 0.25, we can assume that two of the circuits will be working after 10 years.

6.9. a. i. 4.75%.
ii. 20.33%.
iii. 20.33%.
iv. 74.92%.
b. ≈ 24 students.

6.11. a. i. 0.1711.
ii. 0.9713.
iii. 0.0602.
b. i. 0.1736.
ii. 0.9706.
iii. 0.0612.

6.13. 8.832 years (≈ 8 years and 10 months).

6.15. a. i. 0.9783.
ii. 0.9664.
iii. ≈ 0.6128 (using interpolation).
iv. 0.7108.
b. 0.6826, which is in agreement with the Empirical Rule.

6.17. a. 0.0885.
b. 0.0040.
c. 0.9075.

Chapter 7

7.1. a. For example, starting in column 1, row 15; go row-wise (across) taking 3-digit random numbers:

071 048 081 105 070 085 127 075 046 026
011 010 115 043 127.

b. For example, starting in column 1, row 1; go column-wise (down) taking 3-digit random numbers. *Since 007 appears twice, it is ignored the second time.

104 094 103 071 023 010 070 024 007 053
005 007* 097 145 089 019.

7.3. For example, starting in column 3, row 6, take 4-digit numbers row-wise (across). The first digit indicates the day (0 is the 10th day) and the second 3-digit number will be the half a minute. *Numbers greater than 960 are ignored.

1–100	8–427	5–127	7–565	3–498
1–860	2–706	5–990*	6–551	5–053
2–191	6–818	2–544	3–944	2–880
9–956	2–729	0–556	4–206	9–994*
9–887	2–310	1–671	1–941	8–738
4–401	3–488			

7.5.

1,1	1,3	1,5	1,7	1,9
3,1	3,3	3,5	3,7	3,9
5,1	5,3	5,5	5,7	5,9
7,1	7,3	7,5	7,7	7,9
9,1	9,3	9,5	9,7	9,9

$\mu = 5.0$ $\mu_{\bar{x}} = 5.0$ The means are the same.

$\sigma^2 = 8.0$ $\sigma = 2.83$

$\sigma_{\bar{x}}^2 = 4.0$ $\sigma_{\bar{x}}^2 = 2.0.$

The standard deviation of all possible sample means is smaller.

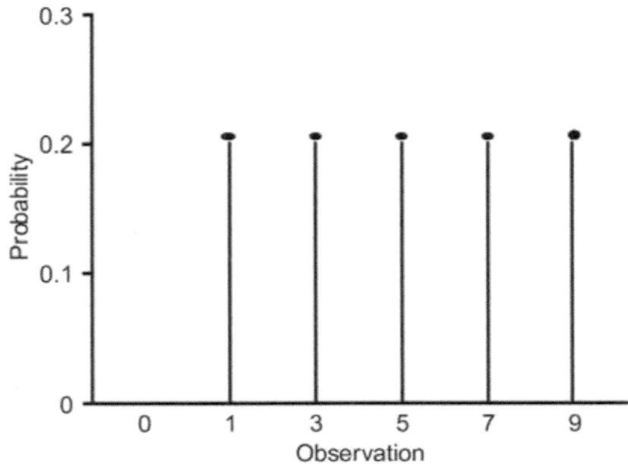

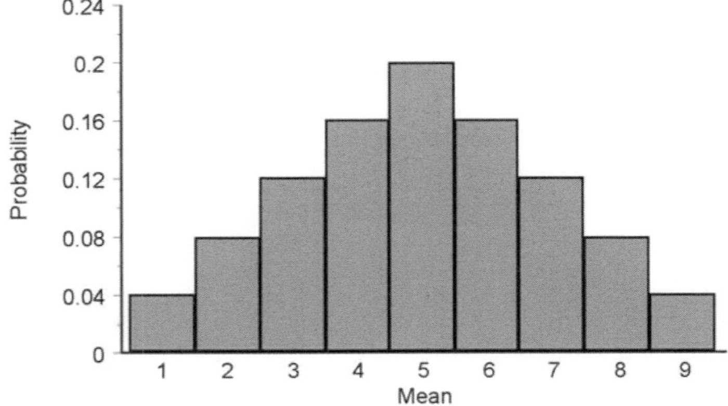

for $n = 3$ μ_x would be the same, 5.0, and $\sigma_{\bar{x}} = \sqrt{\frac{8}{3}} \approx 1.63.$

7.7. a. fpc does not apply.
b. fpc applies.
c. fpc does not apply.
d. fpc does not apply.
e. fpc does not apply.

7.9. 5 or 6 (since n = 5.20).

7.11. ≈ 0.025.

7.13. a. 0.1867.
b. 0.0384.
c. 0.7749.
0.05646, which represents the spread of all possible proportions based on n = 40 from this population.

7.15. a. 0.0197.
b. 0.0708.
c. 0.8106.
3403.4, which represents the spread of all possible differences between the two sample means, $\bar{x}_1$ and $\bar{x}_2$, based on samples of n_1 and n_2 taken, respectively, from the two populations studied.

7.17. a. ≈ 0.05.
b. ≈ 0.025.
8.49, which represents the spread of all possible differences between the two sample means, $\bar{x}_1$ and $\bar{x}_2$, based on samples of n_1 and n_2 taken, respectively, from the two populations studied.

7.19. a. ≈ 0.10.
b. ≈ 0.025.
28.60, which represents the spread of all possible $\bar{d}$ values calculated from these two populations based on 10 pairs of observations.

7.21. a. 0.7019.
b. 0.1446.
0.0943, which represents the spread of all possible $\hat{p}_1 - \hat{p}_2$ values taken from these two populations based on sample sizes of $n_1 = 40$ and $n_2 = 50$.

7.23. a. ≈ 0.14.
b. ≈ 0.09.
c. ≈ 0.81.

7.25. a. ≈ 0.10.
b. ≈ 0.20.
c. i. $P(0.353 < \frac{s_1^2}{s_2^2} < 4.16) = 0.95$.

ii. $P(0.269 < \frac{s_1^2}{s_2^2} < 5.19) = 0.98$.

d. i. 0.05.
ii. ≈ 0.009.

Chapter 8

8.1. Since the mean of all possible means is $\mu_x = \frac{\sum_{i=1}^{25} \bar{x}_i}{25} = 5.0$, which is equal to μ, it is unbiased. $E(\bar{x}) = \mu$.

8.3. Since the mean of all possible s'^2 is 4.0 instead of 8.0, it is biased. It can be shown mathematically that the bias is $\frac{\sigma^2}{n}$. $E(s'^2) \neq \sigma^2$.

8.5.
a. $P(957.36 < \mu < 992.64) = 0.95$.
$P(951.78 < \mu < 998.22) = 0.99$.
The 99% confidence interval is wider.
b. 95% $e = 17.64$.
99% $e = 23.22$.
The 99% margin of error is larger.
c. ≈ 35.
d. $P(962.52 < \mu < 987.47) = 0.95$.
As n increases, the width of the confidence interval decreases.

8.7.
a. $P(26.77 < \mu < 35.17) = 0.95$.
b. 21 (by iteration).
c. 32.
More samples (32 *versus* 21) are required for a higher level of confidence.

8.9.
a. $P(0.670 < p < 0.810) = 0.90$.
$P(0.656 < p < 0.824) = 0.95$.
For the higher level of confidence, the interval is wider.
b. ≈ 462.

8.11.
a. $P(-3.17 < \mu_1 - \mu_2 < -1.23) = 0.95$.
$P(-3.53 < \mu_1 - \mu_2 < -0.87) = 0.99$.
For a higher level of confidence, the interval is wider.
b. Since zero is not included in the two intervals and they are negative, it can be assumed that $\mu_2 > \mu_1$.
c. Both n_1 and $n_2 \approx 9$.

8.13. Using $v = 24$: $P(12.76 < \mu_1 - \mu_2 < 46.04) = 0.95$
$P(6.78 < \mu_1 - \mu_2 < 52.02) = 0.99$.
Using $v = 14$: $P(12.11 < \mu_1 - \mu_2 < 46.69) = 0.95$
$P(5.32 < \mu_1 - \mu_2 < 53.48) = 0.99$.
Since zero is not included in any of the intervals above, it can be concluded that controlled grazing produces higher weight gain than continuous grazing. Note also that the confidence intervals are somewhat wider for 14 degrees of freedom than for 24 degrees of freedom.

8.15. $P(-0.267 < p_1 - p_2 < 0.107) = 0.95$.
$P(-0.326 < p_1 - p_2 < 0.166) = 0.99$.
Since zero is included in the two intervals, it can be assumed that the two unknown population proportions are the same.

8.17. $P(22.01 < \sigma^2 < 126.18) = 0.95.$
$P(17.99 < \sigma^2 < 185.39) = 0.99.$
For a higher level of confidence, the interval is wider.

8.19. $P(0.0135 < \sigma^2 < 0.0953) = 0.95.$
The probability is 0.95 that the unknown population variance is within the interval stated above.

8.21.
$$P\left(1.29 < \frac{\sigma_1^2}{\sigma_2^2} < 7.70\right) = 0.90$$
$$P\left(1.08 < \frac{\sigma_1^2}{\sigma_2^2} < 9.19\right) = 0.95$$

The interval is wider for a higher level of confidence.
Since one is not included in the two intervals, the assumption of unequal variances in Exercise 8.13 was justified.

Chapter 9

9.1.

a. $\bar{x}_c = 0.479.$
The critical region is above this point at $\bar{x}_c > 0.479.$

b. For $\mu_1 = 0.49$: type II error = 0.2514.
For $\mu_1 = 0.53$: type II error $\approx$ 0.0009.

c. $\bar{x}_c = 0.472.$
The critical region is above this point at $\bar{x}_c > 0.472.$
For $\mu_1 = 0.49$: type II error = 0.0778.
For $\mu_1 = 0.53$: type II error $\approx$ 0.0000.

9.3.

a. H_0: $p = 0.8.$
H_1: $p < 0.8.$
The critical region is on the left side of a distribution centred at 0.8.

b. H_0: $\mu = 25.2.$
H_1: $\mu \neq 25.2.$
The critical regions are on both ends of a distribution centred at 25.2.

c. H_0: $p = 0.1.$
H_1: $p > 0.1.$
The critical region is on the right side of a distribution centred at 0.1.

d. H_0: $p = 0.07.$
H_1: $p > 0.07.$
The critical region is on the right side of a distribution centred at 0.07.

e. H_0: $\mu = 18.0.$
H_1: $\mu \neq 18.0.$
The critical regions are on both ends of a distribution centred at 18.

9.5. H_0: $\mu = 150$ $\quad H_1$: $\mu \neq 150$ $\quad \alpha = 0.05.$
$z_{0.05} = \pm 1.96$ $\quad z = -1.19.$
Since $-1.96 < -1.19 < 1.96$, 'accept' H_0.
The unknown population mean is not significantly different from the set 150 ml.

9.7. $H_0: \mu = 600 \quad H_1: \mu < 600 \quad \alpha = 0.01.$
$t_{(14)0.01} = -2.62 \quad t_{(14)} = -3.35.$
Since $-3.35 < -2.62$, reject H_0.
The unknown population mean of lumber prices is significantly less than US$600.

9.9. $H_0: \mu = 32.0 \quad H_1: \mu \neq 32.0 \quad \alpha = 0.05.$
$t_{(11)0.025} = \pm 2.20 \quad t_{(11)} = -0.54.$
Since $-2.20 < -0.54 < 2.20$, 'accept' H_0.
The unknown population mean is not significantly different from 32.0 cm. The 95% confidence interval calculated in Exercise 8.7 is equivalent to the results obtained here. The confidence limits are the critical values expressed in centimetres.

9.11. $H_0: p = 0.20 \quad H_1: p > 0.20 \quad \alpha = 0.05.$
$z_{0.05} = 1.645 \quad z = 0.73.$
Since $0.73 < 1.645$, 'accept' H_0.
The manufacturer's claim can be accepted. The unknown population proportion is not significantly different from 0.20.

9.13. $H_0: \sigma^2 = 1400 \quad H_1: \sigma^2 \neq 1400 \quad \alpha = 0.01.$
$\chi^2_{(14)0.005} = 4.07 \quad \chi^2_{(14)} = 16.56.$
$\chi^2_{(14)0.995} = 31.3.$
Since $4.07 < 16.56 < 31.3$, 'accept' H_0.
The variance of lumber prices is not significantly different from 1400.

9.15. $H_0: \sigma^2 = 0.01 \quad H_1: \sigma^2 > 0.01 \quad \alpha = 0.05.$
$\chi^2_{(9)0.95} = 16.9 \quad \chi^2_{(9)} = 25.74.$
Since $25.74 > 16.9$, reject H_0.
The unknown standard deviation of the weights for the rainbow trout population is significantly greater than 0.1 kg. These results are consistent with the results in Exercise 8.19.

9.17. $H_0: \mu_1 - \mu_2 = 0 \quad H_1: \mu_1 - \mu_2 \neq 0 \quad \alpha = 0.01.$
$z_{0.005} = \pm 2.58 \quad z = 2.17.$
Since $-2.58 < 2.17 < 2.58$, 'accept' H_0.
It is reasonable to assume that the two unknown population means are equal. The 99% confidence interval calculated in Exercise 8.10 is equivalent to this test. The confidence limits (–1381.3 and 16,007.3) are the critical values expressed in kilometres.

9.19. $H_0: \mu_1 - \mu_2 = 0 \quad H_1: \mu_1 - \mu_2 \neq 0 \quad \alpha = 0.05$ case 1.
$t_{(\infty)0.025} = z_{0.025} = \pm 1.96 \quad z = -3.98.$
Since $-3.98 < -1.96$, reject H_0.
The two unknown population means of the tensile strengths of the two commercial fishing lines are not equal. Since zero is not included in the 95% confidence interval (Exercise 8.12), this indicates the same thing. The 95% confidence interval is equivalent to the above two-tailed test.

9.21. $H_0: \mu_1 - \mu_2 = 0$ $H_1: \mu_1 - \mu_2 \neq 0$ $\alpha = 0.01$ case 2.
$t_{(17)0.005} = \pm 2.90$ $t_{(17)} = -4.80$.
Since $-4.80 < -2.90$, reject H_0.
The two unknown population means are significantly different. Since it is a two-tailed test, it is equivalent to the 99% confidence interval used in Exercise 8.11a.

9.23. $H_0: \mu_1 - \mu_2 = 0$ $H_1: \mu_1 - \mu_2 \neq 0$ $\alpha = 0.01$ case 3.
$t_{(14)0.005} = \pm 2.98$ $t_{(v)} = 3.64$.
$t_{(24)0.005} = \pm 2.80$.
Since $3.64 > 2.98$ (or 2.80), reject H_0.
The two unknown population means are significantly different. Since this is a two-tailed test, it is equivalent to the 99% confidence interval calculated in Exercise 8.13 (confidence limits are the same as the critical values).

9.25. $H_0: \mu_1 - \mu_2 = 0$ $H_1: \mu_1 - \mu_2 \neq 0$ $\alpha = 0.05$.
$t_{(7)0.025} = \pm 2.36$ $t_{(7)} = 15.89$.
Since $15.89 > 2.36$, reject H_0.
The unknown population means of the 2-year increments are significantly different. Since the above test is a two-tailed test, it is equivalent to the 95% confidence interval calculated in Exercise 8.14.

9.27. $H_0: p_1 - p_2 = 0$ $H_1: p_1 - p_2 < 0$ $\alpha = 0.05$.
$z_{0.05} = -1.645$ $z = -0.821$.
Since $-0.821 > -1.645$, 'accept' H_0.
The effect of soil preparation on the rate of regeneration for Area II is not superior to that of Area I. Since the above is a one-tailed test, it is not comparable to the confidence interval calculated in Exercise 8.14.

9.29. $H_o: \dfrac{\sigma_2^2}{\sigma_1^2} = 5.0$ $H_1: \dfrac{\sigma_2^2}{\sigma_1^2} < 5.0$ $\alpha = 0.01$.

$F_{(6,4)0.01} = 15.21$ $F_{(6,4)} = 2.38$.
Since $2.38 < 15.21$, 'accept' H_0.
The unknown variation of weight gain in Diet 2 is at most 5 times the variation of Diet 1.

9.31. 9.22.

$H_o: \dfrac{\sigma_2^2}{\sigma_1^2} = 1.0$ $H_1: \dfrac{\sigma_2^2}{\sigma_1^2} \neq 1.0$ $\alpha = 0.10$.

$F_{(6,5)0.05} = 4.95$ $F_{(6,5)} = 0.866$.
$F_{(6,5)0.95} = 0.22$.
Since $0.22 < 0.866 < 4.95$, 'accept' H_0.
The two unknown population variances are not significantly different.

9.24.

$H_o: \dfrac{\sigma_2^2}{\sigma_1^2} = 1.0$ $H_1: \dfrac{\sigma_2^2}{\sigma_1^2} \neq 1.0$ $\alpha = 0.05$.

$F_{(4,6)0.025} = 6.23 \qquad F_{(4,6)} = 0.0839.$
$F_{(4,6)0.975} = 0.109.$
Since $0.0839 < 0.109$, reject H_0.
The two unknown population variances are significantly different.

Chapter 10

10.1. H_0: $O_i = E_i$ for all *i*s *or* the distribution is uniform.
H_1: $O_i \neq E_i$ for at least one *i* *or* the distribution is not uniform.
$\alpha = 0.01$.
$\chi^2_{(9)0.01} = 21.7 \qquad \chi^2_{(9)} = 6.60.$
Since $6.60 < 21.7$, 'accept' H_0.
It can be assumed that the distribution of random numbers is uniform.

10.3. a. A binomial distribution is suggested by the nature of the observations.
b. H_0: $O_i = E_i$ for all *i*s *or* the distribution is binomial.
H_1: $O_i \neq E_i$ for at least one *i* *or* the distribution is not binomial.
$\alpha = 0.01$.
$\chi^2_{(3)0.01} = 11.3 \qquad \chi^2_{(3)} = 0.94.$
Since $0.92 < 11.3$, 'accept' H_0.
The assumption of a binomial distribution is reasonable.

10.5. H_0: $O_{ij} = E_{ij}$ for all *i*s and *j*s *or* the frequency (distribution) of species is independent of type.
H_1: $O_{ij} \neq E_{ij}$ for at least one pair of *i* and *j* *or* the frequency of species is type dependent.
$\alpha = 0.05$.
$\chi^2_{(6)0.05} = 12.6 \qquad \chi^2_{(6)} = 5.27.$
Since $5.27 < 12.6$, 'accept' H_0.
The distribution of the frequency of species is independent of type.

10.7. H_0: $O_{ij} = E_{ij}$ for all *i*s and *j*s *or* the distribution of accidents is plant independent.
H_1: $O_{ij} \neq E_{ij}$ for at least one pair of *i* and *j* *or* the distribution of accidents is plant dependent.
$\alpha = 0.05$.
$\chi^2_{(4)0.05} = 9.49 \qquad \chi^2_{(4)} = 1.27.$
Since $1.27 < 9.49$, 'accept' H_0.
The two distributions of the frequencies of accidents are the same.

10.9. H_0: $O_{ij} = E_{ij}$ for all *i*s and *j*s *or* the distribution of grades are the same for the three areas.
H_1: $O_{ij} \neq E_{ij}$ for at least one pair of *i* and *j* *or* the distribution of grades are not the same for the three areas.
$\alpha = 0.01$.
$\chi^2_{(6)0.01} = 16.8 \qquad \chi^2_{(6)} = 2.96.$
Since $2.96 < 16.8$, 'accept' H_0.
The distributions are area independent; that is, the distributions are the same from area to area.

Chapter 11

11.1. a.

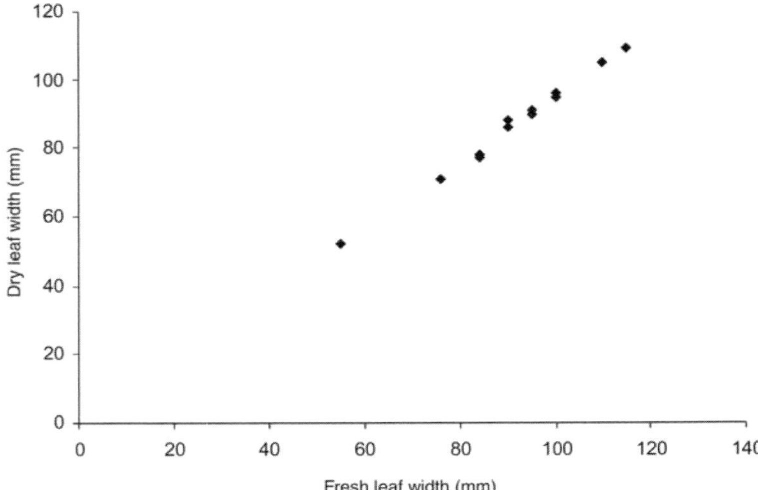

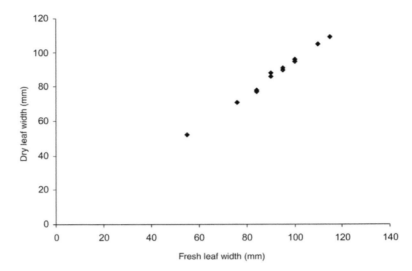

$\hat{y}_i = -2.482 + 0.976x_i$.

b. $H_0: \beta_1 = 0$ or $\dfrac{\sigma^2_{Reg}}{\sigma^2_{Res}} = 1.0$ $\quad H_1: \beta_1 \neq 0$ or $\dfrac{\sigma^2_{Reg}}{\sigma^2_{Res}} \neq 1.0$.

$\alpha = 0.01$.
$F_{(1,10)0.01} = 10.04 \quad F_{(1,10)} = 1421.13$.
Since $1421.13 > 10.04$, reject H_0.
The regression is significant.

c. $r^2 = 0.993$.
99.3% of the variation of the dry leaf width is explained by the independent variable, fresh leaf width.

d. $S_{y \cdot x} = 1.36$.
1.36 is the measure of the spread of the dry leaf width around the regression line.

e. H_0: $\beta_0 = 0 \quad H_1$: $\beta_0 \neq 0$.
$\alpha = 0.05$.
$t_{(10)0.025} = \pm 2.23 \quad t_{(10)} = -1.04$.
Since $-2.23 < -1.04 < 2.23$, 'accept' H_0.
The intercept is not significantly different from zero.

f. H_0: $\beta_1 = 1.0 \quad H_1$: $\beta_1 \neq 1.0$.
$\alpha = 0.05$.
$t_{(10)0.025} = \pm 2.23 \quad t_{(10)} = -0.92$.
Since $-2.23 < -0.92 < 2.23$, 'accept' H_0.
The slope is not significantly different from 1.0. The slope indicates the change in (y) dry leaf width for every unit change of (x), the fresh leaf weight.

g. $P(94.11 < \mu_{y_{100}} < 96.13) = 0.95$.

11.3. a.

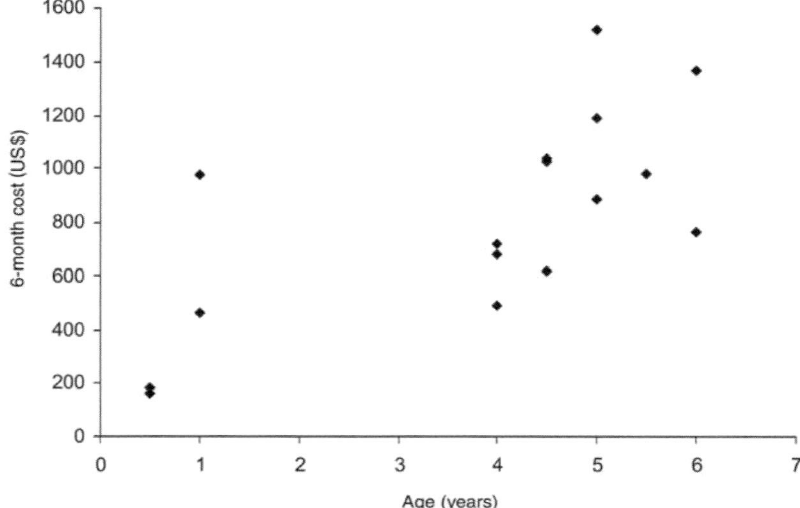

$\hat{y}_i = 300.00 + 136.27x_i$.

b. $H_0: \beta_1 = 0$ or $\dfrac{\sigma^2_{Reg}}{\sigma^2_{Res}} = 1.0$ $\quad H_1: \beta_1 \neq 0$ or $\dfrac{\sigma^2_{Reg}}{\sigma^2_{Res}} \neq 1.0$.

$\alpha = 0.01$.
$F_{(1,14)0.01} = 8.86$ $\quad F_{(1,14)} = 12.20$.
Since $12.20 > 8.86$, reject H_0.
The regression is significant.

c. $r^2 = 0.466$.
46.6% of the variation of the cost in maintenance can be explained by the age of CNC moulders.

d. $S_{y \cdot x} = 291.77$.
291.77 is the measure of the spread of 6-month maintenance costs around the regression line.

e. H_0: $\beta_0 = 0$ $\quad H_1$: $\beta_0 > 0$.
$\alpha = 0.05$.
$t_{(14)0.05} = 1.76$ $\quad t_{(14)} = 1.81$.
Since $1.81 > 1.76$, reject H_0.
The intercept is not significantly greater than zero. The intercept in this example is the cost of maintaining the CNC moulders when they are zero years old (brand new).

f. H_0: $\beta_1 = 240$ $\quad H_1$: $\beta_1 \neq 240$.
$\alpha = 0.05$.
$t_{(14)0.025} = \pm 2.14$ $\quad t_{(14)} = -2.66$.
Since $-2.66 < -2.14$, reject H_0.
The slope is significantly different (lower) than 240. For one unit change of age, the maintenance costs change by US$136.27.

g. i. $P(375.28 < \mu_{y_{2.5}}$ 906.08) = 0.99.
ii. $P(765.50 < \mu_{y_{5.5}} < 1342.48) = 0.99$.
The width of the intervals vary for different values of x because of the 'confidence belt' effect (note that the interval would be at a minimum at $\bar{x}$).

h. $H_0: \frac{\sigma^2_{Lf}}{\sigma^2_{Pe}} = 1.0 \qquad H_1: \frac{\sigma^2_{Lf}}{\sigma^2_{Pe}} > 1.0.$
$\alpha = 0.05$.
$F_{(5,9)0.01} = 3.48 \qquad F_{(5,9)} = 1.44$.
Since 1.44 < 3.48, 'accept' H_0.
There is no lack of fit.

11.5. a. $r \approx 0.976$.
b. $H_0: \rho = 0 \qquad H_1: \rho \neq 0$.
$\alpha = 0.05$.
$t_{(10)0.025} = \pm 2.23 \qquad t_{(10)} = 14.17$.
Since 14.17 > 2.23, reject H_0.
The correlation is significant.

11.7. a. $\bar{y}_i = -0.4914 - 0.3454x_1 + 0.0984x_2$.
b. $H_0: \beta_1 = \beta_2 = 0 \qquad H_1$: at least one is different.
$\alpha = 0.05$.
$F_{(2,7)0.05} = 4.74 \qquad F_{(2,7)} = 20.98$.
Since 20.98 > 4.74, reject H_0.
The multiple regression is significant.
c. $R^2 = 0.857$.
85.7% of the variation of tree volumes can be explained by crown diameter and tree height.
d. $S_{y.x_1,x_2} = 0.2777$.
0.2777 is the measure of spread of volumes around the regression surface.

Chapter 12

12.1. a. $H_0: \mu_{1.} = \mu_{2.} = \mu_{3.} \qquad H_1$: at least one is different.
$\alpha = 0.05$.

Source	*DF*	*SS*	*MS*	*F*	$F_{(2,9)0.05}$
Group-to-group	2	0.068600	0.034300	22.33	4.26
Within group	9	0.013825	0.001536		
Total	11	0.082425			

Since 22.33 > 4.26, reject H_0.
At least one of the three means is different.
b. $S_{\bar{y}_{i.}} \approx 0.01960$.
0.01960 is the measure of the spread of all possible means, based on four observations from any of the three populations described above.

c. $S_{\bar{y}_{p.} - \bar{y}_{r.}} \approx 0.02771$.
0.02771 is the measure of the spread of differences of any two means based on four observations taken from the above populations.

12.3. a. H_0: $\mu_{A.} = \mu_{B.} = \mu_{C.}$ H_1: at least one is different.
$\alpha = 0.05$.

Source	DF	SS	MS	F	$F_{(2,9)0.05}$
Group-to-group	2	57.50	28.750	7.99	4.26
Within group	9	32.39	3.599		
Total	11	89.89			

Since 7.99 > 4.26, reject H_0.
At least one of the three means is different.

b. $P(14.08 < \mu_{B.} < 19.60) = 0.99$.

c. $P(-7.60 < \mu_{A.} - \mu_{B.} < -1.34) = 0.95$.
Since zero is not included in the interval, it can be assumed that the two unknown population means are different.

12.5. $CD \approx 0.375$.

$4.78_{(a)}$ $6.04_{(b)}$ $6.08_{(bc)}$ $6.44_{(c)}$

Any two means underlined by the same line, or labelled with the same subscript, are not significantly different.

12.7. H_0: $\sigma_1^2 = \sigma_2^2 = \sigma_3^2$ H_1: at least one is different.
$\alpha = 0.05$.
$\chi^2_{(2)0.05} = 5.99$ $\chi^2_{(2)} = 1.91$.
Since 1.91 < 5.99, 'accept' H_0.
The assumption of equal variances in Exercise 12.1 was justified.

12.9. a. H_0:

1. Treatment: $\mu_{1.} = \mu_{2.} = \mu_{3.} = \mu_{4.}$
2. Species: $\mu_{.1} = \mu_{.2}$
3. Interaction: $\mu_{11.} - \mu_{12.} = \mu_{21.} - \mu_{22.} = \mu_{31.} - \mu_{32.} = \mu_{41.} - \mu_{42.}$
$\mu_{12.} - \mu_{13.} = \mu_{22.} - \mu_{23.} = \mu_{32.} - \mu_{33.} = \mu_{42.} - \mu_{43.}$
$\mu_{11.} - \mu_{13.} = \mu_{21.} - \mu_{23.} = \mu_{31.} - \mu_{33.} = \mu_{41.} - \mu_{43.}$

H_1:

1. At least two of the means are not equal.
2. At least two of the means are not equal.
3. At least two of the differences are not equal.

$\alpha = 0.05$.

Source	DF	SS	MS	F	Critical F
Nitrogen (N)	3	11.56	3.853	45.33	4.07
Species (S)	1	1.38	1.380	16.23	5.32
N × S	3	0.47	0.157	1.84	4.07
Error	8	0.68	0.085		
Total	15	14.09			

Since 1.84 < 4.07, there is no significant interaction;
16.23 > 5.32, the species means are different;
45.33 > 4.07, the nitrogen means are different.

b. $P(5.76 < \mu_{3..} < 6.44) = 0.95$.
c. $P(4.77 < \mu_{.2.} < 5.25) = 0.95$.
d. The confidence interval for the nitrogen means is wider than the interval for species means, because the means for nitrogen are based on fewer observations.

Chapter 14

14.1. $H_0: \tilde{\mu} = 0 \quad H_1: \tilde{\mu} \neq 0 \quad \alpha = 0.05$.
Number of '–' signs = 3 number of '+' signs = 4.
$P(X \leq 3) = 0.500$ *or* $P(X \geq 4) = 0.500$.
Since both probabilities exceed 0.025, 'accept' H_0.
It is reasonable to assume that the median temperature for May in Vancouver is 14°C.

14.3. $H_0: \mu_1 - \mu_2 = 0 \quad H_1: \mu_1 - \mu_2 > 0 \quad \alpha = 0.05$.
Number of '–' signs = 1.
$P(X \leq 1) = 0.035$.
Since this probability is less than 0.05, reject H_0.
The daily maintenance significantly reduced the number of defective parts produced.

14.5. $H_0: \tilde{\mu} = 0 \quad H_1: \tilde{\mu} \neq 0 \quad \alpha = 0.05$.
$w_+ = 18.5 \quad w_- = 9.5$.
$w \leq 2$.
Since 9.5 > 2, 'accept' H_0.
It is reasonable to assume that the median temperature for May in Vancouver is 14°C.

14.7. $H_0: \mu_1 - \mu_2 = 0 \quad H_1: \mu_1 - \mu_2 \neq 0 \quad \alpha = 0.01$.
$w_+ = 108.5 \quad w_- = 81.5$.
$z_{0.005} = \pm 2.58 \quad z = 0.54$.
Since $-2.58 < 0.54 < 2.58$, 'accept' H_0.
The conclusion is the same as it was in Exercise 14.4. The unknown population means of the leader growth can be assumed to be the same during the 2 years.

14.9. $H_0: \mu_1 - \mu_2 = 0 \quad H_1: \mu_1 - \mu_2 > 0 \quad \alpha = 0.05$.
Critical value (5,7) = 6.
$u_1 = 15.5 \quad u_2 = 19.5$.
Since 15.5 > 6, 'accept' H_0.
Diet 2 is not superior to Diet 1. This agrees with the results in Exercise 9.24.

14.11. $H_0: \mu_1 = \mu_2 = \mu_3 \quad H_1$: at least one is different.
$\alpha = 0.05$.
$\chi^2_{(2)0.05} = 5.99 \quad h = 15.16$.
Since 15.16 > 5.99, reject H_0.
At least one of the three unknown population means is different.

14.13. H_0: the sequence is random H_1: the sequence is not random.
$\alpha = 0.05$.
Number of '–' signs = 44 number of '+' signs = 46 $v = 52$.
$z_{0.025} = \pm 1.96$ $z = 1.28$.
Since $-1.96 < 1.28 < 1.96$, 'accept' H_0.
The 90 digits of π listed in Example 14.10 are random relative to their median of 4.5.

14.15. H_0: the sequence is random H_1: the sequence is not random.
$\alpha = 0.05$.
Number of '–' signs = 9 number of '+' signs = 9 $v = 11$.
Critical values: $a = 5$, $b = 13$.
Since $5 < 11 < 13$, 'accept' H_0.
The sequence is random.

14.17. $H_0: \rho_s = 0$ $H_1: \rho_s \neq 0$ $\alpha = 0.01$.
$z_{0.005} = \pm 2.58$ $z = 3.69$.
Since $3.69 > 2.58$, reject H_0.
There is a significant correlation between leader growths for the 2 years.

Chapter 15

15.1. $\bar{X}$ chart:

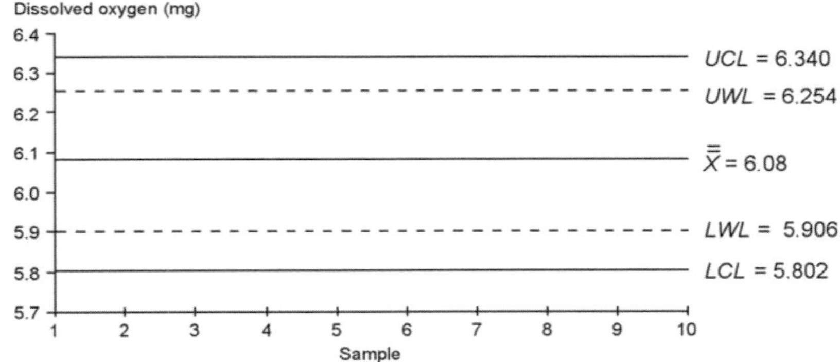

S chart:

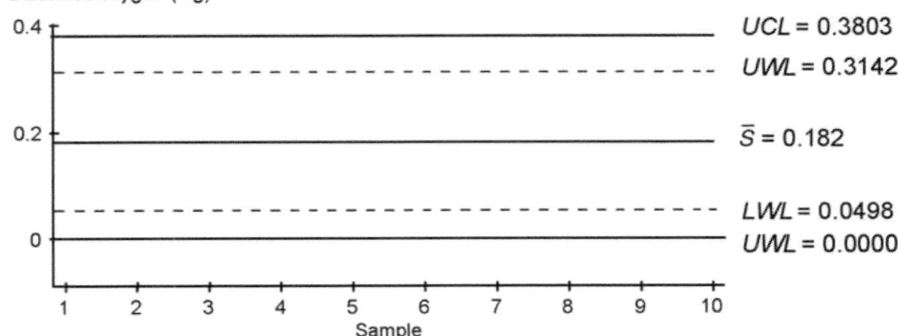

R chart:

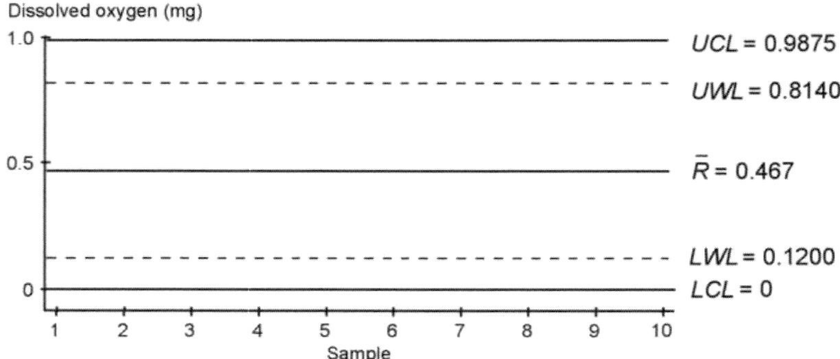

15.3. $\bar{X}$ chart:

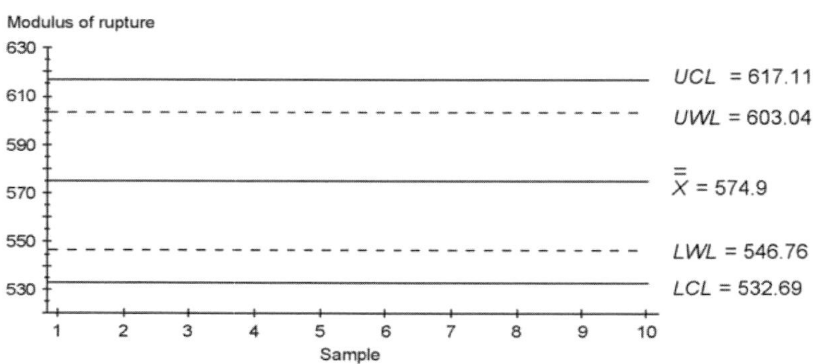

S chart:

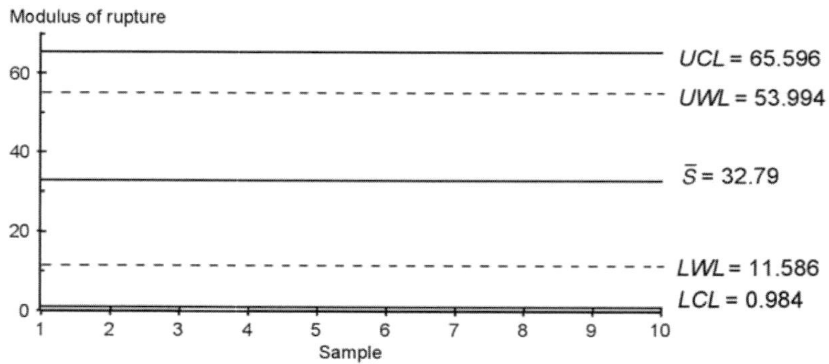

R chart:

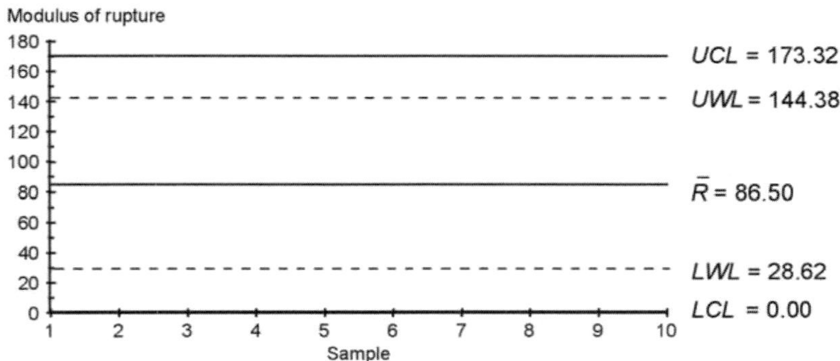

15.5. p chart:

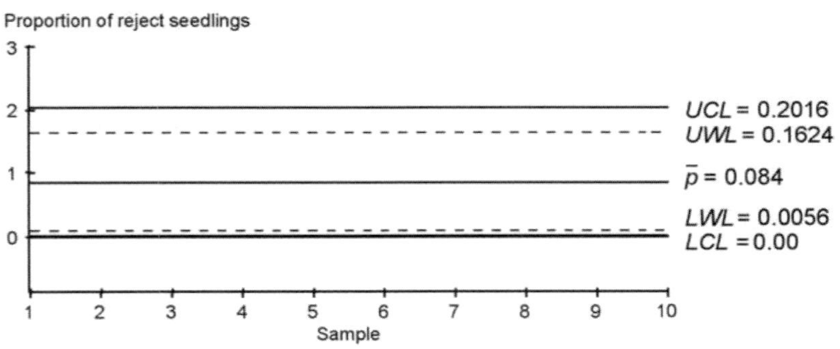

This page intentionally left blank

Appendix A

This page intentionally left blank

Table A.1 Binomial probabilities.

		p										
n	*x*	*0.05*	*0.1*	*0.2*	*0.3*	*0.4*	*0.5*	*0.6*	*0.7*	*0.8*	*0.9*	*0.95*
2	0	0.903	0.810	0.640	0.490	0.360	0.250	0.160	0.090	0.040	0.010	0.003
	1	0.095	0.180	0.320	0.420	0.480	0.500	0.480	0.420	0.320	0.180	0.095
	2	0.003	0.010	0.040	0.090	0.160	0.250	0.360	0.490	0.640	0.810	0.903
3	0	0.857	0.729	0.512	0.343	0.216	0.125	0.064	0.027	0.008	0.001	
	1	0.135	0.243	0.384	0.441	0.432	0.375	0.288	0.189	0.096	0.027	0.007
	2	0.007	0.027	0.096	0.189	0.288	0.375	0.432	0.441	0.384	0.243	0.135
	3		0.001	0.008	0.027	0.064	0.125	0.216	0.343	0.512	0.729	0.857
4	0	0.815	0.656	0.410	0.240	0.130	0.063	0.026	0.008	0.002		
	1	0.171	0.292	0.410	0.412	0.346	0.250	0.154	0.076	0.026	0.004	
	2	0.014	0.049	0.154	0.265	0.346	0.375	0.346	0.265	0.154	0.049	0.014
	3		0.004	0.026	0.076	0.154	0.250	0.346	0.412	0.410	0.292	0.171
	4			0.002	0.008	0.026	0.063	0.130	0.240	0.410	0.656	0.815
5	0	0.774	0.590	0.328	0.168	0.078	0.031	0.010	0.002			
	1	0.204	0.328	0.410	0.360	0.259	0.156	0.077	0.028	0.006		
	2	0.021	0.073	0.205	0.309	0.346	0.313	0.230	0.132	0.051	0.008	0.001
	3	0.001	0.008	0.051	0.132	0.230	0.313	0.346	0.309	0.205	0.073	0.021
	4			0.006	0.028	0.077	0.156	0.259	0.360	0.410	0.328	0.204
	5				0.002	0.010	0.031	0.078	0.168	0.328	0.590	0.774
6	0	0.735	0.531	0.262	0.118	0.047	0.016	0.004	0.001			
	1	0.232	0.354	0.393	0.303	0.187	0.094	0.037	0.010	0.002		
	2	0.031	0.098	0.246	0.324	0.311	0.234	0.138	0.060	0.015	0.001	
	3	0.002	0.015	0.082	0.185	0.276	0.313	0.276	0.185	0.082	0.015	0.002
	4		0.001	0.015	0.060	0.138	0.234	0.311	0.324	0.246	0.098	0.031
	5			0.002	0.010	0.037	0.094	0.187	0.303	0.393	0.354	0.232
	6				0.001	0.004	0.016	0.047	0.118	0.262	0.531	0.735
7	0	0.698	0.478	0.210	0.082	0.028	0.008	0.002				
	1	0.257	0.372	0.367	0.247	0.131	0.055	0.017	0.004			
	2	0.041	0.124	0.275	0.318	0.261	0.164	0.077	0.025	0.004		
	3	0.004	0.023	0.115	0.227	0.290	0.273	0.194	0.097	0.029	0.003	
	4		0.003	0.029	0.097	0.194	0.273	0.290	0.227	0.115	0.023	0.004
	5			0.004	0.025	0.077	0.164	0.261	0.318	0.275	0.124	0.041
	6				0.004	0.017	0.055	0.131	0.247	0.367	0.372	0.257
	7					0.002	0.008	0.028	0.082	0.210	0.478	0.698

Source: Freund, J.E. (1988) *Modern Elementary Statistics*, 7th edn. Prentice-Hall, Inc., Englewood Cliffs, New Jersey. Reproduced with permission.
Note: All values omitted in this table are 0.0005 or less.

Table A.1 Binomial probabilities – *continued*

n	x	p = 0.05	0.1	0.2	0.3	0.4	0.5	0.6	0.7	0.8	0.9	0.95
8	0	0.663	0.430	0.168	0.058	0.017	0.004	0.001				
	1	0.279	0.383	0.336	0.198	0.090	0.031	0.008	0.001			
	2	0.051	0.149	0.294	0.296	0.209	0.109	0.041	0.010	0.001		
	3	0.005	0.033	0.147	0.254	0.279	0.219	0.124	0.047	0.009		
	4		0.005	0.046	0.136	0.232	0.273	0.232	0.136	0.046	0.005	
	5			0.009	0.047	0.124	0.219	0.279	0.254	0.147	0.033	0.005
	6			0.001	0.010	0.041	0.109	0.209	0.296	0.294	0.149	0.051
	7				0.001	0.008	0.031	0.090	0.198	0.336	0.383	0.279
	8					0.001	0.004	0.017	0.058	0.168	0.430	0.663
9	0	0.630	0.387	0.134	0.040	0.010	0.002					
	1	0.299	0.387	0.302	0.156	0.060	0.018	0.004				
	2	0.063	0.172	0.302	0.267	0.161	0.070	0.021	0.004			
	3	0.008	0.045	0.176	0.267	0.251	0.164	0.074	0.021	0.003		
	4	0.001	0.007	0.066	0.172	0.251	0.246	0.167	0.074	0.017	0.001	
	5		0.001	0.017	0.074	0.167	0.246	0.251	0.172	0.066	0.007	0.001
	6			0.003	0.021	0.074	0.164	0.251	0.267	0.176	0.045	0.008
	7				0.004	0.021	0.070	0.161	0.267	0.302	0.172	0.063
	8					0.004	0.018	0.060	0.156	0.302	0.387	0.299
	9						0.002	0.010	0.040	0.134	0.387	0.630
10	0	0.599	0.349	0.107	0.028	0.006	0.001					
	1	0.315	0.387	0.268	0.121	0.040	0.010	0.002				
	2	0.075	0.194	0.302	0.233	0.121	0.044	0.011	0.001			
	3	0.010	0.057	0.201	0.267	0.215	0.117	0.042	0.009	0.001		
	4	0.001	0.011	0.088	0.200	0.251	0.205	0.111	0.037	0.006		
	5		0.001	0.026	0.103	0.201	0.246	0.201	0.103	0.026	0.001	
	6			0.006	0.037	0.111	0.205	0.251	0.200	0.088	0.011	0.001
	7			0.001	0.009	0.042	0.117	0.215	0.267	0.201	0.057	0.010
	8				0.001	0.011	0.044	0.121	0.233	0.302	0.194	0.075
	9					0.002	0.010	0.040	0.121	0.268	0.387	0.315
	10						0.001	0.006	0.0282	0.1074	0.3487	0.5987

11	0	0.569	0.314	0.086	0.020	0.004						
	1	0.329	0.384	0.236	0.093	0.027	0.005	0.001				
	2	0.087	0.213	0.295	0.200	0.089	0.027	0.005	0.001			
	3	0.014	0.071	0.221	0.257	0.177	0.081	0.023	0.004			
	4	0.001	0.016	0.111	0.220	0.236	0.161	0.070	0.017	0.002		
	5		0.002	0.039	0.132	0.221	0.226	0.147	0.057	0.010		
	6			0.010	0.057	0.147	0.226	0.221	0.132	0.039	0.002	
	7			0.002	0.017	0.070	0.161	0.236	0.220	0.111	0.016	0.001
	8				0.004	0.023	0.081	0.177	0.257	0.221	0.071	0.014
	9				0.001	0.005	0.027	0.089	0.200	0.295	0.213	0.087
	10					0.001	0.005	0.027	0.093	0.236	0.384	0.329
	11							0.004	0.020	0.086	0.314	0.569
12	0	0.540	0.282	0.069	0.014	0.002						
	1	0.341	0.377	0.206	0.071	0.017	0.003					
	2	0.099	0.230	0.283	0.168	0.064	0.016	0.002				
	3	0.017	0.085	0.236	0.240	0.142	0.054	0.012	0.001			
	4	0.002	0.021	0.133	0.231	0.213	0.121	0.042	0.008	0.001		
	5		0.004	0.053	0.158	0.227	0.193	0.101	0.029	0.003		
	6			0.016	0.079	0.177	0.226	0.177	0.079	0.016		
	7			0.003	0.029	0.101	0.193	0.227	0.158	0.053	0.004	
	8			0.001	0.008	0.042	0.121	0.213	0.231	0.133	0.021	0.002
	9				0.001	0.012	0.054	0.142	0.240	0.236	0.085	0.017
	10					0.002	0.016	0.064	0.168	0.283	0.230	0.099
	11						0.003	0.017	0.071	0.206	0.377	0.341
	12							0.002	0.0138	0.0687	0.2824	0.5404
13	0	0.513	0.254	0.055	0.010	0.001						
	1	0.351	0.367	0.179	0.054	0.011	0.002					
	2	0.111	0.245	0.268	0.139	0.045	0.010	0.001				
	3	0.021	0.100	0.246	0.218	0.111	0.035	0.006	0.001			
	4	0.003	0.028	0.154	0.234	0.184	0.087	0.024	0.003			
	5		0.006	0.069	0.180	0.221	0.157	0.066	0.014	0.001		
	6		0.001	0.023	0.103	0.197	0.209	0.131	0.044	0.006		
	7			0.006	0.044	0.131	0.209	0.197	0.103	0.023	0.001	
	8			0.001	0.014	0.066	0.157	0.221	0.180	0.069	0.006	
	9				0.003	0.024	0.087	0.184	0.234	0.154	0.028	0.003
	10				0.001	0.006	0.035	0.111	0.218	0.246	0.100	0.021
	11					0.001	0.010	0.045	0.139	0.268	0.245	0.111
	12						0.002	0.011	0.054	0.1787	0.3672	0.3512
	13							0.001	0.0097	0.055	0.2542	0.5133

Table A.1 Binomial probabilities – *continued*

n	*x*	*p*										
		0.05	*0.1*	*0.2*	*0.3*	*0.4*	*0.5*	*0.6*	*0.7*	*0.8*	*0.9*	*0.95*
14	0	0.488	0.229	0.044	0.007	0.001						
	1	0.359	0.356	0.154	0.041	0.007	0.001					
	2	0.123	0.257	0.250	0.113	0.032	0.006	0.001				
	3	0.026	0.114	0.250	0.194	0.085	0.022	0.003				
	4	0.004	0.035	0.172	0.229	0.155	0.061	0.014	0.001			
	5		0.008	0.086	0.196	0.207	0.122	0.041	0.007			
	6		0.001	0.032	0.126	0.207	0.183	0.092	0.023	0.002		
	7			0.009	0.062	0.157	0.209	0.157	0.062	0.009		
	8			0.002	0.023	0.092	0.183	0.207	0.126	0.032	0.001	
	9				0.007	0.041	0.122	0.207	0.196	0.086	0.008	
	10				0.001	0.014	0.061	0.155	0.229	0.172	0.035	0.004
	11					0.003	0.022	0.085	0.194	0.250	0.114	0.026
	12					0.001	0.006	0.032	0.113	0.250	0.257	0.123
	13						0.001	0.007	0.041	0.154	0.356	0.359
	14							0.001	0.007	0.044	0.229	0.488
15	0	0.463	0.206	0.035	0.005							
	1	0.366	0.343	0.132	0.031	0.005						
	2	0.135	0.267	0.231	0.092	0.022	0.003					
	3	0.031	0.129	0.250	0.170	0.063	0.014	0.002				
	4	0.005	0.043	0.188	0.219	0.127	0.042	0.007	0.001			
	5	0.001	0.010	0.103	0.206	0.186	0.092	0.024	0.003			
	6		0.002	0.043	0.147	0.207	0.153	0.061	0.012	0.001		
	7			0.014	0.081	0.177	0.196	0.118	0.035	0.003		
	8			0.003	0.035	0.118	0.196	0.177	0.081	0.014		
	9			0.001	0.012	0.061	0.153	0.207	0.147	0.043	0.002	
	10				0.003	0.024	0.092	0.186	0.206	0.103	0.010	0.001
	11				0.001	0.007	0.042	0.127	0.219	0.188	0.043	0.005
	12					0.002	0.014	0.063	0.170	0.250	0.129	0.031
	13						0.003	0.022	0.092	0.231	0.267	0.135
	14							0.005	0.031	0.132	0.343	0.366
	15								0.005	0.035	0.206	0.463

16	0	0.440	0.185	0.028	0.003							
	1	0.371	0.329	0.113	0.023	0.003						
	2	0.146	0.275	0.211	0.073	0.015	0.002					
	3	0.036	0.142	0.246	0.146	0.047	0.009	0.001				
	4	0.006	0.051	0.200	0.204	0.101	0.028	0.004				
	5	0.001	0.014	0.120	0.210	0.162	0.067	0.014	0.001			
	6		0.003	0.055	0.165	0.198	0.122	0.039	0.006			
	7			0.020	0.101	0.189	0.175	0.084	0.019	0.001		
	8			0.006	0.049	0.142	0.196	0.142	0.049	0.006		
	9			0.001	0.019	0.084	0.175	0.189	0.101	0.020		
	10				0.006	0.039	0.122	0.198	0.165	0.055	0.003	
	11				0.001	0.014	0.067	0.162	0.210	0.120	0.014	0.001
	12					0.004	0.028	0.101	0.204	0.200	0.051	0.006
	13					0.001	0.009	0.047	0.146	0.246	0.142	0.036
	14						0.002	0.015	0.073	0.211	0.275	0.146
	15							0.003	0.023	0.113	0.329	0.371
	16								0.003	0.028	0.185	0.440
17	0	0.418	0.167	0.023	0.002							
	1	0.374	0.315	0.096	0.017	0.002						
	2	0.158	0.280	0.191	0.058	0.010	0.001					
	3	0.041	0.156	0.239	0.125	0.034	0.005					
	4	0.008	0.060	0.209	0.187	0.080	0.018	0.002				
	5	0.001	0.017	0.136	0.208	0.138	0.047	0.008	0.001			
	6		0.004	0.068	0.178	0.184	0.094	0.024	0.003			
	7		0.001	0.027	0.120	0.193	0.148	0.057	0.009			
	8			0.008	0.064	0.161	0.185	0.107	0.028	0.002		
	9			0.002	0.028	0.107	0.185	0.161	0.064	0.008		
	10				0.009	0.057	0.148	0.193	0.120	0.027	0.001	
	11				0.003	0.024	0.094	0.184	0.178	0.068	0.004	
	12				0.001	0.008	0.047	0.138	0.208	0.136	0.017	0.001
	13					0.002	0.018	0.080	0.187	0.209	0.060	0.008
	14						0.005	0.034	0.125	0.239	0.156	0.041
	15						0.001	0.010	0.058	0.191	0.280	0.158
	16							0.002	0.017	0.096	0.315	0.374
	17								0.002	0.023	0.167	0.418

Table A.1 Binomial probabilities – *continued*

n	*x*	*p* 0.05	0.1	0.2	0.3	0.4	0.5	0.6	0.7	0.8	0.9	0.95
18	0	0.397	0.150	0.018	0.002							
	1	0.376	0.300	0.081	0.013	0.001						
	2	0.168	0.284	0.172	0.046	0.007	0.001					
	3	0.047	0.168	0.230	0.105	0.025	0.003					
	4	0.009	0.070	0.215	0.168	0.061	0.012	0.001				
	5	0.001	0.022	0.151	0.202	0.115	0.033	0.004				
	6		0.005	0.082	0.187	0.166	0.071	0.015	0.001			
	7		0.001	0.035	0.138	0.189	0.121	0.037	0.005			
	8			0.012	0.081	0.173	0.167	0.077	0.015	0.001		
	9			0.003	0.039	0.128	0.185	0.128	0.039	0.003		
	10			0.001	0.015	0.077	0.167	0.173	0.081	0.012		
	11				0.005	0.037	0.121	0.189	0.138	0.035	0.001	
	12				0.001	0.015	0.071	0.166	0.187	0.082	0.005	
	13					0.004	0.033	0.115	0.202	0.151	0.022	0.001
	14					0.001	0.012	0.061	0.168	0.215	0.070	0.009
	15						0.003	0.025	0.105	0.230	0.168	0.047
	16						0.001	0.007	0.046	0.172	0.284	0.168
	17							0.001	0.013	0.081	0.300	0.376
	18								0.002	0.018	0.150	0.397

19	0	0.377	0.135	0.014	0.001							
	1	0.377	0.285	0.068	0.009	0.001						
	2	0.179	0.285	0.154	0.036	0.005						
	3	0.053	0.180	0.218	0.087	0.017	0.002					
	4	0.011	0.080	0.218	0.149	0.047	0.007	0.001				
	5	0.002	0.027	0.164	0.192	0.093	0.022	0.002				
	6		0.007	0.095	0.192	0.145	0.052	0.008	0.001			
	7		0.001	0.044	0.153	0.180	0.096	0.024	0.002			
	8			0.017	0.098	0.180	0.144	0.053	0.008			
	9			0.005	0.051	0.146	0.176	0.098	0.022	0.001		
	10			0.001	0.022	0.098	0.176	0.146	0.051	0.005		
	11				0.008	0.053	0.144	0.180	0.098	0.017		
	12				0.002	0.024	0.096	0.180	0.153	0.044	0.001	
	13				0.001	0.008	0.052	0.145	0.192	0.095	0.007	
	14					0.002	0.022	0.093	0.192	0.164	0.027	0.002
	15					0.001	0.007	0.047	0.149	0.218	0.080	0.011
	16						0.002	0.017	0.087	0.218	0.180	0.053
	17							0.005	0.036	0.154	0.285	0.179
	18							0.001	0.009	0.068	0.285	0.377
	19								0.001	0.014	0.135	0.377
20	0	0.358	0.122	0.012	0.001							
	1	0.377	0.270	0.058	0.007							
	2	0.189	0.285	0.137	0.028	0.003						
	3	0.060	0.190	0.205	0.072	0.012	0.001					
	4	0.013	0.090	0.218	0.130	0.035	0.005					
	5	0.002	0.032	0.175	0.179	0.075	0.015	0.001				
	6		0.009	0.109	0.192	0.124	0.037	0.005				
	7		0.002	0.055	0.164	0.166	0.074	0.015	0.001			
	8			0.022	0.114	0.180	0.120	0.035	0.004			
	9			0.007	0.065	0.160	0.160	0.071	0.012			
	10			0.002	0.031	0.117	0.176	0.117	0.031	0.002		
	11				0.012	0.071	0.160	0.160	0.065	0.007		
	12				0.004	0.035	0.120	0.180	0.114	0.022		
	13				0.001	0.015	0.074	0.166	0.164	0.055	0.002	
	14					0.005	0.037	0.124	0.192	0.109	0.009	
	15					0.001	0.015	0.075	0.179	0.175	0.032	0.002
	16						0.005	0.035	0.130	0.218	0.090	0.013
	17						0.001	0.012	0.072	0.205	0.190	0.060
	18							0.003	0.028	0.137	0.285	0.189
	19								0.007	0.058	0.270	0.377
	20								0.001	0.012	0.122	0.358

Table A.2. Poisson probabilities.

x	λ 0.1	0.2	0.3	0.4	0.5	0.6	0.7	0.8	0.9	1.0
0	0.9048	0.8187	0.7408	0.6703	0.6065	0.5488	0.4966	0.4493	0.4066	0.3679
1	0.0905	0.1637	0.2222	0.2681	0.3033	0.3293	0.3476	0.3595	0.3659	0.3679
2	0.0045	0.0164	0.0333	0.0536	0.0758	0.0988	0.1217	0.1438	0.1647	0.1839
3	0.0002	0.0011	0.0033	0.0072	0.0126	0.0198	0.0284	0.0383	0.0494	0.0613
4	0.0000	0.0001	0.0003	0.0007	0.0016	0.0030	0.0050	0.0077	0.0111	0.0153
5	0.0000	0.0000	0.0000	0.0001	0.0002	0.0004	0.0007	0.0012	0.0020	0.0031
6	0.0000	0.0000	0.0000	0.0000	0.0000	0.0000	0.0001	0.0002	0.0003	0.0005
7	0.0000	0.0000	0.0000	0.0000	0.0000	0.0000	0.0000	0.0000	0.0000	0.0001

x	λ 1.1	1.2	1.3	1.4	1.5	1.6	1.7	1.8	1.9	2.0
0	0.3329	0.3012	0.2725	0.2466	0.2231	0.2019	0.1827	0.1653	0.1496	0.1353
1	0.3662	0.3614	0.3543	0.3452	0.3347	0.3230	0.3106	0.2975	0.2842	0.2707
2	0.2014	0.2169	0.2303	0.2417	0.2510	0.2584	0.2640	0.2678	0.2700	0.2707
3	0.0738	0.0867	0.0998	0.1128	0.1255	0.1378	0.1496	0.1607	0.1710	0.1804
4	0.0203	0.0260	0.0324	0.0395	0.0471	0.0551	0.0636	0.0723	0.0812	0.0902
5	0.0045	0.0062	0.0084	0.0111	0.0141	0.0176	0.0216	0.0260	0.0309	0.0361
6	0.0008	0.0012	0.0018	0.0026	0.0035	0.0047	0.0061	0.0078	0.0098	0.0120
7	0.0001	0.0002	0.0003	0.0005	0.0008	0.0011	0.0015	0.0020	0.0027	0.0034
8	0.0000	0.0000	0.0001	0.0001	0.0001	0.0002	0.0003	0.0005	0.0006	0.0009
9	0.0000	0.0000	0.0000	0.0000	0.0000	0.0000	0.0001	0.0001	0.0001	0.0002

x	λ 2.1	2.2	2.3	2.4	2.5	2.6	2.7	2.8	2.9	3.0
0	0.1225	0.1108	0.1003	0.0907	0.0821	0.0743	0.0672	0.0608	0.0550	0.0498
1	0.2572	0.2438	0.2306	0.2177	0.2052	0.1931	0.1815	0.1703	0.1596	0.1494
2	0.2700	0.2681	0.2652	0.2613	0.2565	0.2510	0.2450	0.2384	0.2314	0.2240
3	0.1890	0.1966	0.2033	0.2090	0.2138	0.2176	0.2205	0.2225	0.2237	0.2240
4	0.0992	0.1082	0.1169	0.1254	0.1336	0.1414	0.1488	0.1557	0.1622	0.1680
5	0.0417	0.0476	0.0538	0.0602	0.0668	0.0735	0.0804	0.0872	0.0940	0.1008
6	0.0146	0.0174	0.0206	0.0241	0.0278	0.0319	0.0362	0.0407	0.0455	0.0504
7	0.0044	0.0055	0.0068	0.0083	0.0099	0.0118	0.0139	0.0163	0.0188	0.0216
8	0.0011	0.0015	0.0019	0.0025	0.0031	0.0038	0.0047	0.0057	0.0068	0.0081
9	0.0003	0.0004	0.0005	0.0007	0.0009	0.0011	0.0014	0.0018	0.0022	0.0027
10	0.0001	0.0001	0.0001	0.0002	0.0002	0.0003	0.0004	0.0005	0.0006	0.0008
11	0.0000	0.0000	0.0000	0.0000	0.0000	0.0001	0.0001	0.0001	0.0002	0.0002
12	0.0000	0.0000	0.0000	0.0000	0.0000	0.0000	0.0000	0.0000	0.0000	0.0001

x	λ 3.1	3.2	3.3	3.4	3.5	3.6	3.7	3.8	3.9	4.0
0	0.0450	0.0408	0.0369	0.0334	0.0302	0.0273	0.0247	0.0224	0.0202	0.0183
1	0.1397	0.1304	0.1217	0.1135	0.1057	0.0984	0.0915	0.0850	0.0789	0.0733
2	0.2165	0.2087	0.2008	0.1929	0.1850	0.1771	0.1692	0.1615	0.1539	0.1465
3	0.2237	0.2226	0.2209	0.2186	0.2158	0.2125	0.2087	0.2046	0.2001	0.1954
4	0.1733	0.1781	0.1823	0.1858	0.1888	0.1912	0.1931	0.1944	0.1951	0.1954

Source: From Beyer, W.H. (1986) *Handbook of Tables for Probability and Statistics*, 2nd edn. CRC Press, Boca Raton, Florida. Reproduced with permission.

x	3.1	3.2	3.3	3.4	3.5	3.6	3.7	3.8	3.9	4.0
5	0.1075	0.1140	0.1203	0.1264	0.1322	0.1377	0.1429	0.1477	0.1522	0.1563
6	0.0555	0.0608	0.0662	0.0716	0.0771	0.0826	0.0881	0.0936	0.0989	0.1042
7	0.0246	0.0278	0.0312	0.0348	0.0385	0.0425	0.0466	0.0508	0.0551	0.0595
8	0.0095	0.0111	0.0129	0.0148	0.0169	0.0191	0.0215	0.0241	0.0269	0.0298
9	0.0033	0.0040	0.0047	0.0056	0.0066	0.0076	0.0089	0.0102	0.0116	0.0132
10	0.0010	0.0013	0.0016	0.0019	0.0023	0.0028	0.0033	0.0039	0.0045	0.0053
11	0.0003	0.0004	0.0005	0.0006	0.0007	0.0009	0.0011	0.0013	0.0016	0.0019
12	0.0001	0.0001	0.0001	0.0002	0.0002	0.0003	0.0003	0.0004	0.0005	0.0006
13	0.0000	0.0000	0.0000	0.0000	0.0001	0.0001	0.0001	0.0001	0.0002	0.0002
14	0.0000	0.0000	0.0000	0.0000	0.0000	0.0000	0.0000	0.0000	0.0000	0.0001

x	4.1	4.2	4.3	4.4	4.5	4.6	4.7	4.8	4.9	5.0
0	0.0166	0.0150	0.0136	0.0123	0.0111	0.0101	0.0091	0.0082	0.0074	0.0067
1	0.0679	0.0630	0.0583	0.0540	0.0500	0.0462	0.0427	0.0395	0.0365	0.0337
2	0.1393	0.1323	0.1254	0.1188	0.1125	0.1063	0.1005	0.0948	0.0894	0.0842
3	0.1904	0.1852	0.1798	0.1743	0.1687	0.1631	0.1574	0.1517	0.1460	0.1404
4	0.1951	0.1944	0.1933	0.1917	0.1898	0.1875	0.1849	0.1820	0.1789	0.1755
5	0.1600	0.1633	0.1662	0.1687	0.1708	0.1725	0.1738	0.1747	0.1753	0.1755
6	0.1093	0.1143	0.1191	0.1237	0.1281	0.1323	0.1362	0.1398	0.1432	0.1462
7	0.0640	0.0686	0.0732	0.0778	0.0824	0.0869	0.0914	0.0959	0.1002	0.1044
8	0.0328	0.0360	0.0393	0.0428	0.0463	0.0500	0.0537	0.0575	0.0614	0.0653
9	0.0150	0.0168	0.0188	0.0209	0.0232	0.0255	0.0281	0.0307	0.0334	0.0363
10	0.0061	0.0071	0.0081	0.0092	0.0104	0.0118	0.0132	0.0147	0.0164	0.0181
11	0.0023	0.0027	0.0032	0.0037	0.0043	0.0049	0.0056	0.0064	0.0073	0.0082
12	0.0008	0.0009	0.0011	0.0013	0.0016	0.0019	0.0022	0.0026	0.0030	0.0034
13	0.0002	0.0003	0.0004	0.0005	0.0006	0.0007	0.0008	0.0009	0.0011	0.0013
14	0.0001	0.0001	0.0001	0.0001	0.0002	0.0002	0.0003	0.0003	0.0004	0.0005
15	0.0000	0.0000	0.0000	0.0000	0.0001	0.0001	0.0001	0.0001	0.0001	0.0002

x	5.1	5.2	5.3	5.4	5.5	5.6	5.7	5.8	5.9	6.0
0	0.0061	0.0055	0.0050	0.0045	0.0041	0.0037	0.0033	0.0030	0.0027	0.0025
1	0.0311	0.0287	0.0265	0.0244	0.0225	0.0207	0.0191	0.0176	0.0162	0.0149
2	0.0793	0.0746	0.0701	0.0659	0.0618	0.0580	0.0544	0.0509	0.0477	0.0446
3	0.1348	0.1293	0.1239	0.1185	0.1133	0.1082	0.1033	0.0985	0.0938	0.0892
4	0.1719	0.1681	0.1641	0.1600	0.1558	0.1515	0.1472	0.1428	0.1383	0.1339
5	0.1753	0.1748	0.1740	0.1728	0.1714	0.1697	0.1678	0.1656	0.1632	0.1606
6	0.1490	0.1515	0.1537	0.1555	0.1571	0.1584	0.1594	0.1601	0.1605	0.1606
7	0.1086	0.1125	0.1163	0.1200	0.1234	0.1267	0.1298	0.1326	0.1353	0.1377
8	0.0692	0.0731	0.0771	0.0810	0.0849	0.0887	0.0925	0.0962	0.0998	0.1033
9	0.0392	0.0423	0.0454	0.0486	0.0519	0.0552	0.0586	0.0620	0.0654	0.0688
10	0.0200	0.0220	0.0241	0.0262	0.0285	0.0309	0.0334	0.0359	0.0386	0.0413
11	0.0093	0.0104	0.0116	0.0129	0.0143	0.0157	0.0173	0.0190	0.0207	0.0225
12	0.0039	0.0045	0.0051	0.0058	0.0065	0.0073	0.0082	0.0092	0.0102	0.0113
13	0.0015	0.0018	0.0021	0.0024	0.0028	0.0032	0.0036	0.0041	0.0046	0.0052

Table A.2. Poisson probabilities – *continued*

	λ									
x	*5.1*	*5.2*	*5.3*	*5.4*	*5.5*	*5.6*	*5.7*	*5.8*	*5.9*	*6.0*
14	0.0006	0.0007	0.0008	0.0009	0.0011	0.0013	0.0015	0.0017	0.0019	0.0022
15	0.0002	0.0002	0.0003	0.0003	0.0004	0.0005	0.0006	0.0007	0.0008	0.0009
16	0.0001	0.0001	0.0001	0.0001	0.0001	0.0002	0.0002	0.0002	0.0003	0.0003
17	0.0000	0.0000	0.0000	0.0000	0.0000	0.0001	0.0001	0.0001	0.0001	0.0001

	λ									
x	*6.1*	*6.2*	*6.3*	*6.4*	*6.5*	*6.6*	*6.7*	*6.8*	*6.9*	*7.0*
0	0.0022	0.0020	0.0018	0.0017	0.0015	0.0014	0.0012	0.0011	0.0010	0.0009
1	0.0137	0.0126	0.0116	0.0106	0.0098	0.0090	0.0082	0.0076	0.0070	0.0064
2	0.0417	0.0390	0.0364	0.0340	0.0318	0.0296	0.0276	0.0258	0.0240	0.0223
3	0.0848	0.0806	0.0765	0.0726	0.0688	0.0652	0.0617	0.0584	0.0552	0.0521
4	0.1294	0.1249	0.1205	0.1162	0.1118	0.1076	0.1034	0.0992	0.0952	0.0912
5	0.1579	0.1549	0.1519	0.1487	0.1454	0.1420	0.1385	0.1349	0.1314	0.1277
6	0.1605	0.1601	0.1595	0.1586	0.1575	0.1562	0.1546	0.1529	0.1511	0.1490
7	0.1399	0.1418	0.1435	0.1450	0.1462	0.1472	0.1480	0.1486	0.1489	0.1490
8	0.1066	0.1099	0.1130	0.1160	0.1188	0.1215	0.1240	0.1263	0.1284	0.1304
9	0.0723	0.0757	0.0791	0.0825	0.0858	0.0891	0.0923	0.0954	0.0985	0.1014
10	0.0441	0.0469	0.0498	0.0528	0.0558	0.0588	0.0618	0.0649	0.0679	0.0710
11	0.0244	0.0265	0.0285	0.0307	0.0330	0.0353	0.0377	0.0401	0.0426	0.0452
12	0.0124	0.0137	0.0150	0.0164	0.0179	0.0194	0.0210	0.0227	0.0245	0.0263
13	0.0058	0.0065	0.0073	0.0081	0.0089	0.0099	0.0108	0.0119	0.0130	0.0142
14	0.0025	0.0029	0.0033	0.0037	0.0041	0.0046	0.0052	0.0058	0.0064	0.0071
15	0.0010	0.0012	0.0014	0.0016	0.0018	0.0020	0.0023	0.0026	0.0029	0.0033
16	0.0004	0.0005	0.0005	0.0006	0.0007	0.0008	0.0010	0.0011	0.0013	0.0014
17	0.0001	0.0002	0.0002	0.0002	0.0003	0.0003	0.0004	0.0004	0.0005	0.0006
18	0.0000	0.0001	0.0001	0.0001	0.0001	0.0001	0.0001	0.0002	0.0002	0.0002
19	0.0000	0.0000	0.0000	0.0000	0.0000	0.0000	0.0001	0.0001	0.0001	0.0001

	λ									
x	*7.1*	*7.2*	*7.3*	*7.4*	*7.5*	*7.6*	*7.7*	*7.8*	*7.9*	*8.0*
0	0.0008	0.0007	0.0007	0.0006	0.0006	0.0005	0.0005	0.0004	0.0004	0.0003
1	0.0059	0.0054	0.0049	0.0045	0.0041	0.0038	0.0035	0.0032	0.0029	0.0027
2	0.0208	0.0194	0.0180	0.0167	0.0156	0.0145	0.0134	0.0125	0.0116	0.0107
3	0.0492	0.0464	0.0438	0.0413	0.0389	0.0366	0.0345	0.0324	0.0305	0.0286
4	0.0874	0.0836	0.0799	0.0764	0.0729	0.0696	0.0663	0.0632	0.0602	0.0573
5	0.1241	0.1204	0.1167	0.1130	0.1094	0.1057	0.1021	0.0986	0.0951	0.0916
6	0.1468	0.1445	0.1420	0.1394	0.1367	0.1339	0.1311	0.1282	0.1252	0.1221
7	0.1489	0.1486	0.1481	0.1474	0.1465	0.1454	0.1442	0.1428	0.1413	0.1396
8	0.1321	0.1337	0.1351	0.1363	0.1373	0.1381	0.1388	0.1392	0.1395	0.1396
9	0.1042	0.1070	0.1096	0.1121	0.1144	0.1167	0.1187	0.1207	0.1224	0.1241
10	0.0740	0.0770	0.0800	0.0829	0.0858	0.0887	0.0914	0.0941	0.0967	0.0993
11	0.0478	0.0504	0.0531	0.0558	0.0585	0.0613	0.0640	0.0667	0.0695	0.0722
12	0.0283	0.0303	0.0323	0.0344	0.0366	0.0388	0.0411	0.0434	0.0457	0.0481
13	0.0154	0.0168	0.0181	0.0196	0.0211	0.0227	0.0243	0.0260	0.0278	0.0296
14	0.0078	0.0086	0.0095	0.0104	0.0113	0.0123	0.0134	0.0145	0.0157	0.0169

x	7.1	7.2	7.3	7.4	7.5	7.6	7.7	7.8	7.9	8.0
	λ									
15	0.0037	0.0041	0.0046	0.0051	0.0057	0.0062	0.0069	0.0075	0.0083	0.0090
16	0.0016	0.0019	0.0021	0.0024	0.0026	0.0030	0.0033	0.0037	0.0041	0.0045
17	0.0007	0.0008	0.0009	0.0010	0.0012	0.0013	0.0015	0.0017	0.0019	0.0021
18	0.0003	0.0003	0.0004	0.0004	0.0005	0.0006	0.0006	0.0007	0.0008	0.0009
19	0.0001	0.0001	0.0001	0.0002	0.0002	0.0002	0.0003	0.0003	0.0003	0.0004
20	0.0000	0.0000	0.0001	0.0001	0.0001	0.0001	0.0001	0.0001	0.0001	0.0002
21	0.0000	0.0000	0.0000	0.0000	0.0000	0.0000	0.0000	0.0000	0.0001	0.0001

x	8.1	8.2	8.3	8.4	8.5	8.6	8.7	8.8	8.9	9.0
	λ									
0	0.0003	0.0003	0.0002	0.0002	0.0002	0.0002	0.0002	0.0002	0.0001	0.0001
1	0.0025	0.0023	0.0021	0.0019	0.0017	0.0016	0.0014	0.0013	0.0012	0.0011
2	0.0100	0.0092	0.0086	0.0079	0.0074	0.0068	0.0063	0.0058	0.0054	0.0050
3	0.0269	0.0252	0.0237	0.0222	0.0208	0.0195	0.0183	0.0171	0.0160	0.0150
4	0.0544	0.0517	0.0491	0.0466	0.0443	0.0420	0.0398	0.0377	0.0357	0.0337
5	0.0882	0.0849	0.0816	0.0784	0.0752	0.0722	0.0692	0.0663	0.0635	0.0607
6	0.1191	0.1160	0.1128	0.1097	0.1066	0.1034	0.1003	0.0972	0.0941	0.0911
7	0.1378	0.1358	0.1338	0.1317	0.1294	0.1271	0.1247	0.1222	0.1197	0.1171
8	0.1395	0.1392	0.1388	0.1382	0.1375	0.1366	0.1356	0.1344	0.1332	0.1318
9	0.1256	0.1269	0.1280	0.1290	0.1299	0.1306	0.1311	0.1315	0.1317	0.1318
10	0.1017	0.1040	0.1063	0.1084	0.1104	0.1123	0.1140	0.1157	0.1172	0.1186
11	0.0749	0.0776	0.0802	0.0828	0.0853	0.0878	0.0902	0.0925	0.0948	0.0970
12	0.0505	0.0530	0.0555	0.0579	0.0604	0.0629	0.0654	0.0679	0.0703	0.0728
13	0.0315	0.0334	0.0354	0.0374	0.0395	0.0416	0.0438	0.0459	0.0481	0.0504
14	0.0182	0.0196	0.0210	0.0225	0.0240	0.0256	0.0272	0.0289	0.0306	0.0324
15	0.0098	0.0107	0.0116	0.0126	0.0136	0.0147	0.0158	0.0169	0.0182	0.0194
16	0.0050	0.0055	0.0060	0.0066	0.0072	0.0079	0.0086	0.0093	0.0101	0.0109
17	0.0024	0.0026	0.0029	0.0033	0.0036	0.0040	0.0044	0.0048	0.0053	0.0058
18	0.0011	0.0012	0.0014	0.0015	0.0017	0.0019	0.0021	0.0024	0.0026	0.0029
19	0.0005	0.0005	0.0006	0.0007	0.0008	0.0009	0.0010	0.0011	0.0012	0.0014
20	0.0002	0.0002	0.0002	0.0003	0.0003	0.0004	0.0004	0.0005	0.0005	0.0006
21	0.0001	0.0001	0.0001	0.0001	0.0001	0.0002	0.0002	0.0002	0.0002	0.0003
22	0.0000	0.0000	0.0000	0.0000	0.0001	0.0001	0.0001	0.0001	0.0001	0.0001

x	9.1	9.2	9.3	9.4	9.5	9.6	9.7	9.8	9.9	10.0
	λ									
0	0.0001	0.0001	0.0001	0.0001	0.0001	0.0001	0.0001	0.0001	0.0001	0.0000
1	0.0010	0.0009	0.0009	0.0008	0.0007	0.0007	0.0006	0.0005	0.0005	0.0005
2	0.0046	0.0043	0.0040	0.0037	0.0034	0.0031	0.0029	0.0027	0.0025	0.0023
3	0.0140	0.0131	0.0123	0.0115	0.0107	0.0100	0.0093	0.0087	0.0081	0.0076
4	0.0319	0.0302	0.0285	0.0269	0.0254	0.0240	0.0226	0.0213	0.0201	0.0189
5	0.0581	0.0555	0.0530	0.0506	0.0483	0.0460	0.0439	0.0418	0.0398	0.0378
6	0.0881	0.0851	0.0822	0.0793	0.0764	0.0736	0.0709	0.0682	0.0656	0.0631
7	0.1145	0.1118	0.1091	0.1064	0.1037	0.1010	0.0982	0.0955	0.0928	0.0901
8	0.1302	0.1286	0.1269	0.1251	0.1232	0.1212	0.1191	0.1170	0.1148	0.1126
9	0.1317	0.1315	0.1311	0.1306	0.1300	0.1293	0.1284	0.1274	0.1263	0.1251

Table A.2. Poisson probabilities – *continued*

	λ									
x	*9.1*	*9.2*	*9.3*	*9.4*	*9.5*	*9.6*	*9.7*	*9.8*	*9.9*	*10.0*
10	0.1198	0.1210	0.1219	0.1228	0.1235	0.1241	0.1245	0.1249	0.1250	0.1251
11	0.0991	0.1012	0.1031	0.1049	0.1067	0.1083	0.1098	0.1112	0.1125	0.1137
12	0.0752	0.0776	0.0799	0.0822	0.0844	0.0866	0.0888	0.0908	0.0928	0.0948
13	0.0526	0.0549	0.0572	0.0594	0.0617	0.0640	0.0662	0.0685	0.0707	0.0729
14	0.0342	0.0361	0.0380	0.0399	0.0419	0.0439	0.0459	0.0479	0.0500	0.0521
15	0.0208	0.0221	0.0235	0.0250	0.0265	0.0281	0.0297	0.0313	0.0330	0.0347
16	0.0118	0.0127	0.0137	0.0147	0.0157	0.0168	0.0180	0.0192	0.0204	0.0217
17	0.0063	0.0069	0.0075	0.0081	0.0088	0.0095	0.0103	0.0111	0.0119	0.0128
18	0.0032	0.0035	0.0039	0.0042	0.0046	0.0051	0.0055	0.0060	0.0065	0.0071
19	0.0015	0.0017	0.0019	0.0021	0.0023	0.0026	0.0028	0.0031	0.0034	0.0037
20	0.0007	0.0008	0.0009	0.0010	0.0011	0.0012	0.0014	0.0015	0.0017	0.0019
21	0.0003	0.0003	0.0004	0.0004	0.0005	0.0006	0.0006	0.0007	0.0008	0.0009
22	0.0001	0.0001	0.0002	0.0002	0.0002	0.0002	0.0003	0.0003	0.0004	0.0004
23	0.0000	0.0001	0.0001	0.0001	0.0001	0.0001	0.0001	0.0001	0.0002	0.0002
24	0.0000	0.0000	0.0000	0.0000	0.0000	0.0000	0.0000	0.0001	0.0001	0.0001

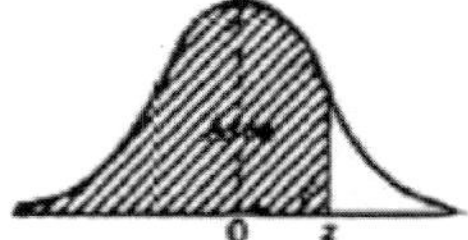

Table A.3. Areas under the normal curve.

z	*0.00*	*0.01*	*0.02*	*0.03*	*0.04*	*0.05*	*0.06*	*0.07*	*0.08*	*0.09*
−3.4	0.0003	0.0003	0.0003	0.0003	0.0003	0.0003	0.0003	0.0003	0.0003	0.0002
−3.3	0.0005	0.0005	0.0005	0.0004	0.0004	0.0004	0.0004	0.0004	0.0004	0.0003
−3.2	0.0007	0.0007	0.0006	0.0006	0.0006	0.0006	0.0006	0.0005	0.0005	0.0005
−3.1	0.0010	0.0009	0.0009	0.0009	0.0008	0.0008	0.0008	0.0008	0.0007	0.0007
−3.0	0.0013	0.0013	0.0013	0.0012	0.0012	0.0011	0.0011	0.0011	0.0010	0.0010
−2.9	0.0019	0.0018	0.0018	0.0017	0.0016	0.0016	0.0015	0.0015	0.0014	0.0014
−2.8	0.0026	0.0025	0.0024	0.0023	0.0023	0.0022	0.0021	0.0021	0.0020	0.0019
−2.7	0.0035	0.0034	0.0033	0.0032	0.0031	0.0030	0.0029	0.0028	0.0027	0.0026
−2.6	0.0047	0.0045	0.0044	0.0043	0.0041	0.0040	0.0039	0.0038	0.0037	0.0036
−2.5	0.0062	0.0060	0.0059	0.0057	0.0055	0.0054	0.0052	0.0051	0.0049	0.0048
−2.4	0.0082	0.0080	0.0078	0.0075	0.0073	0.0071	0.0069	0.0068	0.0066	0.0064
−2.3	0.0107	0.0104	0.0102	0.0099	0.0096	0.0094	0.0091	0.0089	0.0087	0.0084
−2.2	0.0139	0.0136	0.0132	0.0129	0.0125	0.0122	0.0119	0.0116	0.0113	0.0110
−2.1	0.0179	0.0174	0.0170	0.0166	0.0162	0.0158	0.0154	0.0150	0.0146	0.0143
−2.0	0.0228	0.0222	0.0217	0.0212	0.0207	0.0202	0.0197	0.0192	0.0188	0.0183
−1.9	0.0287	0.0281	0.0274	0.0268	0.0262	0.0256	0.0250	0.0244	0.0239	0.0233
−1.8	0.0359	0.0351	0.0344	0.0336	0.0329	0.0322	0.0314	0.0307	0.0301	0.0294
−1.7	0.0446	0.0436	0.0427	0.0418	0.0409	0.0401	0.0392	0.0384	0.0375	0.0367
−1.6	0.0548	0.0537	0.0526	0.0516	0.0505	0.0495	0.0485	0.0475	0.0465	0.0455
−1.5	0.0668	0.0655	0.0643	0.0630	0.0618	0.0606	0.0594	0.0582	0.0571	0.0559
−1.4	0.0808	0.0793	0.0778	0.0764	0.0749	0.0735	0.0721	0.0708	0.0694	0.0681
−1.3	0.0968	0.0951	0.0934	0.0918	0.0901	0.0885	0.0869	0.0853	0.0838	0.0823
−1.2	0.1151	0.1131	0.1112	0.1093	0.1075	0.1056	0.1038	0.1020	0.1003	0.0985
−1.1	0.1357	0.1335	0.1314	0.1292	0.1271	0.1251	0.1230	0.1210	0.1190	0.1170
−1.0	0.1587	0.1562	0.1539	0.1515	0.1492	0.1469	0.1446	0.1423	0.1401	0.1379
−0.9	0.1841	0.1814	0.1788	0.1762	0.1736	0.1711	0.1685	0.1660	0.1635	0.1611
−0.8	0.2119	0.2090	0.2061	0.2033	0.2005	0.1977	0.1949	0.1922	0.1894	0.1867
−0.7	0.2420	0.2389	0.2358	0.2327	0.2296	0.2266	0.2236	0.2206	0.2177	0.2148
−0.6	0.2743	0.2709	0.2676	0.2643	0.2611	0.2578	0.2546	0.2514	0.2483	0.2451
−0.5	0.3085	0.3050	0.3015	0.2981	0.2946	0.2912	0.2877	0.2843	0.2810	0.2776
−0.4	0.3446	0.3409	0.3372	0.3336	0.3300	0.3264	0.3228	0.3192	0.3156	0.3121
−0.3	0.3821	0.3783	0.3745	0.3707	0.3669	0.3632	0.3594	0.3557	0.3520	0.3483
−0.2	0.4207	0.4168	0.4129	0.4090	0.4052	0.4013	0.3974	0.3936	0.3897	0.3859
−0.1	0.4602	0.4562	0.4522	0.4483	0.4443	0.4404	0.4364	0.4325	0.4286	0.4247
0.0	0.5000	0.4960	0.4920	0.4880	0.4840	0.4801	0.4761	0.4721	0.4681	0.4641
0.0	0.5000	0.5040	0.5080	0.5120	0.5160	0.5199	0.5239	0.5279	0.5319	0.5359
0.1	0.5398	0.5438	0.5478	0.5517	0.5557	0.5596	0.5636	0.5675	0.5714	0.5753
0.2	0.5793	0.5832	0.5871	0.5910	0.5948	0.5987	0.6026	0.6064	0.6103	0.6141
0.3	0.6179	0.6217	0.6255	0.6293	0.6331	0.6368	0.6406	0.6443	0.6480	0.6517
0.4	0.6554	0.6591	0.6628	0.6664	0.6700	0.6736	0.6772	0.6808	0.6844	0.6879
0.5	0.6915	0.6950	0.6985	0.7019	0.7054	0.7088	0.7123	0.7157	0.7190	0.7224
0.6	0.7257	0.7291	0.7324	0.7357	0.7389	0.7422	0.7454	0.7486	0.7517	0.7549
0.7	0.7580	0.7611	0.7642	0.7673	0.7704	0.7734	0.7764	0.7794	0.7823	0.7852
0.8	0.7881	0.7910	0.7939	0.7967	0.7995	0.8023	0.8051	0.8078	0.8106	0.8133
0.9	0.8159	0.8186	0.8212	0.8238	0.8264	0.8289	0.8315	0.8340	0.8365	0.8389
1.0	0.8413	0.8438	0.8461	0.8485	0.8508	0.8531	0.8554	0.8577	0.8599	0.8621
1.1	0.8643	0.8665	0.8686	0.8708	0.8729	0.8749	0.8770	0.8790	0.8810	0.8830
1.2	0.8849	0.8869	0.8888	0.8907	0.8925	0.8944	0.8962	0.8980	0.8997	0.9015
1.3	0.9032	0.9049	0.9066	0.9082	0.9099	0.9115	0.9131	0.9147	0.9162	0.9177
1.4	0.9192	0.9207	0.9222	0.9236	0.9251	0.9265	0.9279	0.9292	0.9306	0.9319
1.5	0.9332	0.9345	0.9357	0.9370	0.9382	0.9394	0.9406	0.9418	0.9429	0.9441
1.6	0.9452	0.9463	0.9474	0.9484	0.9495	0.9505	0.9515	0.9525	0.9535	0.9545
1.7	0.9554	0.9564	0.9573	0.9582	0.9591	0.9599	0.9608	0.9616	0.9625	0.9633
1.8	0.9641	0.9649	0.9656	0.9664	0.9671	0.9678	0.9686	0.9693	0.9699	0.9706
1.9	0.9713	0.9719	0.9726	0.9732	0.9738	0.9744	0.9750	0.9756	0.9761	0.9767
2.0	0.9772	0.9778	0.9783	0.9788	0.9793	0.9798	0.9803	0.9808	0.9812	0.9817
2.1	0.9821	0.9826	0.9830	0.9834	0.9838	0.9842	0.9846	0.9850	0.9854	0.9857
2.2	0.9861	0.9864	0.9868	0.9871	0.9875	0.9878	0.9881	0.9884	0.9887	0.9890
2.3	0.9893	0.9896	0.9898	0.9901	0.9904	0.9906	0.9909	0.9911	0.9913	0.9916
2.4	0.9918	0.9920	0.9922	0.9925	0.9927	0.9929	0.9931	0.9932	0.9934	0.9936
2.5	0.9938	0.9940	0.9941	0.9943	0.9945	0.9946	0.9948	0.9949	0.9951	0.9952
2.6	0.9953	0.9955	0.9956	0.9957	0.9959	0.9960	0.9961	0.9962	0.9963	0.9964
2.7	0.9965	0.9966	0.9967	0.9968	0.9969	0.9970	0.9971	0.9972	0.9973	0.9974
2.8	0.9974	0.9975	0.9976	0.9977	0.9977	0.9978	0.9979	0.9979	0.9980	0.9981
2.9	0.9981	0.9982	0.9982	0.9983	0.9984	0.9984	0.9985	0.9985	0.9986	0.9986
3.0	0.9987	0.9987	0.9987	0.9988	0.9988	0.9989	0.9989	0.9989	0.9990	0.9990
3.1	0.9990	0.9991	0.9991	0.9991	0.9992	0.9992	0.9992	0.9992	0.9993	0.9993
3.2	0.9993	0.9993	0.9994	0.9994	0.9994	0.9994	0.9994	0.9995	0.9995	0.9995
3.3	0.9995	0.9995	0.9995	0.9996	0.9996	0.9996	0.9996	0.9996	0.9996	0.9997
3.4	0.9997	0.9997	0.9997	0.9997	0.9997	0.9997	0.9997	0.9997	0.9997	0.9998

Table A.4. Random numbers.

10480	15011	01536	02011	81647	91646	69179	14194	62590	36207	20969	99570	91291	90700
22368	46573	25595	85393	30995	89198	27982	53402	93965	34095	52666	19174	39615	99505
24130	48390	22527	97265	76393	64809	15179	24830	49340	32081	30680	19655	63348	58629
42167	93093	06243	61680	07856	16376	39440	53537	71341	57004	00849	74917	97758	16379
37570	39975	81837	16656	06121	91782	60468	81305	49684	60072	14110	06927	01263	54613
77921	06907	11008	42751	27756	53498	18602	70659	90655	15053	21916	81825	44394	42880
99562	72905	56420	69994	98872	31016	71194	18738	44013	48840	63213	21069	10634	12952
96301	91977	05463	07972	18876	20922	94595	56869	69014	60045	18425	84903	42508	32307
89579	14342	63661	10281	17453	18103	57740	84378	25331	12568	58678	44947	05585	56941
85475	36857	53342	53988	53060	59533	38867	62300	08158	17983	16439	11458	18593	64952
28918	69578	88231	33276	70997	79936	56865	05859	90106	31595	01547	85590	91610	78188
63553	40961	48235	03427	49626	69445	18663	72695	52180	20847	12234	90511	33703	90322
09429	93969	52636	92737	88974	33488	36320	17617	30015	08272	84115	27156	30613	74952
10365	61129	87529	85689	48237	52267	67689	93394	01511	26358	85104	20285	29975	89868
07119	97336	71048	08178	77233	13916	47564	81056	97735	85977	29372	74461	28551	90707
51085	12765	51821	51259	77452	16308	60756	92144	49442	53900	70960	63990	75601	40719
02368	21382	52404	60268	89368	19885	55322	44819	01188	65255	64835	44919	05944	55157
01011	54092	33362	94904	31273	04146	18594	29852	71685	85030	51132	01915	92747	64951
52162	53916	46369	58586	23216	14513	83149	98736	23495	64350	94738	17752	35156	35749
07056	97628	33787	09998	42698	06691	76988	13602	51851	46104	88916	19509	25625	58104
48663	91245	85828	14346	09172	30163	90229	04734	59193	22178	30421	61666	99904	32812
54164	58492	22421	74103	47070	25306	76468	26384	58151	06646	21524	15227	96909	44592
32639	32363	05597	24200	13363	38005	94342	28728	35806	06912	17012	64161	18296	22851
29334	27001	87637	87308	58731	00256	45834	15398	46557	41135	10307	07684	36188	18510
02488	33062	28834	07351	19731	92420	60952	61280	50001	67658	32586	86679	50720	94953
81525	72295	04839	96423	24878	82651	66566	14778	76797	14780	13300	87074	79666	95725
29676	20591	68086	26432	46901	20849	89768	81536	86645	12659	92259	57102	80428	25280
00742	57392	39064	66432	84673	40027	32832	61362	98947	96067	64760	64584	96096	98253
05366	04213	25669	26422	44407	44048	37937	63904	45766	66134	75470	66520	34693	90449
91921	26418	64117	94305	26766	25940	39972	22209	71500	64568	91402	42416	07844	69618
00582	04711	87917	77341	42206	35126	74087	99547	81817	42607	43808	76655	62028	76630
00725	69884	62797	56170	86324	88072	76222	36086	84637	93161	76038	65855	77919	88006
69011	65795	95876	55293	18988	27354	26575	08625	40801	59920	29841	80150	12777	48501
25976	57948	29888	88604	67917	48708	18912	82271	65424	69774	33611	54262	85963	03547
09763	83473	73577	12908	30883	18317	28290	35797	05998	41688	34952	37888	38917	88050
91567	42595	27958	30134	04024	86385	29880	99730	55536	84855	29088	09250	79656	73211
17955	56349	90999	49127	20044	59931	06115	20542	18059	02008	73708	83517	36103	42791
46503	18584	18845	49618	02304	51038	20655	58727	28168	15475	56942	53389	20562	87338
92157	89634	94824	78171	84610	82834	09922	25417	44137	48413	25555	21246	35509	20468
14577	62765	35605	81263	39667	47358	56873	56307	61607	49518	89656	20103	77490	18062
98427	07523	33362	64270	01638	92477	66969	98420	04880	45585	46565	04102	46880	45709
34914	63976	88720	82765	34476	17032	87589	40836	32427	70002	70663	88863	77775	69348
70060	28277	39475	46473	23219	53416	94970	25832	69975	94884	19661	72828	00102	66794
53976	54914	06990	67245	68350	82948	11398	42878	80287	88267	47363	46634	06541	97809
76072	29515	40980	07391	58745	25774	22987	80059	39911	96189	41151	14222	60697	59583
90725	52210	83974	29992	65831	38857	50490	83765	55657	14361	31720	57375	56228	41546
64364	67412	33339	31926	14883	24413	59744	92351	97473	89286	35931	04110	23726	51900
08962	00358	31662	25388	61642	34072	81249	35648	56891	69352	48373	45578	78547	81788
95012	68379	93526	70765	10592	04542	76463	54328	02349	17247	28865	14777	62730	92277
15664	10493	20492	38301	91132	21999	59516	81652	27195	48223	46751	22923	32261	85653

Source: From Beyer, W.H. (1986) *Handbook of Tables for Probability and Statistics*, 2nd edn. CRC Press, Boca Raton, Florida. Reproduced with permission.

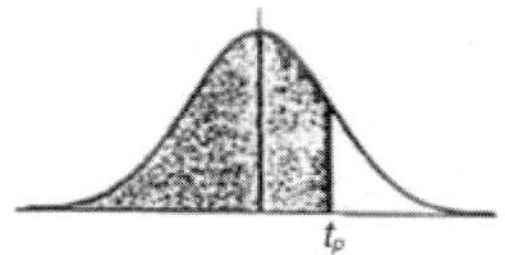

Table A.5. Critical values for the t distribution: percentile values (t_p) for student's t distribution with v degrees of freedom (shaded area = p).

v	$t_{.995}$	$t_{.99}$	$t_{.975}$	$t_{.95}$	$t_{.90}$	$t_{.80}$	$t_{.75}$	$t_{.70}$	$t_{.60}$	$t_{.55}$
1	63.66	31.82	12.71	6.31	3.08	1.376	1.000	0.727	0.325	0.158
2	9.93	6.97	4.30	2.92	1.89	1.061	0.816	0.617	0.289	0.142
3	5.84	4.54	3.18	2.35	1.64	0.978	0.765	0.584	0.277	0.137
4	4.60	3.75	2.78	2.13	1.53	0.941	0.741	0.569	0.271	0.134
5	4.03	3.37	2.57	2.02	1.48	0.920	0.727	0.559	0.267	0.132
6	3.71	3.14	2.45	1.94	1.44	0.906	0.718	0.553	0.265	0.131
7	3.50	3.00	2.37	1.90	1.42	0.896	0.711	0.549	0.263	0.130
8	3.36	2.90	2.31	1.86	1.40	0.889	0.706	0.546	0.262	0.130
9	3.25	2.82	2.26	1.83	1.38	0.883	0.703	0.543	0.261	0.129
10	3.17	2.76	2.23	1.81	1.37	0.879	0.700	0.542	0.260	0.129
11	3.11	2.72	2.20	1.80	1.36	0.876	0.697	0.540	0.260	0.129
12	3.06	2.68	2.18	1.78	1.36	0.873	0.695	0.539	0.259	0.128
13	3.01	2.65	2.16	1.77	1.35	0.870	0.694	0.538	0.259	0.128
14	2.98	2.62	2.15	1.76	1.35	0.868	0.692	0.537	0.258	0.128
15	2.95	2.60	2.13	1.75	1.34	0.866	0.691	0.536	0.258	0.128
16	2.92	2.58	2.12	1.75	1.34	0.865	0.690	0.535	0.258	0.128
17	2.90	2.57	2.11	1.74	1.33	0.863	0.689	0.534	0.257	0.128
18	2.88	2.55	2.10	1.73	1.33	0.862	0.688	0.534	0.257	0.127
19	2.86	2.54	2.09	1.73	1.33	0.861	0.688	0.533	0.257	0.127
20	2.85	2.53	2.09	1.73	1.33	0.860	0.687	0.533	0.257	0.127
21	2.83	2.52	2.08	1.72	1.32	0.859	0.686	0.532	0.257	0.127
22	2.82	2.51	2.07	1.72	1.32	0.858	0.686	0.532	0.256	0.127
23	2.81	2.50	2.07	1.71	1.32	0.858	0.685	0.532	0.256	0.127
24	2.80	2.49	2.06	1.71	1.32	0.857	0.685	0.531	0.256	0.127
25	2.79	2.49	2.06	1.71	1.32	0.856	0.684	0.531	0.256	0.127
26	2.78	2.48	2.06	1.71	1.32	0.856	0.684	0.531	0.256	0.127
27	2.77	2.47	2.05	1.70	1.31	0.855	0.684	0.531	0.256	0.127
28	2.76	2.47	2.05	1.70	1.31	0.855	0.683	0.530	0.256	0.127
29	2.76	2.46	2.05	1.70	1.31	0.854	0.683	0.530	0.256	0.127
30	2.75	2.46	2.04	1.70	1.31	0.854	0.683	0.530	0.256	0.127
40	2.70	2.42	2.02	1.68	1.30	0.851	0.681	0.529	0.255	0.126
60	2.66	2.39	2.00	1.67	1.30	0.848	0.679	0.527	0.254	0.126
120	2.62	2.36	1.98	1.66	1.29	0.845	0.677	0.526	0.254	0.126
∞	2.58	2.33	1.96	1.65	1.28	0.842	0.674	0.524	0.253	0.126

Source: Fisher, R.A. and Yates, F. (1957) *Statistical Tables for Biological, Agricultural, and Medical Research*, 5th edn, Table III. Oliver and Boyd Ltd, Edinburgh. Reproduced with permission of the authors and publisher.

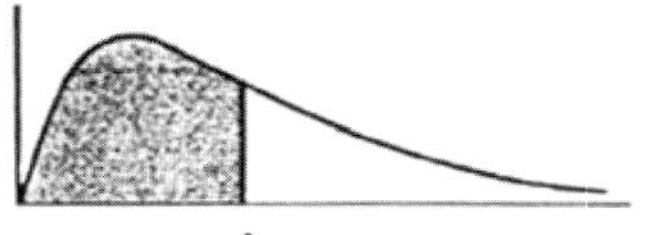

Table A.6. Critical values for the χ^2 distribution: percentile values (χ^2_p) for chi-square distribution with v degrees of freedom (shaded area = p).

v	$\chi^2_{.995}$	$\chi^2_{.99}$	$\chi^2_{.975}$	$\chi^2_{.95}$	$\chi^2_{.90}$	$\chi^2_{.75}$	$\chi^2_{.50}$	$\chi^2_{.25}$	$\chi^2_{.10}$	$\chi^2_{.05}$	$\chi^2_{.025}$	$\chi^2_{.01}$	$\chi^2_{.005}$
1	7.88	6.63	5.02	3.84	2.71	1.32	0.455	0.102	0.016	0.004	0.001	0.000	0.000
2	10.60	9.21	7.38	5.99	4.61	2.77	1.39	0.575	0.211	0.103	0.051	0.020	0.010
3	12.84	11.34	9.35	7.81	6.25	4.11	2.37	1.21	0.584	0.352	0.216	0.115	0.072
4	14.86	13.28	11.14	9.49	7.78	5.39	3.36	1.92	1.06	0.711	0.484	0.297	0.207
5	16.75	15.09	12.83	11.07	9.24	6.63	4.35	2.67	1.61	1.15	0.831	0.554	0.412
6	18.55	16.81	14.45	12.59	10.64	7.84	5.35	3.45	2.20	1.64	1.24	0.872	0.676
7	20.28	18.48	16.01	14.07	12.02	9.04	6.35	4.25	2.83	2.17	1.69	1.24	0.989
8	21.95	20.09	17.53	15.51	13.36	10.22	7.34	5.07	3.49	2.73	2.18	1.65	1.34
9	23.59	21.67	19.02	16.92	14.68	11.39	8.34	5.90	4.17	3.33	2.70	2.09	1.73
10	25.19	23.21	20.48	18.31	15.99	12.55	9.34	6.74	4.87	3.94	3.25	2.56	2.16
11	26.76	24.72	21.92	19.68	17.28	13.70	10.34	7.58	5.58	4.57	3.82	3.05	2.60
12	28.30	26.22	23.34	21.03	18.55	14.85	11.34	8.44	6.30	5.23	4.40	3.57	3.07
13	29.82	27.69	24.74	22.36	19.81	15.98	12.34	9.30	7.04	5.89	5.01	4.11	3.57
14	31.32	29.14	26.12	23.68	21.06	17.12	13.34	10.17	7.79	6.57	5.63	4.66	4.07
15	32.80	30.58	27.49	25.00	22.31	18.25	14.34	11.04	8.55	7.26	6.26	5.23	4.60
16	34.27	32.00	28.85	26.30	23.54	19.37	15.34	11.91	9.31	7.96	6.91	5.81	5.14
17	35.72	33.41	30.19	27.59	24.77	20.49	16.34	12.79	10.09	8.67	7.56	6.41	5.70
18	37.16	34.81	31.53	28.87	25.99	21.60	17.34	13.68	10.86	9.39	8.23	7.01	6.26
19	38.58	36.19	32.85	30.14	27.20	22.72	18.34	14.56	11.65	10.12	8.91	7.63	6.84
20	40.00	37.57	34.17	31.41	28.41	23.83	19.34	15.45	12.44	10.85	9.59	8.26	7.43
21	41.40	38.93	35.48	32.67	29.62	24.93	20.34	16.34	13.24	11.59	10.28	8.90	8.03
22	42.80	40.29	36.78	33.92	30.81	26.04	21.34	17.24	14.04	12.34	10.98	9.54	8.64
23	44.18	41.64	38.08	35.17	32.01	27.14	22.34	18.14	14.85	13.09	11.69	10.20	9.26
24	45.56	42.98	39.36	36.42	33.20	28.24	23.34	19.04	15.66	13.85	12.40	10.86	9.89
25	46.93	44.31	40.65	37.65	34.38	29.34	24.34	19.94	16.47	14.61	13.12	11.52	10.52
26	48.29	45.64	41.92	38.89	35.56	30.43	25.34	20.84	17.29	15.38	13.84	12.20	11.16
27	49.64	46.96	43.19	40.11	36.74	31.53	26.34	21.75	18.11	16.15	14.57	12.88	11.81
28	50.99	48.28	44.46	41.34	37.92	32.62	27.34	22.66	18.94	16.93	15.31	13.56	12.46
29	52.34	49.59	45.72	42.56	39.09	33.71	28.34	23.57	19.77	17.71	16.05	14.26	13.12
30	53.67	50.89	46.98	43.77	40.26	34.80	29.34	24.48	20.60	18.49	16.79	14.95	13.79
40	66.77	63.69	59.34	55.76	51.81	45.62	39.34	33.66	29.05	26.51	24.43	22.16	20.71
50	79.49	76.15	71.42	67.50	63.17	56.33	49.33	42.94	37.69	34.76	32.36	29.71	27.99
60	91.95	88.38	83.30	79.08	74.40	66.98	59.33	52.29	46.46	43.19	40.48	37.48	35.53
70	104.21	100.43	95.02	90.53	85.53	77.58	69.33	61.70	55.33	51.74	48.76	45.44	43.28
80	116.32	112.33	106.63	101.88	96.58	88.13	79.33	71.14	64.28	60.39	57.15	53.54	51.17
90	128.30	124.12	118.14	113.15	107.57	98.65	89.33	80.62	73.29	69.13	65.65	61.75	59.20
100	140.17	135.81	129.56	124.34	118.50	109.14	99.33	90.13	82.36	77.93	74.22	70.06	67.33

Source: Thompson, C.M. (1941) *Table of percentage points of the χ^2 distribution. Biometrika* 32. Reproduced with permission of the author and publisher.

Table A.7. Critical values for the F distribution.

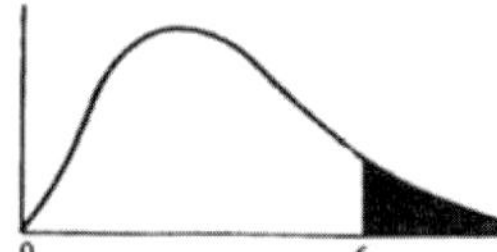

df for Denom.	α	1	2	3	4	5	6	7	8	9	10	12	15	20	24	30	40	60	∞
	0.25	5.83	7.50	8.20	8.58	8.82	8.98	9.10	9.19	9.26	9.32	9.41	9.49	9.58	9.63	9.67	9.71	9.76	9.85
	0.1	39.9	49.5	53.6	55.8	57.2	58.2	58.9	59.4	59.9	60.2	60.7	61.2	61.7	62.0	62.3	62.5	62.8	63.3
	0.05	161.4	199.5	215.7	224.6	230.2	234.0	236.8	238.9	240.5	241.9	243.9	245.9	248.0	249.1	250.1	251.1	252.2	254.3
1	0.025	648	799	864	900	922	937	948	957	963	969	977	985	993	997	1001	1006	1010	1018
	0.01	4052	4999	5403	5625	5764	5859	5928	5981	6022	6056	6106	6157	6209	6235	6261	6287	6313	6366
	0.005	1621*	2000*	2161*	2250*	2306*	2344*	2371*	2393*	2409*	2422*	2443*	2463*	2484*	2494*	2504*	2515*	2525*	2546*
	0.001	4053†	5000†	5404†	5625†	5764†	5859†	5929†	5981†	6023†	6056†	6107†	6158†	6209†	6235†	6261†	6287†	6313†	6366†
	0.25	2.57	3.00	3.15	3.23	3.28	3.31	3.34	3.35	3.37	3.38	3.39	3.41	3.43	3.43	3.44	3.45	3.46	3.48
	0.1	8.53	9.00	9.16	9.24	9.29	9.33	9.35	9.37	9.38	9.39	9.41	9.42	9.44	9.45	9.46	9.47	9.47	9.49
	0.05	18.5	19.0	19.2	19.2	19.3	19.3	19.4	19.4	19.4	19.4	19.4	19.4	19.4	19.5	19.5	19.5	19.5	19.5
2	0.025	38.5	39.0	39.2	39.2	39.3	39.3	39.4	39.4	39.4	39.4	39.4	39.4	39.4	39.5	39.5	39.5	39.5	39.5
	0.01	98.5	99.0	99.2	99.2	99.3	99.3	99.4	99.4	99.4	99.4	99.4	99.4	99.4	99.5	99.5	99.5	99.5	99.5
	0.005	199	199	199	199	199	199	199	199	199	199	199	199	199	199	199	199	199	199
	0.001	999	999	999	999	999	999	999	999	999	999	999	999	999	999	999	999	999	999
	0.25	2.02	2.28	2.36	2.39	2.41	2.42	2.43	2.44	2.44	2.44	2.45	2.46	2.46	2.46	2.47	2.47	2.47	2.47
	0.1	5.54	5.46	5.39	5.34	5.31	5.28	5.27	5.25	5.24	5.23	5.22	5.20	5.18	5.18	5.17	5.16	5.15	5.13
	0.05	10.1	9.55	9.28	9.12	9.01	8.94	8.89	8.85	8.81	8.79	8.74	8.70	8.66	8.64	8.62	8.59	8.57	8.53
3	0.025	17.4	16.0	15.4	15.1	14.9	14.7	14.6	14.5	14.5	14.4	14.3	14.3	14.2	14.1	14.1	14.0	14.0	13.9
	0.01	34.1	30.8	29.5	28.7	28.2	27.9	27.7	27.5	27.3	27.2	27.1	26.9	26.7	26.6	26.5	26.4	26.3	26.1
	0.005	55.6	49.8	47.5	46.2	45.4	44.8	44.4	44.1	43.9	43.7	43.4	43.1	42.8	42.6	42.5	42.3	42.1	41.8
	0.001	167	149	141	137	135	133	132	131	130	129	128	127	126	126	125	125	124	123
	0.25	1.81	2.00	2.05	2.06	2.07	2.08	2.08	2.08	2.08	2.08	2.08	2.08	2.08	2.08	2.08	2.08	2.08	2.08
	0.1	4.54	4.32	4.19	4.11	4.05	4.01	3.98	3.95	3.94	3.92	3.90	3.87	3.84	3.83	3.82	3.80	3.79	3.76
	0.05	7.7	6.94	6.59	6.39	6.26	6.16	6.09	6.04	6.00	5.96	5.91	5.86	5.80	5.77	5.75	5.72	5.69	5.63
4	0.025	12.2	10.6	10.0	9.60	9.36	9.20	9.07	8.98	8.90	8.84	8.75	8.66	8.56	8.51	8.46	8.41	8.36	8.26
	0.01	21.2	18.0	16.7	16.0	15.5	15.2	15.0	14.8	14.7	14.5	14.4	14.2	14.0	13.9	13.8	13.7	13.7	13.5
	0.005	31.3	26.3	24.3	23.2	22.5	22.0	21.6	21.4	21.1	21.0	20.7	20.4	20.2	20.0	19.9	19.8	19.6	19.3
	0.001	74.1	61.2	56.2	53.4	51.7	50.5	49.7	49.0	48.5	48.1	47.4	46.8	46.1	45.8	45.4	45.1	44.7	44.1
	0.25	1.69	1.85	1.88	1.89	1.89	1.89	1.89	1.89	1.89	1.89	1.89	1.89	1.88	1.88	1.88	1.88	1.87	1.87
	0.1	4.06	3.78	3.62	3.52	3.45	3.40	3.37	3.34	3.32	3.30	3.27	3.24	3.21	3.19	3.17	3.16	3.14	3.10
	0.05	6.61	5.79	5.41	5.19	5.05	4.95	4.88	4.82	4.77	4.74	4.68	4.62	4.56	4.53	4.50	4.46	4.43	4.37
5	0.025	10.0	8.43	7.76	7.39	7.15	6.98	6.85	6.76	6.68	6.62	6.52	6.43	6.33	6.28	6.23	6.18	6.12	6.02
	0.01	16.3	13.3	12.1	11.4	11.0	10.7	10.5	10.3	10.2	10.1	9.89	9.72	9.55	9.47	9.38	9.29	9.20	9.02
	0.005	22.8	18.3	16.5	15.6	14.9	14.5	14.2	14.0	13.8	13.6	13.4	13.1	12.9	12.8	12.7	12.5	12.4	12.1
	0.001	47.2	37.1	33.2	31.1	29.8	28.8	28.2	27.6	27.2	26.9	26.4	25.9	25.4	25.1	24.9	24.6	24.3	23.8

* Multiply these values by 10, † multiply these values by 100.

Table A.7. Critical values for the F distribution – *continued*

df for Denom.	α	df for Numerator 1	2	3	4	5	6	7	8	9	10	12	15	20	24	30	40	60	∞
6	0.25	1.62	1.76	1.78	1.79	1.79	1.78	1.78	1.78	1.77	1.77	1.77	1.76	1.76	1.75	1.75	1.75	1.74	1.74
	0.1	3.78	3.46	3.29	3.18	3.11	3.05	3.01	2.98	2.96	2.94	2.90	2.87	2.84	2.82	2.80	2.78	2.76	2.72
	0.05	5.99	5.14	4.76	4.53	4.39	4.28	4.21	4.15	4.10	4.06	4.00	3.94	3.87	3.84	3.81	3.77	3.74	3.67
	0.025	8.81	7.26	6.60	6.23	5.99	5.82	5.70	5.60	5.52	5.46	5.37	5.27	5.17	5.12	5.07	5.01	4.96	4.85
	0.01	13.7	10.9	9.78	9.15	8.75	8.47	8.26	8.10	7.98	7.87	7.72	7.56	7.40	7.31	7.23	7.14	7.06	6.88
	0.005	18.6	14.5	12.9	12.0	11.5	11.1	10.8	10.6	10.4	10.3	10.0	9.81	9.59	9.47	9.36	9.24	9.12	8.88
	0.001	35.5	27.0	23.7	21.9	20.8	20.0	19.5	19.0	18.7	18.4	18.0	17.6	17.1	16.9	16.7	16.4	16.2	15.7
7	0.25	1.57	1.70	1.72	1.72	1.71	1.71	1.70	1.70	1.69	1.69	1.68	1.68	1.67	1.67	1.66	1.66	1.65	1.65
	0.1	3.59	3.26	3.07	2.96	2.88	2.83	2.78	2.75	2.72	2.70	2.67	2.63	2.59	2.58	2.56	2.54	2.51	2.47
	0.05	5.59	4.74	4.35	4.12	3.97	3.87	3.79	3.73	3.68	3.64	3.57	3.51	3.44	3.41	3.38	3.34	3.30	3.23
	0.025	8.07	6.54	5.89	5.52	5.29	5.12	4.99	4.90	4.82	4.76	4.67	4.57	4.47	4.41	4.36	4.31	4.25	4.14
	0.01	12.2	9.55	8.45	7.85	7.46	7.19	6.99	6.84	6.72	6.62	6.47	6.31	6.16	6.07	5.99	5.91	5.82	5.65
	0.005	16.2	12.4	10.9	10.1	9.52	9.16	8.89	8.68	8.51	8.38	8.18	7.97	7.75	7.64	7.53	7.42	7.31	7.08
	0.001	29.2	21.7	18.8	17.2	16.2	15.5	15.0	14.6	14.3	14.1	13.7	13.3	12.9	12.7	12.5	12.3	12.1	11.7
8	0.25	1.54	1.66	1.67	1.66	1.66	1.65	1.64	1.64	1.63	1.63	1.62	1.62	1.61	1.60	1.60	1.59	1.59	1.58
	0.1	3.46	3.11	2.92	2.81	2.73	2.67	2.62	2.59	2.56	2.54	2.50	2.46	2.42	2.40	2.38	2.36	2.34	2.29
	0.05	5.32	4.46	4.07	3.84	3.69	3.58	3.50	3.44	3.39	3.35	3.28	3.22	3.15	3.12	3.08	3.04	3.01	2.93
	0.025	7.57	6.06	5.42	5.05	4.82	4.65	4.53	4.43	4.36	4.30	4.20	4.10	4.00	3.95	3.89	3.84	3.78	3.67
	0.01	11.3	8.6	7.59	7.01	6.63	6.37	6.18	6.03	5.91	5.81	5.67	5.52	5.36	5.28	5.20	5.12	5.03	4.86
	0.005	14.7	11.0	9.60	8.81	8.30	7.95	7.69	7.50	7.34	7.21	7.01	6.81	6.61	6.50	6.40	6.29	6.18	5.95
	0.001	25.4	18.5	15.8	14.4	13.5	12.9	12.4	12.0	11.8	11.5	11.2	10.8	10.5	10.3	10.1	9.92	9.73	9.33
9	0.25	1.51	1.62	1.63	1.63	1.62	1.61	1.60	1.60	1.59	1.59	1.58	1.57	1.56	1.56	1.55	1.54	1.54	1.53
	0.1	3.36	3.01	2.81	2.69	2.61	2.55	2.51	2.47	2.44	2.42	2.38	2.34	2.30	2.28	2.25	2.23	2.21	2.16
	0.05	5.12	4.26	3.86	3.63	3.48	3.37	3.29	3.23	3.18	3.14	3.07	3.01	2.94	2.90	2.86	2.83	2.79	2.71
	0.025	7.21	5.71	5.08	4.72	4.48	4.32	4.20	4.10	4.03	3.96	3.87	3.77	3.67	3.61	3.56	3.51	3.45	3.33
	0.01	10.6	8.02	6.99	6.42	6.06	5.80	5.61	5.47	5.35	5.26	5.11	4.96	4.81	4.73	4.65	4.57	4.48	4.31
	0.005	13.6	10.1	8.72	7.96	7.47	7.13	6.88	6.69	6.54	6.42	6.23	6.03	5.83	5.73	5.62	5.52	5.41	5.19
	0.001	22.9	16.4	13.9	12.6	11.7	11.1	10.7	10.4	10.1	9.89	9.57	9.24	8.90	8.72	8.55	8.37	8.19	7.81
10	0.25	1.49	1.60	1.60	1.59	1.59	1.58	1.57	1.56	1.56	1.55	1.54	1.53	1.52	1.52	1.51	1.51	1.50	1.48
	0.1	3.29	2.92	2.73	2.61	2.52	2.46	2.41	2.38	2.35	2.32	2.28	2.24	2.20	2.18	2.16	2.13	2.11	2.06
	0.05	4.96	4.10	3.71	3.48	3.33	3.22	3.14	3.07	3.02	2.98	2.91	2.85	2.77	2.74	2.70	2.66	2.62	2.54
	0.025	6.94	5.46	4.83	4.47	4.24	4.07	3.95	3.85	3.78	3.72	3.62	3.52	3.42	3.37	3.31	3.26	3.20	3.08
	0.01	10.0	7.56	6.55	5.99	5.64	5.39	5.20	5.06	4.94	4.85	4.71	4.56	4.41	4.33	4.25	4.17	4.08	3.91
	0.005	12.8	9.43	8.08	7.34	6.87	6.54	6.30	6.12	5.97	5.85	5.66	5.47	5.27	5.17	5.07	4.97	4.86	4.64
	0.001	21.0	14.9	12.6	11.3	10.5	9.93	9.52	9.20	8.96	8.75	8.45	8.13	7.80	7.64	7.47	7.30	7.12	6.76

df for Denom.		df for Numerator																	
	α	1	2	3	4	5	6	7	8	9	10	12	15	20	24	30	40	60	∞
	0.25	1.47	1.58	1.58	1.57	1.56	1.55	1.54	1.53	1.53	1.52	1.51	1.50	1.49	1.49	1.48	1.47	1.47	1.45
	0.1	3.23	2.86	2.66	2.54	2.45	2.39	2.34	2.30	2.27	2.25	2.21	2.17	2.12	2.10	2.08	2.05	2.03	1.97
	0.05	4.84	3.98	3.59	3.36	3.20	3.09	3.01	2.95	2.90	2.85	2.79	2.72	2.65	2.61	2.57	2.53	2.49	2.40
11	0.025	6.72	5.26	4.63	4.28	4.04	3.88	3.76	3.66	3.59	3.53	3.43	3.33	3.23	3.17	3.12	3.06	3.00	2.88
	0.01	9.65	7.21	6.22	5.67	5.32	5.07	4.89	4.74	4.63	4.54	4.40	4.25	4.10	4.02	3.94	3.86	3.78	3.60
	0.005	12.2	8.91	7.60	6.88	6.42	6.10	5.86	5.68	5.54	5.42	5.24	5.05	4.86	4.76	4.65	4.55	4.45	4.23
	0.001	19.7	13.8	11.6	10.3	9.58	9.05	8.66	8.35	8.12	7.92	7.63	7.32	7.01	6.85	6.68	6.52	6.35	6.00
	0.25	1.46	1.56	1.56	1.55	1.54	1.53	1.52	1.51	1.51	1.50	1.49	1.48	1.47	1.46	1.45	1.45	1.44	1.42
	0.1	3.18	2.81	2.61	2.48	2.39	2.33	2.28	2.24	2.21	2.19	2.15	2.10	2.06	2.04	2.01	1.99	1.96	1.90
	0.05	4.75	3.89	3.49	3.26	3.11	3.00	2.91	2.85	2.80	2.75	2.69	2.62	2.54	2.51	2.47	2.43	2.38	2.30
12	0.025	6.55	5.10	4.47	4.12	3.89	3.73	3.61	3.51	3.44	3.37	3.28	3.18	3.07	3.02	2.96	2.91	2.85	2.72
	0.01	9.33	6.93	5.95	5.41	5.06	4.82	4.64	4.50	4.39	4.30	4.16	4.01	3.86	3.78	3.70	3.62	3.54	3.36
	0.005	11.8	8.51	7.23	6.52	6.07	5.76	5.52	5.35	5.20	5.09	4.91	4.72	4.53	4.43	4.33	4.23	4.12	3.90
	0.001	18.6	13.0	10.8	9.63	8.89	8.38	8.00	7.71	7.48	7.29	7.00	6.71	6.40	6.25	6.09	5.93	5.76	5.42
	0.25	1.45	1.55	1.55	1.53	1.52	1.51	1.50	1.49	1.49	1.48	1.47	1.46	1.45	1.44	1.43	1.42	1.42	1.40
	0.1	3.14	2.76	2.56	2.43	2.35	2.28	2.23	2.20	2.16	2.14	2.10	2.05	2.01	1.98	1.96	1.93	1.90	1.85
	0.05	4.67	3.81	3.41	3.18	3.03	2.92	2.83	2.77	2.71	2.67	2.60	2.53	2.46	2.42	2.38	2.34	2.30	2.21
13	0.025	6.41	4.97	4.35	4.00	3.77	3.60	3.48	3.39	3.31	3.25	3.15	3.05	2.95	2.89	2.84	2.78	2.72	2.60
	0.01	9.07	6.70	5.74	5.21	4.86	4.62	4.44	4.30	4.19	4.10	3.96	3.82	3.66	3.59	3.51	3.43	3.34	3.17
	0.005	11.4	8.19	6.93	6.23	5.79	5.48	5.25	5.08	4.94	4.82	4.64	4.46	4.27	4.17	4.07	3.97	3.87	3.65
	0.001	17.8	12.3	10.2	9.07	8.35	7.86	7.49	7.21	6.98	6.80	6.52	6.23	5.93	5.78	5.63	5.47	5.30	4.97
	0.25	1.44	1.53	1.53	1.52	1.51	1.50	1.49	1.48	1.47	1.46	1.45	1.44	1.43	1.42	1.41	1.41	1.40	1.38
	0.1	3.10	2.73	2.52	2.39	2.31	2.24	2.19	2.15	2.12	2.10	2.05	2.01	1.96	1.94	1.91	1.89	1.86	1.80
	0.05	4.60	3.74	3.34	3.11	2.96	2.85	2.76	2.70	2.65	2.60	2.53	2.46	2.39	2.35	2.31	2.27	2.22	2.13
14	0.025	6.30	4.86	4.24	3.89	3.66	3.50	3.38	3.29	3.21	3.15	3.05	2.95	2.84	2.79	2.73	2.67	2.61	2.49
	0.01	8.86	6.51	5.56	5.04	4.69	4.46	4.28	4.14	4.03	3.94	3.80	3.66	3.51	3.43	3.35	3.27	3.18	3.00
	0.005	11.1	7.92	6.68	6.00	5.56	5.26	5.03	4.86	4.72	4.60	4.43	4.25	4.06	3.96	3.86	3.76	3.66	3.44
	0.001	17.1	11.8	9.73	8.62	7.92	7.44	7.08	6.80	6.58	6.40	6.13	5.85	5.56	5.41	5.25	5.10	4.94	4.60
	0.25	1.43	1.52	1.52	1.51	1.49	1.48	1.47	1.46	1.46	1.45	1.44	1.43	1.41	1.41	1.40	1.39	1.38	1.36
	0.1	3.07	2.70	2.49	2.36	2.27	2.21	2.16	2.12	2.09	2.06	2.02	1.97	1.92	1.90	1.87	1.85	1.82	1.76
	0.05	4.54	3.68	3.29	3.06	2.90	2.79	2.71	2.64	2.59	2.54	2.48	2.40	2.33	2.29	2.25	2.20	2.16	2.07
15	0.025	6.20	4.77	4.15	3.80	3.58	3.41	3.29	3.20	3.12	3.06	2.96	2.86	2.76	2.70	2.64	2.59	2.52	2.40
	0.01	8.68	6.36	5.42	4.89	4.56	4.32	4.14	4.00	3.89	3.80	3.67	3.52	3.37	3.29	3.21	3.13	3.05	2.87
	0.005	10.8	7.70	6.48	5.80	5.37	5.07	4.85	4.67	4.54	4.42	4.25	4.07	3.88	3.79	3.69	3.58	3.48	3.26
	0.001	16.6	11.3	9.34	8.25	7.57	7.09	6.74	6.47	6.26	6.08	5.81	5.54	5.25	5.10	4.95	4.80	4.64	4.31

Table A.7. Critical values for the F distribution – *continued*

df for Denom.		df for Numerator																	
	α	1	2	3	4	5	6	7	8	9	10	12	15	20	24	30	40	60	∞
16	0.25	1.42	1.51	1.51	1.50	1.48	1.47	1.46	1.45	1.44	1.44	1.43	1.41	1.40	1.39	1.38	1.37	1.36	1.34
	0.1	3.05	2.67	2.46	2.33	2.24	2.18	2.13	2.09	2.06	2.03	1.99	1.94	1.89	1.87	1.84	1.81	1.78	1.72
	0.05	4.49	3.63	3.24	3.01	2.85	2.74	2.66	2.59	2.54	2.49	2.42	2.35	2.28	2.24	2.19	2.15	2.11	2.01
	0.025	6.12	4.69	4.08	3.73	3.50	3.34	3.22	3.12	3.05	2.99	2.89	2.79	2.68	2.63	2.57	2.51	2.45	2.32
	0.01	8.53	6.23	5.29	4.77	4.44	4.20	4.03	3.89	3.78	3.69	3.55	3.41	3.26	3.18	3.10	3.02	2.93	2.75
	0.005	10.6	7.51	6.30	5.64	5.21	4.91	4.69	4.52	4.38	4.27	4.10	3.92	3.73	3.64	3.54	3.44	3.33	3.11
	0.001	16.1	11.0	9.01	7.94	7.27	6.80	6.46	6.19	5.98	5.81	5.55	5.27	4.99	4.85	4.70	4.54	4.39	4.06
17	0.25	1.42	1.51	1.50	1.49	1.47	1.46	1.45	1.44	1.43	1.43	1.41	1.40	1.39	1.38	1.37	1.36	1.35	1.33
	0.1	3.03	2.64	2.44	2.31	2.22	2.15	2.10	2.06	2.03	2.00	1.96	1.91	1.86	1.84	1.81	1.78	1.75	1.69
	0.05	4.45	3.59	3.20	2.96	2.81	2.70	2.61	2.55	2.49	2.45	2.38	2.31	2.23	2.19	2.15	2.10	2.06	1.96
	0.025	6.04	4.62	4.01	3.66	3.44	3.28	3.16	3.06	2.98	2.92	2.82	2.72	2.62	2.56	2.50	2.44	2.38	2.25
	0.01	8.40	6.11	5.18	4.67	4.34	4.10	3.93	3.79	3.68	3.59	3.46	3.31	3.16	3.08	3.00	2.92	2.83	2.65
	0.005	10.4	7.35	6.16	5.50	5.07	4.78	4.56	4.39	4.25	4.14	3.97	3.79	3.61	3.51	3.41	3.31	3.21	2.98
	0.001	15.7	10.7	8.73	7.68	7.02	6.56	6.22	5.96	5.75	5.58	5.32	5.05	4.78	4.63	4.48	4.33	4.18	3.85
18	0.25	1.41	1.50	1.49	1.48	1.46	1.45	1.44	1.43	1.42	1.42	1.40	1.39	1.38	1.37	1.36	1.35	1.34	1.32
	0.1	3.01	2.62	2.42	2.29	2.20	2.13	2.08	2.04	2.00	1.98	1.93	1.89	1.84	1.81	1.78	1.75	1.72	1.66
	0.05	4.41	3.55	3.16	2.93	2.77	2.66	2.58	2.51	2.46	2.41	2.34	2.27	2.19	2.15	2.11	2.06	2.02	1.92
	0.025	5.98	4.56	3.95	3.61	3.38	3.22	3.10	3.01	2.93	2.87	2.77	2.67	2.56	2.50	2.44	2.38	2.32	2.19
	0.01	8.29	6.01	5.09	4.58	4.25	4.01	3.84	3.71	3.60	3.51	3.37	3.23	3.08	3.00	2.92	2.84	2.75	2.57
	0.005	10.2	7.21	6.03	5.37	4.96	4.66	4.44	4.28	4.14	4.03	3.86	3.68	3.50	3.40	3.30	3.20	3.10	2.87
	0.001	15.4	10.4	8.49	7.46	6.81	6.35	6.02	5.76	5.56	5.39	5.13	4.87	4.59	4.45	4.30	4.15	4.00	3.67
19	0.25	1.41	1.49	1.49	1.47	1.46	1.44	1.43	1.42	1.41	1.41	1.40	1.38	1.37	1.36	1.35	1.34	1.33	1.30
	0.1	2.99	2.61	2.40	2.27	2.18	2.11	2.06	2.02	1.98	1.96	1.91	1.86	1.81	1.79	1.76	1.73	1.70	1.63
	0.05	4.38	3.52	3.13	2.90	2.74	2.63	2.54	2.48	2.42	2.38	2.31	2.23	2.16	2.11	2.07	2.03	1.98	1.88
	0.025	5.92	4.51	3.90	3.56	3.33	3.17	3.05	2.96	2.88	2.82	2.72	2.62	2.51	2.45	2.39	2.33	2.27	2.13
	0.01	8.18	5.93	5.01	4.50	4.17	3.94	3.77	3.63	3.52	3.43	3.30	3.15	3.00	2.92	2.84	2.76	2.67	2.49
	0.005	10.1	7.09	5.92	5.27	4.85	4.56	4.34	4.18	4.04	3.93	3.76	3.59	3.40	3.31	3.21	3.11	3.00	2.78
	0.001	15.1	10.2	8.28	7.27	6.62	6.18	5.85	5.59	5.39	5.22	4.97	4.70	4.43	4.29	4.14	3.99	3.84	3.51
20	0.25	1.40	1.49	1.48	1.47	1.45	1.44	1.43	1.42	1.41	1.40	1.39	1.37	1.36	1.35	1.34	1.33	1.32	1.29
	0.1	2.97	2.59	2.38	2.25	2.16	2.09	2.04	2.00	1.96	1.94	1.89	1.84	1.79	1.77	1.74	1.71	1.68	1.61
	0.05	4.35	3.49	3.10	2.87	2.71	2.60	2.51	2.45	2.39	2.35	2.28	2.20	2.12	2.08	2.04	1.99	1.95	1.84
	0.025	5.87	4.46	3.86	3.51	3.29	3.13	3.01	2.91	2.84	2.77	2.68	2.57	2.46	2.41	2.35	2.29	2.22	2.09
	0.01	8.10	5.85	4.94	4.43	4.10	3.87	3.70	3.56	3.46	3.37	3.23	3.09	2.94	2.86	2.78	2.69	2.61	2.42
	0.005	9.94	6.99	5.82	5.17	4.76	4.47	4.26	4.09	3.96	3.85	3.68	3.50	3.32	3.22	3.12	3.02	2.92	2.69
	0.001	14.8	10.0	8.10	7.10	6.46	6.02	5.69	5.44	5.24	5.08	4.82	4.56	4.29	4.15	4.00	3.86	3.70	3.38

df for Denom.	α	1	2	3	4	5	6	7	8	9	10	12	15	20	24	30	40	60	∞
		df for Numerator																	
22	0.25	1.40	1.48	1.47	1.45	1.44	1.42	1.41	1.40	1.39	1.39	1.37	1.36	1.34	1.33	1.32	1.31	1.30	1.28
	0.1	2.95	2.56	2.35	2.22	2.13	2.06	2.01	1.97	1.93	1.90	1.86	1.81	1.76	1.73	1.70	1.67	1.64	1.57
	0.05	4.30	3.44	3.05	2.82	2.66	2.55	2.46	2.40	2.34	2.30	2.23	2.15	2.07	2.03	1.98	1.94	1.89	1.78
	0.025	5.79	4.38	3.78	3.44	3.22	3.05	2.93	2.84	2.76	2.70	2.60	2.50	2.39	2.33	2.27	2.21	2.14	2.00
	0.01	7.95	5.72	4.82	4.31	3.99	3.76	3.59	3.45	3.35	3.26	3.12	2.98	2.83	2.75	2.67	2.58	2.50	2.31
	0.005	9.73	6.81	5.65	5.02	4.61	4.32	4.11	3.94	3.81	3.70	3.54	3.36	3.18	3.08	2.98	2.88	2.77	2.55
	0.001	14.4	9.61	7.80	6.81	6.19	5.76	5.44	5.19	4.99	4.83	4.58	4.33	4.06	3.92	3.78	3.63	3.48	3.15
24	0.25	1.39	1.47	1.46	1.44	1.43	1.41	1.40	1.39	1.38	1.38	1.36	1.35	1.33	1.32	1.31	1.30	1.29	1.26
	0.1	2.93	2.54	2.33	2.19	2.10	2.04	1.98	1.94	1.91	1.88	1.83	1.78	1.73	1.70	1.67	1.64	1.61	1.53
	0.05	4.26	3.40	3.01	2.78	2.62	2.51	2.42	2.36	2.30	2.25	2.18	2.11	2.03	1.98	1.94	1.89	1.84	1.73
	0.025	5.72	4.32	3.72	3.38	3.15	2.99	2.87	2.78	2.70	2.64	2.54	2.44	2.33	2.27	2.21	2.15	2.08	1.94
	0.01	7.82	5.61	4.72	4.22	3.90	3.67	3.50	3.36	3.26	3.17	3.03	2.89	2.74	2.66	2.58	2.49	2.40	2.21
	0.005	9.55	6.66	5.52	4.89	4.49	4.20	3.99	3.83	3.69	3.59	3.42	3.25	3.06	2.97	2.87	2.77	2.66	2.43
	0.001	14.0	9.34	7.55	6.59	5.98	5.55	5.23	4.99	4.80	4.64	4.39	4.14	3.87	3.74	3.59	3.45	3.29	2.97
30	0.25	1.38	1.45	1.44	1.42	1.41	1.39	1.38	1.37	1.36	1.35	1.34	1.32	1.30	1.29	1.28	1.27	1.26	1.23
	0.1	2.88	2.49	2.28	2.14	2.05	1.98	1.93	1.88	1.85	1.82	1.77	1.72	1.67	1.64	1.61	1.57	1.54	1.46
	0.05	4.17	3.32	2.92	2.69	2.53	2.42	2.33	2.27	2.21	2.16	2.09	2.01	1.93	1.89	1.84	1.79	1.74	1.62
	0.025	5.57	4.18	3.59	3.25	3.03	2.87	2.75	2.65	2.57	2.51	2.41	2.31	2.20	2.14	2.07	2.01	1.94	1.79
	0.01	7.56	5.39	4.51	4.02	3.70	3.47	3.30	3.17	3.07	2.98	2.84	2.70	2.55	2.47	2.39	2.30	2.21	2.01
	0.005	9.18	6.35	5.24	4.62	4.23	3.95	3.74	3.58	3.45	3.34	3.18	3.01	2.82	2.73	2.63	2.52	2.42	2.18
	0.001	13.3	8.77	7.05	6.12	5.53	5.12	4.82	4.58	4.39	4.24	4.00	3.75	3.49	3.36	3.22	3.07	2.92	2.59
60	0.25	1.35	1.42	1.41	1.38	1.37	1.35	1.33	1.32	1.31	1.30	1.29	1.27	1.25	1.24	1.22	1.21	1.19	1.15
	0.1	2.79	2.39	2.18	2.04	1.95	1.87	1.82	1.77	1.74	1.71	1.66	1.60	1.54	1.51	1.48	1.44	1.40	1.29
	0.05	4.00	3.15	2.76	2.53	2.37	2.25	2.17	2.10	2.04	1.99	1.92	1.84	1.75	1.70	1.65	1.59	1.53	1.39
	0.025	5.29	3.93	3.34	3.01	2.79	2.63	2.51	2.41	2.33	2.27	2.17	2.06	1.94	1.88	1.82	1.74	1.67	1.48
	0.01	7.08	4.98	4.13	3.65	3.34	3.12	2.95	2.82	2.72	2.63	2.50	2.35	2.20	2.12	2.03	1.94	1.84	1.60
	0.005	8.49	5.79	4.73	4.14	3.76	3.49	3.29	3.13	3.01	2.90	2.74	2.57	2.39	2.29	2.19	2.08	1.96	1.69
	0.001	12.0	7.77	6.17	5.31	4.76	4.37	4.09	3.86	3.69	3.54	3.32	3.08	2.83	2.69	2.55	2.41	2.25	1.89
120	0.25	1.34	1.40	1.39	1.37	1.35	1.33	1.31	1.30	1.29	1.28	1.26	1.24	1.22	1.21	1.19	1.18	1.16	1.10
	0.1	2.75	2.35	2.13	1.99	1.90	1.82	1.77	1.72	1.68	1.65	1.60	1.55	1.48	1.45	1.41	1.37	1.32	1.19
	0.05	3.92	3.07	2.68	2.45	2.29	2.18	2.09	2.02	1.96	1.91	1.83	1.75	1.66	1.61	1.55	1.50	1.43	1.25
	0.025	5.15	3.80	3.23	2.89	2.67	2.52	2.39	2.30	2.22	2.16	2.05	1.94	1.82	1.76	1.69	1.61	1.53	1.31
	0.01	6.85	4.79	3.95	3.48	3.17	2.96	2.79	2.66	2.56	2.47	2.34	2.19	2.03	1.95	1.86	1.76	1.66	1.38
	0.005	8.18	5.54	4.50	3.92	3.55	3.28	3.09	2.93	2.81	2.71	2.54	2.37	2.19	2.09	1.98	1.87	1.75	1.43
	0.001	11.4	7.32	5.78	4.95	4.42	4.04	3.77	3.55	3.38	3.24	3.02	2.78	2.53	2.40	2.26	2.11	1.95	1.54

Table A.7. Critical values for the F distribution – *continued*

df for Denom.		df for Numerator																	
	α	1	2	3	4	5	6	7	8	9	10	12	15	20	24	30	40	60	∞
	0.25	1.32	1.39	1.37	1.35	1.33	1.31	1.29	1.28	1.27	1.25	1.24	1.22	1.19	1.18	1.16	1.14	1.12	1.00
	0.1	2.71	2.30	2.08	1.94	1.85	1.77	1.72	1.67	1.63	1.60	1.55	1.49	1.42	1.38	1.34	1.30	1.24	1.00
	0.05	3.84	3.00	2.60	2.37	2.21	2.10	2.01	1.94	1.88	1.83	1.75	1.67	1.57	1.52	1.46	1.39	1.32	1.00
∞	0.025	5.02	3.69	3.12	2.79	2.57	2.41	2.29	2.19	2.11	2.05	1.94	1.83	1.71	1.64	1.57	1.48	1.39	1.00
	0.01	6.63	4.61	3.78	3.32	3.02	2.80	2.64	2.51	2.41	2.32	2.18	2.04	1.88	1.79	1.70	1.59	1.47	1.00
	0.005	7.88	5.30	4.28	3.72	3.35	3.09	2.90	2.74	2.62	2.52	2.36	2.19	2.00	1.90	1.79	1.67	1.53	1.00
	0.001	10.8	6.91	5.42	4.62	4.10	3.74	3.47	3.27	3.10	2.96	2.74	2.51	2.27	2.13	1.99	1.84	1.66	1.00

Table A.8 Critical values for the r distribution, probability level = 0.05.

	Number of independent variables																				
df	1	2	3	4	5	6	7	8	9	10	11	12	14	16	20	24	30	40	50	75	100
1	0.997	0.999	0.999	0.999	1.000	1.000	1.000	1.000	1.000	1.000	1.000	1.000	1.000	1.000	1.000	1.000	1.000	1.000	1.000	1.000	1.000
2	0.950	0.975	0.983	0.987	0.990	0.991	0.993	0.994	0.994	0.995	0.995	0.996	0.996	0.997	0.997	0.998	0.998	0.999	0.999	0.999	0.999
3	0.878	0.930	0.950	0.961	0.968	0.973	0.977	0.979	0.982	0.983	0.985	0.986	0.988	0.989	0.991	0.993	0.994	0.996	0.997	0.998	0.998
4	0.811	0.881	0.912	0.930	0.942	0.950	0.956	0.961	0.965	0.968	0.971	0.973	0.977	0.979	0.983	0.986	0.989	0.991	0.993	0.995	0.996
5	0.754	0.836	0.874	0.898	0.914	0.925	0.934	0.941	0.946	0.951	0.955	0.958	0.964	0.968	0.974	0.978	0.982	0.986	0.989	0.993	0.994
6	0.707	0.795	0.839	0.867	0.886	0.900	0.911	0.920	0.927	0.933	0.938	0.943	0.950	0.955	0.963	0.969	0.975	0.981	0.984	0.989	0.992
7	0.666	0.758	0.807	0.838	0.860	0.876	0.889	0.900	0.909	0.916	0.922	0.927	0.936	0.943	0.953	0.960	0.967	0.975	0.980	0.986	0.989
8	0.632	0.726	0.777	0.811	0.835	0.854	0.868	0.880	0.890	0.898	0.906	0.912	0.922	0.930	0.942	0.950	0.959	0.969	0.975	0.983	0.987
9	0.602	0.697	0.750	0.786	0.812	0.832	0.848	0.861	0.872	0.882	0.890	0.897	0.908	0.917	0.931	0.941	0.951	0.962	0.969	0.979	0.984
10	0.576	0.671	0.726	0.763	0.790	0.812	0.829	0.843	0.855	0.865	0.874	0.882	0.895	0.905	0.920	0.932	0.943	0.956	0.964	0.975	0.981
11	0.553	0.648	0.703	0.741	0.770	0.792	0.811	0.826	0.839	0.850	0.859	0.867	0.882	0.893	0.910	0.922	0.936	0.950	0.959	0.972	0.978
12	0.532	0.627	0.683	0.722	0.751	0.774	0.793	0.809	0.823	0.835	0.845	0.854	0.869	0.881	0.900	0.913	0.928	0.943	0.953	0.968	0.975
13	0.514	0.608	0.664	0.703	0.733	0.757	0.777	0.794	0.808	0.820	0.831	0.840	0.856	0.869	0.889	0.904	0.920	0.937	0.948	0.964	0.972
14	0.497	0.590	0.646	0.686	0.717	0.741	0.762	0.779	0.794	0.806	0.818	0.828	0.844	0.858	0.879	0.895	0.912	0.931	0.943	0.960	0.969
15	0.482	0.574	0.630	0.670	0.701	0.726	0.747	0.765	0.780	0.793	0.805	0.815	0.833	0.847	0.870	0.886	0.904	0.924	0.938	0.956	0.966
16	0.468	0.559	0.615	0.655	0.687	0.712	0.733	0.751	0.767	0.780	0.793	0.803	0.822	0.837	0.860	0.878	0.897	0.918	0.932	0.953	0.963
17	0.456	0.545	0.601	0.641	0.673	0.698	0.720	0.738	0.754	0.768	0.781	0.792	0.811	0.826	0.851	0.869	0.890	0.912	0.927	0.949	0.960
18	0.444	0.532	0.587	0.628	0.660	0.686	0.707	0.726	0.742	0.757	0.769	0.781	0.800	0.816	0.842	0.861	0.882	0.906	0.922	0.945	0.957
19	0.433	0.520	0.575	0.615	0.647	0.673	0.696	0.714	0.731	0.746	0.759	0.770	0.790	0.807	0.833	0.853	0.875	0.900	0.917	0.941	0.954
20	0.423	0.509	0.563	0.604	0.636	0.662	0.684	0.703	0.720	0.735	0.748	0.760	0.780	0.797	0.825	0.845	0.868	0.894	0.912	0.937	0.951
21	0.413	0.498	0.552	0.593	0.624	0.651	0.673	0.693	0.710	0.725	0.738	0.750	0.771	0.788	0.816	0.837	0.861	0.888	0.906	0.933	0.948
22	0.404	0.488	0.542	0.582	0.614	0.640	0.663	0.682	0.699	0.715	0.728	0.741	0.762	0.780	0.808	0.830	0.854	0.883	0.901	0.930	0.945
23	0.396	0.479	0.532	0.572	0.604	0.630	0.653	0.673	0.690	0.705	0.719	0.731	0.753	0.771	0.800	0.823	0.848	0.877	0.897	0.926	0.942
24	0.388	0.470	0.523	0.562	0.594	0.621	0.643	0.663	0.681	0.696	0.710	0.722	0.744	0.763	0.793	0.815	0.841	0.871	0.892	0.922	0.939
25	0.381	0.462	0.514	0.553	0.585	0.612	0.634	0.654	0.672	0.687	0.701	0.714	0.736	0.755	0.785	0.808	0.835	0.866	0.887	0.919	0.936
26	0.374	0.454	0.506	0.545	0.576	0.603	0.626	0.645	0.663	0.679	0.693	0.706	0.728	0.747	0.778	0.802	0.829	0.860	0.882	0.915	0.933
27	0.367	0.446	0.498	0.536	0.568	0.594	0.617	0.637	0.655	0.670	0.685	0.698	0.720	0.739	0.771	0.795	0.823	0.855	0.877	0.911	0.930
28	0.361	0.439	0.490	0.529	0.560	0.586	0.609	0.629	0.647	0.662	0.677	0.690	0.713	0.732	0.764	0.788	0.817	0.850	0.873	0.908	0.928
29	0.355	0.432	0.483	0.521	0.552	0.579	0.601	0.621	0.639	0.655	0.669	0.682	0.705	0.725	0.757	0.782	0.811	0.845	0.868	0.904	0.925
30	0.349	0.425	0.476	0.514	0.545	0.571	0.594	0.614	0.631	0.647	0.662	0.675	0.698	0.718	0.750	0.776	0.805	0.840	0.864	0.901	0.922
31	0.344	0.419	0.469	0.507	0.538	0.564	0.587	0.607	0.624	0.640	0.655	0.668	0.691	0.711	0.744	0.769	0.799	0.835	0.859	0.897	0.919
32	0.339	0.413	0.462	0.500	0.531	0.557	0.580	0.600	0.617	0.633	0.648	0.661	0.684	0.705	0.738	0.763	0.794	0.830	0.855	0.894	0.916
34	0.329	0.402	0.450	0.488	0.518	0.544	0.566	0.586	0.604	0.620	0.635	0.648	0.672	0.692	0.725	0.752	0.783	0.820	0.846	0.887	0.910
36	0.320	0.392	0.439	0.476	0.506	0.532	0.554	0.574	0.591	0.608	0.622	0.636	0.659	0.680	0.714	0.741	0.773	0.811	0.838	0.880	0.905
38	0.312	0.382	0.429	0.465	0.495	0.520	0.542	0.562	0.580	0.596	0.610	0.624	0.648	0.668	0.703	0.730	0.763	0.802	0.829	0.873	0.899

Table A.8 Critical values for the r distribution, probability level = 0.05 – *continued*

df	Number of independent variables																				
	1	2	3	4	5	6	7	8	9	10	11	12	14	16	20	24	30	40	50	75	100
40	0.304	0.373	0.419	0.455	0.484	0.509	0.531	0.551	0.569	0.585	0.599	0.613	0.637	0.657	0.692	0.720	0.753	0.793	0.821	0.867	0.894
42	0.297	0.365	0.410	0.445	0.474	0.499	0.521	0.541	0.558	0.574	0.589	0.602	0.626	0.647	0.682	0.710	0.744	0.784	0.814	0.861	0.888
44	0.291	0.357	0.401	0.436	0.465	0.490	0.511	0.531	0.548	0.564	0.579	0.592	0.616	0.637	0.672	0.701	0.735	0.776	0.806	0.854	0.883
46	0.285	0.349	0.393	0.428	0.456	0.481	0.502	0.521	0.539	0.555	0.569	0.583	0.607	0.628	0.663	0.692	0.726	0.768	0.799	0.848	0.878
48	0.279	0.343	0.386	0.420	0.448	0.472	0.493	0.513	0.530	0.546	0.560	0.573	0.598	0.619	0.654	0.683	0.717	0.760	0.791	0.842	0.873
50	0.273	0.336	0.379	0.412	0.440	0.464	0.485	0.504	0.521	0.537	0.551	0.565	0.589	0.610	0.645	0.674	0.709	0.753	0.784	0.836	0.868
55	0.261	0.321	0.362	0.395	0.422	0.445	0.466	0.485	0.502	0.517	0.531	0.545	0.568	0.590	0.625	0.654	0.690	0.735	0.768	0.822	0.855
60	0.250	0.308	0.348	0.380	0.406	0.429	0.449	0.467	0.484	0.499	0.513	0.526	0.550	0.571	0.607	0.636	0.672	0.718	0.752	0.808	0.844
65	0.240	0.297	0.335	0.366	0.392	0.414	0.434	0.452	0.468	0.483	0.497	0.510	0.533	0.554	0.590	0.619	0.656	0.702	0.737	0.795	0.832
70	0.232	0.286	0.324	0.354	0.379	0.401	0.420	0.438	0.454	0.469	0.482	0.495	0.518	0.539	0.574	0.604	0.640	0.687	0.723	0.783	0.821
80	0.217	0.269	0.304	0.332	0.356	0.377	0.396	0.413	0.428	0.443	0.456	0.469	0.491	0.511	0.546	0.576	0.613	0.660	0.697	0.760	0.800
100	0.195	0.241	0.274	0.299	0.321	0.341	0.358	0.374	0.388	0.402	0.414	0.426	0.448	0.467	0.501	0.530	0.566	0.614	0.652	0.718	0.763
125	0.174	0.216	0.246	0.269	0.289	0.307	0.323	0.338	0.351	0.364	0.376	0.387	0.407	0.425	0.458	0.485	0.521	0.568	0.606	0.675	0.722
150	0.159	0.198	0.225	0.247	0.265	0.282	0.297	0.311	0.323	0.335	0.346	0.356	0.375	0.393	0.424	0.450	0.485	0.531	0.569	0.639	0.688
200	0.138	0.172	0.196	0.215	0.231	0.246	0.259	0.271	0.282	0.293	0.303	0.312	0.330	0.345	0.374	0.398	0.430	0.475	0.511	0.580	0.631
400	0.098	0.122	0.139	0.153	0.165	0.176	0.185	0.194	0.203	0.210	0.218	0.225	0.238	0.250	0.272	0.291	0.317	0.353	0.384	0.445	0.493
1000	0.062	0.077	0.088	0.097	0.105	0.112	0.118	0.124	0.129	0.134	0.139	0.144	0.153	0.161	0.175	0.188	0.206	0.231	0.253	0.298	0.334

Table A.8 Critical values for the r distribution, probability level = 0.01.

	Number of independent variables																				
df	1	2	3	4	5	6	7	8	9	10	11	12	14	16	20	24	30	40	50	75	100
1	1.000	1.000	1.000	1.000	1.000	1.000	1.000	1.000	1.000	1.000	1.000	1.000	1.000	1.000	1.000	1.000	1.000	1.000	1.000	1.000	1.000
2	0.990	0.995	0.997	0.997	0.998	0.998	0.999	0.999	0.999	0.999	0.999	0.999	0.999	0.999	0.999	1.000	1.000	1.000	1.000	1.000	1.000
3	0.959	0.977	0.983	0.987	0.990	0.991	0.992	0.993	0.994	0.995	0.995	0.995	0.996	0.997	0.997	0.998	0.998	0.999	0.999	0.999	0.999
4	0.917	0.949	0.962	0.970	0.975	0.979	0.981	0.984	0.985	0.987	0.988	0.989	0.990	0.991	0.993	0.994	0.995	0.996	0.997	0.998	0.999
5	0.875	0.917	0.937	0.949	0.957	0.963	0.967	0.971	0.974	0.976	0.978	0.980	0.982	0.984	0.987	0.989	0.991	0.993	0.995	0.996	0.997
6	0.834	0.886	0.911	0.927	0.938	0.946	0.952	0.957	0.961	0.964	0.967	0.969	0.973	0.976	0.980	0.983	0.986	0.990	0.992	0.994	0.996
7	0.798	0.855	0.885	0.904	0.918	0.928	0.935	0.942	0.947	0.951	0.955	0.958	0.963	0.967	0.973	0.977	0.981	0.986	0.988	0.992	0.994
8	0.765	0.827	0.860	0.882	0.898	0.909	0.919	0.926	0.932	0.938	0.942	0.946	0.952	0.957	0.965	0.970	0.975	0.981	0.985	0.989	0.992
9	0.735	0.800	0.837	0.861	0.878	0.891	0.902	0.911	0.918	0.924	0.929	0.934	0.941	0.947	0.956	0.963	0.969	0.976	0.981	0.987	0.990
10	0.708	0.776	0.814	0.840	0.859	0.874	0.886	0.895	0.904	0.911	0.916	0.922	0.930	0.937	0.948	0.955	0.963	0.971	0.977	0.984	0.988
11	0.684	0.753	0.793	0.821	0.841	0.857	0.870	0.881	0.889	0.897	0.904	0.910	0.919	0.927	0.939	0.947	0.956	0.966	0.972	0.981	0.985
12	0.661	0.732	0.773	0.802	0.824	0.841	0.855	0.866	0.876	0.884	0.891	0.898	0.909	0.917	0.930	0.940	0.950	0.961	0.968	0.978	0.983
13	0.641	0.712	0.755	0.785	0.807	0.825	0.840	0.852	0.862	0.871	0.879	0.886	0.898	0.907	0.922	0.932	0.943	0.956	0.964	0.975	0.981
14	0.623	0.694	0.737	0.768	0.791	0.810	0.825	0.838	0.849	0.859	0.867	0.875	0.887	0.897	0.913	0.924	0.937	0.950	0.959	0.972	0.978
15	0.606	0.677	0.721	0.752	0.776	0.796	0.812	0.825	0.837	0.847	0.856	0.864	0.877	0.888	0.904	0.917	0.930	0.945	0.955	0.968	0.976
16	0.590	0.662	0.706	0.738	0.762	0.782	0.799	0.813	0.825	0.835	0.844	0.853	0.867	0.878	0.896	0.909	0.924	0.940	0.950	0.965	0.973
17	0.575	0.647	0.691	0.724	0.749	0.769	0.786	0.801	0.813	0.824	0.834	0.842	0.857	0.869	0.888	0.902	0.917	0.934	0.946	0.962	0.971
18	0.561	0.633	0.678	0.710	0.736	0.757	0.774	0.789	0.802	0.813	0.823	0.832	0.847	0.860	0.880	0.894	0.911	0.929	0.941	0.959	0.968
19	0.549	0.620	0.665	0.697	0.723	0.745	0.762	0.778	0.791	0.802	0.813	0.822	0.838	0.851	0.872	0.887	0.904	0.924	0.936	0.955	0.965
20	0.537	0.607	0.652	0.685	0.712	0.733	0.751	0.767	0.780	0.792	0.803	0.812	0.829	0.842	0.864	0.880	0.898	0.918	0.932	0.952	0.963
21	0.526	0.596	0.641	0.674	0.700	0.722	0.740	0.756	0.770	0.782	0.793	0.803	0.820	0.834	0.856	0.873	0.892	0.913	0.927	0.949	0.960
22	0.515	0.585	0.630	0.663	0.690	0.711	0.730	0.746	0.760	0.773	0.784	0.794	0.811	0.825	0.848	0.866	0.886	0.908	0.923	0.945	0.957
23	0.505	0.574	0.619	0.653	0.679	0.701	0.720	0.736	0.751	0.763	0.775	0.785	0.803	0.817	0.841	0.859	0.880	0.903	0.919	0.942	0.955
24	0.496	0.565	0.609	0.643	0.669	0.692	0.711	0.727	0.741	0.754	0.766	0.776	0.794	0.810	0.834	0.852	0.874	0.898	0.914	0.939	0.952
25	0.487	0.555	0.600	0.633	0.660	0.682	0.701	0.718	0.733	0.746	0.757	0.768	0.786	0.802	0.827	0.846	0.868	0.893	0.910	0.935	0.949
26	0.479	0.546	0.590	0.624	0.651	0.673	0.692	0.709	0.724	0.737	0.749	0.760	0.778	0.794	0.820	0.839	0.862	0.888	0.905	0.932	0.947
27	0.471	0.538	0.582	0.615	0.642	0.664	0.684	0.701	0.716	0.729	0.741	0.752	0.771	0.787	0.813	0.833	0.856	0.883	0.901	0.929	0.944
28	0.463	0.529	0.573	0.607	0.633	0.656	0.676	0.693	0.708	0.721	0.733	0.744	0.763	0.780	0.806	0.827	0.850	0.878	0.897	0.925	0.942
29	0.456	0.522	0.565	0.598	0.625	0.648	0.668	0.685	0.700	0.713	0.726	0.737	0.756	0.773	0.800	0.821	0.845	0.873	0.893	0.922	0.939
30	0.449	0.514	0.558	0.591	0.618	0.640	0.660	0.677	0.692	0.706	0.718	0.729	0.749	0.766	0.793	0.815	0.839	0.868	0.888	0.919	0.936
31	0.442	0.507	0.550	0.583	0.610	0.633	0.652	0.670	0.685	0.699	0.711	0.722	0.742	0.759	0.787	0.809	0.834	0.864	0.884	0.916	0.934
32	0.436	0.500	0.543	0.576	0.603	0.625	0.645	0.662	0.678	0.692	0.704	0.716	0.736	0.753	0.781	0.803	0.829	0.859	0.880	0.912	0.931
34	0.424	0.487	0.530	0.562	0.589	0.612	0.631	0.649	0.664	0.678	0.691	0.702	0.723	0.740	0.769	0.792	0.818	0.850	0.872	0.906	0.926
36	0.413	0.475	0.517	0.549	0.576	0.599	0.618	0.636	0.651	0.665	0.678	0.690	0.711	0.728	0.758	0.781	0.808	0.841	0.864	0.900	0.921
38	0.403	0.464	0.505	0.537	0.564	0.586	0.606	0.623	0.639	0.653	0.666	0.678	0.699	0.717	0.747	0.771	0.799	0.832	0.856	0.894	0.916

Table A.8 Critical values for the r distribution, probability level = 0.01 – *continued*

	Number of independent variables																				
df	1	2	3	4	5	6	7	8	9	10	11	12	14	16	20	24	30	40	50	75	100
40	0.393	0.454	0.494	0.526	0.552	0.575	0.595	0.612	0.628	0.642	0.655	0.667	0.688	0.706	0.736	0.761	0.789	0.824	0.849	0.888	0.910
42	0.384	0.444	0.484	0.516	0.542	0.564	0.584	0.601	0.617	0.631	0.644	0.656	0.677	0.696	0.726	0.751	0.780	0.816	0.841	0.882	0.905
44	0.376	0.435	0.474	0.506	0.532	0.554	0.573	0.591	0.606	0.620	0.633	0.645	0.667	0.685	0.717	0.742	0.771	0.808	0.834	0.876	0.900
46	0.368	0.426	0.465	0.496	0.522	0.544	0.563	0.581	0.596	0.611	0.624	0.636	0.657	0.676	0.707	0.733	0.763	0.800	0.827	0.870	0.896
48	0.361	0.418	0.457	0.487	0.513	0.535	0.554	0.571	0.587	0.601	0.614	0.626	0.648	0.667	0.698	0.724	0.755	0.792	0.820	0.864	0.891
50	0.354	0.410	0.449	0.479	0.504	0.526	0.545	0.562	0.578	0.592	0.605	0.617	0.639	0.658	0.689	0.715	0.746	0.785	0.813	0.858	0.886
55	0.339	0.393	0.430	0.460	0.484	0.506	0.525	0.542	0.557	0.571	0.584	0.596	0.618	0.637	0.669	0.695	0.727	0.767	0.797	0.845	0.874
60	0.325	0.377	0.414	0.442	0.467	0.488	0.506	0.523	0.538	0.552	0.565	0.577	0.599	0.618	0.650	0.677	0.710	0.751	0.781	0.832	0.863
65	0.313	0.363	0.399	0.427	0.451	0.471	0.490	0.506	0.521	0.535	0.548	0.560	0.581	0.600	0.633	0.660	0.693	0.735	0.766	0.819	0.852
70	0.302	0.351	0.386	0.413	0.436	0.456	0.475	0.491	0.506	0.519	0.532	0.544	0.565	0.584	0.617	0.644	0.678	0.720	0.752	0.807	0.841
80	0.283	0.330	0.363	0.389	0.411	0.431	0.448	0.464	0.478	0.492	0.504	0.516	0.537	0.556	0.588	0.615	0.649	0.693	0.726	0.784	0.821
100	0.254	0.297	0.327	0.351	0.372	0.390	0.406	0.421	0.435	0.447	0.459	0.470	0.491	0.509	0.541	0.568	0.602	0.647	0.682	0.743	0.784
125	0.228	0.267	0.294	0.316	0.335	0.352	0.367	0.381	0.394	0.406	0.417	0.428	0.447	0.464	0.495	0.521	0.555	0.600	0.635	0.700	0.744
150	0.208	0.244	0.269	0.290	0.308	0.324	0.338	0.351	0.363	0.374	0.385	0.395	0.413	0.430	0.459	0.485	0.517	0.562	0.597	0.663	0.709
200	0.181	0.212	0.235	0.253	0.269	0.283	0.295	0.307	0.318	0.328	0.338	0.347	0.363	0.379	0.406	0.430	0.460	0.503	0.538	0.604	0.652
400	0.128	0.151	0.167	0.180	0.192	0.202	0.212	0.221	0.229	0.236	0.244	0.250	0.263	0.275	0.296	0.315	0.340	0.376	0.406	0.466	0.512
1000	0.081	0.096	0.106	0.115	0.122	0.129	0.135	0.141	0.146	0.151	0.156	0.160	0.169	0.177	0.191	0.204	0.221	0.246	0.268	0.312	0.349

Table A.9. Critical values for the Bonferroni t Statistic: $t_{a,k,\nu}$ for $P(|t| \geq t_{a,k,\nu}) = \alpha$.

$\alpha = 0.05$

ν	$k = 2$	3	4	5	6	7	8	9	10	11	12	13	14	15	20
5	3.16	3.53	3.81	4.03	4.22	4.38	4.53	4.66	4.77	4.88	4.98	5.08	5.16	5.25	5.60
6	2.97	3.29	3.52	3.71	3.86	4.00	4.12	4.22	4.32	4.40	4.49	4.56	4.63	4.70	4.98
7	2.84	3.13	3.34	3.50	3.64	3.75	3.86	3.95	4.03	4.10	4.17	4.24	4.30	4.36	4.59
8	2.75	3.02	3.21	3.36	3.48	3.58	3.68	3.76	3.83	3.90	3.96	4.02	4.07	4.12	4.33
9	2.69	2.93	3.11	3.25	3.36	3.46	3.55	3.62	3.69	3.75	3.81	3.86	3.91	3.95	4.15
10	2.63	2.87	3.04	3.17	3.28	3.37	3.45	3.52	3.58	3.64	3.69	3.74	3.79	3.83	4.00
11	2.59	2.82	2.98	3.11	3.21	3.29	3.37	3.44	3.50	3.55	3.60	3.65	3.69	3.73	3.89
12	2.56	2.78	2.93	3.05	3.15	3.24	3.31	3.37	3.43	3.48	3.53	3.57	3.61	3.65	3.81
13	2.53	2.75	2.90	3.01	3.11	3.19	3.26	3.32	3.37	3.42	3.47	3.51	3.55	3.58	3.73
14	2.51	2.72	2.86	2.98	3.07	3.15	3.21	3.27	3.33	3.37	3.42	3.46	3.49	3.53	3.67
15	2.49	2.69	2.84	2.95	3.04	3.11	3.18	3.23	3.29	3.33	3.37	3.41	3.45	3.48	3.62
16	2.47	2.67	2.81	2.92	3.01	3.08	3.15	3.20	3.25	3.30	3.34	3.38	3.41	3.44	3.58
17	2.46	2.65	2.79	2.90	2.98	3.06	3.12	3.17	3.22	3.27	3.31	3.34	3.38	3.41	3.54
18	2.45	2.64	2.77	2.88	2.96	3.03	3.09	3.15	3.20	3.24	3.28	3.32	3.35	3.38	3.51
19	2.43	2.63	2.76	2.86	2.94	3.01	3.07	3.13	3.17	3.22	3.25	3.29	3.32	3.35	3.48
20	2.42	2.61	2.74	2.85	2.93	3.00	3.06	3.11	3.15	3.20	3.23	3.27	3.30	3.33	3.46
25	2.38	2.57	2.69	2.79	2.86	2.93	2.99	3.03	3.08	3.12	3.15	3.19	3.22	3.24	3.36
30	2.36	2.54	2.66	2.75	2.82	2.89	2.94	2.99	3.03	3.07	3.10	3.13	3.16	3.19	3.30
40	2.33	2.50	2.62	2.70	2.78	2.84	2.89	2.93	2.97	3.01	3.04	3.07	3.10	3.12	3.23
60	2.30	2.46	2.58	2.66	2.73	2.79	2.83	2.88	2.91	2.95	2.98	3.01	3.03	3.06	3.16
120	2.27	2.43	2.54	2.62	2.68	2.74	2.78	2.82	2.86	2.89	2.92	2.95	2.97	3.00	3.09
∞	2.24	2.39	2.50	2.58	2.64	2.69	2.73	2.77	2.81	2.84	2.87	2.89	2.91	2.94	3.02

$\alpha = 0.01$

ν	$k = 2$	3	4	5	6	7	8	9	10	11	12	13	14	15	20
5	4.77	5.25	5.60	5.89	6.14	6.35	6.54	6.71	6.87	7.01	7.15	7.27	7.39	7.50	7.98
6	4.32	4.70	4.98	5.21	5.40	5.56	5.71	5.84	5.96	6.07	6.17	6.26	6.35	6.43	6.79
7	4.03	4.36	4.59	4.79	4.94	5.08	5.20	5.31	5.41	5.50	5.58	5.66	5.73	5.80	6.08
8	3.83	4.12	4.33	4.50	4.64	4.76	4.86	4.96	5.04	5.12	5.19	5.25	5.32	5.37	5.62
9	3.69	3.95	4.15	4.30	4.42	4.53	4.62	4.71	4.78	4.85	4.91	4.97	5.02	5.08	5.29
10	3.58	3.83	4.00	4.14	4.26	4.36	4.44	4.52	4.59	4.65	4.71	4.76	4.81	4.85	5.05
11	3.50	3.73	3.89	4.02	4.13	4.22	4.30	4.37	4.44	4.49	4.55	4.60	4.64	4.68	4.86
12	3.43	3.65	3.81	3.93	4.03	4.12	4.19	4.26	4.32	4.37	4.42	4.47	4.51	4.55	4.72
13	3.37	3.58	3.73	3.85	3.95	4.03	4.10	4.16	4.22	4.27	4.32	4.36	4.40	4.44	4.60
14	3.33	3.53	3.67	3.79	3.88	3.96	4.03	4.09	4.14	4.19	4.23	4.28	4.31	4.35	4.50
15	3.29	3.48	3.62	3.73	3.82	3.90	3.96	4.02	4.07	4.12	4.16	4.20	4.24	4.27	4.42
16	3.25	3.44	3.58	3.69	3.77	3.85	3.91	3.96	4.01	4.06	4.10	4.14	4.18	4.21	4.35
17	3.22	3.41	3.54	3.65	3.73	3.80	3.86	3.92	3.97	4.01	4.05	4.09	4.12	4.15	4.29
18	3.20	3.38	3.51	3.61	3.69	3.76	3.82	3.87	3.92	3.96	4.00	4.04	4.07	4.10	4.23
19	3.17	3.35	3.48	3.58	3.66	3.73	3.79	3.84	3.88	3.93	3.96	4.00	4.03	4.06	4.19
20	3.15	3.33	3.46	3.55	3.63	3.70	3.75	3.80	3.85	3.89	3.93	3.96	3.99	4.02	4.15
25	3.08	3.24	3.36	3.45	3.52	3.58	3.64	3.68	3.73	3.76	3.80	3.83	3.86	3.88	4.00
30	3.03	3.19	3.30	3.39	3.45	3.51	3.56	3.61	3.65	3.68	3.71	3.74	3.77	3.80	3.90
40	2.97	3.12	3.23	3.31	3.37	3.43	3.47	3.51	3.55	3.58	3.61	3.64	3.67	3.69	3.79
60	2.91	3.06	3.16	3.23	3.29	3.34	3.39	3.43	3.46	3.49	3.52	3.54	3.57	3.59	3.68
120	2.86	3.00	3.09	3.16	3.22	3.26	3.31	3.34	3.37	3.40	3.43	3.45	3.47	3.49	3.58
∞	2.81	2.94	3.02	3.09	3.14	3.19	3.23	3.26	3.29	3.32	3.34	3.36	3.39	3.40	3.48

k, number of comparisons.

Source: Bailey, B.J.R. (1977) Tables of the Bonferroni *t* statistics. *Journal of the American Statistical Association* 72, 459–478. Reproduced with permission from the *Journal of the American Statistical Association*. Copyright 1977 by the American Statistical Association. All rights reserved.

Table A.10. Critical values for the Wilcoxon signed rank test.

n	One-sided $\alpha = 0.01$ Two-sided $\alpha = 0.02$	One-sided $\alpha = 0.025$ Two-sided $\alpha = 0.05$	One-sided $\alpha = 0.05$ Two-sided $\alpha = 0.10$
5			1
6		1	2
7	0	2	4
8	2	4	6
9	3	6	8
10	5	8	11
11	7	11	14
12	10	14	17
13	13	17	21
14	16	21	26
15	20	25	30
16	24	30	36
17	28	35	41
18	33	40	47
19	38	46	54
20	43	52	60
21	49	59	68
22	56	66	75
23	62	73	83
24	69	81	92
25	77	90	101
26	85	98	110
27	93	107	120
28	102	117	130
29	111	127	141
30	120	137	152

Source: Wilcoxon, F. and Wilcox, R.A. (1964) *Some Rapid Approximate Statistical Procedures*. American Cyanamid Company, Pearl River, New York. Reproduced with permission.

Table A.11. Critical values for the Wilcoxon rank sum test.

One-tailed test at $\alpha = 0.001$ or two-tailed test at $\alpha = 0.002$

	n_2														
n_1	6	7	8	9	10	11	12	13	14	15	16	17	18	19	20
1															
2															
3												0	0	0	0
4					0	0	0	1	1	1	2	2	3	3	3
5		0	0	1	1	2	2	3	3	4	5	5	6	7	7
6	0	1	2	2	3	4	4	5	6	7	8	9	10	11	12
7		2	3	3	5	6	7	8	9	10	11	13	14	15	16
8			5	5	6	8	9	11	12	14	15	17	18	20	21
9				7	8	10	12	14	15	17	19	21	23	25	26
10					10	12	14	17	19	21	23	25	27	29	32
11						15	17	20	22	24	27	29	32	34	37
12							20	23	25	28	31	34	37	40	42
13								26	29	32	35	38	42	45	48
14									32	36	39	43	46	50	54
15										40	43	47	51	55	59
16											48	52	56	60	65
17												57	61	66	70
18													66	71	76
19														77	82
20															88

One-tailed test at $\alpha = 0.01$ or two-tailed test at $\alpha = 0.02$

	n_2															
n_1	5	6	7	8	9	10	11	12	13	14	15	16	17	18	19	20
1																
2									0	0	0	0	0	0	1	1
3			0	0	1	1	1	2	2	2	3	3	4	4	4	5
4	0	1	1	2	3	3	4	5	5	6	7	7	8	9	9	10
5	1	2	3	4	5	6	7	8	9	10	11	12	13	14	15	16
6		3	4	6	7	8	9	11	12	13	15	16	18	19	20	22
7			6	8	9	11	12	14	16	17	19	21	23	24	26	28
8				10	11	13	15	17	20	22	24	26	28	30	32	34
9					14	16	18	21	23	26	28	31	33	36	38	40
10						19	22	24	27	30	33	36	38	41	44	47
11							25	28	31	34	37	41	44	47	50	53
12								31	35	38	42	46	49	53	56	60
13									39	43	47	51	55	59	63	67
14										47	51	56	60	65	69	73
15											56	61	66	70	75	80
16												66	71	76	82	87
17													77	82	88	93
18														88	94	100
19															101	107
20																114

Source: Based in part on Tables 1, 3, 5 and 7 of Auble, D. (1953) Extended tables for the Mann–Whitney statistic. *Bulletin of the Institute of Educational Research at Indiana University* 1(2). Reproduced with permission of the director.

Table A.11. Critical Values for the Wilcoxon Rank Sum Test – *continued*

One-tailed test at $\alpha = 0.025$ or two-tailed test at $\alpha = 0.05$

n_1 \ n_2	4	5	6	7	8	9	10	11	12	13	14	15	16	17	18	19	20
1																	
2					0	0	0	0	1	1	1	1	1	2	2	2	2
3		0	1	1	2	2	3	3	4	4	5	5	6	6	7	7	8
4	0	1	2	3	4	4	5	6	7	8	9	10	11	11	12	13	13
5		2	3	5	6	7	8	9	11	12	13	14	15	17	18	19	20
6			5	6	8	10	11	13	14	16	17	19	21	22	24	25	27
7				8	10	12	14	16	18	20	22	24	26	28	30	32	34
8					13	15	17	19	22	24	26	29	31	34	36	38	41
9						17	20	23	26	28	31	34	37	39	42	45	48
10							23	26	29	33	36	39	42	45	48	52	55
11								30	33	37	40	44	47	51	55	58	62
12									37	41	45	49	53	57	61	65	69
13										45	50	54	59	63	67	72	76
14											55	59	64	67	74	78	83
15												64	70	75	80	85	90
16													75	81	86	92	98
17														87	93	99	105
18															99	106	112
19																113	119
20																	127

One-tailed test at $\alpha = 0.05$ or two-tailed test at $\alpha = 0.10$

n_1 \ n_2	3	4	5	6	7	8	9	10	11	12	13	14	15	16	17	18	19	20
1																	0	0
2			0	0	0	1	1	1	1	2	2	3	3	3	3	4	4	4
3	0	0	1	2	2	3	4	4	5	5	6	7	7	8	9	9	10	11
4		1	2	3	4	5	6	7	8	9	10	11	12	14	15	16	17	18
5			4	5	6	8	9	11	12	13	15	16	18	19	20	22	23	25
6				7	8	10	12	14	16	17	19	21	23	25	26	28	30	32
7					11	13	15	17	19	21	24	26	28	30	33	35	37	39
8						15	18	20	23	26	28	31	33	36	39	41	44	47
9							21	24	27	30	33	36	39	42	45	48	51	54
10								27	31	34	37	41	44	48	51	55	58	62
11									34	38	42	46	50	54	57	61	65	69
12										42	47	51	55	60	64	68	72	77
13											51	56	61	65	70	75	80	84
14												61	66	71	77	82	87	92
15													72	77	83	88	94	100
16														83	89	95	101	107
17															96	102	109	115
18																109	116	123
19																	123	130
20																		138

Source: Based in part on Tables 1,3,5 and 7 of Auble, D. (1953) Extended tables for the Mann–Whitney statistic. *Bulletin of the Institute of Educational Research at Indiana University* 1(2). Reproduced with permission of the director.

Table A.12. Critical Values for the Runs Test; $P(V \leq v^*$ when H_0 is true).

(n_1,n_2)	v^* 2	3	4	5	6	7	8	9	10
(2, 3)	0.200	0.500	0.900	1.000					
(2, 4)	0.133	0.400	0.800	1.000					
(2, 5)	0.095	0.333	0.714	1.000					
(2, 6)	0.071	0.286	0.643	1.000					
(2, 7)	0.056	0.250	0.583	1.000					
(2, 8)	0.044	0.222	0.533	1.000					
(2, 9)	0.036	0.200	0.491	1.000					
(2, 10)	0.030	0.182	0.455	1.000					
(3, 3)	0.100	0.300	0.700	0.900	1.000				
(3, 4)	0.057	0.200	0.543	0.800	0.971	1.000			
(3, 5)	0.036	0.143	0.429	0.714	0.929	1.000			
(3, 6)	0.024	0.107	0.345	0.643	0.881	1.000			
(3, 7)	0.017	0.083	0.283	0.583	0.833	1.000			
(3, 8)	0.012	0.067	0.236	0.533	0.788	1.000			
(3, 9)	0.009	0.055	0.200	0.491	0.745	1.000			
(3, 10)	0.007	0.045	0.171	0.455	0.706	1.000			
(4, 4)	0.029	0.114	0.371	0.629	0.886	0.971	1.000		
(4, 5)	0.016	0.071	0.262	0.500	0.786	0.929	0.992	1.000	
(4, 6)	0.010	0.048	0.190	0.405	0.690	0.881	0.976	1.000	
(4, 7)	0.006	0.033	0.142	0.333	0.606	0.833	0.954	1.000	
(4, 8)	0.004	0.024	0.109	0.279	0.533	0.788	0.929	1.000	
(4, 9)	0.003	0.018	0.085	0.236	0.471	0.745	0.902	1.000	
(4, 10)	0.002	0.014	0.068	0.203	0.419	0.706	0.874	1.000	
(5, 5)	0.008	0.040	0.167	0.357	0.643	0.833	0.960	0.992	1.000
(5, 6)	0.004	0.024	0.110	0.262	0.522	0.738	0.911	0.976	0.998
(5, 7)	0.003	0.015	0.076	0.197	0.424	0.652	0.854	0.955	0.992
(5, 8)	0.002	0.010	0.054	0.152	0.347	0.576	0.793	0.929	0.984
(5, 9)	0.001	0.007	0.039	0.119	0.287	0.510	0.734	0.902	0.972
(5, 10)	0.001	0.005	0.029	0.095	0.239	0.455	0.678	0.874	0.958
(6, 6)	0.002	0.013	0.067	0.175	0.392	0.608	0.825	0.933	0.987
(6, 7)	0.001	0.008	0.043	0.121	0.296	0.500	0.733	0.879	0.966
(6, 8)	0.001	0.005	0.028	0.086	0.226	0.413	0.646	0.821	0.937
(6, 9)	0.000	0.003	0.019	0.063	0.175	0.343	0.566	0.762	0.902
(6, 10)	0.000	0.002	0.013	0.047	0.137	0.288	0.497	0.706	0.864
(7, 7)	0.001	0.004	0.025	0.078	0.209	0.383	0.617	0.791	0.922
(7, 8)	0.000	0.002	0.015	0.051	0.149	0.296	0.514	0.704	0.867
(7, 9)	0.000	0.001	0.010	0.035	0.108	0.231	0.427	0.622	0.806
(7, 10)	0.000	0.001	0.006	0.024	0.080	0.182	0.355	0.549	0.743
(8, 8)	0.000	0.001	0.009	0.032	0.100	0.214	0.405	0.595	0.786
(8, 9)	0.000	0.001	0.005	0.020	0.069	0.157	0.319	0.500	0.702
(8, 10)	0.000	0.000	0.003	0.013	0.048	0.117	0.251	0.419	0.621
(9, 9)	0.000	0.000	0.003	0.012	0.044	0.109	0.238	0.399	0.601
(9, 10)	0.000	0.000	0.002	0.008	0.029	0.077	0.179	0.319	0.510
(10,10)	0.000	0.000	0.001	0.004	0.019	0.051	0.128	0.242	0.414

Source: Eisenhart, C. and Swed, F. (1943) Tables for testing randomness of grouping in a sequence of alternatives. *Annals of Mathematical Statistics 14*. Reproduced with permission of the editor.

Table A.12. Critical Values for the Runs Test. $P(V \le v^*$ when H_0 is true) – *continued*

	v^*									
(n_1,n_2)	11	12	13	14	15	16	17	18	19	20
(2, 3)										
(2, 4)										
(2, 5)										
(2, 6)										
(2, 7)										
(2, 8)										
(2, 9)										
(2, 10)										
(3, 3)										
(3, 4)										
(3, 5)										
(3, 6)										
(3, 7)										
(3, 8)										
(3, 9)										
(3, 10)										
(4, 4)										
(4, 5)										
(4, 6)										
(4, 7)										
(4, 8)										
(4, 9)										
(4, 10)										
(5, 5)										
(5, 6)	1.000									
(5, 7)	1.000									
(5, 8)	1.000									
(5, 9)	1.000									
(5, 10)	1.000									
(6, 6)	0.998	1.000								
(6, 7)	0.992	0.999	1.000							
(6, 8)	0.984	0.998	1.000							
(6, 9)	0.972	0.994	1.000							
(6, 10)	0.958	0.990	1.000							
(7, 7)	0.975	0.996	0.999	1.000						
(7, 8)	0.949	0.988	0.998	1.000	1.000					
(7, 9)	0.916	0.975	0.994	0.999	1.000					
(7, 10)	0.879	0.957	0.990	0.998	1.000					
(8, 8)	0.900	0.968	0.991	0.999	1.000	1.000				
(8, 9)	0.843	0.939	0.980	0.996	0.999	1.000	1.000			
(8, 10)	0.782	0.903	0.964	0.990	0.998	1.000	1.000			
(9, 9)	0.762	0.891	0.956	0.988	0.997	1.000	1.000	1.000		
(9, 10)	0.681	0.834	0.923	0.974	0.992	0.999	1.000	1.000	1.000	
(10,10)	0.586	0.758	0.872	0.949	0.981	0.996	0.999	1.000	1.000	1.000

Table A.13. Critical values for Spearman's rank correlation coefficient test.

n	$\alpha = 0.05$	$\alpha = 0.025$	$\alpha = 0.01$	$\alpha = 0.005$
5	0.900	—	—	—
6	0.829	0.886	0.943	—
7	0.714	0.786	0.893	—
8	0.643	0.738	0.833	0.881
9	0.600	0.683	0.783	0.833
10	0.564	0.648	0.745	0.794
11	0.523	0.623	0.736	0.818
12	0.497	0.591	0.703	0.780
13	0.475	0.566	0.673	0.745
14	0.457	0.545	0.646	0.716
15	0.441	0.525	0.623	0.689
16	0.425	0.507	0.601	0.666
17	0.412	0.490	0.582	0.645
18	0.399	0.476	0.564	0.625
19	0.388	0.462	0.549	0.608
20	0.377	0.450	0.534	0.591
21	0.368	0.438	0.521	0.576
22	0.359	0.428	0.508	0.562
23	0.351	0.418	0.496	0.549
24	0.343	0.409	0.485	0.537
25	0.336	0.400	0.475	0.526
26	0.329	0.392	0.465	0.515
27	0.323	0.385	0.456	0.505
28	0.317	0.377	0.448	0.496
29	0.311	0.370	0.440	0.487
30	0.305	0.364	0.432	0.478

Source: Olds, E.G. (1938) Distribution of sums of squares of rank differences for small samples. *Annals of Mathematical Statistics* 9. Reproduced with permission of the editor.

This page intentionally left blank

Appendix B

Summation Notation

In mathematics and statistics, the Greek letter Σ (capital sigma) is used to indicate a summation. For example:

$$\sum_{i=1}^{5} x_i$$

reads 'the summation of x_i, i going from 1 to 5', where 1 is the lower limit and 5 is the upper limit of the summation. Therefore:

$$\sum_{i=1}^{5} x_i = x_1 + x_2 + x_3 + x_4 + x_5.$$

In general, $\sum_{i=1}^{5} x_i$ means that we replace i by 1, then by 2, ... and so on up to n, and add up the numerical values of $x_1, x_2, ..., x_n$.

If the sum of the squares of n observations is needed, we write:

$$\sum_{i=1}^{n} x_i^2 = x_1^2 + x_2^2 + \ldots + x_n^2,$$

which says 'add the squares of the observations from 1 to n inclusive.' The sum of the products of two variables, X and Y, from 1 to n is:

$$\sum_{i=1}^{n} x_i x_y = x_1 y_1 + x_2 y_2 + \cdots\cdots + x_n y_n.$$

Partial sums, sums of squares and sums of products can be represented as:

$$\sum_{i=3}^{5} x_i = x_3 + x_4 + x_5,$$

or

$$\sum_{i=6}^{n} x_i^2 = x_6^2 + x_7^2 + \cdots + x_n^2,$$

or

$$\sum_{i=2}^{7} x_i y_i = x_2 y_2 + x_3 y_3 + \ldots + x_6 y_6 + x_7 y_7.$$

Note the following when a constant is summed:
from 1 to 5:

$$\sum_{i=1}^{5} a = a + a + a + a + a = 5a$$

from 1 to n:

$$\sum_{i=1}^{n} a = a + a + a + \cdots + a = na$$

and from 3 to 5:

$$\sum_{i=3}^{5} a = a + a + a = 3a$$

Also, if a variable, x, is multiplied by a constant, a:

$$\sum_{i=1}^{n} ax_i = ax_1 + ax_2 + \cdots + ax_n = a(x_1 + x_2 + \cdots + x_n) = a\sum_{i=1}^{n} x_i,$$

so that:

$$\sum_{i=1}^{n} (x_i - a) = x_1 - a + x_2 - a + \cdots + x_n - a = \sum_{i=1}^{n} x_i - na$$

and:

$$\sum_{i=1}^{n} (x_i - a)^2 = \sum_{i=1}^{n} (x_i^2 - 2x_i a + a^2) = \sum_{i=1}^{n} x_i^2 - 2a\sum_{i=1}^{n} x_i + na^2.$$

Any letter can be used as a subscript. For example,

$$\sum_{i=1}^{n} x_i = \sum_{j=1}^{n} x_j = x_1 + x_2 + \cdots + x_n.$$

Observations arranged in rows and columns are usually described by two subscripts:

12.5	14.3	11.5	8.2
6.1	12.1	15.3	14.4
10.2	11.3	13.6	14.8

In general:

x_{11}	x_{12}	x_{13}	x_{14}
x_{21}	x_{22}	x_{23}	x_{24}
x_{31}	x_{32}	x_{33}	x_{34}

The sum of these values is:

$$\sum_{i=1}^{3} \sum_{j=1}^{4} x_{ij} = x_{11} + x_{12} + x_{13} + \cdots + x_{33} + x_{34}$$

and their sum of squares is:

$$\sum_{i=1}^{3} \sum_{j=1}^{4} x_{ij}^2 = x_{11}^2 + x_{12}^2 + x_{13}^3 + \cdots + x_{33}^2 + x_{34}^2.$$

In general:

$$\sum_{i=1}^{r} \sum_{j=1}^{c} x_{ij} = x_{11} + x_{12} + \cdots + x_{rc},$$

where r = number of rows and c = number of columns.

Glossary

acceptance region: the range of values for a sample statistic where the null hypothesis is not rejected.

addition rule: a probability rule based on the union of events. For two events A and B, the addition rule is denoted by: $P(A \cup B) = P(A) + P(B) - P(A \cap B)$.

alternative hypothesis: a statement which is contradictory to the null hypothesis, denoted by H_1.

arithmetic average: see **mean.**

attribute charts: statistical process control charts used for monitoring attribute data, including p charts.

attribute data: in statistical process control, production-related data that require an operational definition of acceptable and defective products.

average: see **mean.**

bar graphs: graphical tools used to present information summarized in categorical frequency distributions or ungrouped frequency distributions created for discrete variables. Since the horizontal axis is not a continuous random variable, the bars do not touch each other.

Bayes' Theorem: a logical proposition used to solve conditional probability problems that generally occur in reverse order of time. Bayes' Theorem gives the conditional probability of the random variable A given B in terms of the marginal probability distribution of A alone and the conditional probability distribution of variable B given A.

bias: the amount by which a sample estimate systematically under/over-estimates the true value of a parameter. Bias can occur, for example, when equipment used for recording measurements are not calibrated properly.

bimodal: a population or sample with two modes.

bivariate distribution: *see* **joint probability distribution.**

bivariate frequency distribution: the joint, simultaneous distribution of two variables.

bivariate normal distribution: a joint statistical distribution of two random variables which may or may not be correlated, and where each has a normal marginal distribution.

blocks: groups of smaller, more uniform experimental units used in experimental designs if the experimental units, area, time or material are not homogeneous.

blocking: *see* **blocks.**

categorical frequency distributions: frequency distributions used to place qualitative, ordinal or nominal level variables into specific categories.

categorical variables: *see* **qualitative variables.**

Central Limit Theorem: one of the most important theorems in statistics, formalizing the relationship between a specific parameter of a population and its estimate (statistic). This theorem posits that when the sample size (n) is sufficiently large ($n \geq 30$), the sampling distribution of sample means approaches a normal distribution with a mean equaling the population mean and the standard deviation equaling the standard error of the mean.

Chebyshev's Theorem: a theorem which can be applied to samples or populations of any kind, and states that at least the fraction $(1 - 1/k^2)$ of the observations must lie within k standard deviations of the mean, regardless of the shape of the distribution of the data (where k is any constant greater than one).

chi-square (χ^2) distribution: a positively skewed, positive-valued distribution that describes the sampling distribution of the variances. It has a mean of $n - 1$ and approaches the normal distribution at larger sample sizes.

circular permutation: the number of permutations of n distinct subjects positioned in a circle, denoted by P_c.

class boundaries: the values occurring halfway between the upper class limit of one interval and the lower class limit of the next interval in a frequency distribution.

class frequency: the number of observations that fall in a particular class in a frequency distribution.

class intervals: *see* **classes.**

class limits: the smallest and largest possible values that can fall into a given class in a frequency distribution.

class mark: *see* **class midpoint.**

class midpoint: the average of the upper and lower class limits, or upper and lower class boundaries, of a class in a frequency distribution.

class width: the difference between the upper and lower class boundaries of a given class in a frequency distribution.

classes: the various bounded groupings (generally with similar intervals) defined for a frequency distribution within which data observations are placed.

classical probability: probability calculated as the ratio of the number of outcomes favourable to a particular event versus the number of possible outcomes in a sample space.

coefficient of variation: the standard deviation expressed as a percentage of the mean.

collectively exhaustive: a quality of events where the sum of the probabilities for all possible events in the sample space equals unity.

combination: the number of possible outcomes when order is not important. Commonly denoted by $\binom{n}{r}$ or ${}_nC_r$ and often stated as 'n choose r'.

complement: the event containing all the elements of the sample space that are not contained in the event. The complement of B is denoted by B'.

completely randomized design: the simplest of the experimental designs wherein treatments are randomly assigned to each experimental unit (in time or space).

compound event: an event that consists of two or more simple events.

conditional distribution: the distribution of a random variable given that other variables have certain specified values.

conditional probability: a redefined sample space, where a given event, B, has occurred, and we are interested in understanding the effect of this information on the probability of event A occurring. The conditional probability of event A given that event B has occurred is denoted by $P(A|B)$.

confidence interval: for a given confidence level (or degree of confidence), the interval between the lower confidence limit (LCL) and upper confidence limit (UCL). *See* **confidence limits.**

confidence level: the quantity, $(1 - \alpha)100\%$, which describes the degree of statistical certainty that can be attached to an observed statistic. The most frequently used values of α are 0.10, 0.05 and 0.01, resulting in 90%, 95% and 99% confidence intervals, respectively.

confidence limits: upper (UCL) and lower (LCL) bounds of the interval where the probability of finding the true parameter, θ, is set at a confidence value, $1 - \alpha$. The probability that we will find the true population parameter between LCL and UCL is $1 - \alpha$: $P(\text{LCL} < \theta < \text{UCL}) = 1 - \alpha$.

consistent: a quality of an estimator such that as the sample size, n, approaches infinity, the value of the estimator approaches the value of the population parameter. An unbiased estimator is consistent if, as $n \to \infty$, var $(\hat{\theta}) \to 0$ and $\hat{\theta} \to \theta$.

continuity correction: a constant applied to a random variable (usually equal to half of the unit of measurement for continuous variables).

continuous random variable: a random variable defined over a continuous sample space, where the probability of any exact value is always zero.

continuous sample space: a sample space that contains an infinite and uncountable number of outcomes.

continuous variable: a quantitative variable that can take on all possible values over a specific interval.

control chart: a graphical device used in statistical process control to determine whether a production process is in or out of control based on sampled data.

control chart constants: conversion and correction factors used in the production of statistical process control charts.

corrected sum of squares: a measure of spread equal to the sum of squared deviations of each observation from the mean, so named because each observation is 'corrected for' the mean before it is squared.

covariance: the measure of joint variation between two random variables. Covariance may be zero (when two random variables are independent), positive (when the value of the variables increases together), or negative (when the value of one variable increases, the value of the other variable decreases).

critical region: the range of values for a sample statistic where the null hypothesis is rejected.

critical value: a selected arbitrary value along a statistical distribution, below or above which the null hypothesis is rejected.

cumulative frequency: the frequency of all observations less than a particular value of a random variable (for a frequency distribution, the upper class boundary of a given class). Often referred to as the 'less than frequency'.

data: pieces of information collected on subjects or items from a population that form the building blocks of statistics.

deciles: divisions of the frequency distribution into ten equal groups that correspond to the 10th, 20th, ..., and 90th percentiles.

degree of confidence: *see* **confidence level.**

degrees of freedom: the number of unrestricted observations used to calculate a statistic.

dependent populations: random variables that occur in pairs and where the response value of one variable is at least partly a function of the response of the other.

dependent samples: sampled observations that occur in pairs and where the response value of one sample is at least partly a function of the response of the other.

descriptive statistics: a branch of statistics dealing with the collection, organization and presentation of information, and the calculation of some measures (statistics) which describe the information.

discrete random variable: a random variable defined over discrete sample space.

discrete sample space: a sample space that contains a finite number of elements. A discrete sample space can be unending, but countable.

discrete variables: quantitative variables which take on whole numbers only and usually result from counting (tallying) items.

disjoint: *see* **mutually exclusive.**

distribution-free tests: see **non-parametric tests.**

efficient: the quality of the unbiased estimator of a given parameter, θ, having the smallest variance.

element: a single outcome of an experiment within a given sample space.

empirical probability: the likelihood of an event happening based on experiments for which all possible outcomes and the number of outcomes favouring the event are not known exactly, but have generally been observed.

Empirical Rule: a rule which states that approximately 68%, 95% and 99.7% of the observations from a normal distribution will lie within one, two or three standard deviations of the mean, respectively.

estimate: *see* **point estimate.**

estimation: the process of estimating the values of parameters based on measured or empirical data.

estimator: a function used to estimate an unknown parameter from observed data.

event: a subset or portion of the elements in a sample space.

expected value: the theoretical mean of a probability distribution, denoted by $E(X)$, interpreted as the long-term average that is 'expected' if an experiment is conducted repeatedly.

experimental design: a means of collecting data in which one or more of the factors affecting the variable(s) of interest are controlled, with the purpose of investigating how these controlled factors affect the variable(s) of interest.

experimental error: the pooled variation among experimental units receiving the same treatment in an experimental design.

experimental study: *see* **experimental design**

exponential distribution: the continuous counterpart to the Poisson distribution. The exponential distribution describes the elapsed times between occurrences of consecutive events as a function of the mean elapsed time.

***F* distribution:** a distribution which describes the ratio of two independent χ^2-values, where each is divided by its degrees of freedom. There exist many such curves, but each is positively skewed and positive-valued.

finite population: a population consisting of a fixed, countable number of elements, which can be, if necessary, listed.

finite population correction factor: a multiplicative adjustment used in the calculation of the standard error of the mean when the sample size is large relative to the population size, specifically when $n < 0.05N$.

frequency distribution: a systematic arrangement of data to describe a variable, where observations (raw data) are ordered or grouped into classes, and the frequency of observations is tallied and presented in tabular form. Frequency distributions can be categorical, ungrouped, or grouped.

frequency polygon: a graphical display of a frequency distribution constructed by plotting frequency (or relative frequency) against class mark (or value of the random variable in the case of ungrouped data), and then joining each point by a sequence of line segments. To close the polygon, an 'imaginary' class midpoint with zero frequency is added to both ends of the distribution.

geometric distribution: a discrete probability function which possesses all the properties of a binomial experiment except that trials are repeated until the first success occurs. The geometric random variable, X, represents the number of repeated independent trials required to produce the first success, the probability of which is p.

geometric experiment: *see* **geometric distribution**.

geometric mean: a special form of the mean that is used for ratio data like population growth, rates of change, economic indicators, etc. The geometric mean of n observations is the nth root of the product of the n observations.

grand mean: a special application of the weighting procedure used to find the overall combined mean of several groups of data when the mean of each individual group is known.

grouped frequency distribution: a frequency distribution usually used to summarize continuous (interval or ratio scale) variables.

***H*-test:** *see* **Kruskal-Wallis test**

harmonic mean: a special form of mean used for data where one element remains constant but another changes. The harmonic mean is calculated as the reciprocal of the mean of the reciprocals of the individual values.

histogram: a graphical tool for presenting the grouped frequency distribution of a continuous variable. Like a bar graph, the middle of each bar is the class midpoint; however, histograms do not contain spaces between bars so that bars touch at class boundaries.

hypergeometric distribution: a discrete probability distribution that has two possible outcomes, but where the probability of subsequent events are dependent upon previous outcomes. In other words, the probability of success from trial to trial is not constant and the successive trials (made without replacement from a finite population) are not independent.

hypothesis: a statement or claim made about a parameter or a certain characteristic of a population.

hypothesis testing: a procedure in applied statistics for determining whether a statement or claim made about a parameter or a certain characteristic of a population is plausible, based on some sample data collected from the population.

independence: two events are statistically independent if the probability of one event is not affected by the occurrence or nonoccurrence of the other event.

independent populations: two populations are statistically independent if the distribution of values in one population is not affected by the values in the other population.

inferential statistics: a branch of statistics dealing with the generalization of information obtained in a sample to an entire population. Common procedures include estimation, hypothesis testing, determining relationships and prediction.

infinite population: a population where (in theory) there is no limit to the number of possible observations (or measurements). In sampling, the word 'infinite' is used rather loosely and is used to refer to a population with a large number of possible measurements.

intersection: for two events, A and B, the event that contains all the elements common to both A and B. The intersection is denoted by $A \cap B$.

interval estimate: *see* **confidence interval.**

interval estimation: the process of determining a confidence interval; that is, an interval within which we expect to find the unknown population parameter.

interval scale: a scale of measurement with the same properties as the ordinal scale, but where the data are always quantitative and the differences between data values are meaningful.

inverse cumulative frequency: the frequency of all values greater than a particular value of a random variable (for a frequency distribution, the lower class boundary of a given class). Often referred to as the 'more than frequency'.

inverse relative cumulative frequencies: inverse cumulative frequencies expressed as percentages (or proportions) of the total frequencies.

joint probability distribution: for two random variables, X and Y, the probability distribution of X and Y together.

joint probability function: joint probability expressed as a function of the random variables X and Y. The function is denoted by $f(x,y)$, which represents the probability that X assumes the value x at the same time Y assumes the value y.

Kruskal-Wallis test: a non-parametric test used to compare three or more unknown population means.

Latin square design: an experimental design used when the natural variation between experimental units cannot be reduced by simple blocking alone and the variation of the experimental units are removed in two directions.

layout: the placement of treatments on experimental units in an experimental design.

level of significance: the size of type I error (α). The value is arbitrary in that it is selected by the person carrying out the statistical test, but 0.1, 0.05 or 0.01 are generally used.

lower class limit: the smallest possible value that can fall into a given class in a grouped frequency distribution.

lower confidence limit (LCL): *see* **confidence limit.**

lower control limit: the lower limit on a statistical process control chart, beyond which production processes are said to be out of control.

lower warning limit: a lower limit on a statistical process control chart which is used to draw attention to potential production-related problems.

Mann-Whitney U-test: *see* **Wilcoxon rank sum test.**

marginal probability: the probability of some event, regardless of the outcome of other events. For a joint probability distribution $f(x,y)$, the marginal probability

$f(x)$ results from constructing a probability distribution for X over all possible values of Y.

mathematical expectation: *see* **expected value.**

mathematical expectation of a random variable: *see* **population mean of a random variable.**

mean: a measure of central tendency that is calculated by dividing the sum of the observations by the number of observations.

mean deviation: a measure of variation, calculated as the average of the absolute values of the deviations of each of the observations from the sample or population mean.

mean of a random variable: the weighted average of all possible outcomes of a random variable, where the weights are the probabilities of the respective outcomes.

mean square: *see* **variance.**

median: the middle value when a set of observations is arranged in increasing or decreasing order of magnitude, dividing the frequency distribution into two equal groups and corresponding to the 50th percentile. The median is the preferred measure of central location when extreme values are present.

midrange: a measure of central tendency defined as the average of the minimum and maximum values.

mode: a measure of central tendency defined as the most frequently occurring value in a sample or a population. Some data sets may have more than one mode (e.g. when several values occur with the greatest frequency) and others may have no mode at all.

multimodal: a population or sample with more than two modes.

multinomial distribution: a discrete probability distribution having all the properties of a binomial distribution, except that more than two outcomes are possible from each trial.

multiplication rule: a counting rule used to calculate the total number of outcomes for a sample space or event. The rule states that if a random experiment has a sequence of two steps, in which there are n_1 possible outcomes for the first step and n_2 for the second, the total number of outcomes is the product of the two numbers ($n_1 \times n_2$).

multivariate hypergeometric distribution: a probability distribution having all the properties of a hypergeometric distribution, except there are more than two possible outcomes.

mutually exclusive: a quality ascribed to two or more events which have no common intersecting elements (i.e., when one event occurs the others cannot). For two mutually exclusive events, A and B, $A \cap B = \varnothing$.

negative binomial distribution: a discrete probability distribution which is an extension of the binomial and geometric distribution, describing the situation where trials are repeated until a fixed number of successes, k, occurs.

nominal scale: a scale of measurement where numbers or categories are used to classify, name or label an individual or attribute, but the numbers or categories have no specific order or importance.

non-critical region: *see* **acceptance region.**

non-parametric test: a statistical test that makes no assumptions about the distribution or the parameters of the distribution from which observations are drawn.

non-sampling error: errors arising during the course of data collection that are not due to sampling. This includes errors from non-responses, improper coding, instrument miscalibration, etc.

normal distribution: a continuous, symmetrical, bell-shaped distribution whose shape and position are determined by the mean and standard deviation. Many of the most important theories in statistical inference are based on the normal distribution, also often referred to as the Gaussian distribution or the Laplacian distribution.

null hypothesis: a statement about a characteristic of the population assumed to be true, denoted by H_0.

null space: an event containing no elements in a given sample space.

observational study: a study where investigators observe without altering or influencing the variable under study.

odds: a term used in subjective probability, often seen in gambling, sporting events, and horse racing, which refers to the ratio of the probability of an event occurring versus the probability of the event not occurring.

ogive: a graphical tool representing cumulative or inverse cumulative frequencies, plotted in a similar manner to a frequency polygon. The cumulative frequencies are plotted against the upper (cumulative) or lower (inverse cumulative) class boundaries and joined by line segments. Also known as a cumulative frequency or inverse cumulative frequency graph.

one-tailed tests: a hypothesis test which can be refuted in only one direction, i.e., the inequality in the alternative hypothesis is generally 'less than' or 'greater than' some value.

open class: in a grouped frequency distribution, when the first (or last) class has no lower (or upper) limit, to accommodate a very few (one or two) extreme observations in the data set.

operating characteristic (OC) curve: a curve describing how the values of β (the probability of 'accepting' the null hypothesis when it is false) change over a range of values of μ, n and/or α.

ordinal scale: a scale of measurement similar to the nominal scale, but where the order or rank of the categories is meaningful.

outcome: the result of an experiment.

outliers: extreme values in a data set.

p chart: an attribute control chart used in statistical process control for monitoring the sample proportion of defective products.

***p*-value:** the smallest level of significance at which H_0 will be rejected. Depending on the direction of the test, the p-value indicates the probability of obtaining a value in the sampling distribution of the test statistic less than or greater than the calculated test statistic.

parameters: the characteristics of a population, usually denoted with Greek letters (e.g. μ, σ).

parametric tests: statistical testing methods that use values which uniquely define a probability distribution and involve testing estimates of parameter values.

percentile: a measure indicating the position of an observation within a data set (not the same as a percentage). In general, the pth *percentile* is the value such that p per cent of the items in the data set fall at or below that value.

permutation: the number of possible outcomes when order is important. Commonly denoted by ${}_nP_r$.

permutation of similar objects: a special kind of permutation used when some of the objects, among the n objects, are not distinguishable.

pie chart: a graphical presentation of a variable relative to a totality using a circle divided into sectors representing each category's frequency proportional in size to the total.

point estimate: a single numeric estimate of a population parameter calculated from the information in a sample.

point estimation: see **point estimate**.

Poisson distribution: a discrete probability distribution describing independent events that occur in a fixed time (or space) with a known average rate.

Poisson experiments: a series of trials or tests where the variable of interest follows a Poisson distribution.

population: the entire collection of items/subjects possessing certain common characteristics about which information is being sought.

population mean: the mean of all elements in a population.

posterior probabilities: reversed conditional probabilities used in Bayes' Theorem.

power of a test: the probability that a test will reject the null hypothesis when it is in fact false.

prediction: the value of the dependent variable obtained from a regression equation using a particular value of the independent variable.

prior probability: a conditional probability based on previously observed frequencies in a sample space or event.

probability: (i) the branch of mathematics incorporating the most important set of concepts used in statistics; (ii) the measure of likelihood of the occurrence or nonoccurrence of an event. The probability of an event, A, is denoted by $P(A)$ and can be classical, empirical, or subjective.

probability density: a function associated with a probability distribution that specifies how the values of a random variable are distributed over its possible range.

probability distribution: for a given random variable, the list of all possible outcomes and their associated probabilities.

probability function: a formula (or mathematical expression) expressing probabilities associated with given values of a random variable.

properties of probability: (i) for any given event A, the probability of A must be between zero and one; (ii) the sum of the probabilities of all possible events in a sample space must equal one; and (3) the sum of the probabilities of A and its complement, A', must equal one.

qualitative survey methods: behavioural survey methods which are exploratory in nature and are generally used to gain insight into a research problem or for theory development.

qualitative variables: variables which can be placed into distinct categories according to some characteristic.

quality control: *see* **statistical process control**.

quantitative survey methods: behavioural survey methods which employ rigorous sampling methods and make it possible to draw inferences about populations.

quantitative variables: variables which are numerical in nature and indicate 'how many' or 'how much' or 'how big' on a numeric scale.

quartiles: percentiles which divide a frequency distribution into four equal groups corresponding to the 25th, 50th, and 75th percentiles.

R chart: a variable control chart used in statistical process control for measuring and monitoring sample ranges.

random number: a number that is determined entirely by chance from some specified distribution, without bias and without correlations between successive numbers.

random variable: a variable whose value is determined by the outcome of a random experiment, denoted by capital letters, such as X, Y or Z.

randomized complete block design: an experimental design wherein each treatment is applied to one experimental unit within each block, and treatments are randomly allotted to the experimental units independently within each block.

range: the simplest measure of variation, calculated as the difference between the highest and lowest values in a data set.

ratio scale: a scale of measurement similar to the interval scale, but where zero means 'none', and therefore, the ratio of two variables becomes meaningful.

rejection region: *see* **critical region.**

relative cumulative frequencies: cumulative frequencies expressed as percentages (or proportions) of the total frequencies.

replication: applying the same treatment to more than one experimental unit within an experimental design.

response variable: the variable of interest in an experimental design.

runs rule: a systematic procedure used in statistical process control to determine whether a process is out of control based on a pattern of consecutive measurements.

runs test: a non-parametric method for testing if observations are drawn in random order.

S chart: a variable control chart used in statistical process control for measuring and monitoring sample standard deviations.

sample: a portion or subset of the population.

sample mean: the mean of all elements measured in a sample.

sample point: *see* **element.**

sample space: an event containing all possible outcomes of an experiment, denoted by (S).

sample survey: collection of information from a population through interviews or the application of questionnaires to a sample from the group.

sampling: the collection of data from a subset of the population leading to prediction, or inferences about the entire population. There is no attempt to control the variable(s) of interest, rather a given situation is merely observed.

sampling distribution: the probability distribution of a statistic, e.g., a sample mean, the difference between two means, a sample proportion, the difference between two proportions, a single variance, or the ratio of two variances.

sampling distribution of the differences between two means: the probability distribution for the random variable describing the differences between two independent sample means.

sampling distribution of the mean: the probability distribution for the random variable describing sample means.

sampling distribution of the statistic: see **sampling distribution.**

sampling error: uncertainty which occurs because observations arising from samples tend to deviate from one sample to another (a natural consequence of taking samples).

sampling with replacement: selection from a population such that each element can appear in the sample as often as it is selected (the element is replaced every time it is sampled). If a sample is selected with replacement, there are N^n possible samples.

sampling without replacement: selection from a population such that each element of a population can only be selected once (the element is not replaced when it is sampled). If a sample is selected without replacement, there are ${}_NC_n$ possible samples.

scale of measurement: a classification that refers to the nature of information contained within a random variable and indicates what types of statistical analyses are appropriate, e.g., nominal ordinal, interval, or ratio scales.

shape: a quality of a distribution described by its frequency histogram or bar graph. In the case of a Normal distribution, shape is defined by the variance, σ^2 (or standard deviation, σ).

sign test: a non-parametric test of the median value of a single population that uses plus and minus signs to identify differences between observations and their median.

significance level: *see* **level of significance.**

simple event: an event which contains only one element of a sample space.

simple random sample: *see* **simple random sampling.**

simple random sampling: a sample selection method in which observations are drawn randomly from a population and each sampling unit (or group of sampling units) has the same probability of being chosen.

skewed: a quality of a frequency distribution that lacks symmetry with respect to a central vertical axis through the distribution. Frequency distributions may be skewed positively (i.e., have a long right tail) or negatively (i.e., have a long left tail).

Spearman's rank correlation test: a non-parametric test used to test the significance of a sample correlation coefficient based on ranks known as Spearman's rank correlation coefficient.

standard deviation: a measure of variation in the same units as the original observations (and the mean) which is the square root of the variance, denoted by σ or σ_x from a population and s or s_x from a sample.

standard error of the mean: the standard deviation of the sample means for a given sample size.

standard error of the statistic: the standard deviation of a statistic for a given sample size. It measures the spread of all possible values of a statistic.

standard normal distribution: a normal distribution with a mean of zero and variance of one. A random variable, X, is transformed into a standard normal random variable, Z, in order to use standard normal probability tables.

standard score: the relative position of an observation within a particular data set expressed in terms of the mean and standard deviation.

statistical estimation: *see* **estimation.**

statistical hypothesis: *see* **hypothesis.**

statistical inference: *see* **inferential statistics.**

statistical process control: statistical procedures for measuring production-related metrics and monitoring them on control charts.

statistical quality control: *see* **statistical process control.**

statistics: (i) the science of collecting, organizing, analysing and interpreting information; (ii) numbers that describe characteristics of a sample from a population. Statistics are usually denoted with Roman letters (e.g., x, p).

stratified random sampling: a sampling method in which the sampling units (individual measurements) in a population are grouped together to form a stratum on the basis of similarity of some characteristic or characteristics and each group or stratum is treated as an individual population.

Student's *t* distribution: *see* ***t* distribution.**

Sturges' Rule: a formula used to determine the number of classes in a grouped frequency distribution.

subjective probabilities: probabilities based solely on an individual's experiences, or 'educated guesses', and not substantiated by exact scientific evidence.

subset: a group of elements, C, that are also elements of another (larger) event, A. When C is a subset of A, it is denoted by ($C \subset A$).

sum of squares of the deviations from the mean: *see* **corrected sum of squares.**

symmetric: a quality of a distribution where a central vertical axis separates the distribution into two identical (mirror image) or near-identical parts.

systematic sampling: a sampling method in which the sampling units are numbered from 1 to N, and n units are selected using a regular interval.

***t* distribution**: the probability distribution of Student's t statistic. The t distribution is a symmetrical (about zero), bell-shaped curve. Its standard deviation depends on the sample size, and will always be somewhat higher than one.

test of hypothesis: *see* **hypothesis testing.**

test statistic: a statistic computed from sample data which is compared to a critical value to determine the outcome of a hypothesis test.

treatments: factors that are controlled or kept at fixed levels in order to estimate their effect in experimental designs.

tree diagram: a systematic procedure for graphically listing all possible outcomes in a sample space or an event.

trimmed mean: a special form of the mean, calculated after removing the upper and lower 5% of the ranked data, used in cases when very small or large values are apparent.

two-stage sampling: sample selection which takes place in two distinct phases. First primary units are selected which are divisible into multiple secondary units, then samples are selected from these secondary units.

two-tailed tests: a hypothesis test which can be refuted in two directions, i.e, the inequality in the alternative hypothesis is generally 'not equal to' some value.

type I error: the probability of rejecting H_0 when it is true, denoted by α. The value of α is decided on by the person conducting the test and is equal to the area under the curve in the rejection region.

type II error: the probability of not rejecting ('accepting') H_0 when it is false, denoted by β. The value of β is rarely known to us because its value depends on knowledge that we generally do not possess, namely the true value of the population parameter, sample size and the size of α (level of significance).

unbiased: the quality of a sample estimator when the mean of its sampling distribution is equal to the population parameter. An unbiased estimate of the true population parameter occurs when $E(\hat{\theta}) = \theta$.

ungrouped frequency distributions: frequency distributions used to summarize discrete quantitative variables using each unique value of the random variable.

uniform distribution: a discrete or continuous probability distribution whereby the probability of every outcome is the same.

uniform probability distribution: see **uniform distribution.**

uniform random variable: a random variable which follows a uniform distribution.

union: for two given events, A and B, the event that contains all of the elements in A or in B, including elements common to both. The union is denoted by $A \cup B$.

upper class limit: the largest possible value that can fall into a given class in a grouped frequency distribution.

upper confidence limit (UCL): *see* **confidence limit.**

upper control limit: the upper limit on a statistical process control chart, beyond which production processes are said to be out of control.

upper warning limit: an upper limit on a statistical process control chart which is used to draw attention to potential production-related problems.

variable charts: statistical process control charts used for monitoring variable data, including $\bar{X}$ charts, R charts and S charts.

variable data: in statistical process control, measured quantitative production-related data.

variance: a measure of variation equal to the corrected sum of squares divided by its degrees of freedom.

variance ratio test: a statistical test that determines if the ratio of two variances is significantly different from a constant, usually one.

Venn diagram: a picture of events as they relate to each other within a sample space, especially useful where compound (multiple) events are concerned. The sample space is shown as the interior of a rectangle and the events are identified (often as circles) as specified regions inside the rectangle.

weighted mean: a special formulation of the arithmetic mean used to find the average of a number of values, attaching more importance to some values than to others by assigning different weights to the n observations (representing their relative contribution to the overall average).

Wilcoxon rank sum test: a non-parametric test for comparing two unknown population means.

Wilcoxon signed rank test: a non-parametric test of the median value of a single population that uses plus and minus signs to identify differences between observations and their median.

$\bar{X}$ chart: a variable control chart used in statistical process control for measuring and monitoring sample means.

***Z* distribution:** a standard normal distribution.

***Z*-transformation:** see **standard normal distribution**

This page intentionally left blank

Index

Page numbers in *italics* indicate figures and tables. Page numbers followed by 'n' indicate footnotes.